SYSTEMS ANALYSIS IN URBAN POLICY-MAKING AND PLANNING

NATO CONFERENCE SERIES

I   Ecology
II Systems Science
III Human Factors
IV Marine Sciences
V   Air—Sea Interactions
VI Materials Science

II SYSTEMS SCIENCE

Volume 1   Transportation Planning for a Better Environment
            Edited by Peter Stringer and H. Wenzel

Volume 2   Arctic Systems
            Edited by P J Amaria, A. A Bruneau, and P. A. Lapp

Volume 3   Environmental Assessment of Socioeconomic Systems
            Edited by Dietrich F Burkhardt and William H. Ittelson

Volume 4   Earth Observation for Resource Management and
            Environmental Control
            Edited by Donald J Clough and Lawrence W. Morley

Volume 5   Applied General Systems Research: Recent Developments
            and Trends
            Edited by George J. Klir

Volume 6   Evaluating New Telecommunications Services
            Edited by Martin C. J Elton, William A. Lucas, and
            David W Conrath

Volume 7   Manpower Planning and Organization Design
            Edited by Donald T. Bryant and Richard J. Niehaus

Volume 8   Search Theory and Applications
            Edited by K. Brian Haley and Lawrence D. Stone

Volume 9   Energy Policy Planning
            Edited by B. A. Bayraktar, E A. Cherniavsky, M. A. Laughton, and
            L. E Ruff

Volume 10   Applied Operations Research in Fishing
            Edited by K. Brian Haley

Volume 11   Work, Organizations, and Technological Change
            Edited by Gerhard Mensch and Richard J. Niehaus

Volume 12   Systems Analysis in Urban Policy-Making and Planning
            Edited by Michael Batty and Bruce Hutchinson

Volume 13   Homotopy Methods and Global Convergence
            Edited by B. Curtis Eaves, Floyd J. Gould,
            Heinz-Otto Peitgen, and Michael J. Todd

SYSTEMS ANALYSIS IN URBAN POLICY-MAKING AND PLANNING

Edited by

Michael Batty

University of Wales
Institute of Science and Technology
Cardiff, United Kingdom

and

Bruce Hutchinson

University of Waterloo
Waterloo, Ontario, Canada

Published in cooperation with NATO Scientific Affairs Division

PLENUM PRESS · NEW YORK AND LONDON

Library of Congress Cataloging in Publication Data

Main entry under title:

Systems analysis in urban policy-making and planning.

(NATO conference series. II, Systems science; v. 12)
   "Proceedings of a NATO Advanced Research Institute on Systems Analysis in Urban
Policy-Making and Planning, held September 21–27, 1980, at New College, Oxford
University, Oxford, Eng."—Verso t.p.
   Includes bibliographical references and index.
   1. Urban policy—Congresses. 2. City planning—Congresses. 3. Systems analysis—
Congresses. 4. Land use, Urban—Congresses. I. Batty, Michael. II. Hutchinson,
Bruce. III. NATO Advanced Research Institute on Systems Analysis in Urban Policy-
Making and Planning (1980: Oxford University) IV Title. V. Series.
RT107.S96 1982   307'.12   82-13172
ISBN-13: 978-1-4613-3562-7   e-ISBN-13: 978-1-4613-3560-3
DOI: 10.1007/978-1-4613-3560-3

Proceedings of a NATO Advanced Research Institute on Systems Analysis in Urban
Policy-Making and Planning, held September 21–27, 1980, at New College, Oxford
University, Oxford, England

© 1983 Plenum Press, New York
A Division of Plenum Publishing Corporation
233 Spring Street, New York, N.Y. 10013
Softcover reprint of the hardcover 1st edition 1983

ACKNOWLEDGEMENTS

The manuscript for this book was prepared on a MICOM Word Processing Unit at the Department of Civil Engineering, University of Waterloo. Miss Rosemary McCleary typed the entire manuscript, contributed substantially to the grammatical quality of the text and was responsible for the splendid layout. Her assistance is gratefully acknowledged. Any errors or omissions remain, of course, the sole responsibility of the Editors.

Many other people made the NATO Advanced Research Institute and this book which emanated from it possible. We wish to acknowledge all those who were involved, whether or not they find mention here. In particular, Mrs. Irene Steffler of the University fo Waterloo and Mrs. Beryl Collins of the University of Wales Institute of Science and Technology (UWIST) were involved in a variety of organizational and secretarial matters throughout the period during which the Conference was planned and held, and during the preparation of this book. Michael Breheny of the University of Reading and Richard Spooner of UWIST contributed immensely to the smooth running of the Conference itself, and we gratefully acknowledge their help.

PREFACE

In September 1980, the Special Programme Panel on Systems Sciences of the North Atlantic Treaty Organization (NATO) sponsored an Advanced Research Institute (ARI) on "Systems Analysis in Urban Policy-Making and Planning" which was held at New College, University of Oxford, from 21st to 27th September. This week-long meeting brought together 35 invited delegates from most countries of the NATO Alliance to discuss the impact which systems analysis has had and is likely to have on urban affairs. The manuscript was submitted to the publisher in June of 1982.

Although the goal of the ARI was to assess the impact of urban systems analysis as seen through the eyes of those closely involved in such work, the meeting also addressed opportunities for future research and development, and therefore in this book we have attempted to synthesize discussions at the meeting with this in mind. But before we describe the structure of this book, it is worth recounting in a little more detail the intentions and organization of the meeting, for this has had an important effect on the type of papers produced here, the way they have been written, and the issues they address.

The structure of the week-long ARI was developed by an Organizing Committee consisting of Michael Batty of the University of Wales, UK, Bruce Hutchinson of the University of Waterloo, Ontario and Douglas Wright, then with the Provincial Government of Ontario, and now with the University of Waterloo. In addition, Donald Clough of the University of Waterloo, delegate to and former Chairman of the NATO Systems Sciences Panel, provided much helpful advice on the organization of the ARI. Bulent Bayraktar, Secretary of the NATO Systems Sciences Panel, also helped in establishing an appropriate focus for the meeting and in inviting a representative list of delegates.

This book represents the outcome of the meeting and as such, the papers included are broadly representative of the ideas discussed at the meeting. Indeed, all the papers included here were circulated prior to the meeting in draft form and were used to direct discussion. None of the papers, with the exception of those by Vickers, Ridley and Cole which were invited presentations, were formally delivered at the meeting, but all the papers were revised in the light of the discussions. We thus refer to these papers as "background papers", and as they constitute the bulk of this volume, they should be considered in the light of the introduction which presents the original intentions of the ARI, and the Conclusions which present its recommendations.

The book is organized as follows. In the introduction, the backgound to the meeting is sketched, the scope of the papers briefly reviewed, and the paper by Vickers which was delivered as the opening address, is presented. Vickers' paper introduced many of the themes developed here, and as such, it represents a useful focus on the subsequent contributions. The background papers are then presented, grouped into six parts. Finally, the conclusions comprise a paper by Batty and Spooner, outlining the results of a survey of the participants' attitudes towards certain issues in the field, and a paper written by the editors which synthesizes the recommendations emerging from the meeting.

Although we have imposed a fairly strict order on the papers by grouping them into distinct parts, there are many common themes cutting across these contributions; intending readers might there-fore best consider each of these papers as statements of each indi-vidual author's view of the contribution, actual or intended, of urban systems analysis to planning. Indeed, this was the stated intention of the organizers in asking participants to write these papers. Accordingly, these papers may be read in any order for in essence they represent "snapshots" of opinions concerning the state of the art in urban systems analysis, among experts working in a field which has emerged during the last two or three decades.

Michael Batty

Waterloo, Ontario   Bruce Hutchinson

April, 1982

CONTENTS

INTRODUCTION

Systems Analysis in Urban Policy-Making and Planning:
Background to the NATO Advanced Research Institute   3

Introduction to the Papers   13

The Poverty of Problem-Solving   17
        Geoffrey Vickers

PART 1:   CRITICAL ASSESSMENTS

The Technology of Urban Systems Modelling: Nagging
Questions, Sagging Hopes and Reasons for Continued
Research   31
        Michael A. Goldberg

The Effectiveness of Urban Transport Systems Analysis   53
        Bruce Hutchinson

Systematic Methods in Strategic Land Use Planning:
Reflections on Recent British Experience   69
        Peter Batey

Strategic Land Use Planning: An Evaluation of Procedural
Methodology   89
        Ian Masser

Systems Analysis in Planning: A Critique of Critiques   107
        Michael Breheny

PART 2:   SYSTEMS ANALYSIS IN DIVERSE CONTEXTS

The Impact of Systems Analysis on Urban Planning: The
West German Experience   125
        Michael Wegener

On the Use of Strategic Planning Models in Iberian Cities   153
        Pedro Geraldes

Systems Analysis in a Developing Country: The Case in
Turkey   169
        Gunduz Ulusoy

Systems Analysis in Urban Policy-Making and Planning:
The Munich Experience   177
        Klaus Schussmann

French Local Planning Practice   193
        Robert Laurini

PART 3:   APPLICATIONS OF SYSTEMS MODELS

Early Warning Systems for Urban Policy-Making and Planning   207
        John Dickey

Multiregional Population Analysis for Urban and Regional
Planning   227
        Frans Willekens

The Sao Paulo Metropolitan Study: A Case Study of the
Effectiveness of Urban Systems Analysis   243
        Marcial Echenique

Agency Policy Requirements and System Design   271
        William Goldner

PART 4:   POLICY ANALYSIS

The Failure of Model Use for Policy Analysis in Regional
Planning   293
        Janet Rothenberg Pack

Applications of Corporate Planning in Urban Policy
Analysis and Decision-Making   309
        Juri Pill

Political Realities and the Implementation of Urban
Transport Policies   327
        John Bonsall

PART 4:   Continued

Tyne and Wear and Hong Kong: Can Systems Analysis
Tackle the Realities of Decision-Making?   341
    Tony M. Ridley

Systems Concepts in Government Planning: Experience and
Prospects   363
    Douglas T. Wright

PART 5:   PHILOSOPHIES OF SYSTEMS ANALYSIS

Systems Analysis and Urban Policy   379
    E. S. Savas

Reflections on and Implications of Systems Analysis as a
Sociological Phenomenon   391
    Ida R. Hoos

Models, Metaphors and the State of Knowledge   407
    Sam Cole

On Systems Theory in Urban Planning: An Assessment   423
    Michael Batty

What Reasons for Rationality?   In Search of a Future for
Rational Methods in Urban Planning   449
    Helen Couclelis

PART 6:   EMERGING THEMES

Positive and Normative Aspects of Modelling Large-Scale
Social Systems   475
    Britton Harris

Planning and Decision-Making in Human Systems:
Modelling Self-Organization   491
    Peter Allen

Technology Considerations in Urban Systems Analysis   525
    William L. Garrison

A Plea for Planning-Oriented Research   545
    Henk Voogd

CONCLUSIONS

Attitudes towards Urban Systems Analysis: A Survey of
Participants   567
        Michael Batty and Richard Spooner

The Future of Urban Systems Analysis: Issues, Themes and
Recommendations of the Institute   583

The Authors and Participants in the Institute   601

Author Index   607

Subject Index   613

INTRODUCTION

SYSTEMS ANALYSIS IN URBAN POLICY-MAKING AND PLANNING:

BACKGROUND TO THE NATO ADVANCED RESEARCH INSTITUTE

INTRODUCTION

This paper outlines the kinds of issues which guided the organization of the Advanced Research Institute. The goal of the Institute was to assess the impacts of systems analysis on urban policy-making and planning, to discuss the success or failure of such ventures, and to attempt some evaluation of the experience both historically and in terms of the lessons to be learned for future work.

In a wider context, the purpose of the seminar was to explore the age-old problem of the relationship between scientific endeavour and social action; that is, to use systems analysis as a window through which to view the dilemmas, paradoxes, insights and ambiguities which plague the use of any science or rational method in a social context. To this end, the seminar was not focused primarily on the technical issues of systems analysis, but on the use and impact of such technical analysis in a wider political environment.

Accordingly, the definition of systems analysis which guided the organizing committee was fairly liberal and included any rational analysis procedure that contributed to the understanding of urban problems and the prescription of possible solutions to these problems. The term "systems analysis" refers to systematic procedures which are at least partly explicit, and when applied to policy-making and planning enable scientists and decision-makers to do more than simply "muddle-through". Different styles of analysis have developed during the last twenty to thirty years which reflect the various aspects of work which are included under the umbrella term of urban policy-making and planning. The ability to develop scien-

3

tific analysis varies quite widely between different aspects of
planning. Moreover, the emphasis to these different aspects demands
different techniques which embrace the overall scope of systems
analysis. For example, short-term problems of planning involving
the organization of public utilities and facilities are quite well-
defined, and are well-suited to the application of operational
research methods. In contrast, longer-term problems of population,
transport and housing are more suited to the development of economic
systems analysis and dynamic simulations. The organization of these
areas of technical work themselves in terms of budgeting and scheduling, involves the development of information and management
systems. The field is large and an attempt was made to be clear
about its scope at the outset of the Institute.

More specifically, the work of the Advanced Research Institute
was structured with the aid of the following questions:

> What has been the experience of systems analysis
> in this area during the last two decades?
>
> How relevant have been the techniques and appli-
> cations?
>
> What has been learned about the theory of systems
> analysis and the ways of applying it?
>
> Are the difficulties and contradictions which
> characterize the field simply manifestations of
> some broader issue about the relevance of science
> to human affairs?
>
> How can analysis be improved?
>
> What are the current and likely future issues?
>
> What are the prospects?
>
> What should be done?

Clearly it was not assumed that all of these questions could be
answered, at least in any definitive way, but in attempting to evaluate systems analysis in this way, the hope was that the seminar
would begin to raise the collective awareness of the delegates concerning these issues. An attempt was made to select delegates to
the Institute with a healthy balance of academic and non-academic
experience, theoretical insight and practical knowledge, intellectual rigour and political sensitivity to the series of questions
around which the seminar was structured.

One final distinction was made and this relates to the distinction between the subject areas comprising systems analysis in urban
policy-making and planning, and the methodological themes involving

analysis and planning per se. In essence, the domain of urban planning can be separated into quite distinct areas, the prerogative of different professional groups and different disciplinary perspectives, but cutting across these are common analytical and methodological approaches. The Organizing Committee proposed that the various subject areas be represented in some way in the seminar through the experience and skills of the various participants, but that the prime focus of the seminar would be on extracting commonalities and differences in the approach and impact of various types of systems analysis.

THE SCOPE OF THE INSTITUTE

At the risk of oversimplification the Organizing Committee defined five broad clusters of activity which form the substantive basis of systems analysis in urban policy-making and planning. These five areas can be considered as separate subject areas but each area is also distinct in terms of its professional skills, methodological approaches and possibly its impact on practice. These areas are listed as follows:

1. Public Services Planning: involving rather well-defined short run problems such as fire and police protection, health care, educational provision and location, the management of existing transport services, and the resource allocation process involving such delivery systems. In this area, systems analysis has been developed in quite straightforward ways using existing OR techniques and simulation methods.

2. Physical Network Planning: involving somewhat longer run problems than Item 1 above, often with a strategic emphasis but concerning fairly well-defined network structures in cities such as transport, water supply, sewerage, electricity and so on. Systems analysis has involved the development of more special purpose models and techniques than Item 1 above, and these are embedded in a somewhat looser planning context.

3. Built-Form Planning: involving the design of adapted spaces in cities, typically housing and service functions. Such planning is short- to medium-term, involving less direct systems analytic techniques than Items 1 or 2 above but being quite strongly related to Strategic Land Use Planning which is accomplished at a higher level.

4. Strategic Land Use Planning: involving much longer range problems where the emphasis is typically on urban spatial organization and the informing theory is largely economic in emphasis. Systems analysis has developed in terms of special purpose computer models, more strongly mirroring economic theory than those discussed so far. Such problems are quite ill-defined and the range of analytic techniques is wide.

5.   Corporate Planning: involving a higher level decision-making
     function than those areas sketched above, but typically coor-
     dinating the resource allocation process and budgeting require-
     ments of the shorter-term demands made by Items 1, 2, 3 and 4
     above.

     These subject areas, apart possibly from Corporate Planning,
are largely organized in terms of the application of techniques to
specific problems within appropriate domains.   Corporate Planning
exists as a higher level function, but it too is concerned with a
fairly narrow focus on resource allocation which lends itself to
specific techniques of analysis.   From all these areas, two distinct
methodological characteristics emerge.   The first relates to analysis.   Clearly, the way in which systems analysis is pursued, the
theories which are used as a basis for the appropriate systems
models, and the specific techniques in use can be compared and
assessed.   Second, there is the relationship of these subject areas
to planning or design; in effect this is the relationship to the
wider political and organizational environment.   In terms of both
the methodology of analysis and of planning, there are a series of
themes which can be drawn out.   These can be broadly structured as
follows, noting that the division into methodological themes is
somewhat arbitrary in comparison to that made for subject areas; for
the subject area division relates to how the field is actually
organized, whereas the methodological division relates to the themes
that were explored in the seminar.   Part of the seminar was to
generate common methodological assessments of systems analysis.
Five broad themes can be distinguished relating to analysis and
planning, and these are:

1.   Methods of Analysis: involving the specific techniques applied,
     theories on which such techniques are based, and the scien-
     tific-experimental method used in their selection and valida-
     tion.

2.   The Use of Analysis in Planning: involving the juxtaposition of
     the systems analytic process with the narrow technical process
     of design and planning; and the contrast of positive analysis
     with normative.

3.   The Decision-Making Environment: involving the organizational
     and political environment in which systems analysis is used,
     and the contradictory demands and aspirations of technicians,
     scientists, bureaucrats and politicians.

4.   The Philosophic Basis of Systems Analysis: involving the link
     between theory and practice, thus leading to the assessment of
     systems analysis in urban policy-making and planning as a spe-
     cific case of the impact and relationship of science to
     society.

5. Analysis of and Speculation on Future Decisions: involving an
    assessment of the overall experience and designs for the appro-
    priate types of systems analysis in the appropriate contexts
    and environments.

The Advanced Research Institute attempted to address both subject areas and methodological themes, and the clearest way to conceive its scope is to imagine a concatenation of the five areas with
the five methodological themes. This then provided the scope of the
seminar, but before the concatenation of areas and themes is discussed it is worth digressing slightly to try to give more flavour
to the seminar which the Organizing Committee had in mind.

EXISTING ASSESSMENTS OF SYSTEMS ANALYSIS

In a short paper such as this it is impossible to do more than
sketch the key issues which characterize the field. Moreover, in a
field so diverse as this one, everyone will have a slightly different interpretation as to the significance of these issues, and
everyone will perceive the organization of the field differently.
However, one way of providing a clearer idea of what the organizers
had in mind is to discuss the subject areas and methodological
themes in terms of the pertinent literature already devoted to these
issues. Thus, in this section the issues raised above are clarified
in terms of what the organizers considered to be the literature most
relevant to the issues already raised. In the following exposition
an attempt is made to be as brief as possible, emphasizing only the
books, not papers available, and with the selection having been made
on the basis of the predilections of the Organizing Committee.

The five subject areas are easiest to identify in terms of
existing literature because systems analysis has been developed as
part of these areas. The co-directors of the ARI originate from the
Physical Network Planning - Strategic Land Use Planning areas and
have both written about these topics. The book on transport planning by Hutchinson (1974) gives a comprehensive account of the
modelling styles involved, that by Batty (1976) on land use modelling is slightly narrower, focusing on applications. The book by
Beltrami (1977) on public systems analysis gives a good account of
the models in the public services - delivery systems area, and it
seemed to the organizers that these areas of modelling, reflecting
the previous subject areas of physical network planning, strategic
land use planning and public services planning are the best defined
of any of those considered. In terms of built form planning, the
book by Martin and March (1972) is representative, but this field is
less organized than those just described. So is that of corporate
planning where the work of Wildavsky (1974) is well-known. But in
these latter two areas, the issues are slightly broader and the
subject areas less well-defined.

These books deal variously with the substance of these subject areas, as well as with the techniques of systems analysis, that is model-building techniques. But these contributions do not really deal directly with any of the methodological issues concerned with planning with the exception of Wildavsky's. There are quite a few books which deal with philosophic/theoretical rationale for systems analysis in planning: a good example is that by Chadwick (1971). But in terms of the Institute, of more pertinent interest were those contributions which have attempted some assessment of the application of systems analysis, and have emphasized the wider methodological issues. The history of systems analysis documenting the successes and failures in these fields is dealt with by several authors, in particular by Hoos (1972) for corporate planning and more generally, and by Boyce, Day and McDonald (1970) for strategic land use planning. In terms of the critiques which have been developed, these can be seen as emphasizing theoretical problems and/or organizational problems: that is, as emphasizing the analytic aspects of systems analysis, or the social context in which it fits. Drake's (1973) book on transport modelling projects, Brewer's (1973) book on particular land use modelling experiences in Pittsburgh and San Francisco, and Greenberger, Crenson and Crissey's (1976) book on several experiences in socio-economic modelling are all excellent examples of the kind of assessment which the Organizing Committee hoped to make as a result of the seminar. Another essential component in the assessment is work such as that by Pack (1978) which emphasizes the actual scale and amount of systems analysis in urban policy-making.

There are many more contributions to the field in the form of other books, papers and reports which are important, many of which are identified in the papers presented by delegates. These books provide something of the flavour that was expected to be generated during the seminar. To conclude this brief excursion into the literature, it is worth noting the major themes which emerge from these existing assessments. In applying systems analysis to planning, there is always a mismatch between the theory available and that required to inform about the problem. Moreover, this appears to be compounded by the different motivations displayed by analysts and decision-makers. There is also an organizational problem in that the appropriate environments are seldom available for the proper application of such techniques. There are resource problems based on manpower, computer capacity and data limitations. And there are problems posed by investing time and money in techniques which are eventually judged to be inappropriate. Finally, there are problems reflecting different perceptions of what is possible in the scientific and political domains. All these and more emerge from the critiques and formed the heart of the material to be assimilated, discussed and evaluated during the seminar.

ORGANIZATION OF THE INSTITUTE

The week-long seminar was structured around a series of major and minor questions with various sub-sets of the delegates asked to address these questions with respect to the principal subject areas identified previously, and to report their findings to a plenary session for further discussion. These questions are:

What has been the experience in applying rational analysis techniques to urban problems?

> What issues have been addressed and what types of techniques have been used?

> Were decisions actually based fully or partially on the results of these analyses?

> What disciplines were involved in the development and application of these techniques?

Why are the results frequently ignored from the perspective of the analyst?

> Is it because of data limitations?

> Is it because urban phenomena are very complex and models are trivial representations?

> Is there a fundamental mismatch between analysis time and decision-making time?

What have we learned from trying to implement urban systems analyses?

> What has been the experience in the organizational/political area?

> What conditions or restrains the responses of organizations?

> What have we learned in the areas of data, education and the computing area and why?

> Are there attitudinal problems and is there a basic incompatibility between the attitudes of the analyst and the administrator/decision-maker?

> Are there differences in experience across cultures?

Are the difficulties in the urban field simply a manifestation of the broader issue about the relevance of
science to human affairs?

Is there a crisis of confidence in urban management?

Is this crisis unique to the urban field?

Are the techniques simply an intellectual fabri-
cation?

What are the current and expected urban issues and
what processes should be followed to respond to these
issues?

Are there differences in the perceptions of urban
problems between institutions and groups?

What uncertainties are there about defining
future issues?

Should we plan for the future or adapt to
changes?

What things can be realistically achieved by the
public sector?

What should be done and what should the various actors
do?

Can the formal analysis of urban problems be made
more effective through the development of better
techniques?

Will developments in computer technology and
better data bases improve the effectiveness of
systems analyses?

Will organizational changes help to solve the
problems and improve the effectiveness of systems
analysis?

CONCLUSIONS

As mentioned in the Introduction, this is a large area with
extremely diverse aspects, and in the event, it was not expected
that a completely balanced picture could be provided. Some interests were bound to be represented more than others. Moreover, none
of the questions can be answered definitively, if at all, but at
least it appeared to be a suitable way of raising awareness about

the problem, and of generating some mutual understanding between scientists and decision-makers. The intent was that the seminar would provide an opportunity for scientists and decision-makers to cooperate in making a useful assessment of the state of the art and the prospects for future research and application. The observations and conclusions which emerged from the meeting are described in a subsequent section of this book.

REFERENCES

Batty, M., 1976, Urban Modelling: Algorithms, Calibrations, Predictions, Cambridge University Press, Cambridge, UK.

Beltrami, E., 1977, Models for Public Systems Analysis, Academic Press, New York.

Boyce, D., Day, N. and McDonald, C., 1970, Metropolitan Plan Making, Regional Science Research Institute, Philadelphia, Penn.

Brewer, G.D., 1973, Politicians, Bureaucrats and the Consultant: A Study of Urban Problem-Solving, Basic Books Inc., New York.

Chadwick, G.F., 1971, A Systems View of Planning, Pergamon Press, Oxford, UK.

Drake, J.W., 1973, The Administration of Transportation Modeling Projects, D.C. Heath, Lexington, Mass.

Greenberger, M., Crenson, M.A. and Crissey, B.L., 1976, Models in the Policy Process: Public Decision-Making in the Computer Era, Russell Sage Foundation, New York.

Hoos, I., 1972, Systems Analysis in Public Policy: A Critique, University of California Press, Berkeley, California.

Hutchinson, B.G., 1974, Principles of Urban Transport Systems Planning, McGraw-Hill Book Company, New York.

Martin, L. and March, L., (Eds.), 1972, Urban Space and Structures, Cambridge University Press, Cambridge, UK.

Pack, J.R., 1978, Urban Models: Diffusion and Policy Application, Regional Science Research Institute, Philadelphia, Penn.

Wildavsky, A., 1974, The Politics of the Budgetary Process, Little, Brown and Company, Boston, Mass.

INTRODUCTION TO THE PAPERS

The first paper was presented by Sir Geoffrey Vickers at the opening session of the meeting and is entitled "The Poverty of Problem-Solving". His principal point is that systems analysis is not a technique for solving problems but rather an approach to understanding situations; that understanding may or may not be followed by some action involving a detailed second round of analysis. Systems analysis is viewed as offering a conceptual framework that helps make sense of what one knows already. The theme proposed by Vickers emerged frequently during the discussions throughout the week of the Advanced Research Institute. It is also a theme which emerges in many of the papers and thus Vickers' paper has been included as part of the introduction to give some perspective to the arguments which follow.

The remaining papers have been grouped into six parts, reflecting their emphasis and enabling the reader to direct his or her attention to particular types of critique. All the papers included reflect the state of the art in urban systems analysis and all are critical in some respects, for the intention of the meeting was to develop critical assessments. In fact, none of the papers dwell on techniques but all attempt some critique of past experiences and some speculation on future directions for the development of this field. Some authors have chosen to produce broad surveys of the experiences, others more specific assessments based on particular techniques, applications and political contexts. Some have dwelt on the culture of systems analysis and its use in policy-making, some have attempted a more fundamental critique in terms of its philosophy, while others have chosen to emphasize the extent to which systems analysis is responding to new ideas concerning urban problems. Yet these papers are not specifically about the future role of urban systems analysis for it was the meeting itself, not the background papers which sought to emphasize this. Accordingly, recommendations about future developments are developed in the concluding paper.

The first part includes five papers which develop general critical assessments of experiences in urban systems analysis, reviewing the suitability of techniques, the appropriateness of organizational and political environments in which such analyses have taken place, and the relevance of the methods used. These papers use the experiences with land use and transportation planning to emphasize these themes. Goldberg examines the experience in urban systems modelling in North America while Hutchinson explores the experience with transport systems analysis in several countries. Batey and Masser both review the British experience with systematic methods in planning, and finally Breheny develops a broad based critique from the more general perspective of urban and regional planning.

Five papers are included in the second part which also presents experiences from specific contexts, covering both techniques and applications. Wegener begins by providing a comprehensive account of the West German experience while Geraldes reviews experiences with land use-transport models in several regions of Spain and Portugal. Ulusoy then illustrates some of the difficulties in implementing systems analysis in a developing country, Turkey, while in complete contrast, Schussmann describes the elaborate systems of planning models developed for the City of Munich. Finally, Laurini describes the character of French local planning practice and points out that it is more concerned with the resolution of planning conflicts than with formal analyses.

The third part deals with more specific applications of systems models, largely urban models, which emphasize technique but also the ways in which model-builders have responded to the requirements of policy-makers. Dickey describes an application of a Systems Dynamics type model to the forecasting of area-specific socio-economic conditions in the Miami region for social planning purposes. Willekens then presents experiences with the development and use of multi-regional population models, emphasizing ways in which techniques need to be adapted to specific conditions, and Echenique discusses the application of a land use-transport model to the formulation and evaluation of transport policy options in Sao Paulo, Brazil. Finally, Goldner outlines some twenty years of experience with the development, application and adaptation of spatial models in the San Francisco Bay area, introducing explicitly the policy context.

The fourth part includes five papers which describe experiences with systems analysis from the perspectives of policy advisors and policy-makers. Pack describes the results of a comprehensive survey of the use and impact of planning models in US planning agencies, while Pill outlines the experience with techniques for corporate planning in the Toronto Transit Commission. Bonsall describes the experience in using transport systems analysis methods in the formulation of transport policy in the Ottawa-Carleton region of Canada, while Ridley compares his experiences in developing new rapid transit facilities for the Newcastle-upon-Tyne, UK, and Hong Kong regions. Finally, Wright describes his experience with systematic methods in a corporate context, as a senior civil servant dealing with a variety of social policy areas in the Government of Ontario.

The emphasis of the papers in part five begins to shift back towards more general critiques, where the papers have a slightly more philosophic orientation. Savas, using his experience in New York City, poses a variety of questions concerning the role of the analyst in a policy context, and Hoos continues by arguing that systems analysis in its present form severely constrains the search for innovative solutions to complex problems. Cole, Batty and Couclelis all point to the preoccupation of systems analysis with modelling, and the lack of attention paid in most studies to the problem context. Couclelis in particular hints that new developments in ways of thinking about systems are required if formal analysis is to have greater relevance in the near future.

Part six identifies some of the potential opportunities for new techniques and new approaches. Harris points to the unification which is taking place between positive and normative styles of systems thinking while Allen examines the potential of treating urban systems in particular, and systems in general, as dissipative structures. Garrison discusses the need for understanding the fundamental characteristics of the technology that systems analysis is attempting to manage, and Voogd concludes with the plea that future research should address processes of planning rather than techniques to be used in such processes.

The conclusions contain two papers which emerge directly from the meeting itself. First, Batty and Spooner undertook a survey of the attitudes of participants to factors affecting the development of urban systems analysis, and the results are presented in a short paper. Finally, the editors have synthesized the ideas generated by the participants during the meeting, in a paper reflecting the themes discussed, as well as providing recommendations for the future of urban systems analysis.

THE POVERTY OF PROBLEM SOLVING

Geoffrey Vickers

The Grange
Manor Road
Goring, RG8 9EA
United Kingdom

THE PROPER ROLE OF SYSTEMS ANALYSIS

Most of the papers presented to this conference seem to me to disclose one or more of three misconceptions about what systems analysis should be and might be. These, I shall suggest, sometimes exaggerate, sometimes restrict and always distort what should be regarded as the proper contribution to policy-making to be expected of the profession of the systems analyst — an important profession, though not a new one, except for the new technique which is now at its disposal.

The first of these misconceptions is the idea that systems analysis is primarily a technique for solving problems. It should rather, I suggest, be regarded as a means of understanding situations. Understanding may or may not be followed by the hope that something can be done to make the situation better or to stop it getting worse. And if so, a second and different round of analysis will follow. The analyst is then required to analyze over some future time span the hypothetical situation which would be created if the hypothetical action were taken. He may even be required to compare the probable outcomes of more than one hypothetical response or himself to suggest and explore the effect of more than one or more possible responses. And at this stage he may regard himself, if not as a problem solver, at least as one concerned with exploring the possibility of a solution. But a "second round" by no means always follows. Once the situation is understood it is often apparent without further analysis both what if anything can be done, and what needs to be done. The luxury of choosing between alternatives is by no means always open. In any case, to focus on problem solving is to divert attention from the far more important function of problem definition and to confuse the continuing process of system regulation with the episodic activity of seeking specific goals and the much more frequent and radically different activity of averting specific threats.

Of course people do not seek to deepen their understanding of a situation unless it is causing them some concern, but this of itself does not necessarily create a problem, still less a soluble one.

Concerns may be of two kinds — perception of some present state which awakens anxiety or of some imaginable future state which awakens aspiration. But it does not follow that the concerned person can do anything either to abate his anxiety or to realize his hope. Through most of human history most human ills have been regarded not as problems but as part of the human condition. Some still are; others should be. It is important to know the difference.

Nor does government or management consist primarily in solving problems. It consists in regulating systems. Systems analysis means to me the analysis of a system. And the system to be analyzed has first to be defined by reference to the concern which makes it of interest to the policy-maker.

This defining of the situation to be analyzed is a crucial preliminary task of the systems analyst. He must include all those factors, which are so important in the context of that concern, that none can be omitted without making nonsense of the others. (Water pollution, for example, is not a very important factor to someone planning a system of water-borne transport, but it is vital if his concern is a trout hatchery or the distribution of drinking water.) Some concerns relate to situations which are much simpler to analyze than others. Some relate to situations which are impossible to analyze because we do not know the crucial factors involved, or because we do not agree about them. Some relate to situations which we can analyze but not model in any quantitative way. The systems analyst should know these differences better than anyone; and his client senses them and is influenced by them in deciding what analyses to commission. It is no accident that these papers are almost wholly about transportation problems. None is concerned with crime or unemployment or housing policy, or (with one partial exception) with education or with the treatment of cultural minorities. Yet these are major concerns of city governments and threaten dangers at least as great as overcrowded roads and subways.

I do not criticize these omissions. They acknowledge the limitations of systems analysis or at least of its current concepts of "modelling", which are limited as well as enlarged by concern with quantitative computerized models so obsessive as almost to extinguish belief in the unaided powers of human judgement on which human governance has relied, not always unsuccessfully, for many millenia before the present generation. It is this obsession, as Dr.

Archibald has pointed out (Archibald, 1979), which has virtually confined the phrase "systems analysis" to "systems modellers", and brought into use the wider phrase "policy analysis" to cover all analyses made by whatever means to help the policy-maker.

I do, however, criticize the paradoxical eclipse of the concept of system itself. Contributors always use the adjective systematic rather than systemic. One paper defines systems analysis as "any rational procedure for understanding urban problems and predicting and prescribing possible solutions" and further defines these as "systematic procedures which are at least partly explicit". The terms "rational" and "systematic" remain undefined except that the procedures which they involve must be "at least partly explicit". There is scarcely any indication that "systems" (open systems for our purposes) are a specific kind of phenomena, a broad class indeed but one which has important characteristics in common and that the study of them has greatly clarified and modified our conception of the nature and purpose of government and to a lesser extent of business management.

SYSTEMS AND THEIR REGULATION

Open systems have four characteristics which I wish to stress, familiar though they are, because they profoundly affect the nature of government and hence the objects of systems analysis.

First, an open system is a form more enduring than its constituents. The old philosopher who insisted that we cannot step twice into the same river seems to me to have been wilfully perverse. For he doubtless knew as well as we that the name of a river is the name of a form, not the name of a particular collection of water which happens to be flowing through it at any moment of time. The first concern of any systems analyst is to identify the factors which are preserving this form and to determine whether they are equal to their task. For these forms can change or oscillate between extremes or even dissolve. The dimension of stability-instability is inseparable from the concept of form. The form of a river is largely given by external factors, chiefly the contours of its catchment area. The form of an organism is largely given by internal factors, some of which are set to follow a course of most complex development, whilst others, such as those which preserve the internal temperature of so-called warm-blooded animals (and countless other internal relations, chemical and physical) are set to preserve internal stasis, despite external change. The form of a society is determined partly by external pressures and opportunities and partly by internal cultural structures ranging from explicit laws to the countless subtle conventions which deter or control the spread of deviance.

Systems, then, are bundles of inter-acting relations, internal and external. The internal ones enable each to maintain coherence

even through change. The external ones regulate the system's relation as a whole with its milieu including all the other systems of
which it forms part. This is as true of a city government as of a
cat. A city government contains departments, most of which are
responsible for maintaining some set of relations within acceptable,
or at least viable, limits -- the relation of sewers to sewerage,
schools to schoolchildren, roads to traffic and a dozen others.
Other departments are concerned with the relations of the government
as a whole with its surround, notably the relation of its revenue to
its expenditure. All these departments compete with each other for
scarce resources and some involve sometimes inconsistent activities. And each set of relations has a limit beyond which its breakdown will have sudden, acute and possibly irreversible repercussions
on all the others. (How few city activities can continue if the
sewers stop working!) The primary task of government is to keep all
those relations, internal and external, within their permissible
ranges - and hence to detect and correct instabilities before they
become overwhelming. Its secondary task is to alter some or all of
these relationships if and when it can in a way which is regarded by
all who cannot be ignored as being on the whole more acceptable to
them.

All this was true of times far more stable than our own. Today
we see everywhere evidence of instability approaching breakdown.
Inflation, unemployment, multiplying populations, increasing cost of
energy and raw materials, degeneration of the biosphere, famine and
endemic war all signal the breakdown of systems necessary to the
survival of the race or at least of most of the major political
systems by which human life is at present sustained on the planet.
A systems view is necessary and it is useful even when it can be
neither computerized nor quantified.

Everyone knows, for example, that if Britain continues to
increase the populations of its prisons which are already grossly
overcrowded even by the 19th century standards to which many of them
were built, something will pass a critical point and the system will
break down. Perhaps all the prison governors will resign. Perhaps
all the warders will walk out. Perhaps the felons will burn the
places down. Perhaps .... No one can predict precisely when the
system will break down, or what will trigger the breakdown, or what
form it will take. But the event is sufficiently certain to set a
host of different "problems" at different levels of the hierarchy -
to prison governors regulating as best they may the relations of
felons to prisons and especially to their warders and to each other;
to the police in deciding what prosecutions to bring; to judges
deciding what prison sentences to award; to the Home Office considering the organization of the service, the building of new prisons
and alternative forms of custodial and non-custodial care; and to
sociologists and others seeking to understand the generation of
crime. All require analysis of many situations. Probably none
would be aided by a computerized model. Yet I see no reason to
suppose that any of them are beyond the understanding of a human
mind.

Whether such understanding would be attained by a wholly rational process depends on the meaning given to that term but almost any currently accepted meaning would be obviously inadequate for at least four familiar reasons. First the judgments made by men about men are inescapably affected by the human experience of those who make the judgments and are usually affected for the better. Most of our vocabulary concerning human experience would have no meaning if it were not enlightened by experience of our own. Secondly these judgments involve appreciations of value as well as fact; for if no human value were involved there would be no concern and no analysis. Thirdly human experience and action takes place in specific contexts and is radically affected by them. There is no means of formulating laws appropriate to each of an indefinite number of contexts or laws of any substantial use which are so general as to be indifferent to context. Fourthly the situation as validly analyzed by each actor in a situation is different from the analysis made by all the others, even though each actor is aware of those others and takes account of them. For example, the situation posed by the collapse in demand for steel, looks different and is different for a steel worker, a plant manager, a steel marketeer and a general manager, even though each succeeds by an effort of imagination in understanding what the appreciations of others mean to them and incorporating as much of it as is relevant to him. Concerns are highly personal and unless the essential parts of the analysis made by each are common to all, no concerted action is to be expected. The technologist of pre-1914, turning one physical state into another to the general delight and amazement of his fellows and as indifferent to its systemic effects as they were is the most unsuitable of all imaginable soils in which to grow the systems analyst of today.

REGULATION AS RESPONSE TO CHANGE

My experience of what I understand by systems analysis has not been primarily in the area of urban government but I will summarize it briefly to make the point. It begins on 12th November 1918 — long before anyone was talking systems language, but not before competent policy-makers were thinking systemically. On that day — the morning after the Armistice which ended the first World War — the general commanding the division in which I was serving in France called together his battalion commanders and addressed them in substance as follows.

> "Gentlemen, have you considered how you are going to manage without the Germans? Reflect on the changed situation in which this armistice has left you.
>
> For more than four years everything we have done has been dictated by the German army as everything it has done has been dictated by the

Allies. And our efforts to destabilize each other have produced a situation not merely stable but self-stabilizing. Until these last few months every attack by one side has petered out from inability to sustain its own momentum, at least as much from the resistance of the other side. And now after only a few months of mobile warfare the inherent weaknesses have broken surface, political agreement has stopped the war — and where are we?

For more than four years France has been divided into an occupied zone — occupied by the Germans — and a free zone, defended by us. Now all the Germans will be gone in a few weeks but we shall be here for months. For the policy is to demobilize our army individually from here. Meantime we shall be the "occupiers". The French who were behind the German lines will be free to get their working life back into its familiar pattern. The French on our side will still have hundreds of thousands of idle, armed, alien men in their barns and houses. Think what that will be like for them.

And this demobilization policy. Men who have been out here for years may wait for months before the factories where they have been promised their old jobs can be turned round for civilian work again. But boys who have been out less than a year may be the first to go home to places in universities which are empty and waiting for them. All good logical sense. But think what it will be like for the old timers?

And how will they pass their time? Toughening up for a next battle that isn't going to be fought? Learning skills to win a war that has already been won? They can't play football all day. Yet your continually shrinking force will be here all through a Normandy winter.

The conclusions are obvious. The external relations that matter to you now are no longer with the Germans or even with me but with the French civilian population and with the demobilization authorities at home. The first must be fostered as never before. The second must be trusted and endured. Neither will happen unless your men understand and respond to the changes I am describing. Please see that they do. Internally you have to find ways for them to spend their

time bearably, if possible enjoyably, even usefully. You are sure to find lots of internal
resources if you look for them — skills which
people have and will gladly teach; skills which
others will be glad to learn. Present the situation to your men as a problem and engage them in
solving it.   And please start today.   Today
everyone is just listening to the silence.
Tomorrow they will begin asking questions about
what happens next. I want them to come up with
the answers I'm giving you. This is a transient
situation but it will last for months and a lot
can go wrong in a few months. Start now."

Across more than sixty years I remember that man with respect
and admiration. He told his hearers nothing they did not know but
he sent them away with an understanding of the situation and what it
demanded of them which none, I think, had when they arrived. He did
not weigh alternatives. The policy was clear once the situation was
understood.

Did he act rationally? Certainly not irrationally. And yet
what he said was not derived simply by logical deduction from proved
or assumed postulates. It was primarily an exercise in empathy. He
had never been a French farmer or a conscripted toolmaker but his
experience of being human, though gained in the restricted context
of a British pre-first-war regular soldier, was sufficient for him
to see the situation not only as it was for him but as it would be
seen by people whose position and experience were remote from his
own. This is an important element in human judgement. Has it any
bearing on the analysis of urban problems? What about those high
rise, low rent apartment blocks of which at least two have had to be
demolished by explosives within 30 years of their construction
because no one would live in them? Presumably those who decided on
them and planned them calculated the costs and benefits in terms of
site economy, increased traffic density, containment of urban
sprawl, relative cost of erection and maintenance and other such
quantifiable features. But it remains an essential criterion of
success of any policy that it should be acceptable to those for
whose benefit it is designed. The insights of General Campbell
would perhaps have averted one of the most dramatic boobs in the
history of urban planning.

My experiences of systems analysis seem always to have been
gained in the context of some sudden instability. But so have
yours. Even transport problems become urgent only when the demand
for mobility passes or is seen to be about to pass the facilities
for mobility to an extent sufficiently acute to be deemed unacceptable by those affected. This threshold of the unacceptable is highly
subjective but no less important for that. And it tends to be
reached suddenly. For worsening relations tend to escalate and the
threshold between the bearable and the unbearable is hard to antici-

pate until it is crossed. I recall the European glut of coke in the depression between the wars when the collapse in the demand for steel left the makers of the "hard" coke used in blast furnaces with the alternatives of extinguishing their ovens or throwing their production onto the domestic market, already served in those days by the coke produced as a by-product by the gas industry. Their consumers, like the German army in 1918, had suddenly gone away. The only overriding certainty was that some agreement between the stranded coke producers of five countries would be less disastrous for them than a "free for all" battle. I recall the even more elusive problems of the short term money market when, following the first bank failures, confidence evaporated - confidence of banks in each other, confidence of industrial clients in banks, confidence of everyone in the banking system. Of all the constituents in any system of relations, confidence is the most basic and the hardest to renew.

This is no place to tell the resultant stories. They had a common theme -- the restoration of stable relationships and of the trust by which all relationships are sustained. They had a common factor -- shared understanding of what had destabilized those relations and what were the limits within which self-sustaining balance could be again restored. They had a common outcome -- shared revision, usually downwards of the threshold of the acceptable.

They also shared some element of encouragement; for they showed that the self-regulating capacity of these complex systems was very great. This lesson was to be intensified for me a few years later when I found myself concerned with economic intelligence during the second World War. It had been expected that the increased interdependence of the world for technological skills and materials would make economic limitations far more important and economic targets far more vulnerable. Again it was found that politico-economic systems are far more adaptive than had been expected -- provided always that the will to sustain them remained unimpaired.

When in the years after the war I read some of the founding fathers of cybernetics and systems theory -- Wiener, Bertalannfy, Ross Ashby -- I felt much as I had felt thirty years before when that general analysed what the armistice "meant" for us. I remember still the intellectual excitement of acquiring a conceptual framework which made sense of much that I already knew and promised better understanding of whatever I might learn thereafter. Far more important, it seemed to promise a basis for common understanding of a manifestly unstable world and hence hope for common action and acceptance of action on a scale which would manifestly be needed but had not previously been in sight. Another thirty years on that hope is heavily clouded.

Two landmarks taken almost at random will make the point. In the early 1950's the State of California commissioned several "systems analyses" of different fields of governmental activity. I

write from memories I cannot now confirm and hope that I do not
understate the vision behind those enquiries. But if I recall the
affair aright, one enquiry concerned the disposal of "trash" -- but
not the generation of trash. I was shocked. Could a systemic view
take the generation of trash as a given and concern itself only with
disposing of a nuisance which was expanding exponentially? Apparently it could. I could think of many reasons why the enquiry's
terms of reference should be thus truncated but none of them were
encouraging.

A decade later the British report on Traffic in Towns
(Buchanan et al, 1963) was more encouraging. This report (the
Buchanan report) began by gently rewriting its terms of reference
which were to consider the "problem" of vehicular traffic in towns.
This, said the Committee was no problem -- only a symptom of the
fact that modern towns generated more vehicular traffic than their
often mediaeval layout could contain. They could be redesigned in
four dimensions (including the allocation of temporal time bands) so
as to accommodate more traffic. The report illustrated a host of
unfamiliar innovations. Equally their activities could be regulated
so as to generate less traffic. But urban design and urban activities, not merely roads and traffic, were the minimal factors for
inclusion in the situation to be analysed. And the criteria for
assessing any hypothetical action for improving the situation were
equally widened, for any change would surely affect not only the
flow of traffic but also pedestrian access, pedestrian safety, amen-
ity, parking and the whole quality of urban life.

I was encouraged. This was no mere technical report. It was
education in the art of governing and of being governed. I would
like to see it a text book in all secondary schools.

For there is no more important element in education today than
the understanding of the systems which we form and by which we are
both sustained and constrained. And these systems are formed not
wholly or even chiefly of roads and buildings but of the standards
of expectation entertained by ordinary men and women. These standards especially in the West are today, as I believe, exaggerated and
conflicting to an unparalleled degree. Technological euphoria comparable in its potency and unreality to a Polynesian cargo cult is
combined with eclipse and confusion of ethical standards and even of
the relevance of ethical standards to a degree which seems to me
wholly inconsistent with the demands made on every individual and
every society by living in an overcrowded world of limited resources. My view may be unduly dark but even if it is, there must
remain a radical inconsistency between the traditional attitude of
technology and the realities of government.

"The difficult we do at once. The impossible takes a little
longer." So boasted the technologist in his finest hour. None but
the craziest autocrat has ever claimed such power in the field of
government, least of all in the government of peoples among whom

even small minorities have today powers of veto and destruction of which even majorities did not dream in days gone by. And on the other hand, the field of "knowledge" (obscure and ambiguous word!) which the technologist claims today is curiously limited. For it is increasingly confined to processes which can be transferred to a computer, still usually defined as "logical processes which can be fully described" (and therefore programmed) whereas even such confused epistemology as we have credits us with far wider mental capacities than this.

THE MEANING OF RATIONALITY

Thus I return to a topic on which I have already touched, the meaning of rationality. Almost all the papers at this conference declare their allegiance to the god of rationality but its meaning seems to vary as much as those of most gods to their various worshippers. For all it has a strong negative connotation -- the reverse of what is irrational or, worse still, anti-rational. But can there be mental activities which are neither rational nor irrational? For many the proposition seems to be regarded as a contradiction in terms. Yet it was not always so and it is less so today than it was a hundred years ago.

This is no place for an essay on epistemology but it should be possible to make four points without fear of serious contradiction and it may be useful to do so.

First the power to reason deductively by logical steps from known premises or from hypotheses to their necessary or probable results is a useful and distinctive property of the human mind for which the word rationality might well be reserved, as in fact it often is.

Brain scientists distinguish a separate mental capacity, involving the manipulation of form and context, figure and ground and sensitive to context rather than to cause and effect. They identify this as the source of creativity, including the origination of those scientific hypotheses which are subsequently tested by deductive processes. They thus legitimize what for many is a fact of direct experience, that thinking involves a dialogue between at least two mental faculties of which one is tacit, though its activities can be recognized. Since most brain scientists are neuro-physiologists, much interest has attached to the location of these two faculties in different hemispheres of the brain. Recent work suggests that either location can in time learn to perform at least some of the functions of the other if the other is wholly removed. For the epistemologist these questions of location and of adaptation are unimportant. What matters is the confirmation that we possess at least two "cognitive styles" (Galin, 1974), which supplement each other.

Thirdly, nothing we yet know "scientifically" about the working of the brain accounts adequately for the "mental maps" which we all build up from experience and reflection (including the experience of communication with other human beings) and which we constantly use for three not always consistent purposes -- to guide effective action; to sustain successful communication with each other; and to make meaningful and bearable all incoming experience, including our memory of past experience. It would none the less be more than perverse to try to ignore such maps since without them we could not carry on any of these three essential activities. Nor should we try to exclude their contents from the corpus of our knowledge, even though we know that they are fragmentary and not necessarily shared by others. Nor should we ignore our ability to understand or imagine the mental maps of others. For though such understanding is bound to be partial, sometimes wrong and often misleading, it immensely exceeds anything we could construct without it.

Fourthly what I have called "concern" (which includes the whole field of what is commonly called values) is essential to the accumulation, use and revision of knowledge and can neither be kept in a separate compartment nor reduced to conditioned or unconditioned responses or to purposeful sequences which leave the origin of the purpose unexplained. The most important impact of control theory on our understanding of the higher levels of human motivation may well be that it legitimizes the concept of internal standards capable of generating signals of match and mis-match which in turn give coherence and quality to human behaviour (Vickers, 1973). Shared standards of this kind have always been necessary to the coherence of human societies. They are the prime sources of both commitment and constraint and their decay is the most sinister feature of the current human predicament, especially in the West.

They are not, of course, solely a personal artifact. Human beings are both social and acculturable creatures and such regularities as their societies show are inter-subjective artifacts. Most philosophers of science today would I think agree that even the immense system of natural science is fundamentally only a inter-subjective artifact. It may be hoped then that the 19th century concept of knowledge as either subjective or objective (the first illusory, the second real) is at last widening to make room for the inter-subjective dimension. Policy makers have never been able to avoid it. Their advisers should not try to do so.

The domain of the inter-subjective can no longer be ignored, even though it may mislead. It conditions and enables all our understanding, scientific as well as political and ethical and aesthetic, our understanding of ourselves and each other even more than our understanding of the natural world. The analyst with his computer has as much right in that world as anyone else but at the moment perhaps more influence. We do not want him to make it wholly in his own image.

REFERENCES

Archibald, A., 1979, Policy Analysis and Social Science. Proceedings
of the Annual Meeting of the Association for Canadian Studies,
Toronto, Ontario.

Buchanan, C. et al, 1963, Traffic in Towns, Penguin Books, London,
UK.

Galin, D., 1974, "Implications for Psychiatry of Left and Right Hem-
isphere Specialisation", Archives of General Psychiatry, 31, 572-
583.

Vickers, G., 1973, Motivation Theory - A Cybernetic Contribution,
Behavioral Science, 18, 242-249. Reprinted in G. Vickers, 1980,
Responsibility; Its Sources and Limits, Intersystems Publications,
Seaside, California.

EDITORS' NOTE

This paper is an expanded version of the invited lecture presented
to the NATO Advanced Research Institute.

PART 1

CRITICAL ASSESSMENTS

THE TECHNOLOGY OF URBAN SYSTEMS MODELLING: NAGGING QUESTIONS, SAGGING HOPES, AND REASONS FOR CONTINUED RESEARCH

Michael A. Goldberg

Faculty of Commerce
University of British Columbia
Vancouver, British Columbia, V6T 1Y8
Canada

INTRODUCTION: BACK TO SOME BASIC QUESTIONS

In attempting to assess the past several decades of urban system modelling, the prospective assessor must revert to some basic questions about the purposes of such modelling activities. This line of reasoning leads to a more fundamental series of queries about the nature of the systems we are attempting to mimic through models. Accordingly, we must examine the urban system and its evolution over the past thirty years or so if we are to have some basis for exploring the utility of models in simulating these systems.

THE CHANGING NATURE OF THE URBAN SYSTEM: A RETROSPECTIVE PROSPECTIVE

The first modelling efforts were conceived in the 1950's, and it is to this decade that we must first look for insights into the present state of modelling. What follows are rather rough, but also roughly reliable, thumbnail sketches of three periods of interest: the 1950's and the 1960's and the formative years of modelling activity; the 1970's when modelling grew into its present state; and the 1980's and their likely demands on modelling, planning and policymaking. We follow the US model as it gave rise to most large modelling efforts.

The 1950's and 1960's: The Go-Go Years

These two decades represent a highwater mark in optimism and enthusiasm, and the belief that large expenditures could solve most, if not all, problems. Growth and construction/demolition held the key to solving urban problems.

The following, nicely sums up the rapid evolution of American suburbs and the resulting emergence of a range of urban problems loosely termed "the urban crisis". The set of consequences which resulted from a short-run post-war housing policy quickly manifested itself in a number of unexpected ways, as will be seen later. Returning to the scenario that gave us the American city of the 1950's and 1960's:

> "By the end of the 1940's, urban areas were suf-
> fering from severe housing shortages, due largely
> to war-stimulated in-migration to the city, plus
> increasing urban birth rates. One approach to
> the housing crisis was to employ new building
> techniques that made possible the development of
> mass suburban housing tracts. This technological
> response was facilitated by important social
> organizational changes in financing, such as the
> GI Bill and liberalized FHA loans. And, the lack
> of established local governmental regulations
> within emerging suburbs -- such as lenient
> building codes and minimal requirements for
> sewers, fire protection, lights, and so on --
> also facilitated extremely rapid housing growth.
>
> The impact on the physical environment of rapid
> suburban development is not hard to discern.
> Indiscriminate land clearance destroyed the
> existing physical ecology while shortcuts, such
> as lack of sewers, polluted even distant areas.
> Burgeoning suburban population growth, almost
> exclusively dependent on the technology of the
> automobile, resulted in the proliferation of
> highways. Increased reliance on the automobile,
> together with the emphasis on single-family unit
> construction, resulted in the massive disappear-
> ance of open space surrounding the central city.
> Thus, the immediate housing crisis was met
> largely by turning to technology with little
> consideration of the effects of rapid population
> growth on the environment and social organiza-
> tion."[1]

An enormous host of consequences were thus set in motion. Suburban growth, aided by massive highway construction, low cost financing of new construction, and minimal servicing requirements, allowed suburbs to grow, really encouraging their growth.[2] Central cities in the meantime were left increasingly with less affluent households, traffic congestion, declining tax bases, aging physical capital, etc. Slum clearance, intra-urban highway programmes, urban renewal, and public housing programmes did not stem the tide. Some critics assert that these programmes only worsened the situation.[3]

Attempts to deal with these growing problems, and growing perceptions of these phenomena as problems, were firmly rooted in the American tradition: additional monies, new technologies, and new programmes could turn the situation around. Despite the magnitude of the problems and the heightened public awareness, there was still an underlying optimism that construction and renewal expenditures could reverse the trend.

What were the underlying socio-economic conditions that gave rise to such a complex of interacting problems and expenditure solutions? The decades of the 1950's and 1960's could first of all be characterized by relatively high levels of economic growth and relatively stable price levels. In addition, population growth, particularly in the 1950's, was buoyant. Household formation was dominated by traditional households (husband and wife with children soon to follow). As families grew, there was enormous pressure for additional single-family housing units and these, in general, were developed in suburban locations. Mobility was also very high, not just within metropolitan areas but also among urban regions of the United States. Unemployment was low, optimism high.

In short, the 1950's and especially the first half of the 1960's were typified by suburban growth, decline of the central city, especially in older eastern and mid-western cities, and a general feeling that massive capital expenditures and aid programmes could overcome the twin problems of opening up the suburbs to supply needed single-family housing and simultaneously re-doing the central city. In reflecting on the "metropolitan enigma" in 1968, James Q. Wilson remarked:

> "If people object to pumping millions of dollars
> into the central cities, the alternative --
> unless we are to do nothing -- is to draw the
> people out of the central slums and put them in
> new environments. Looked at another way, if we
> wish to maintain both our suburban lifestyles and
> our concern for the future of the central city,
> it is going to cost us a lot of money. Having
> your cake and eating it, too, is rarely accomplished but when accomplished, expensive."[4]

While largely an American problem, this perception of central city decay, suburban growth and the need to take vigorous action, spread elsewhere, often uncritically. In Canada, the late 1960's saw several pivotal works that spoke to Canada's urban problems.[5] Nor was the Atlantic Ocean a suitable barrier to stall the spread of such "urban crisis thinking". In the United Kingdom the most ambitious programme to overcome urban blight took shape in the form of the new town programme.[6] Continental European cities also were

subjected to re-evaluation influenced by the American "model".[7]
Even Australia, halfway around the globe from the United States, did
not escape unscathed.[8]

It could be argued, of course, that the US model did fit quite
well, and there was in fact an epidemic of urban decay and suburban
expansion. Such an argument would have some credibility, if looking
back from the perspective of 1980 we could identify widespread
central city decay outside the US. We cannot.[9] This is so despite
the fact that significant suburban growth did take place in the
major cities of Canada, Europe and Australia, yet the American urban
crisis failed to materialize. What is important for our present
purposes, however, is that during the 1950's and 1960's many urban
policymakers and researchers were responding as if the US urban
crisis were upon them.[10] It is this myth, not the reality, that is
important for our purposes, since the myth gave rise to policy and
research initiatives, of which urban systems modelling was a component. In a world of rapid economic growth, price stability,
generally low unemployment, and perceived large urban needs, the
monies flowed freely, to study urban areas, to make policy for them,
and to build, rebuild and expand them. Such a spree had to come to
a close. It did starting in the 1970's.

The 1970's: The So-So Years

Even before the close of the 1960's it was becoming apparent
that massive doses of money were not achieving their stated goals of
revitalizing and repairing the central city and providing new and
reasonably priced residential and work opportunities in the suburbs. The groundswell of environmental concerns from the 1960's
erupted into important political forces within and among US metropolitan areas. Inflation soared from an average of near three percent during the 1960's to more than eight percent during the
1970's. Unemployment climbed from three to four percent during the
1960's to six to eight percent during the 1970's. Growth was
replaced by stagnation. Family formation was of less interest than
household formation, as non-traditional households proliferated
(eg. single-parents, unrelated individuals living together, one
person households, and married couples with no children). Metropolitan growth slowed or turned to decline while non-metropolitan areas
breathed with new life.[11]

Growth promotion, highway building and suburban expansion began
to be superceded by programmes of growth management and control.[12]
Housing production slowed dramatically. Re-use and renovation
became both economically feasible and acceptable in the market-
place. In short, the buoyancy, optimism, spending and growth of the
earlier decades came to a surprisingly abrupt halt, to be replaced
by significant changes in demographic structure, social attitudes,
and economic reality. This was not the decade to launch either
expensive new urban policies or research programmes. Emphasis began

shifting from growth and new construction, to better management of existing capital resources.[13]   The urban system had indeed changed in the United States, as had the perceptions of the system and the strategies for solving urban problems.

While the foregoing scenario is rooted in US experience, similar forces were at work elsewhere.   The entire developed world was affected by rising oil prices, economic slowdown, growing environmental concern, inflation, unemployment, and a decline in the blind belief that we could spend our way out of urban and social problems.   The money was no longer available, and the record of past "successes" discouraging.   Urban systems were everywhere affected by these general societal forces.   In Britain and Western Europe programmes of urban containment to prevent sprawl and US-type problems from eroding large cities, were beginning to be replaced by the realization that major central cities were not growing, nor were the major metropolitan areas.   Growth controls and population redistribution were not needed, or, if needed, needed in reverse: to attract people back to the major cities.[14]   The dynamics of the urban system had shifted dramatically in virtually all developed countries. Enter the future: the 1980's and beyond.

The 1980's: Questions, Questions and More Questions

The past is prologue, it is said.   Which past?: The 1950's and 1960's, or the 1970's, or some heretofore unexperienced past?   At this writing, we are fourteen months into the 1980's, not much to go on, but some clues nonetheless.

It appears extremely unlikely that there will be a revival of the thinking and the excesses of the 1950's and 1960's.   The decade of the 80's is most likely, from the present perspective, to look like the close of the 1970's, but more so.   Inflation seems to be a reasonable assumption for much of the decade, particularly, the first half.   Low levels of general economic growth appear in order as well.   The population will grow slowly, there will be minimal immigration, and thus, considerable aging of the population.   Capital should remain in reasonably tight supply.   Energy and selected key materials are also likely to remain constrained.   Massive government capital construction projects cannot be expected to come to pass in such an environment.

Bringing this general overview to bear on the urban system in the United States, implies the following.   Continued suburbanization is not very probable.   Re-use, renovation, rehabilitation, etc., are more reasonable expectations for the shape of urban development than is new construction.   Mobility within and among regions of the US can be expected to decline in view of reasonably high unemployment, high energy costs, sluggish growth generally.   The continued growth of the so-called "sun-belt" states (eg. the US south and west) should slow somewhat, as should the movement to non-metropolitan

urban areas (eg. those under 50,000 inhabitants). Central cities
with stocks of inexpensive old housing can be foreseen to continue
to attract households and experience upgrading (gentrification,
white-painting, etc.). In general, the patterns that began in the
1970's will be filled in during the 1980's. Stability (or stagnation depending upon whether you view the situation as good or bad)
should typify the state of the urban system during the decade. The
Reagan administration is unlikely to begin pouring new funds into
old cities.

Some rather large unknowns loom, however. Are recent demographic patterns to continue? Will the young households of the 1970's
really abstain from having children, or were they merely postponing
children until the 1980's? Will divorce rates and re-marriages
remain at present high levels? Will older people return to the
labour force, or will they accept retirement gladly?

The demographic unknowns are the critical ones. Should there
develop a mini-baby boom, then school enrollments will climb toward
the end of the decade. Central city school systems could experience
growth, if central city households view the central city as a sound
place to raise families. If not, a new wave of suburbanization
could ensue. The concentration of older populations in and near
central cities could also bring new strains and opportunities for
locating economic activities that use their skills and that serve
the elderly. If older buildings and neighbourhoods remain in vogue,
we can expect significant improvements in central city neighbourhoods both physically and economically. This raises other questions
about where the present urban poor will (can?) locate. Some of
these problems are already beginning to make themselves felt. [15]

In summary, the 1980's are likely to be different from the preceding three decades. They will force questions upon us that we
have not had to ask before (such as: how do we accommodate the urban
poor, a mini-baby boom, re-use of buildings, etc.). Given the
expected slow pace of the decade however, there should not be very
much of the urgency and crisis feeling that accompanied urban problems in previous decades. It should provide a breathing space and
an opportunity to accommodate to change at a reasonable and healthy
pace, if we are willing to accept that the decade will be different
in character from previous post-war decades.

The outlook for the 1980's sketched out above, should have some
relevance for cities outside the US as well. Similar forecasts
would be reasonable from our present knowledge for Canada's urban
system, and probably for cities in Britain and Western Europe. So,
times are changing and urban systems can be expected to change with
the times. The implications for urban systems modelling are all too
obvious, models and modelling must change as well.

Recapitulation: What Sort(s) of Urban Systems Are We to Model?

All of the foregoing really brings us to the central question of this paper: what sort(s) of urban systems are we to model, now and into the 1980's? Are we to rely on models derived for US cities and urban regions? Are we to model the urban system of the 1950's and 1960's (growth, construction, dispersion, etc.)? Is the situation of the 1970's more reasonable as a model (sluggish growth, growth controls, etc.), or are we entering a new era in the evolution of our cities and of our urban systems, requiring not only new models, but new strategies for developing the models as well?

Once armed with an appropriate expectation of the nature of the urban system that is the subject of our modelling work, we are in a position to move on and to develop appropriate and effective mimics of urban processes and of the urban system. Given these rather basic questions, it is the purpose of the present paper to explore the following:

1.   To review, very briefly, the past evolution of urban systems models and modelling techniques;

2.   To review, also very briefly, some of the successes, failures and calls for improvements in modelling techniques;

3.   To assess, through comparison with trends in urban development, emerging trends in modelling;

4.   To raise some basic questions about and for the future of models/modelling and to suggest alternatives, refinements and areas for additional research.

WHERE HAVE WE BEEN, HOW HAVE WE DONE?

A review of the record of model-building and model-using activity over the past two decades reveals that while there has been a great deal of activity, much success, and much refinement, there has also been much disappointment and a lingering feeling that models and modelling have not lived up to early promises and hopes.

A (Very) Brief Historical Overview

Excellent review articles exist on the subject of urban systems modelling, and it is not the intent of this overview to survey the literature in any depth.[16] Rather, it is intended to provide some context and background for evaluating models in light of their historical development and use.

The first urban simulation models were developed in the US in the early 1960's most frequently under the auspices of either federally-aided community renewal programmes (CRP's) or metropolitan transportation plans. These model-building programmes were designed to deal with the twin-horned dilemma sketched earlier of simultaneous suburbanization (thus transportation plans for the region) and central city decline (thus the renewal focus). With large-scale expenditures to cure the urban crisis in its various dimensions came large-scale modelling to provide a policy and impact testing device to simulate the effects of proposed policies and programmes (including the do-nothing alternative). By the close of the 1960's, the promises of both the models and of the various renewal and transportation policies/programmes were largely unmet. Central cities were still in decline, congestion was still on the rise, as was pollution, urban sprawl, environmental degradation, and the costs of urban growth. Models provided little help in stemming the tide and in improving the situation materially.

Enter the 1970's and the emergence of growth controls, scarcity of land and energy, and a different planning environment. What sort of models came to the fore, and what sorts of developments did we see to cope with these emerging changes in the urban system? Unfortunately, little change in modelling took place. A great deal of technical refinement occurred, along with much more sophisticated and efficient algorithms, but the models themselves did not change. Why? A number of factors come to mind. After a decade of development and innovation, much of the initial excitement in modelling had doubtless worn off. Money and energy had begun to be diverted away from large-scale transportation and renewal projects, and thus, a good deal of the demand for urban systems models softened. The early performance of models had not helped stimulate demand either. Finally, models came under increasingly harsh questioning by people in the field, not merely by professional planners and politicians. All of the above argue strongly for changes in basic approaches to model-building, changes which, however, were not generally forthcoming. The large-scale urban systems models of the 1960's were hardly calibrated, and certainly not fully refined when the 1970's and their quite different environment struck. Modellers were still dealing with normal "mop-up" and "fill-in" work left over from the heyday of modelling during the previous decade. Little energy, demand or resources were available for letting go of the old and launching headlong into a new round of quite different modelling efforts.

As noted, part of the lack of enthusiasm for the task of moving urban systems models into the 1970's derived from the earlier experiences with models and model use, an area worth exploring critically before continuing to suggest ways to bring model-building into the 1980's. A look at past problems is definitely in order.

Past Problems and Pitfalls

A considerable part of the problems that models faced derived from some rather unreasonable expectations about just what models

were capable of dealing with effectively. To put all this in perspective, we should begin by looking at the urban environments that
models could reasonably imitate. One comes to mind immediately,
that of rapidly growing and outwardly expanding urban regions. In
short, our US model for the 1950's and 1960's. In such a setting,
the complex and intertwined micro-economic details of competition
among land uses, of re-use of centrally located land and buildings,
and of site and neighbourhood characteristics, can reasonably be
overlooked in favour of broader considerations of accessibility,
land availability, and trunk infrastructure placement of roads,
water and sewers. The record of modelling in such circumstances is
not bad, and provides a legitimate base for optimism on the univer-
sality of modelling in a broader span of urban circumstances.[17]

However, when we turn to more complex situations, the record
also turns -- for the worse. Models have yet to deal adequately
with such difficult urban phenomena as neighbourhood change and
succession in developed areas, commercial and industrial decline,
intrametropolitan office location, public facility location, and
recreation and entertainment activity location, to name the most
obvious.[18]

During the past decade, a number of criticisms have, legitimately in my opinion, been levelled at models and modelling. What
follows is a synthesis of these criticisms, virtually all of which
are as current today as when they were written. It is contended
that models have failed to achieve their lofty objectives for such
reasons as:[19]

1.   Excessive Costs: The costs of developing, refining, and using
      urban systems models can run into millions of dollars, with
      several hundreds of thousands of dollars being a reasonable
      minimum. Data gathering and development costs are enormous in
      most instances, and until very recently, the costs of using the
      models was also frequently excessive.[20]

2.   "Comprehensive incomprehensibility":[21] The size and complexity
      of most urban systems models preclude their comprehension by
      all but their builders and a handful of technical experts with
      the time and resources to delve into the internal structure of
      these "black boxes".[22] A tool that is not easily understood is
      one that is not easily or probably skillfully used.

3.   Relative Absence of Meaningful Policy Inputs and Outputs:
      Model-building seldom proceeded backward from policy problem to
      model-building as Lee suggested it should.[23] Rather, model-
      builders built models to simulate, as best they could, the
      urban region of interest. Policy variables were not paramount,
      either in the minds of the modellers or in the structure of the
      models. Such an oversight made models, even the best of them
      in terms of accuracy, quite difficult to use.

4. <u>Lack of Access to Modellers and Models</u>: Excessive emphasis was
    placed on the development of operational models often developed
    in a void, divorced from potential model users. Such a clois-
    tered modelling process could not serve the cause of modelling
    very well.   Critical suggestions from users, policymakers,
    citizens and urban professionals could have gone a long way in
    heading off many of the difficulties encountered in putting
    models to actual use (see Point 3 immediately above for an
    example). The need for greater balance between the <u>process</u> of
    modelling and the <u>product(s)</u> of model-building (eg. models), is
    an important point to which we will return later.

5. <u>Unfulfilled Promises</u>: Model-building has not, and is not likely
    to in the near future, live up to the early expectations of
    either modellers or of model users. This failure to meet
    initial expectations has soured many on urban systems modelling
    as a technique for planning, impact analysis, and forecasting.
    Such negative expectations act to dampen enthusiasm for further
    experimentation, development and use. Skepticism and cynicism
    replaced (unfounded) optimism.

A more detailed review of model performance can be had elsewhere (see Note 17). For present purposes, the point to be stressed
is a simple one: while great effort has been expended and great
refinements have been made in models, the record to date is considerably less impressive than the early literature of the 1960's
would have led us to expect after nearly twenty years of modelling.
Despite past disappointments, or perhaps because of them, a number
of changes are on the horizon that auger quite well, I suspect, for
the immediate future of modelling, albeit in a quite different form
from past large-scale comprehensive efforts.

EMERGING SOURCES OF CHANGE IN MODELS, MODELLERS AND MODELLING

The foregoing could either be interpreted as forecasting the
doom and end for modelling, or as sources of wisdom and experience
upon which to build alternative modelling activities. I choose to
think it provides us with the necessary perspective to launch new
and more appropriate efforts, grounded in two decades of exciting
experimentation, and hopefully, learning.

Possible Responses: Build On Successes, Learn From Failures

Failure provides both a <u>raison d'etre</u> for learning and a substantive body of knowledge to learn from. Success, on the other
hand, can be terribly stifling and lead to complacency and stagnation. Modelling fortunately has had a wonderfully healthy mix of
both failure and success. To this point I have stressed the disappointments and failures. Now we can balance the accounts somewhat
by looking at the successes of the past two decades which provide

the building-blocks potentially for renewed and alternative approaches.

Salvagable parts and usable elements of present (and past) models would likely include the following:

1. A Stock of Technical Achievements and Operational Models: Over the past two decades a number of operational models and sub-models have been developed and refined. The Lowry Model has been elaborated and linked with a transportation model and applied in a number of settings.[24] Entropy models have been expanded considerably over their early applications to residential location.[25] Huff-type models have been applied with great regularity and success in the retail location field,[26] and matrix methods have been developed for a number of urban sub-systems.[27]

2. A Stock of Model-Building Expertise: Paralleling the development of models has been the development of a large stock of knowledgeable modellers. Their experience in the development, refinement and application of urban systems models is probably the greatest asset we can bring into the 1980's and beyond.

3. A Stock of Large-Scale Data Bases and Data Handling Systems: One of the early and consistent complaints by model-builders was the lack of suitable data bases upon which data-hungry large-scale models could be built. This situation has changed dramatically during the 1970's when very large data banks and retrieval systems were developed and made widely available. As a result, data shortages do not represent the stumbling block that they once did, though disaggregated land use and activity data are still not abundant. Such data bases do exist, and of equal importance, so do the means for retrieving and manipulating them.

4. A Stock of Fast, Flexible and Economical Hardware and Software: Another supposed bottle-neck in the model-building process was the slowness, limited storage capacity and cost of using computer systems. It was reasoned that if these technical limitations could be removed, then the path would be smooth on the way toward the full development and elaboration of usable and realistic models. The past ten years have seen truly awesome developments in this area. Machine computing speeds have increased several orders of magnitude with additional improvements on the horizon. Storage capacities have increased as dramatically as has storage flexibility with the advent of discs, belts and the pending use of bubble memories. Originally designed to do masses of calculation with relatively limited data sets (eg. solving systems of differential equations), computing machinery has evolved to the point where it can equally well handle relatively simple calculations with masses of data (eg. the usual urban systems model). In addi-

tion to these computing and data improvements, the ability to
communicate with computers and programmes has increased many-
fold, both in terms of quality of communication and quantity.
The existence of elaborate, and now routine, interactive computer graphic programmes for display on CRT's and for printing
out simulations, has greatly enhanced our ability to use models
and disseminate results in a meaningful form. The existence of
time-sharing, portable terminals, hard-wired CRT consoles, and
telecommunication data links, all serve to make computing
machinery more usable for modelling purposes and provide the
potential for making models significantly more accessible than
ever before. Lastly, elaborate and sophisticated supervisory
software programmes have been developed that allow very complex
models to be developed and run with relative ease and with
great economy and efficiency.[28]

In light of such an impressive list of achievements and usable
building-blocks, the future of model-building should be far from
grim. The question remains: to what sorts of models can these
advances be applied, to enable modelling to cope with changes in
urban systems during the 1980's? Some suggestions follow.

To cope with changes in the urban system itself, and to build
on the successes and failures of past modelling efforts, a number of
suggestions for future models come to mind.[29]

1. Build Smaller Models: It would appear that small models hold
    the promise of overcoming a number of previously-noted weak-
    nesses of past efforts. They would necessarily be less cost-
    ly. They would also be much simpler conceptually, and as a
    result, much more accessible to potential users. Moreover, it
    would be significantly easier to involve potential users in the
    design process if that process were simplified, and if the
    resulting models were more readily understood and comprehen-
    sible, as indeed smaller models could be.

2. Build Special-Purpose Models: By focusing model-building activ-
    ity on well-defined issues, costs can be cut substantially, and
    policy issues can be dealt with directly. It was noted earlier
    that a frequent criticism of extant models was their seeming
    inability to treat policy issues adequately, either as inputs
    or outputs. By starting with the issue and developing a
    highly-focused model, such shortcomings should be overcome.
    Even though such models are not likely to be able to deal with
    systemic issues, the modeller, and more importantly, the model-
    user, must keep such broader questions fully in mind. (This is
    a worthwhile caveat for policy analysts in general who deal
    with highly specific issues and policies. Housing policies do
    affect land use and vice versa just to illustrate the point.)

3. Build Modular Models: Given the above-mentioned emphasis on
    smaller and more specialized models, there is a potential for

great gain if these models could be developed in the context of a more comprehensive framework.[30] The existence of highly sophisticated and powerful simulation supervisory software packages make such a modelling strategy feasible. By developing models as modules in a (potentially) larger context, a balance can begin to be struck between highly focused and more general questions, without necessarily sacrificing cost economy and/or operational utility.

4. Take Advantage of Technological Advances in Computing Machinery: Recent advances in computer hardware have not really been adequately, let alone cleverly, exploited by modellers. Of great potential are the dazzling array of interactive real-time computer graphics routines. Such hardware and software allow users to get visual pictures (maps, charts, bar graphs, tables, etc.) as the model operates. Moreover, they allow intervention by the user during model operation, thus opening up the possibility of much greater involvement by users. With greater involvement will doubtless come better models and better uses. These technological advances allow modellers, in theory at least, to overcome much of isolation that has characterized modelling in the past. Remote terminals also enable models to be physically brought to locations that would have been inaccessible previously. The whole constellation of technical achievements in computer hardware and software bring new possibilities for model use, development and re-development. Models (the product) and modelling (the process) should take significant advantage of these technical developments to bring modellers and users closer together, and to greatly improve use and dialogue.

THE PLANNING PROCESS: A MAJOR FORCE FOR, AND SOURCE OF, CHANGE

The foregoing sources of change are all essentially internal to model-building experiences, and derive rather directly from past criticisms, achievements and failures. Perhaps the most powerful source of change (and force for change) comes not from modelling per se (which is, after all, a means and not an end), but rather from the larger planning and political processes for which models were to provide critical and unique inputs. However, the planning and decision-making processes for urban areas have changed dramatically during the 1970's, responding in part to the changes that took place in the urban system. With changed societal priorities (eg. growth control instead of growth promotion) and with changed resource availability (shortage as opposed to abundance), the planning process has undergone significant changes. The implications for models and modelling are very profound.

In a challenging and thoughtful piece on urban transportation and land use planning, Douglass Lee made the following characterization to distinguish between the "new" and the "old" planning processes:

"The "old" process in this somewhat overdrawn
dichotomy can be described as long range, comprehensive, top-down, end state, closed option planning, based on the engineer-architectonic
approach that requires a detailed, fixed end
product from which everything else is subsequently determined, the whole predicated on the belief
that it is possible to forecast future events.
The alternative, or the emerging "new" process,
is characterized as short range, incremental,
politically open, and multi-optioned in the sense
of narrowing but not eliminating choice. Methodologies and techniques for the emerging paradigm
have not been settled upon, but the intent of
sketch planning and quick response analytic
procedures is in this direction. The shift,
technically, is clearly well underway, but there
is still a long way to go."[31]

With such changes on the horizon, we can expect pressures
brought to bear on modelling to provide the kind of short run,
incremental, etc., knowledge that the emerging planning process
requires. These pressures are not at all inconsistent with the
earlier demands by critics for different kinds of models and modelling. They are also entirely consistent with the emerging pattern
of actual model use which has tended toward short run specialized
forecasting instead of the comprehensive impact testing which was
one of the most frequently cited initial objectives of model-
building.[32] Finally, they are completely consistent with the suggestion set out above for alternative approaches to model-building.

Thus, we have, in a sense, come full circle. Starting with
changes in the urban system over the past thirty or so years, we
have noted the need for corresponding change in models and model-
building processes. The need is augmented by the emergence of a
"new" planning process, one that is also more consistent with emerging changes in the urban system. Accordingly, the stage is set for
change.

POISED TO LEAP INTO THE 1980'S: AN "EX ANTE" GUIDE TO THE DECADE FOR
PROSPECTIVE MODELLERS

In conclusion, I will quickly review the argument made to this
point in the paper, and then extend it somewhat to suggest ways in
which model-building can capitalize on the spirit of change that
surrounds the planning process in the large. The demands of a
changing urban system and a changing urban planning process present
unique and exciting challenges and opportunities for urban system
modelling and can provide the spark that is needed to touch off a
round of innovation and excitement such as that which typified more
pioneering efforts in the early 1960's.

Accommodating Changing Urban Systems

The emergence during the 1970's of a very different kind of urban system and very different kinds of societal concerns and constraints placed great strains on existing modes of urban planning, as we have already seen. These changes in the nature and purposes of urban planning in turn place great demands and strains on traditional planning technologies. Perhaps the most elaborate of these technologies was none other than the focus of our discussion, urban systems modelling. As planning changes, so must the tools upon which it relies.

All of this implies that models and modelling must be clearly seen to be a part of a larger planning environment, which itself is part of larger societal concerns. In short, modelling (and planning too, for that matter) must be seen as means, not as autonomous ends. Modelling is an irreducibly applied art/science. As such, it must not lose sight of its role in planning and decision-making. It is a part of, not apart from, these larger issues. Figure 1 below attempts to summarize this idea.

The figure sets out schematically the flows of information between technicians and their techniques (modellers and models) and the decision-makers who rely on this technical information for taking decisions. Information flows up to decision-makers (including the citizenry who are the ultimate cause of all of this planning and decision-making activity). Information needs flow down to the technicians and their tools. While acknowledging that this is something of an idealization of how planning and modelling should occur, it is not terribly at odds with the reality, in most instances.

Embedding models and modelling in such a structure points up sharply their respective roles. It provides some perspective for modellers who get overwhelmed by their models and their imitation worlds (new Frankensteins?). It also provides a launching pad for the "new" models demanded by the "new" planning process, which is itself subservient to and dependent upon an ever changing political and social environment, represented in a sense by the totality of elements in Figure 1. The question that looms is, "How can models and modelling make the suggested changes in scale, emphasis and use?" Accordingly, we turn to these questions next.

MOVING MODELLING INTO THE 1980'S: SOME SPLITS THAT NEED HEALING

In describing the "new" versus the "old" planning process above, Lee set out a number of dualistic dichotomies to make his point. Included were: long-run versus short-run; top-down versus bottom-up; closed-option planning versus multi-option planning; comprehensive versus incremental; and product-oriented versus process-oriented. To this list we might add some others that need to be considered explicitly. For example: theory versus practice;

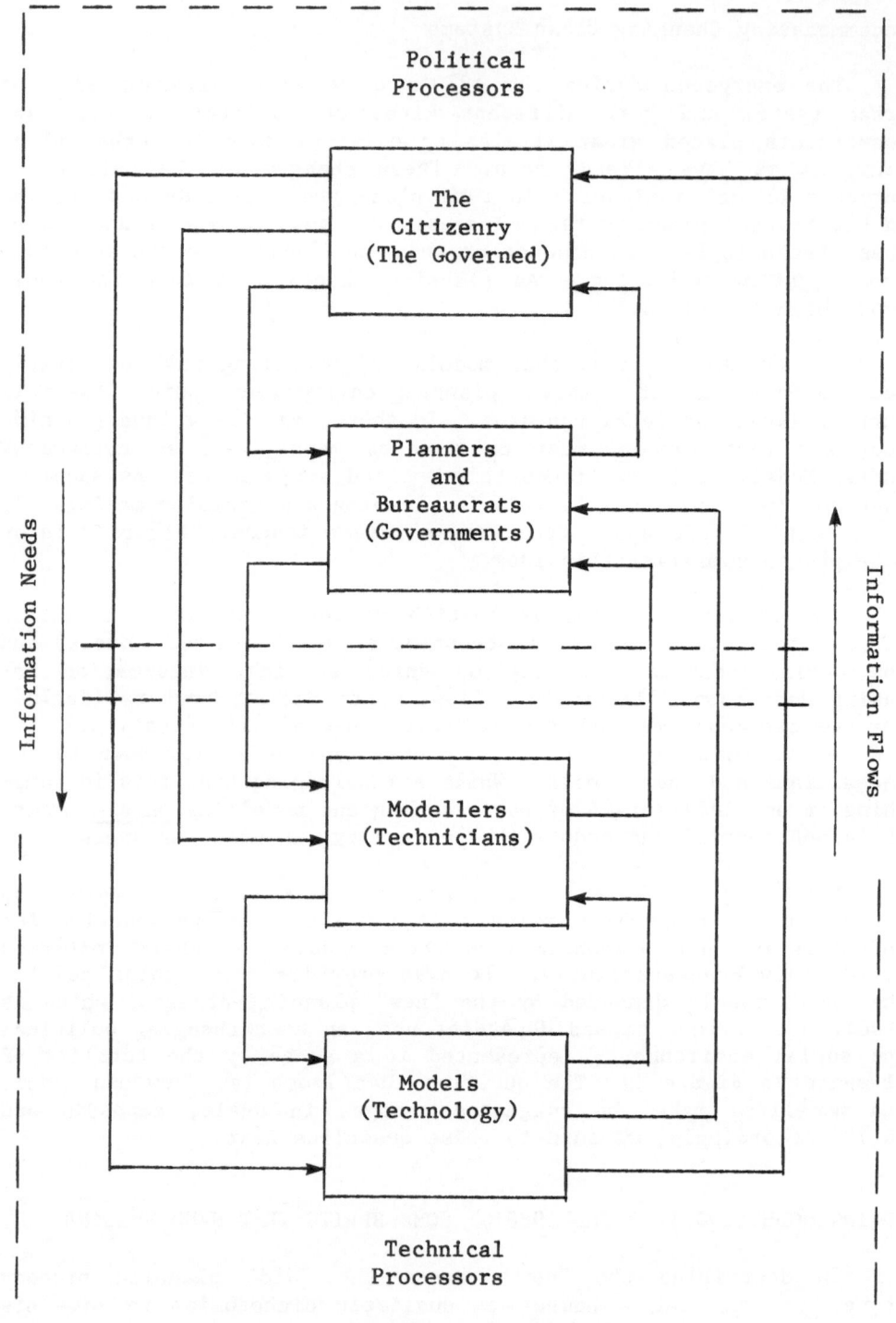

Figure 1 - Models, Modelling and Societal Decision-Making
(Adapted from Goldberg, 1978:7)

analytic versus synthetic; holistic versus partial; inductive versus deductive; and experience/intuition versus abstract/rational. Such dualisms pervade virtually all human thought and action. They cannot be conveniently eliminated by simply ignoring, or defining away, the dual element(s) that is (are) not to our liking. Rather, the existence of such opposing approaches to knowledge and action must be acknowledged explicitly. The appropriate role of each element can only be fully appreciated by placing it in juxtaposition with its opposing force. For instance, theory can only be fully grasped when put to practical test. Similarly, the cleverness and skill of many practitioners is best seen in the context of generalizations about their behaviour (ie. theories). (The skill and technical expertise of commodity traders, for example, appears quite awesome when their efforts are placed in the context of efficient market theory, and it is seen how closely they operate (unknowingly) in accordance with powerful propositions of microeconomic theory).

The point that needs making here is simply that we can no longer meaningfully talk of theory and practice, process and product, etc., as if each of these opposite approaches to modelling and decision-making could actually exist independent of the other. Instead we must begin to appreciate that good theory is practical and good practice is governed by sound (if often implicit) theory. Thus, if we are to move model-building ahead to cope with the emerging needs of the urban systems of the 1980's and beyond, we must adopt a more flexible and accepting view of model-building. Models built on the intuition of the practitioner, have as much to offer if well done, as do models deduced from economic and spatial abstractions. In different settings, different approaches and different types of models are likely to be appropriate. Remember, from Figure 1, that model-building is a means to broader and more fundamental societal needs and objectives. The needs and objectives go a long way toward defining the context and the types of models that will be needed. In a changing environment such as that we have experienced during the past ten years or so, we need diversity. Innovation borne of past experience has much to commend it. So do the mopping-up and refining operations that have characterized much in the modelling area over the recent past. Again, we need both.

By building on refinements of the past, and setting out on innovative and high risk sortees in the future, a new era of model-building can begin. One filled with excitement, adventure and promise, yet tempered by past difficulties and a proper appreciation of the complexity of the systems we attempt to model. A synthesis of the tried and true and the daring, of the intuitive and of the analytical, and of the abstract and concrete, holds the key to future success (and failure). A modelling process that encourages diversity of models (products) and that is itself dynamic and open-ended, can provide the environment within which we can develop models to cope with the dramatic changes we have experienced in the form, structure and behaviour of urban systems in the Western world.

CONCLUSIONS: A NOTE OF TEMPERED OPTIMISM

Past experience provides the surest footing for future model-building efforts. To benefit most from the richness of these past two decades of modelling, we as modellers need to broaden our perspective considerably to admit new and seemingly risky approaches to modelling urban systems, and approaches toward which we might have deep-seated distrust, either borne out of experience or borne out of our particular research styles and preferences (prejudices).

To successfully handle the diversity and complexity of urban systems, especially during these times of transition, we must foster diversity and experimentation in our own work and in that of our colleagues. We need to balance the respective skills of researchers and practitioners, of specialists and generalists, and of technicians and decision-makers. The process-product dichotomy needs to be viewed instead as a cycle. We need models as devices to stimulate discussion and provide information for decision-making, but we must also clearly recognize that such models are ideally part of an ongoing process of model development/evaluation/redevelopment.

Modelling has fallen into the doldrums of late. To get it moving again, we need to reflect on where we have been, what we have accomplished, where we have failed and what needs to be done for the ever-changing future. The present paper is offered in this spirit, to provide a focus for discussion and perhaps as a result, a firmer foundation for future efforts.

NOTES AND REFERENCES

1. Palen, J.J. and Flaming, K.H., 1972, (Eds.), Urban America: Conflict and Change, Praeger Publishers, New York.

2. The role of highways in urban decentralization is nicely documented by de Leon, P. and Enns, J., 1973, The Impact of Highways upon Metropolitan Dispersion: St. Louis, The RAND Corporation, Santa Monica, Calif.

3. A nice compendium of urban renewal criticisms and analyses is: Wilson, J.Q., 1966, (Ed.), Urban Renewal: The Record and the Controversy, The MIT Press, Cambridge, Mass. Also see, Bellush, J. and Hausknecht, M., 1967, (Eds.), Urban Renewal: People, Politics and Planning, Anchor Books, Garden City, New York.

4. Wilson, J.Q., 1968, (Ed.), The Metropolitan Enigma, Harvard University Press, Cambridge, Mass.

5. The report by Harvey Lithwick, commissioned by the Canadian government makes this point strongly. See, Lithwick, N.H., 1970, Urban Canada: Problems and Prospects, Central Mortgage and Housing Corporation, Ottawa, Ontario.

6. The new town program is summed up and analyzed in Schaffer, F., 1972, The New Town Story, Paladin Books, London. For a more descriptive and architectural view, see Osborn F.J. and Whittick, A., 1969, The New Towns: The Answer to Megalopolis, The MIT Press, Cambridge, Mass.

7. Similar pressures and thinking was at work in many of Europe's largest cities. See, for example, Hall, P., 1977, The World Cities, McGraw-Hill Book Co., New York.

8. Neutze, M., 1977, Urban Development in Australia, George Allen and Unwin, Sydney, Australia.

9. This point is stressed in Goldberg, M.A. and Mercer, J., 1981, Canadian and U.S. Cities: Basic Differences, Possible Explanations, and their Meaning for Public Policy, Papers of the Regional Science Association, in the press. A perusal of Hall, op. cit., and Neutze, op. cit., will also show that there was in fact little reason to believe that the US model holds outside the US.

10. Lithwick, op. cit., and Goldberg and Mercer, op. cit.

11. See Hansen, N.M., 1973, The Future of Nonmetropolitan America: Studies in the Reversal of Rural and Small Town Population Decline, Lexington Books, D.C. Heath, Lexington, Mass. Also see, B.J.L. Berry, 1976, (Ed.), Urbanization and Counter-Urbanization, Sage Publications, Beverly Hills, Calif.

12. For a four volume compendium on the growth control literature, see Management and Control of Urban Growth, 1975, The Urban Land Institute, Washington, DC.

13. This shift has been most apparent in the transportation planning area where construction ruled the planning process up through the mid-1970's. The emphasis on Transportation System Management (TSM in the transportation planning jargon) as opposed to new construction signals this mind switch.

14. The Fourth Advanced Studies Institute in Regional Science, held at Siegen, West Germany in August, 1978 was organized around this theme of regional stagnation and decline in developed countries. The proceedings of the Institute document such concerns in a number of Western settings. See Buhr, W. and Friedrich, P., Planning for Stagnating Regions, Universitat Siegen, Siegen, West Germany, forthcoming.

15. For example, fighting over condominium conversion of apartment rental units has been observed in virtually every US city. The apparent trend of middle income households with no children living at home moving into central locations has put tremendous demand pressure on central city housing markets, and has encouraged developers to renovate sound, but old, structures and sell them as condominiums. The social consequences of such renovation and "gentrification" are as yet unknown, though there is cause for concern, as some cities such as Washington DC have experienced significant declines in their black population (in Washington it is estimated that the black population has fallen to about 60 percent from its high in the 1970's of nearly 77 percent). Where do the former residents find suitable housing, and what happens to the whole constellation of services that provided to former poor residents? These kinds of questions are beginning to be asked. It is ironic that self-regeneration of the American central city should represent a problem after all of these decades of large-scale federal spending to rehabilitate it.

16. Among the excellent reviews of the modelling literature see: Batty, M., 1972, Recent Developments in Land Use Modelling: A Review of British Research, Urban Studies, 9, 151-177; Goldner, W., 1971, The Lowry Model Heritage, Journal of the American Institute of Planners, 37, 100-111; Putman, S., 1975, Urban Land Use and Transportation Models: A State of the Art Summary, Transportation Research, 9, 187-202; and Wilson, A.G., 1974, Urban and Regional Models in Geography and Planning, John Wiley, Chichester, UK.

17. See Putnam, S., 1976, Laboratory Testing of Predictive Land-Use Models: Some Comparisons, US Department of Transportation, Washington, DC; and Goldberg, M.A., 1978, Developer Behaviour and Urban Growth: Analysis and Synthesis, in L.S. Bourne and J.R. Hitchcock, (Eds.), Urban Housing Markets: Recent Directions in Research and Policy, University of Toronto Press, Toronto, Ontario.

18. Similar points are made at somewhat greater length and with somewhat greater documentation in Goldberg, M.A., 1977, Simulating Cities: Process, Product, and Prognosis, Journal of the American Institute of Planners, 43, 148-157.

19. Much of the following discussion parallels Goldberg, M.A., 1979, Modelers, Muddlers, and Multitudes: Establishing a Balanced Transportation Planning Process, A Paper presented to the 58th Annual Meeting of the Transportation Research Board, Washington, DC.

20. According to Lee, the prohibitive costs entailed in using the San Francisco CRP Model effectively led to the abandonment of model despite a $500,000 investment by the City. See Lee, D.B., Jr., 1973, Requiem for Large-Scale Models, Journal of the American Institute of Planners, 39, 163-178.

21. See Goldberg, 1979, op. cit.

22. Lee, op. cit., 167.

23. Lee, op. cit., 176. Also see Voelker, A.H., 1975, Some Pitfalls of Land-Use Modelling, Oak Ridge National Laboratory, Oak Ridge, Tenn.

24. See Putman, op. cit.

25. See Wilson, op. cit.

26. Huff, D.L., 1963, A Probability Analysis of Shopping Center Trading Areas, Land Economics, 53, 81-90.

27. See Wilson, A.G., Rees, P.H. and Leigh, C.M., 1977, (Eds.), Models of Cities and Regions, John Wiley, Chichester, UK, especially Chapters 3 and 4.

28. Hueftlein, L.D. and Ash, D.A., 1974, The User's Manual for the Simulation Supervisor, Resource Science Centre, University of British Columbia, Vancouver, BC.

29. This discussion draws heavily upon and largely synthesizes similar comments from Lee, op. cit., and Goldberg, 1979, op. cit.

30. Such a strategy was recommended early on by Nobel Laureate Herbert Simon in setting out ways to deal with complexity. See Simon, H.A., 1962, The Architecture of Complexity, Proceedings of the American Philosophical Society, 106, 467-482.

31. Lee, D.B., Jr., 1977, Improving Communication Among Researchers, Professionals and Policy Makers in Land Use and Transportation Planning, US Department of Transportation, Washington, DC.

32. In her review of model use in planning agencies, Janet Pack found that users tended heavily toward using simple models for forecasting purposes. Comprehensive impact testing and policy evaluation was not a preferred use. See Pack, J.R., 1975, The Use of Urban Models: Report on a Survey of Planning Organizations, Journal of the American Institute of Planners, 41, 191-199.

THE EFFECTIVENESS OF URBAN TRANSPORT SYSTEMS ANALYSIS

Bruce Hutchinson

Department of Civil Engineering
University of Waterloo
Waterloo, Ontario, N2L 3G1
Canada

Metropolitan transport systems analysis methods had their origins in several large-scale urban transport planning studies conducted during the late 1950's and early 1960's in a number of North American cities such as Detroit, Chicago and Toronto. The principal objective of these studies was to identify medium to long-run capital investments in new transport capacity. The stimuli for the adoption of this objective were sustained urban growth and the rapid increases in car ownership. The planning and analysis process developed in these original studies has been adopted in hundreds of planning studies performed throughout the world during the 1960's. In the developed world these studies usually recommended heavy investments in freeway systems, while in the developing world recommendations for large capital expenditures in public transport capacity usually emerged from the planning process.

In the late 1960's and early 1970's the mixed blessings of the capital investment option as the solution to the urban transport problem were experienced. Large increases in freeway capacity tended to exacerbate the problems of traffic congestion and environmental degradation. The increased mobility of car owners provided by the freeway systems stimulated the development of low density urban areas with unstructured travel demands, resulting in substantial reductions in the role and effectiveness of public transport. Transit patronage decreased and many transit properties were unable to provide adequate levels of service to newly developing residential and employment areas, compromising severely the mobility of non-car owners. Attempts to reverse these trends towards excessive dependence on the private car through new investments in public transport capacity were also of limited effectiveness, except in those transport corridors in expanding urban areas in which new transit-related development could occur.

During the late 1960's and 1970's the focus of urban transport policy shifted, with the principal emphases becoming the fuller exploitation of existing facilities and the equity with which all urban groups were served by transport. The shift to these non-capital intensive policy options was stimulated by reductions in urban growth rates in many large urban areas and the lower rates of economic growth in many countries.

These changes in urban transport policy emphasis have been accompanied by some changes in the analysis tools used for policy analysis. In fact, urban transport systems analysis has reached a critical point where the routine application of techniques developed in the earlier studies has been widely rejected by transport plan-ners, but generally accepted methods to replace these earlier tech-niques have not been developed. A wide range of transport policy decisions continue to be taken in spite of very little effective prior policy analysis using formal techniques. Serious questions arise as to the continuing role of formal transport systems analysis techniques in urban transport policy analysis.

The basic objective of this paper is to provide a critical review of the effectiveness of transport systems analysis methods, and to explore the reasons for their successes and failures. The final part of the paper explores potentially productive areas of development.

RANGE OF POTENTIAL POLICY RESPONSES

The particular focii of urban transport policy that exist in individual urban areas and countries at any particular point in time vary widely and reflect differences in rates of urban growth, changes in automobile ownership, financial resources available for urban transport investment, the spatial structures of communities, social attitudes and political philosophies. In spite of these differences, it is possible to detect common themes in urban trans-port policy responses and Table 1 provides a summary.

The policy options listed in Table 1 are grouped under three major classifications and these are: (i) urban spatial management where the principal transport-related concern is in encouraging patterns of travel demand that may be served more efficiently by public transport; (ii) transport demand modification where the emphasis is on altering the timing and overall magnitude of demand; and (iii) policies directed towards the supply-side of transport. Typical examples are provided within each classification and it should be noted that the policies range in implementation time from the shorter-run to the longer-run policies. The policy options listed in Table 1 really represent fairly minor corrections to some of the unsatisfactory features of existing transport systems rather than significant changes in the direction of transport policy.

TABLE 1

POTENTIAL URBAN TRANSPORT POLICY RESPONSES

URBAN SPATIAL MANAGEMENT

- zoning by-laws and building codes

- local area planning for land use and transport

- phasing the development of residential and employment areas to achieve structured travel demands

- structure-planning including satellite towns

DEMAND MODIFICATION

- flexitime and shortened work weeks

- pricing of public transport including passes, peak/off-peak fare differentials, etc.

- pricing of automobile travel through taxes, supplementary licenses, parking charges

- car pooling, van pooling, para-transit

- speed restrictions on urban freeways and the urban sections of inter-city freeways

SUPPLY RELATED

- improved public transport services and the integration and coordination of services

- integration of public transport and private road transport through park-and-ride, etc.

- priority schemes for public transport and high occupancy vehicles

- special transport services for particular groups

- restricted auto access and auto-free zones

- provision of improved bikeways and pedestrian facilities

- new investments in public transport and road network capacities

TRANSPORT SYSTEM ANALYSIS AND POLICY DECISIONS

Two basic idealizations of decision-making may be identified
which in a sense represent the opposite ends of a continuum. At one
end is the intuitive approach where the Gestalt of information,
including the insights provided by systems analysis, results in a
particular policy direction being taken. At the other end is the
analytical approach whereby a set of alternative policies is identified, some logical and explicit analytical process is used to
explore the potential outcomes of each policy, and the outcomes are
evaluated to arrive at a preferred course of action.

Many of the freeway and fixed right-of-way public transport
plans developed in the 1960's for North American cities were formulated directly from the results of formal transport systems analyses. Expected vehicle and passenger volumes in major travel corridors were estimated and freeway and transit capacities sufficient
to handle the expected volumes over a twenty-year period were proposed. The rejection, or drastic modification of most of these
early freeway and rapid transit plans also led to concerns about the
utility of the earlier analysis methods and their more recent derivatives. One consequence of this adverse reaction to demand forecasting based transport planning has been the frequent adoption of
transport policies without prior formal analyses.

There are many reasons for this shift in urban transport policy
decision-making from the analytical end of the spectrum to the
intuitive end. A major reason has been the failure to separate
clearly the issue of what is worth doing from the issue of how to do
it most efficiently. In many cases, ends and means were confused
with decision-makers and their policy advisors being reluctant to
articulate clearly desirable transport goals. Transport systems
analyses have been effective in those urban areas in which transport
goals have been clearly and realistically established, and technical
analyses have been restricted to exploring alternative answers to
the second issue. A related problem is that the information
required from systems analyses to help in choosing between alternative policies has often not been carefully identified. As a result
the analyses have often not been relevant to the alternative policies being considered.

A second, and perhaps the most important reason, has been that
the paradigms of the transport analyst have not been consistent with
political decision-making styles. Decisions usually have a strong
short-run, incremental and fragmentary character, while transport
analysis has tended to focus on the longer-run comprehensive types
of policy packages that could yield substantial changes in travel
behaviour. For example, many analyses have demonstrated that the
most productive methods of obtaining substantial shifts in transport
mode usage are land use development control and traffic restraint
policies, yet policies of this type have gained little political
support. Analysts frequently point to these differences as evidence

of the irrationality of political decision-making rather than attempting to develop analysis methods that are more compatible with decision-making styles.

A third reason for this shift has been the imperfections in the analysis techniques themselves which have inhibited policy advisors in advocating more strongly potentially productive policy changes. Many of the model imperfections are related to the complexities of the urban system, the role of the transport sector within the urban system, and the responses of both to changes in public policy and other exogenous factors. The transport sector interacts strongly with many other urban sectors and actions in the transport sector frequently have counter-intuitive impacts. For example, the provision of high mobility freeway corridors in North American cities had important impacts on both residential and employment location behaviour, and the actual patterns of travel demand that emerged were substantially different from those estimated in the supporting analytical studies.

Available transport models focus mainly on the transport sector, yet the principal characteristics of transport behaviour are frequently determined by factors exogenous to the transport sector. The subsequent sections of this paper focus on the capabilities of available transport models. A first step in attempting to improve the relevance of formal analysis methods is to examine the capabilities of existing transport models as policy analysis tools.

MODELLING STYLES

Urban travel is simply a reflection of the underlying human activities, the ways in which these are organized in space and time, and the transport resources available to facilitate interactions between spatially separated groups of activities. Three basic groups of information are required in the analysis of policy options of the type listed in Table 1 and these are:

1. The number of trips between origins and destinations by transport mode, route within a mode and the times at which trips are taken.

2. The internal resources required by various types of households to satisfy their mobility needs for various types of policy environments.

3. The ways in which trip-making behaviour and household mobility are influenced by changes in public policy.

Four broad types of transport modelling style have emerged over the past twenty-five years and these are usually referred to as multi-stage aggregate models, land use-activity allocation models, disaggregate behavioural models, and household-based activity models.

The first three are essentially concerned with estimating transport demand while the fourth style focuses primarily on the mobility-transport needs of different types of households. None of these four modelling styles are comprehensive in the sense that they are capable of capturing all of the characteristics mentioned above. They are partial models and are capable of responding to a very limited range of policies of the type listed in Table 1.

Multi-Stage Aggregate Models

Multi-stage aggregate models developed during the late 1950's and 1960's when analytical techniques were required to help in locating regional scale transport facilities and in dimensioning their traffic carrying capacities. They attempt to represent a number of aspects of transport decision-making behaviour by the well-known sequence of sub-models of trip generation -- trip distribution -- modal split -- traffic assignment. The term aggregate identifies the fact that these models attempt to explain the average behaviour of groups of urban trip-makers moving between urban sub-areas.

Most of the multi-stage transport models developed for North American planning studies were not formulated with respect to a consistent theoretical perspective. The structure of the individual sub-models and their inter-connections emerged principally from empirical analyses of cross sectional samples of travel behaviour. A variety of disciplines provided the inspiration for these analyses.

In Great Britain the maximum entropy work of Wilson (1968) had an important impact on the structure of the multi-stage aggregate models developed in that country. Wilson's major contribution was to demonstrate that consistent sequences of sub-models could be generated from a limited number of explicit a priori assumptions about the aggregate behaviour of groups of trip makers. Wilson's work stimulated a tremendous amount of theoretical work and some of this work has been reflected in operational models. Senior and Williams (1977), Williams and Senior (1977) and Williams (1979) provide good reviews of much of this work.

Land Use-Activity Allocation Models

The multi-stage aggregate models require as input a completely specified spatial pattern of human activities, and proceed to calculate transport demands for alternative transport system configurations. Concerns about the joint nature of land development and transport led to the development of land use-activity allocation models capable of estimating simultaneously future development patterns and transport flows. Most of the operational models really represent extensions of the multi-stage aggregate models where gravity-type allocation functions are used to estimate the spatial patterns of development. Batty (1976) provides a good summary of the work in this area, while Putman (1973, 1974), Ayeni (1975, 1976), Mackett (1976a, 1976b) and Said and Hutchinson (1980) provide more recent examples.

Disaggregate Behavioural Models

In the late 1960's, the ideas of generalized travel cost and trip dis-utility, as incorporated in the logit model, began to emerge largely as a result of the need to understand better the modal choice decisions of urban trip makers (Stopher, 1969 and de Donnea, 1971). The basic idea was that trip-makers consider a number of modal attributes such as time, cost and service frequency in the choice of one mode over another. These were aggregate models in the sense that they attempted to explain the average behaviour of groups of trip-makers and recognized implicitly that choices had a random component which reflected, because of the averaging process, differences in preferences and the stimuli affecting choice.

In the 1970's random utility-based models began to develop which focused on the behaviour of individual trip-makers, and because of this they became known as disaggregate models, (Domencich and McFadden, 1975; Harris and Tanner, 1975). With this approach individual trip-makers are assumed to weight the various attributes of travel differently, where the weights are assumed to be randomly distributed according to particular probability distributions. Utility functions are developed for each mode and dis-utility coefficients are estimated so as to maximize the ability of the model to explain the modal choice decisions of individuals. A major contribution to this work has been the development of efficient techniques for estimating the models from very modest samples of individual behaviour. These techniques have been extended beyond the modal choice situation and offer a utility-based theory of choice allowing destination, time of day, car ownership decisions, etc., to be modelled independently, sequentially or simultaneously (Ben-Akiva and Lerman, 1977; Ben-Akiva and Atherton, 1977; Lerman, 1976; Ruiter and Ben-Akiva, 1978; Ben-Akiva, Manski and Sherman, 1980).

Household-Based Activity Models

During the 1970's, transport policy in many countries became increasingly concerned with the extent to which urban transport services satisfied the needs of all social groups, particularly those without routine access to a private car. It was argued that transport needs of all urban residents could not be equitably established simply by satisfying estimated demands. Transport needs could only be established through an understanding of the activity patterns of different household groups. With this approach the activities in which individuals wish to engage are constrained in space and time, and the activity patterns depend on household structure, and the transport resources within and external to the households. The primary goal of many of these activity-based approaches was to understand how households of various types might adjust to changes in the external environment, including transport policy.

One of the most comprehensive studies of this approach has been that reported by Jones (1979), and Clarke et al (1980). The major contributions of this work have been:

1. To establish that household structure, principally in terms of the number and ages of children, is the major determinant of activity patterns, and therefore, of travel.

2. To identify that it is from the inter-relationships between the members of the household, their activity patterns and transport resources that the constraints emerge.

3. To indicate that it is possible to establish a simulation method to predict how the different household types might respond to changes in transport policy.

This approach requires detailed surveys of travel by various household groups and no attempt is made to estimate a model in the traditional sense.

THE INTEGRITY OF MODELS

In assessing the capabilities of available models as policy analysis tools it is useful to identify a number of characteristics that should be evaluated and these include:

1. The model structure in terms of its underlying theory and its relationship to public policy variables such as transport service characteristics, fares, etc.

2. The exogenous variables required as input such as population, employment, transport network characteristics, etc., and the extent to which these dictate model predictions.

3. The power of the calibrated model to explain observed variations in trip-making behaviour, and in some cases, transport-related location behaviour.

4. The estimated parameter magnitudes, such as the deterrence function parameters of gravity type allocation functions and the dis-utility coefficients of disaggregate models, and their variation and interpretation across a range of urban areas and policy contexts.

5.   The ability of a model to estimate the observed behaviour
      responses of trip-makers to new changes in public policy.

THE CAPABILITIES OF AVAILABLE MODELS

     In spite of the tremendous amount of theoretical work on multi-
stage aggregate transport models during the past fifteen years, and
their on-going use in many transport studies, substantial reservations exist about their continued use as policy analysis tools.
Some of the concerns include the continued absence of a sound
behaviourally-orientated theoretical basis for all sub-models,
inconsistencies in the inputs required by the different sub-models,
the inabilities of certain sub-models to explain observed behaviours, the very limited connections of the model set to the transport policy environment, the dominance of the exogenous inputs in
determining model outcomes, and the loss of important information on
individual travel behaviour through the use of the average behaviour
of groups of individuals.

     Perhaps the greatest deficiency of multi-stage aggregate models
is the inability of the pivotal trip distribution sub-model to
explain observed spatial interaction behaviour.   While hundreds of
gravity-type spatial interaction models have been calibrated in
various studies, there have been few exhaustive empirical tests of
the explanatory powers of the model.   The systematic evaluation of
these models has been hindered by the unavailability of good data
bases, and the very high costs of acquiring these data bases.   Most
data bases have been derived from sample sizes of 2 to 5 percent and
while these may have been adequate for parameter estimation, they
have certainly not been large enough for assessing the goodness of
fit of model-estimated trip interchange magnitudes.

     Sikdar and Hutchinson (1980) using a 100 percent sample of the
journey to work collected in Edmonton, Alberta have shown that the
explanatory power of existing maximum entropy doubly constrained
models is extremely poor.   Using a simulation approach proposed by
Smith and Hutchinson (1980) they have shown that existing cross-
sectional trip distribution models have the same explanatory power
as a trip table produced by multiplying the cells of the observed
trip table by random errors with a range of ± 75 percent.   These
observations are supported by the work of Hutchinson and Smith
(1979) who examined the capabilities of existing trip distribution
models to explain work trip patterns observed in thirty Canadian
cities in the 1971 Census of Canada.   It was concluded on the basis
of these Canadian studies that the continued use of existing trip
distribution models as the central phase of a multi-stage aggregate
model could not be justified.

     A second major criticism of the aggregate trip distribution
models is the difficulty in developing rational interpretations of
the model parameters.   The theoretical view of spatial interaction
offered by the entropy maximizing models suggests that the deter-

rence function parameter may be interpreted as an elasticity to trip
cost changes and the balancing factors are frequently given a shadow
price interpretation. These behavioural-type interpretations of the
calibration parameters are difficult to sustain from the evidence
obtained in the above-mentioned studies.   Hutchinson and Smith
(1979) obtained trends in the magnitudes of the deterrence parameters across the thirty Canadian census areas and across model
types which contradict the elasticity interpretation given above.
The observed variation in parameter magnitudes really reflected the
way in which the areas had been zoned for the purposes of collecting
census data with parameter magnitudes increasing as the dominance of
the central area as an employment zone increased.   Sikdar and
Hutchinson (1980) showed that the balancing factors of doubly
constrained gravity models cannot be given any other interpretation
other than arbitrary parameters which ensure that the trip end
constraints are satisfied.

The calibration procedure produces a balancing factor product
that is unique rather than individual balancing factor magnitudes
with the destination end factors being adjusted arbitrarily to
accommodate for trip end over- and under-estimations which are
largely unrelated to trip costs.   Aggregate models may also be criticized very strongly in terms of some of the other characteristics
listed previously.

Most of the operational land use-activity allocation models
represent extensions of the multi-stage aggregate modelling tradition and many of the criticisms raised in the previous paragraphs
also apply. Model activity allocations are very dependent upon the
exogenously specified constraints with the only behavioural element
being the travel deterrence term incorporated in most of the allocation functions.

While most studies have produced allocation models that are
capable of reproducing observed activity vectors and the broad patterns of transport flows, the models really represent descriptions
of observed behaviour. Available data bases are usually inadequate
for model calibration and certainly not good enough for independent
evaluation of model performance.

Perhaps the most serious criticism of available models is their
very limited connection to the public policy environment. Alternative public development policies are usually input into the models
through the zone-specific constraints on the total amounts of activity that can occur in a zone, with changes in transport services
being incorporated through changes in travel cost measures.   At
best, available models are capable of providing internally consistent allocations of activities to location opportunities that are
specified exogenously to the model. The critical need is for models
that capture the ways in which location opportunities are created,
not just in new areas of development, but more importantly in the
already developed areas through re-development.

The superiority of the disaggregate behavioural models over the multi-stage aggregate models has been suggested on a number of grounds. An acceptable theoretical base, a capability of responding to a much higher range of policy options that are currently important, more efficient utilization of data, and parameters that can be interpreted more rationally are some of the supporting arguments. The identification of the relevant attributes of the utility functions and the estimation of the dis-utility coefficients are statistical estimation problems, and like all estimation problems significant theoretical assumptions have to be made in order to derive an appropriate estimation procedure. A variety of utility functions may be estimated and the relative powers of each function in explaining choice are used to select the most appropriate function. Clearly, conclusions about the validity of the utility function require that the theoretical assumptions be satisfied by the empirical observations. Like other statistically estimated coefficients, the relative magnitudes of the dis-utility coefficients are inversely related to the variance of the utility distributions from which they were estimated and this variability forms the basis for the significance testing. Perhaps the key assumption of these models is that the attributes incorporated in the utility functions are independent with each having a constant dis-utility coefficient over the range of attributes of interest.

While many disaggregate models of the logit type have been estimated, particularly in modal choice situations, and have been shown to explain the choices of the sample of trip-makers used for calibration, substantial reservations exist about their abilities to capture the underlying behaviour of individuals other than in a statistical sense. Many of the attributes of transport modes incorporated in the utility functions such as times, costs and distances are spatially correlated with each other, as well as with the socio-economic characteristics of trip-makers and land use variables. High socio-economic status is typically associated with the outer suburbs, lower levels of public transport service and long journeys to work. Linear utility functions estimated for choice situations have to be interpreted with a great deal of care, particularly if they are to be applied in contexts in which the co-variances between the utility function attributes are not the same.

Spatial interaction behaviour in urban areas is governed by two broad sets of factors and these are factors exogenous to the transport system such as the locations of housing and job opportunities, and factors that relate directly to the transport system such as service and cost. It is difficult to justify the use of models estimated from behaviour that is jointly determined by these two sets of factors and to assign the cause of the behaviour only to those attributes incorporated into the utility function. The responses of trip-makers to changes in transport policy are unlikely to be captured by models that are really statistical analyses of a cross section. Appropriate data sets that capture the changes in behaviour of trip-makers resulting from changes in transport are not available.

The utility coefficients of the attributes of the utility function are usually interpreted as the marginal valuations of changes
in the magnitudes of the attributes. This assumption that the utility of a decision may be determined by the sum of a set of utility
components implies that the partial derivatives of the utility function may be interpreted as cardinal utilities. Most economists are
not prepared to make this assumption in their models of consumer
behaviour. Finally, the assumption that these marginal valuations
are constant with changes in attributes is in conflict with the idea
of decreasing marginal utility, particularly when attempts are being
made in many countries to develop policies which embody non-marginal
changes.

Other problems that are experienced in the application of
disaggregate models are the problems created by the independence of
irrelevant alternatives assumption and the problems of aggregating
the results of disaggregate analyses to represent the response of a
population to policy changes. These problems are discussed widely
in the literature (Charles River Associates, 1977; and Daly, 1979).

Household-based activity approaches are not constrained by
imposed model structures and really represent some fairly subjective
generalizations about household behaviour. Their sphere of application is restricted primarily to the short-run analysis of fairly
conventional changes in transport policy. The extension of these
techniques to larger-scale applications clearly depends on the
availability of fairly elaborate data bases describing the structure
of existing households.

CONCLUDING REMARKS

It seems fairly clear that the transport modelling capabilities
that have been developed during the past twenty years have fairly
modest capabilities as aids to policy analysis. All of the four
modelling styles described previously are essentially descriptive in
nature despite the often extravagant claims by their proponents. At
best, some models seem to be capable of reproducing a few of the
important dimensions of travel observed at one point in time. While
the multi-stage aggregate models have been used on a fairly routine
basis to estimate future traffic volumes, their ability to explain
travel patterns seems to be so poor that their continued use in the
development of policy advice is difficult to justify.

The land use-activity allocation models that have been
developed also have very limited capabilities as policy analysis
tools. Their valid sphere of application seems to be restricted to
a land use and activity allocation accounting function where the
insights provided by the model results depend on the skill with
which the land development constraints and exogenous inputs have
been specified.

The more recent disaggregate and household activity modelling initiatives have fairly limited scopes of application. Little evidence has been assembled to demonstrate that available disaggregate models have any real capability of estimating the responses to urban residents to changes in transport policy. While the many disaggregate modelling exercises and the few household activity studies have provided important insights into the mechanisms of transport-related behaviour, they have not reached the stage where they might be used for large-scale analyses of policy change.

The traditional scientific-based response to this problem would be to argue for an improved theory of cities, and through this the development of more relevant and effective policy analysis tools. While this improved understanding might contribute to the development of better analysis tools, the most pressing need is for the development of analysis paradigms that are more compatible with political decision-making styles. Rather than focusing mainly on transport behaviour, analyses should concentrate on understanding the major economic and social processes that give rise to transport behaviour.

Urban transport behaviour is governed primarily by the constraints on that behaviour imposed by sectors external to the transport sector. Understanding housing markets, isolating trends in employment location, identifying demographic trends and life style changes, and understanding how these relate to transport behaviour would seem to be more productive avenues to support policy analyses.

A number of issues are emerging in North American cities that require the attention of policy advisors and policy-makers and these include:

1. Increasing energy costs and the impact on residential, job and retail markets and the impacts on transport demand.

2. The aging of the population and its impacts on transport needs and other markets such as housing, health services, and so on.

3. The changing employment bases of urban areas and the implications for fixed route, centrally focused transport facilities.

4. Changes in family structure and life style and the related impacts on transport and housing markets.

Available modelling techniques are not capable of responding to these issues, where most of the issues relate to changes in the transport system environment rather than in changes in the transport system itself. Currently observed behaviour is unlikely to be the best indicator of future responses under these circumstances, and the focus of analysis will have to change from revealed to stated preferences. It seems clear that current analytical techniques will have little impact on the policy issues of the 1980's.

REFERENCES

Ayeni, M., 1975, A Predictive Model of Urban Stock and Activity: 1. Theoretical Considerations, Environment and Planning A, 7, 965-980.

Ayeni, M., 1976, A Predictive Model of Urban Stock and Activity: 2. Empirical Development, Environment and Planning A, 8, 59-78.

Batty, M., 1976, Urban Modelling, Cambridge University Press, Cambridge, UK.

Ben-Akiva, M. and Atherton, T., 1977, Methodology for Short Range Travel Demand Predictions, Journal of Transport Economics and Policy, 9, 224.

Ben-Akiva, M. and Lerman, S.R., 1977, Disaggregate Travel and Mobility Choice Models and Measures of Accessibility, Proceedings: 3rd Conference on Behavioural Travel Modelling, Tanunda, Australia.

Ben-Akiva, M., Manski, C.F. and Sherman, L., 1980, A Behavioural Approach to Modelling Household Motor Vehicle Ownership and Applications to Aggregate Policy Analysis, Proceedings: World Conference on Transportation Research, London, UK.

Clarke, M.I., Dix, M.C., Jones, P.M. and Heggie, I.G., 1980, Recent Advances in Activity-Travel Analysis, PTRC Summer Meeting, University of Warwick, UK.

Charles River Associates, 1977, Disaggregate Travel Demand Models, II, Final Report Project 8-13, NCHRP, Washington DC.

Daly, A., 1979, Behavioural Travel Modelling - Some European Experience, Paper to Transport Studies Group, Department of Geography, University of Reading, Reading, UK.

de Donnea, F.X., 1971, The Determinants of Transport Mode Choice in Dutch Cities, Rotterdam University Press, Rotterdam, The Netherlands.

Domencich, T. and McFadden, D., 1975, Urban Travel Demand: A Behavioural Analysis, North Holland Publishing Company, Amsterdam.

Harris, A.J. and Tanner, J.C., 1975, Transport Demand Models Based on Personal Characteristics, Transport and Road Research Laboratory, Supplementary Report, GSUC, UK.

Hutchinson, B.G. and Smith, D.P., 1979, Empirical Studies of the Journey to Work in Urban Canada, Canadian Journal of Civil Engineering, 6, 308-318.

Jones, P.M., 1979, HATS - A Technique for Investigating Household Decisions, Environment and Planning A, 11, 59-70.

Lerman, S.R., 1976, Location, Housing, Automobile Ownership and Mode to Work: A Joint Choice Model, Transportation Research Record, 610, 6.

Mackett, R., 1976a, A Dynamic Integrated Activity Allocation-Transportation Model for West Yorkshire, WP 40, School of Geography, University of Leeds, UK.

Mackett, R., 1976b, The Theoretical Structure of a Dynamic Urban Activity and Stock Allocation Model, WP 135, School of Geography, University of Leeds, UK.

Putman, S.H., 1973, The Interdependence of Transportation Development and Land Development, Institute of Environmental Studies, Department of City and Regional Planning, University of Pennsylvania, Philadelphia, Penn.

Putman, S.H., 1974, Preliminary Results from an Integrated Transportation and Land Use Package, Transportation, 3, 193-224.

Ruiter, E. and Ben-Akiva, M., 1978, A System of Disaggregate Travel Demand Models: Structure, Components, Models, and Application Procedure, Transportation Research Record, 673, 121-128.

Said, G. and Hutchinson, B.G., 1980, A Transport Policy Related Urban Systems Model for the Toronto Region, Annual Conference, Roads and Transportation Association of Canada, Ottawa, Ontario.

Senior, M.L. and Williams, H.W.C.L., 1977, Model Based Transport Policy Assessment I, WP-195, School of Geography, University of Leeds, UK.

Sikdar, P.K. and Hutchinson, B.G., 1980, Empirical Studies of Work Trip Distribution Models, Transportation Research A, 15, 233-243.

Smith, D.P. and Hutchinson, B.G., 1980, Goodness of Fit Statistics for Trip Distribution Models, Transportation Research A, 15, 295-303.

Stopher, P.R., 1969, A Probability Model of Travel Mode Choice for the Work Journey, Highway Research Record, 283, Washington, DC.

Williams, H.W.C.L. and Senior, M.L., 1977, Model Based Transport Policy Assessment II, WP-196, School of Geography, University of Leeds, UK.

Williams, H.W.C.L., 1979, Travel Demand Forecasting - An Overview of Theoretical Developments, Paper to Transport Studies Group, Department of Geography, University of Reading, Reading, UK.

Wilson, A.G., 1967, A Statistical Theory of Spatial Distribution Models, Transportation Research, 1, 253-269.

SYSTEMATIC METHODS IN STRATEGIC LAND USE PLANNING: SOME REFLECTIONS

ON RECENT BRITISH EXPERIENCE

Peter Batey

Department of Civic Design
University of Liverpool
Liverpool, L69 3BX
United Kingdom

INTRODUCTION

This paper is concerned with the impact that systems analysis has had upon one particular mode of urban policy-making: strategic land use planning. Drawing upon British experience, the paper explores the way in which the relationship between planning methods and strategic planning practice has evolved since systems concepts were first introduced less than twenty years ago.

Although land use planning in Britain has a comparatively long history, it is only since the mid-1960's that a significant degree of sophistication has been achieved in planning methodology. Before this, planning methods tended to be highly informal "rules-of-thumb", frequently without proper documentation, and used as part of a simple, survey-analysis-plan process. As a result, it was often difficult to obtain consistency from one plan to another, as well as being almost impossible to establish the reasoning behind the various policies making up a plan. At least at the strategic level, the picture has changed, and nowadays, land use planning is based upon more formal, replicable, methods. While some of these methods are computer-based and utilise quantitative data, these features are less important than the fact that the methods are <u>systematic</u>. This means that, like the planning process itself, a method can be expressed as a series of logical steps.

The next section of the paper provides an historical account of the development of systematic methods within British planning practice. Rather than concentrating on the technical details of individual methods, which are fully documented elsewhere (Wilson, 1974; Batey, 1978; Batey and Breheny, 1978), the emphasis here is on the

planning context within which methods are applied. The paper shows
that changes in this context have had an important influence upon
the adoption and use of particular types of method. It also indicates that the adoption of new methods can often depend on a sensitive relationship between academics and planning practitioners: a
"live" demonstration project is generally the most effective way of
convincing practitioners of the usefulness of new methods and methodologies. Some specific attempts to improve communications between
academics and planning practitioners are the subject of Appendix 1
where a brief account is given of the aims and activities of the
Workshop on Regional Science Methods in Structure Planning.

In the third section, the paper looks more critically at the
practical application of systematic methods in British planning. As
well as drawing attention to the technical and theoretical problems
which face users of systematic methods in planning, this section
reviews the main organizational constraints which limit the adoption
and use of methods. The section concludes that while there was once
a tendency to think of methods in isolation from specific planning
issues, this is no longer the case: the choice of methods is now
based on a careful consideration of several factors, most important
of which is immediate relevance to a planning problem.

SYSTEMATIC METHODS AND THE CHANGING CONTEXT OF BRITISH LAND USE
PLANNING: 1965-1980

The Advent of the Systems Approach: 1965-1969

Since the late 1940's, local planning authorities in Britain
have been required to prepare land use development plans, and to
submit these plans to central government for approval. The 1947
Planning Act obliged local authorities to prepare detailed land use
allocation maps setting out the proposed pattern of development over
a twenty-year period, and gave them the power to control development
through the issuing (or withholding) of planning permission.

In the mid-1960's, this system of plan-making was reviewed by
the Planning Advisory Group, a government-appointed committee whose
main recommendations (Planning Advisory Group, 1965) were embodied
in the 1968 Planning Act. This Act reformed the system of development plans, but without altering the powers available to local authorities to control the development of land. The reforms had two
main purposes: to differentiate between those aspects of local
development that were strategic[1] and those that were not; and to
enable greater efficiency and speed to be achieved in the preparation, consultation and, most important, central government approval
of plans. Two types of development plan were introduced under the
1968 Act: structure plans intended to deal with matters of strategic
importance; and local plans which it was hoped would cover smaller-
scale and shorter-term planning issues. In contrast to the earlier
development plans, structure plans would consist of written state-

ments of land use policy, with (generalized) maps and diagrams playing only a supporting role. Local plans would be more similar to the old development plans and, unlike structure plans, would not normally require central government approval.

Although the initial impetus for these reforms came from planners and civil servants dissatisfied with the administration of the development plan system (Roberts, 1976), it was not long before the proposals contained in the Planning Advisory Group's report began to be associated with new ideas about the nature of the planning process. A small number of British academics had been taking a close interest in technical developments in North American planning during the early 1960's, particularly in relation to the land use transportation studies then in vogue. One academic, J.B. McLoughlin, was especially influential in his efforts to show how structure plans could follow these new approaches. McLoughlin drew together a large number of ideas based on his observations of American planning practice, and on his wide reading of management theory (McLoughlin, 1967; 1969). He summarized his own view of planning as the "systems approach":

> "Planning is not centrally concerned with the design of artifacts, but with a continuing process that begins with the identification of social goals and the attempt to realise these through the guidance of change in the environment. At all times the system will be monitored to show the effects of recent decisions and how these relate to the course being steered."

> McLoughlin (1965b:339)

McLoughlin popularised this approach to planning in a series of articles in the Journal of the Town Planning Institute during the mid-1960's (McLoughlin, 1965a, 1965b, 1966), and thus ensured that at least a small proportion of professional planners in Britain became familiar with rudimentary systems concepts. However, perhaps McLouglin's main contribution was to demonstrate how the systems approach might be implemented in planning practice, by directing the first sub-regional planning study, of Leicester and Leicestershire (1969) in the English East Midlands. This, and other similar studies, acted as prototypes for the structure planning exercises which followed at the end of the 1960's.

The sub-regional studies, like the American land-use transportation studies earlier in the 1960's (Boyce, Day and McDonald, 1970), were largely concerned with the allocation of demographic and economic growth. They were carried out by small teams of planners working outside the day-to-day pressures of a busy local authority planning office. This environment provided planners with the stimulus to experiment with new planning methodologies, and with computer-based analytical techniques. The studies were conducted within a strict timetable -- usually over a two-year period -- and this meant that in applying new methods, planners were compelled to be more pragmatic than their American counterparts. The fact that these studies showed that worthwhile results could be obtained comparatively quickly, and at a much higher level of technical sophistication than had hitherto been possible, persuaded many British planners of the value of a systems approach in strategic planning studies.

Sub-regional studies attracted a large amount of attention in the professional journals during the late 1960's and early 1970's, and, as a result, many planners became fascinated with the techniques accompanying the systems approach, to such an extent that, at least for a brief period, it seemed that no study was complete without its own Lowry model[2] or potential surface analysis[3]. This was a time when some sections of the planning profession in Britain were particularly receptive to the introduction of new techniques, and when groups of academics and planning practitioners believed there was mutual benefit in cooperating on the development and application of these techniques.[4] Attention was not confined to methods of analysis and forecasting procedures: gradually planners became interested in systematic methods which could be applied at other stages in the planning process, methods such as the goals achievement matrix (Hill, 1968), a popular evaluation method during this period.[5]

Systematic Methods in Structure Planning: 1970-1980

Work on structure planning began slowly in the late 1960's and, by the time of local government reorganisation in 1974, only in the West Midlands and in South Hampshire had significant progress been made (Smith, 1974). These structure plans were prepared against the background of population and employment growth, and there was no doubt in planners' minds that the most important role of the structure plan should be that of recommending locations for growth. Like the sub-regional studies, therefore, the early structure plans were primarily concerned with spatial issues.

Taken as a whole, this first group of structure plans failed to achieve the level of technical sophistication evident in the sub-regional studies. Some fell a long way short of it.[6] Despite efforts by central government to encourage the use of systematic methods (Department of the Environment, 1973), the impetus developed previously was lost. Several factors can be put forward to explain this "failure" to maintain progress in the development and application of new methods. Langley et al (1975), who examined analytical and predictive methods in the first twenty structure plans, suggest that the failure was due to the nature of the institutional setting of the plans. Sub-regional plans had been produced in almost "laboratory" conditions, while structure plans were being prepared as statutory documents by hard-pressed local government officers who generally lacked the time or expertise to carry out innovative work. Also important was the fact that some of the earlier plans were prepared extremely quickly for reasons of political expediency. Some of the local planning authorities in the West Midlands, for example, wanted to leave the incoming post-reorganisation local authorities with certain policy commitments, and a "submitted" structure plan was regarded as an appropriate means of doing this.

The period 1972-74 witnessed a number of important related developments affecting the use of systematic methods in British strategic planning. There were the first signs of a reaction against the "overly-rational" features of the systems approach (McDougall, 1973). This coincided with an increasing interest in the social and economic, as opposed to physical, aspects of planning, and with scepticism about the ability of planners to produce long-term plans and cope with the inherent uncertainties in the environment. These developments, along with the sudden downturn in anticipated growth, led to variants on the so-called rational planning process. There was an increasing interest in the pragmatic "incrementalist" approach, advocated by Thorburn, County Planning Officer of East Sussex (Thorburn, 1975), and supported indirectly by a Department of the Environment Circular (98/74) which urged local authorities to concentrate on "key issues" in structure plan preparation.

The result of these developments was to create by the mid-1970's, an enormous diversity in the use made of systematic methods by different local authorities. In a few cases, their methods of plan-making differed only slightly from those used to prepare 1947 Act development plans.[7] At the other extreme, some local authorities had developed elaborate frameworks linking together several forecasting models, or were using the full range of generation and evaluation tools originally developed in the sub-regional studies, albeit in a more mature and flexible fashion.[8]

A particularly striking contrast in the approach adopted to the use of systematic methods is that between shire and metropolitan counties. In the shire counties, mainly containing rural and smaller urban areas, the nature of the planning problems tackled and the likelihood of some growth enabled some of the earlier forecasting and spatial methods to be used. However, in metropolitan counties, population and employment decline had set in by the early 1970's (if not before), and the major structure planning issues were not spatial, but social and economic. New methods were needed to handle the complex relationships between these issues, particularly in the generation and evaluation of structure plan policies. One method designed to deal with such problems was AIDA, the Analysis of Interconnected Decision Areas. AIDA was developed originally as a tool in architectural design (Luckman, 1967), and in the late 1960's it had been used with some success in local planning (Friend et al, 1970). Its main virtues as a structure planning tool were that it was relatively easy to understand, it enabled complex problems with

many possible "solutions" to be studied systematically, and it could
be used, whatever quantity of data was available. AIDA had the
added advantage of fitting neatly into a cyclical, continuous process of plan-making of the kind that many local authorities were
trying to follow.

AIDA provides a comparatively rare example of a "new" (to strategic planning) planning method which found widespread acceptance
among practitioners during the mid-1970's. It may be instructive,
therefore, to examine briefly how this came about. In 1974, central
government, concerned about the length of time it was taking to
prepare structure plans, commissioned the Institute for Operational
Research (IOR) to conduct an "action research" project aimed at
exploring ways in which AIDA and related techniques could be used to
assist in strategic policy formulation (Sutton, Hickling and Friend,
1977). Local authority involvement in the project was of two
kinds. On the one hand, a small group of local authorities offered
to act as "guinea pigs" for the development of the AIDA methodology:
these local authorities worked closely with IOR personnel who were
sometimes seconded for short periods to the local planning departments. On the other hand, a second group of local authorities, who
had expressed a general interest in the AIDA project, were kept
informed of its progress by a series of seminars and briefings.
These meetings served to convey the basic ideas of AIDA, and confirmed that the methodology could be applied in practice (a vital
step, as McLoughlin had found in promoting the systems approach
during the late 1960's). As a result, several local authorities
drew upon the general principles of AIDA, and applied the technique
in their own structure-planning exercises, modifying it where necessary to suit their own local requirements. In both shire and metropolitan counties, AIDA fulfilled an important role in bridging the
gap between detailed survey and analysis and strategic policy-
making.[9]

The intensive phase of structure <u>plan-making</u> during the first
two-thirds of the 1970's acted in some ways to deflect planners'
attention away from planning as a continuous process, even though
this was one of the prerequisites of McLoughlin's systems approach.
Perhaps because it had to be formally submitted to central government, there was a tendency to think of the structure plan as an end-
-state plan. One important consequence was that systematic methods
of monitoring, aimed at keeping the plan up-to-date and relevant,
tended to be ignored until the plan was submitted. This was only to
be expected, given the way in which planners' interest in systematic
methods had evolved during the ten years or so since the 1968 Planning Act. Generally, local authority interest in methods assumed a
linear process, paralleling the basic plan-making process. Initially, the dominant concern was with methods of analysis, procedures for forecasting activity levels and spatial allocation
methods. As a majority of structure planning teams progressed from
survey and analytical work to policy formulation, so interest in
methods of plan generation and evaluation began to grow, reaching a

peak in the mid-1970's. Eventually, towards the end of the decade, attention turned to monitoring. Prior to this, only a handful of local authorities, including Hertfordshire, East Sussex and Hampshire, had taken any serious steps to establish formal monitoring procedures.[10]

Plan-making activity, as such, had begun to decline in importance by 1980, as most local planning authorities reached the stage of submitting a structure plan for central government approval. Staffing levels in local planning authorities, especially in structure planning teams, started to decline in response to cuts in public expenditure. Planners who remained in these teams, after spending a considerable amount of time "processing" the plan so that it became an acceptable statutory document, proceeded to work either on the implementation of policies, or on the monitoring and review of policies. Monitoring activity was usually closely associated with the establishment of systematic procedures for collecting, storing and manipulating planning data. These information systems represented an attempt to bring together relevant planning information, if possible, for a common set of spatial units, and to present it in a clear and concise form to policy-makers. The presentation and interpretation of this information by planners, an activity sometimes referred to as policy analysis, generally made little use of sophisticated methods of analysis of the kind that had been popular during the previous decade.[11] Where a method satisfied a well-defined need (eg. a population projection method supplying forecasts to other local government departments), it was likely to remain in use. Other methods, perhaps because they had always been regarded as experimental and research-oriented, were neglected by practitioners, and attracted interest only amongst academic planners. With notable exceptions, such as work on the integration of forecasting (Breheny and Roberts, 1978), there were few attempts among local authorities to adopt new, and perhaps more appropriate methods. Although planners were more prepared to select methods suited to local circumstances, and were more aware of time, organizational and data constraints than ten years previously, these methods were almost always chosen from among the body of "tried and tested" methods with which they were already familiar.

SHORTCOMINGS IN THE DEVELOPMENT AND APPLICATION OF SYSTEMATIC METHODS

The Practical Application of Methods

If there is one basic criticism that can be made of the early applications of systematic methods in British planning, it is that too much of the enthusiasm for these methods was channelled in the wrong direction. Rather than considering the practical context within which methods were to be applied, planners allowed themselves to become immersed in technical detail: what sort of calibration procedures to use, how best to measure the goodness-of-fit, and so

on. It is very rare to find any assessment of how effective the
methods proved to be, of how politicians or chief officers reacted
to the results, or the commitment the use of the methods represented
in terms of, say, time or staff resources.

This obsession with the technology of methods, to the exclusion
of practical matters, carried over into planning education. By the
end of the 1960's, most planning schools in Britain had begun to
include systematic planning methods as part of the teaching sylla-
bus, and while it was quite common for such courses to contain
numerical practical exercises, there was a tendency to teach only
the "mechanics" of methods. Never was sufficient attention given to
the factors influencing the choice of appropriate methods.

There are, in fact, three important issues which need to be
considered by any planner contemplating the use of systematic
methods in planning: how to obtain the best possible match between
methods and planning problems; how best to communicate the results
obtained to those involved in the planning process (and, in addition, how best to organise the inputs that these individuals and
groups make to certain methods); and how to ensure that results are
provided while they are still relevant, and can still inform policy
decisions.

During the early days of the systems approach, the choice of
planning methods was often far from rational, and owed much to
fashion, a point stressed by D.M. Smith in relation to the work of
geographers at that time:

> "During the quantitative revolution the focus of
> attention was on methods rather than on subject
> matter...Applied human geography in regional
> planning and urban development was often merely
> an opportunity for the application of some
> favoured technique which may have been quite
> ill-suited to the problem at hand."
>
> Smith (1977:4)

By the mid-1970's, there were still difficulties in matching
methods and problems, but for different reasons. Given the pace at
which planning problems (or, more correctly, their perception) were
changing, it was hardly surprising that there was a time lag between
the recognition of new planning problems, and the development of
appropriate methods. The fall-off in demographic and economic
growth, and the strengthening of links between physical planning and
financial budgeting created a need for new methods which has still
to be adequately met. In the meantime, many planners have either
abandoned the use of all but the simplest of methods, or continued
to use methods which are no longer appropriate.

The problem of communicating the results of applying planning methods has been recognized only comparatively recently. As long as strategic land-use planning was viewed largely as a technical process, the task of presenting the principles of a method, together with the results it provided, was not regarded as an issue. If anything, planners involved in a modelling exercise liked to cultivate the image of being "experts", and saw no reason to remove the "mystery" surrounding their work. However, when the use of systematic methods began to be more fully integrated into the work of a structure planning team, the need to communicate to a wider audience became apparent. To some extent, this change of outlook represents a response to the concern expressed by certain sections of the general public, but the wider audience also includes professional colleagues — planners and others — since it has become clear that in order to implement a strategic plan, the cooperation and involvement of other departments and agencies is vital.

Also of crucial importance is the timing of the results obtained from the use of methods. Excessive amounts of time spent in data collection, and in setting up modelling exercises on a computer have frequently meant that planners have been unable to produce information when it is needed in the decision-making process. Timing promises to be of particular significance in relation to the use of methods in the monitoring and review of strategic plans: there is a danger of opting for very elaborate computerized information systems which may take several years to become fully operational - and which may rely upon data series which are seriously delayed in publication.[12] The risk with a "grand design" approach is that the absence of usable results in the interim may lead policy-makers to abandon the whole system in favour of less sophisticated, but more flexible, alternative procedures.

Organizational Constraints

One of the strongest influences on the use made of systematic methods concerns the nature of the organization within which strategic land use planning is taking place. A number of studies (eg. Jefferson, 1973 and Wade, 1971) in the early 1970's examined factors affecting the adoption of new planning methods by local authorities. Most planning authorities were (and, in many cases, still are) characterized by a fairly pragmatic, ad hoc approach to management, involving a preoccupation with day-to-day tasks, and leaving little scope or encouragement for innovative work. The sub-regional studies, referred to earlier, were an exception, since a separate unit was established and given a free hand to organize its work. There were other cases of local authorities which were more receptive to the introduction of new methods. Cheshire, a county in North West England, is one example. Cheshire experienced a long period of relative inactivity as far as plan-making was concerned, at the end of the 1960's. This offered scope for the establishment of joint research projects with outside bodies, such as the Centre for Environmental Studies (who cooperated in the building of a Lowry model), and the Local Government Operational Research Unit (who contributed a computerized procedure for evaluating land use plans). Neither of these projects was particularly successful, however, and as a result, the methods were never used in the subsequent Cheshire structure planning exercise.

Because technical expertise was scarce, the initial development of new methods in planning departments was often heavily dependent upon particular individuals. Sometimes this meant that when a key person left an authority, the development work ceased, since no replacement could be found with the requisite skills and experience. In any case, given the poor level of documentation which existed in most planning departments, it would be difficult for a new member of staff, however knowledgeable, to take up the work where his predecessor left off. By the end of the 1970's, the development and use of methods still relied upon small groups of staff, but, perhaps because of lower levels of staff mobility, projects stood a greater chance of being seen through to completion.

Some planning departments, however, have adopted a deliberate policy of not employing specialist staff. Merseyside County, for example, was set up with a relatively small planning department, the intention being to commission from consultants, work that could not be done in the department. In three years after the county was established (in 1974, upon local government reorganization), several pieces of analytical work were carried out by consultants, including a small area population forecasting procedure, a multi-level forecasting framework, a social area analysis and an input-output study. The county appears to have had difficulties in absorbing the results of this technical work: only the social area analysis and the population forecasting procedure have had any lasting impact upon the department's work. In the case of the other projects, staff turnover appears to have been a factor in the failure to carry forward the input-output study, while the development of a more straightforward rival approach within the department reduced the chances of the multi-level forecasting framework being adopted.

Finally, in most planning authorities there can be found what Batty (1976) calls "organizational inertia". Even though the average length of service in planning departments is low, when compared with other sections of local government, there is still likely to be suspicion among senior members of staff when attempts are made to introduce new methods and methodologies. Partly this reflects differences in the initial training of senior and junior staff. Understandably, it may also reflect the natural caution exercised by senior officers in adopting fashionable, but risky, technical innovations.

Technical Problems

Earlier in this paper, the point was made that until the mid-1970's, most of the effort in developing and applying new systematic methods had been directed towards methods of analysis; methods which assist in understanding the urban system or parts thereof; the product of planning. Methods concerned with the planning process -- with policy generation, evaluation and monitoring had received much less attention.

Of the various types of methods of analysis, most progress had been made in the fields of transport and population: most strategic planning authorities in Britain had, by the end of the 1970's, acquired some experience in the use of these methods. In housing market analysis and in studies of the local economy, however, the methods used are comparatively crude. Although considerable attention is now being given to policy formulation in these areas, the supporting analysis tends to be weak.

The problem of linking together various individual pieces of analysis has also been neglected by British planners. Barras, in drawing conclusions from a comprehensive study of analysis in twenty structure plans (see Barras and Broadbent, 1979), noted that:

> "While the analysis of separate "subject" areas
> (eg. population, employment and housing) is
> usually quite substantive, there is inadequate
> integration of these partial analyses via the
> key supply and demand linkages in the system
> (eg. the labour market links between "popula-
> tion" and "employment" or the housing market
> links between "population" and "housing"."

Barras (1978:297)

Only in some of the more recent structure plans have attempts been made to bring together analyses and forecasts in a consistent manner: in Gloucestershire, for example, an integrated forecasting system has been developed which allows the interplay of supply and demand factors to be taken into account (Breheny and Roberts, 1978).

Sometimes there can be quite severe problems when methods are used in conjunction with one another. Often a comparatively primitive method is used to provide inputs for a more sophisticated method. A well-known example is the case where "land use" inputs -- spatial distributions of population and employment -- are being prepared for a transport model. It is common for the employment and population components to be derived independently of one another. Although the area-wide control totals for these activities may be derived using a formal method, the allocation to fine spatial zones will usually be carried out "by hand": a return to the informal, difficult-to-replicate, methods of twenty or more years ago. Not sufficient is known about the effects that such poor quality land use inputs are likely to have upon the various items of output obtained from a transport model.

The final example of technical problems which face planners in applying systematic methods concerns the operational decisions taken during a given process of analysis. This point can best be illustrated by reference to an example: social area analysis based upon multivariate statistical methods. The results of such an exercise, often expressed in terms of a classification of spatial units, have been shown by Openshaw and Gillard (1978) to be highly sensitive to the initial choice of attributes or variables, whether these variables are standardized, whether the principal components are rotated, how many principal components are used as inputs to cluster analysis and the method of clustering adopted, inter alia. Awareness of the effects of these choices is very limited, both among planners and other professional staff who make use of the results.

The same type of problem is encountered in population forecasting where a variety of assumptions have to be made about fertility rates, survival rates and migration. While different assumptions will obviously produce different results, planners have often succumbed to the understandable temptation of presenting results as point estimates, rather than as ranges, on the grounds that such ranges might introduce ambiguity among the users of the results.

Theoretical Shortcomings

The attitude amongst British planners towards the application of strategic planning methods has always been the pragmatic one that if a method appears to work and produces reasonable results -- accurate predictions or acceptable strategies -- it is worth using, regardless of whether it has any sound theoretical basis. The most popular methods have been those that can be made to work without delay, and within the limitations of existing data.

During the early days of the systems approach, many planners naively believed that the new systems-based methods were value-free and objective; model results were thought to be correct and unambiguous, and little, if any, attention was given to the fact that some form of bias was bound to influence the interpretation of these results. A planner interpreting the results of a travel demand model, for example, ten years ago would probably arrive at one conclusion: more road space was needed. Nowadays the same planner would almost certainly recognize that a variety of interpretations is possible, policies are needed to reduce projected travel demand, or more road space is needed. The early naivety and over-enthusiasm has, to a large extent, been tempered, and more reasonable views are now held about the role of methods. No longer is the planning process seen merely as a technical process. More care is taken in the choice of methods, and in the presentation of results and assumptions, and although there are still important practical, technical and organizational problems to be addressed, skill and experience in the use of methods is now far more widespread among British planners.

APPENDIX 1: FOSTERING LINKS BETWEEN ACADEMICS AND PLANNING PRACTITIONERS: THE WORKSHOP ON REGIONAL SCIENCE METHODS IN STRUCTURE PLANNING, 1975-1980

The Workshop on Regional Science Methods in Structure Planning was established in 1975, as one of a number of workshops within the British Section of the Regional Science Association. A group of members had been conscious for some time that much of the research carried out in the field of regional science had focused too narrowly on the mathematical representation of urban and regional systems, and on the elaboration of theory. The application of regional science methods to "real world" planning problems had tended to be overlooked and, in particular, little attention had been paid to the practical difficulties associated with the use of these methods by people working in local authorities. At the same time, many local authorities had recently entered an intensive phase of policy-making involving the preparation of structure plans. In spite of their many limitations, structure plans did have the effect of making local authority planners consider more carefully the kinds of method and methodology appropriate for strategic land use planning. Because of this clear local authority interest in, and commitment to, one form of strategic planning, the convenors decided initially to limit the Workshop's primary field of concern to the application of regional science methods in structure planning. Later, as other forms of strategic planning assumed greater importance in local authority work, the field of interest of the Workshop broadened. This was accompanied by greater involvement of practitioners from other types of planning agency.

OBJECTIVES

The principal aim of the Workshop during the five years of its existence was to study the practical application of systematic methods in strategic planning. This aim was translated into two main objectives:

1.   To provide a forum for critical discussion of methods by practitioners and academics.

2.   To assemble information on current practice, and to disseminate this information to people working in the field.

The Workshop attempted to meet these objectives in three main ways: Workshop sessions, Workshop publications and international meetings. A brief account of each of these activities is given below.

WORKSHOP SESSIONS

A total of nineteen Workshop sessions were held, each based around a specific theme. Workshop sessions were organised at a variety of different venues throughout Britain, and each was attended by twenty to twenty-five people. Attendance was by invitation only, so that the convenors could achieve a balance between practitioners and academics. People were invited whose interests coincided with the theme of a session.

Three or four speakers, drawn either from research institutions or from local authorities, were invited to prepare a short paper on a topic appropriate to the theme of the session. Each session lasted for one whole day, and this generally provided sufficient time for the papers to be presented and fully discussed.

The finance needed to run the Workshop sessions (and, indeed, the other Workshop activities) was minimal, usually amounting to no more than thirty pounds per session. Participants provided their own travelling expenses and no charge was made for attendance, other than to cover refreshments. Moving the venue of the Workshop meant that the organisers were able to spread travelling costs among a wider group of people, and also to attract some participants who might otherwise have been deterred from attending. Usually it was possible to obtain free accommodation, either in the offices of a local authority or in a university department.

Themes adopted for Workshop sessions have included: Census Analysis and the Planning Process, Land Use and Transport Interaction, Methods of Resource Allocation, Forecasting Frameworks for Strategic Planning, Activity Allocation Models: A Review of their Role in Planning, Methods for the Study of Rural Deprivation, Plan Generation and Evaluation Methods, Subjective Social Indicators and Input-Output Analysis and its Applications in Strategic Planning.

WORKSHOP PUBLICATIONS

During the first two years of its existence, the Workshop compiled and published a news-sheet which was circulated, free of charge, to all structure-planning local authorities, and to other interested individuals. The news-sheet reported on the proceedings of Workshop sessions, and also provided a register of recent and current work on systematic methods appropriate to structure planning. The news-sheet served a valuable role over the short period during which it was produced. Eventually, however, it was discontinued in favour of the more formal method of publication in journals.

The Workshop has been responsible for a number of publications, including:

Two half-issues of Town Planning Review, 1978, later reprinted as Systematic Methods in British Planning Practice (reflecting the work discussed at Workshop sessions 1975-77).

A collection of three papers in Environment and Planning A, 1979, based on papers given at a Workshop session in 1979 on Activity Allocation Models.

Two papers in Papers of the Regional Science Association, 1980, reflecting the Workshop's interest in problems of forecasting in strategic planning.

Other papers, on plan generation and on input-output analysis, have appeared in the London Papers in Regional Science series.

Probably the most successful feature of the Workshop's publications was the fact that practitioners, as well as academics, were encouraged to document their use and development of methods. In certain cases, the Workshop papers represent the only comprehensive documentation of particular applications of methods.

INTERNATIONAL MEETINGS

In 1978, and again in 1980, a group of Workshop members took part in an international workshop meeting in the Netherlands. Each of these meetings was attended by Dutch and German planners and academics, as well as by the British group. At the first meeting, the programme was organised so that each national group was allocated a day in which to present papers on the application of systematic methods. This permitted useful comparisons to be made of practice (and the relationship between research and practice) in the three countries. The second meeting was organised on a different basis, with the programme divided between "procedural" and "substantive" planning methods.

This series of international meetings promises to continue in future years. It has enabled valuable links to be established between Workshop groups in each of the participating countries who share a concern to improve communication between planning practitioners and academics in relation to the appropriate development and application of systematic methods.

NOTES

1. According to Solesbury (1974:98), strategic planning decisions may give rise to "important marginal changes in the amounts or conditions of activities, land uses, accommodations, movements, townscapes, landscapes or other environmental attributes". The impact of these decisions "may be direct or it may be indirect, consequential on chain reactions which cumulatively have major impact".

2. See, for example, Batty (1970).

3. See, for example, Coventry-Solihull-Warwickshire Sub-Regional Study Team (1971).

4. An example of this cooperation is the special issue of Regional Studies (part of Volume 3, 1969) which focuses on quantitative methods in urban and regional planning.

5. A special issue of Regional Studies (part of Volume 4, 1970) is devoted to evaluation methodology.

6. See Barras and Broadbent (1979) for a detailed critique of the analysis carried out in twenty of the early structure plans.

7. Breheny and Roberts (1978) describe how an integrated forecasting system has been established as part of structure plan work in Gloucestershire.

8. See, for example, Hertfordshire County Council (1976).

9. See Bather, Williams and Sutton (1976) for an example of the application of AIDA to a shire county, and Greater Manchester Council (1977) for an example of its application to a metropolitan county.

10. See Perry and Chamberlain (1977) – Hertfordshire, Duc (1977) – East Sussex, and Francis (1980) – Hampshire, for examples of local authorities with well-established monitoring systems.

11. A good example of this type of work may be seen in the Proceedings of the Policy Analysis in Urban and Regional Planning Seminar, containing papers presented at the annual meetings of PTRC.

12. Data series, such as the Annual Census of Employment and the Census of Population, both of importance to strategic land use planning, have been subject to long delays in publication in recent years: on occasions, the delay has been as much as three years.

REFERENCES

Barras, R., 1978, A Resource Allocation Framework for Strategic Planning, Town Planning Review, 49, 296-305.

Barras, R. and Broadbent, T.A., 1979, The Analysis in English Structure Plans, Urban Studies, 16, 1-18.

Batey, P.W.J., 1978, Planning Techniques in Practice: A Digest of Techniques Used in Recent British Planning Practice, Working Paper 10, Department of Civic Design, University of Liverpool, Liverpool, UK.

Batey, P.W.J. and Breheny, M.J., 1978, Methods in Strategic Planning: A Descriptive Review, Town Planning Review, 49, 259-273.

Bather, N., Williams, C. and Sutton, A., 1976, Strategic Choice in Practice: The West Berkshire Structure Plan Experience, Geographical Paper 50, University of Reading, Reading, UK.

Batty, M., 1970, An Activity Allocation Model for the Nottingham-shire-Derbyshire Sub-Region, Regional Studies, 4, 307-332.

Batty, M., 1976, Models, Methods and Rationality in Urban and Regional Planning: Developments Since 1960, Area, 8, 93-97.

Boyce, D., Day, N. and McDonald, C., 1970, Metropolitan Plan Making, Regional Science Research Institute, Philadelphia, Penn.

Breheny, M.J. and Roberts, A.J., 1978, An Integrated Forecasting System for Structure Planning, Town Planning Review 49, 306-318.

Coventry-Solihull-Warwickshire Sub-Regional Study Team, 1971, The Report on the Sub-Regional Planning Study, Coventry City Council, Coventry, UK.

Department of the Environment, 1973, Using Predictive Models in Structure Planning, HMSO, London.

Duc, T., 1977, Aspects of Monitoring and Review in East Sussex, Paper presented at the eighth session of the Workshop on Regional Science Methods in Structure Planning, Wakefield, West Yorkshire, UK.

Francis, K., 1980, The Monitoring of Structure Plans in the 1980's, Paper presented at the Second International Workshop on Strategic Planning, Delft University of Technology, Delft, The Netherlands.

Friend, J., Wedgwood-Oppenheim, F. et al, 1970, The LOGIMP Experiment: A Collaborative Exercise in the Application of a New Approach to Local Planning Problems, Centre for Environmental Studies, London.

Greater Manchester Council, 1977, Structure Plan Alternative Strategies: Main Report, Manchester, UK.

Hertfordshire County Council, 1976, Hertfordshire County Structure Plan: Report of Studies, Hertford, UK.

Hill, M., 1968, A Goals-Achievement Matrix for Evaluating Alternative Plans, Journal of the American Institute of Planners, 34, 19-19.

Jefferson, R., 1973, Planning and the Innovation Process, Pergamon Press, Oxford, UK.

Langley, P., Masters, R., McCreery, P. and Robertson, A., 1975, The Use of Analytical and Predictive Techniques in Structure Planning, unpublished paper, School of Environmental Studies, University College, London.

Leicester City Council and Leicestershire County Council, 1969, Leicester and Leicestershire Sub-Regional Study, Leicester, UK.

Luckman, J., 1967, An Approach to the Management of Design, Operational Research Quarterly, 19, 345-358.

McDougall, G., 1973, The Systems Approach to Planning: A Critique, Socio-Economic Planning Sciences, 7, 79-90.

McLoughlin, J.B., 1965a, The Planning Profession: New Directions, Journal of the Town Planning Institute, 51, 258-261.

McLoughlin, J.B., 1965b, Notes on the Nature of Physical Change: Towards a View of Physical Planning, Journal of the Town Planning Institute, 51, 397-400.

McLoughlin, J.B., 1966, The PAG Report: Background and Prospect, Journal of the Town Planning Institute, 52, 257-261.

McLoughlin, J.B., 1967, A Systems Approach to Planning, in Report of the Town and Country Planning Summer School, Belfast, UK.

McLoughlin, J.B., 1969, Urban and Regional Planning: A Systems Approach, Faber and Faber, London.

Openshaw, S. and Gillard, A.A., 1978, On the Stability of the Spatial Classification of Census Enumeration District Data, in P.W.J. Batey, (Ed.), Theory and Method in Urban and Regional Analysis, Pion, London.

Perry, H.A. and Chamberlain, K.J., 1977, Hertfordshire: Monitoring and the On-going Review Process, Proceedings of the Policy Analysis for Urban and Regional Planning Seminar, PTRC, London.

Planning Advisory Group, 1965, The Future of Development Plans, HMSO, London.

Roberts, N.A., 1976, The Reform of Planning Law, MacMillan, London.

Smith, D.L., 1974, The Progress and Style of Structure Planning in England, Local Government Studies, 21, 12-17.

Smith, D.M., 1977, Human Geography: A Welfare Approach, Edward Arnold, London.

Solesbury, W., 1974, Policy in Urban Planning, Pergamon Press, Oxford, UK.

Sutton, A., Hickling, A. and Friend, J., 1977, The Analysis of Policy Options in Structure Plan Preparation: The Strategic Choice Approach, Institute for Operational Research, Coventry, UK.

Thorburn, A., 1975, Regional and Structure Planning in a Time of Uncertainty, in Report of the Town and Country Planning Summer School, Aberystwyth, UK.

Wade, B.F., 1971, Some Factors Affecting the Use of New Techniques in Planning Agencies, Environment and Planning, 3, 109-113.

Wilson, A.G., 1974, Urban and Regional Models in Geography and Planning, John Wiley, Chichester, UK.

STRATEGIC LAND USE PLANNING: AN EVALUATION OF PROCEDURAL METHODOLOGY

Ian Masser

Department of Town and Regional Planning
University of Sheffield
Sheffield, S10 2TN
United Kingdom

INTRODUCTION

This paper reviews experience with rational system based methods in the field of strategic planning during the last twenty years. It draws heavily upon British experience because circumstances in Britain have highlighted some of the issues involved in a particularly dramatic way. This is because of the marked shift which has taken place in Britain since the late sixties from planning styles which emphasized the technical analysis of substantive issues to those which place greater stress on procedural questions in an increasingly political environment. These changes have had important implications for procedural methodology in that the traditional blue-print/master plan approach to planning has been largely replaced by procedures designed to increase the effectiveness of planning and decision-making processes in a changing environment. These reflect the many-sidedness of planning activities and the uncertain circumstances in which decision-making takes place.

THE DEVELOPMENT OF PROCEDURAL METHODOLOGY

Introduction

The most dramatic changes that have taken place in planning methodology during the last twenty years have occurred in the field of procedural methodology. The implications of these changes have had profound implications also on the development of substantive methodology because of the extent to which it has defined the context in which substantive methods can be used in the planning process. Twenty years ago procedural methodology had developed very little beyond the survey-analysis-plan (implementation) syndrome first outlined by Geddes more than fifty years earlier and elaborated in the writings of followers like Abercrombie (1959) and Mumford (1938). Since then it has been transformed by a series of waves of innovation which have built upon the experience that has been gained as a result of a particularly intensive period of plan-making in countries like the US and Great Britain. This process began with the introduction of a new procedural approach in the transportation studies and the land use transportation studies that were carried out in the United States in the late fifties and early sixties. The rational plan-making methodology that was introduced in these studies was consolidated in the second half of the sixties by British planners who based their thinking on an explicit systems approach. From the early seventies the focus of attention shifted from plan-making to questions of monitoring and control which could be represented in a process planning framework. As the seventies progressed the growing diversity of planning activities made it necessary to recognise the existence of many different styles of planning and planning situation. This led to a growing concern with questions of constructing comparative analytical frameworks to enable the evaluation and transfer of experience.

These developments not only reflect the rapid expansion of planning activities during the last twenty years but also the growing interest of academics and researchers from a wide variety of disciplines in planning as a field of study. This interest is partly due to the enormous growth in the social sciences that took place during this period and the emergence of social science research as an activity in its own right supported by public funding agencies such as the British Social Science Research Council (established 1965) and the Centre for Environmental Studies (established 1967). During this period, strategic land use planners have also responded very rapidly to developments that have taken place in other fields. In the sixties procedural methodology was very strongly influenced by the principles of systems analysis and cybernetics which had been introduced in engineering technology and corporate management. At the same time, the increasing number of critical studies of planning and decision-making processes that have been carried out by sociologists and political scientists has stimulated the development of urban and regional planning theory.

The Development and Evaluation of Alternatives

The first major departure from traditional planning thinking came in transportation studies, and more particularly, the land use transportation studies that were carried out in the United States in the late fifties and early sixties. These studies initiated a series of revolutionary developments in substantive methods because of their use of integrated forecasting frameworks and the application of computer-based mathematical models to simulate future states. These innovations had important implications for procedural methodology. The development of forecasting frameworks in these studies was prompted by the desire to examine a range of alternative plans for the future with a view to selecting the best possible course of action for implementation purposes. This standpoint represented a radical departure from earlier thinking which had been based on the assumption that "the Survey naturally leads to the Plan; the study of the present inevitably foreshadowing the future" (Abercrombie, 1959:137).

The spirit of optimism that underlay this thinking greatly increased the potential importance of planning as an activity with its own explicit procedural methodology. The land use transportation plan-making process that emerged from these studies consisted of initial identification of the range of possible future alternatives based on certain key organizing principles related largely to differences in plan form and transportation systems. These alternatives were then developed in some detail so that future states could be simulated by means of either urban development models or simple manual techniques. The results of these simulations were then evaluated both in terms of a more detailed analysis of each alternative and also by comparisons between alternatives. As a result of this evaluation, the final step of the process was for a decision to be made by the professional staff and/or decision-makers regarding which plan or combination of plans should be adopted for policy purposes.

In practice a number of important problems emerged in the course of the land use transportation studies which exposed their limitations with respect to both procedural and substantive methodology. A review of thirteen American land use transportation studies that was carried out by Boyce, Day and McDonald (1970) drew attention to the lack of evidence from these studies to support the fundamental assumption that was implicitly made in them about the nature of the relationship between land use systems and transportation systems. The review found that when transportation alternatives were considered no significant differences were found among land use patterns and when land use alternatives were considered no large differences emerged between transportation systems. Where both varied "the resulting land use patterns and transportation requirements were not sufficiently different to provide a technical basis for policy decisions" (Boyce, Day and McDonald, 1970:84). Although it was argued that these findings were partly due to the difficulties experienced in many studies with respect to the preparation of alternatives for testing purposes and the problems that emerged in making the mathematical models operational, they also drew attention to the problem of evaluation. As long as sizable differences between alternatives were expected, the task of evaluation was seen largely as one of totting up these differences, but the marginal differences between alternatives that were recorded in many of these studies drew attention to the limitations of both efficiency-based techniques like benefit-cost analysis and effectiveness-based techniques like goal-achievement evaluation at the

strategic level of thinking. In the process of these studies, problems of measurement, methods to deal with contradictory and often
conflicting criteria and allocating weights to priorities were grap-
pled with for the first time in strategic land use planning.

Because of these difficulties, Boyce, Day and McDonald concluded that the procedural methods that had been developed for the
land use transportation studies should be used with greater caution
in the future. They suggested that a cyclical approach to plan-
making should be devised which placed greater emphasis on the need
to learn in the course of the policy-formulation. This should take
the form of an iterative process whereby goals and objectives are
progressively clarified during the course of plan-making and would
also help to reduce the gap between plan-making and plan implementation in studies of this kind which often led to a failure to
implement the chosen plan because of differences between plan-making
and plan implementation agencies.

Plan Generation Methodology

From the mid-sixties onwards, the procedural methods that had
been developed during the land use transportation studies phase were
extended and consolidated in a formal plan generation methodology
which overcame some of the main limitations noted above. This was
particularly marked in Great Britain where a series of sub-regional
strategic land use planning studies and the emergence of a new
structure planning system gave an added stimulus to the adoption of
rational methods in plan-making. These developments were both
influenced by an interest amongst British planners in systems analysis in its own right as a starting point for the development of a
theory of planning which is evident in the writings of McLoughlin
(1969) and Chadwick (1971).

In "Urban and Regional Planning: A Systems Approach",
McLoughlin argues that the systems approach is not only a powerful
tool in explaining and understanding human relationships with the
environment but also in the planning and control of these relationships. The basic theme of his book is that "deliberate control of
the man-environment relationship must be firmly based on the systems
view" (McLoughlin, 1969:94). With this in mind he proposes a plan
generation cycle which begins with the formulation of planning goals
in broad terms and then proceeds to the identification of the more
precise objectives which must be achieved in order to move towards
these goals. Given these performance measures, it is possible to
examine alternative courses of action and to evaluate them in terms
of the extent to which they satisfy the specified objectives. On
the basis of this evaluation, a preferred course of action can be
chosen for implementation. It is also recognised that the process
of implementation involves a large number of decisions that result
in continuous changes over time and that planning and control mechanisms must be devised which are able to operate continuously

through a sequence of minor modifications interspersed with major reviews of the selected policy.

In "A Systems View of Planning", Chadwick (1971) argues that this approach provides a basis for a formal theory of the urban and regional planning process and translates the plan-generation cycle that has been described above into an integrated spatial method for regional planning. This method had a profound influence on subsequent applications because of the way in which it dealt with the questions of the representation of objectives. Chadwick showed how objectives could be converted to factors which could be represented in a measurable form in the alternatives that were generated by the adoption of standard criteria related to spatial performance. This is because "planning is concerned with space, whether physical or abstract (and) the solution space can be represented literally in spatial terms" (Chadwick, 1971:286).

By the early seventies several attempts had been made to put Chadwick's method into practice in strategic land use studies. One of the most successful of these, and a model for many later studies, was the Coventry-Solihull-Warwickshire Sub-Regional Planning Study (1971). The presentation of the plan-generation sequence in this study clearly indicates the influence of McLoughlin and Chadwick. In the first place four broad goals are defined in general terms and subsequently elaborated in two types of objective. These take the form of eight essential objectives concerned with overall levels of growth which it was assumed would not vary between alternatives and twenty discriminatory objectives concerned principally with questions of spatial choice which it was assumed would vary between alternatives. Then, as many as possible of the twenty discriminatory objectives were converted to factors that could be measured in spatial terms. The development potential scores were calculated for each of the ten factors that are used in the exercise and then combined together to form an aggregate total using various factor weights reflecting different rankings of the priority that was given to these factors by different groups. From an evaluation of these scores, four alternatives emerged for more detailed examination in relation to all the initial objectives. This was carried out by means of a goals (or more specifically objectives) achievement method which also involved a combination of performance measures and weighting of priorities in order to identify the preferred strategy.

The essential differences between studies of this kind and the land use transportation studies are summarised by the authors of the Coventry-Solihull and Warwickshire study in the following way:

> "We felt that sub-regional forecasting and model-
> ling techniques were in the main as advanced as
> was justified by the quality of available data
> and by the overall logic of sub-regional plan-
> making, and that we should therefore concentrate
> on system and logic to ensure that the sum of the

parts of the study were as sound as the parts
might be separately."

Because of this the entire process of objective specification,
alternative developments and plan evaluation could be followed at
each stage and the process as a whole possessed the property of
"transparency" that Lee (1973) regarded as absent from most land use
transportation studies. In addition, the simple combination of
scoring measures and weighting techniques that was used in the study
enabled all kinds of information to be incorporated in the plan
generation process. This provided a framework for the integration
of the findings of qualitative as well as quantitative studies and
also enabled intuitive views to be taken account of as well as the
findings of systematic research. In this respect, the use of
weighting techniques was particularly valuable in that it allowed
planners to examine the robustness of their alternative strategies
in the face of different rankings of priorities and the chosen plan
was the one which stood up best to these differences.

Despite these advantages, integrated plan generation methods of
the type used in the Coventry-Solihull-Warwickshire study had
serious limitations which restricted their potential usefulness.
Like many other studies that were carried out at the same time, the
Coventry-Solihull-Warwickshire study was concerned principally with
the allocation of an agreed level of population and employment
growth. Possible variations in the scale as well as the spatial
distribution of future population and employment growth in the
region were not directly considered in the plan generation process.
By these means the number of possible alternatives was considerably
reduced but the conclusions of the study can be regarded as valid
only within these limits. It should also be noted that an important
effect of broadening the range of alternatives would be to further
reduce the proportion of objectives that could be converted into
factors suitable for the development potential method. Even in this
restricted application, only half the discriminatory objectives were
represented in the factors used to generate the alternatives and the
other half were mainly involved only at the evaluation stage in the
exercise.

Some of these problems are dealt with by a more general plan
generation methodology which is based on the concept of land use
planning as a process of strategic choice

"in which it is sought to reduce the difficulties
encountered in dealing with current decision
problems by exploring them in a wider strategic
context embracing other related problems of present and future choice."

Friend, Power and Yewlett (1974:23)

This view was one of the main findings from a series of empirical
studies of planning and decision-making processes in British

cities that have been carried out by the Institute for Operations Research since 1965. In this perspective, public planning must be regarded as a field where both managerial and political responsibilities are inherently diffuse in nature and poorly-structured in character because of the uncertainties that surround all decisions.

The main objective of the plan generation methodology that has developed within this framework is to enable the management of these uncertainties. The strategic choice approach differs from that used in studies like Coventry-Solihull-Warwickshire in that it is explicitly decision-oriented and takes account of non-spatial as well as spatial factors. Its potential range of application is also much broader than that used in those studies. It can be used for aspect studies as well as comprehensive exercises because it can be operated simultaneously at several different levels of generalization within the planning process.

The essential features of the strategic choice approach can be summarised as follows. The process of plan/policy generation begins with the definition of decision areas and the specification of the connections between them. These links are represented in the form of a strategy graph (Hickling, 1974). Then a subset or subsets of strongly-connected decision areas is chosen for more detailed investigation as a result of an examination of the structure of the graph in terms of the strength of the linkages between decision areas and criteria related to the urgency or the degree of priority that is given to the need for a decision. In the next stage a number of mutually exclusive decision options are defined for each decision area in the chosen subset and the analysis of interrelated decision areas (AIDA) technique is used to eliminate contradictory or incompatible relations between options in different decision areas. The remaining compatible combinations of decision area options provide the set of feasible alternatives which is then subjected to conventional evaluation methods in order to identify the alternative that best satisfies the objectives of the exercise.

The strategic choice approach is essentially a more general form of the rational plan-generation methodology that has emerged in the course of the development of a systems approach to strategic land use planning. It is flexible in form and applicable to a very wide range of planning situations. It provides a logical framework which makes the best possible use of a wide variety of information and enables individual and organisational standpoints to be taken account of in the plan-making process. This makes it particularly appropriate where plan-making takes place by means of what Lindblom (1965) terms "partisan mutual adjustment" whereby those involved in plan-making must adjust their positions progressively so that they can bargain with each other in order to find an acceptable solution.

The strategic choice approach emphasises the organisational context in which most plan-generation related to key issues takes place. As Wiseman (1978) has pointed out, the process of plan-making and policy formulation within an organization begins with the

establishment of an ad hoc group or working party of those departments that are most closely involved. In public sector agencies the
working party may also include elected representatives and can also
be extended to incorporate representatives of external interests
such as pressure groups or building contractors according to the
type of problem that is under consideration (see, for example,
Dekker and Mastop, 1979). The essential task of the working party
is to identify the decision areas that must be considered and to
explore possible solutions. In situations of this kind the strategic choice approach can be used to ensure that the process of
communal problem formulation and the exploration of alternatives
takes place in as orderly and systematic manner as possible, so that
the working party can make recommendations to its superiors with a
minimum of delay.

Even with these improvements in plan-generation methodology,
major problems are still likely to arise in practice with the evaluation of alternatives, despite the development of more powerful
methods of multi-criteria decision analysis (Nijkamp and van Delft,
1977). It should be noted that the main objective of exercises of
this kind is to find a solution which is acceptable to those concerned with the task of plan-generation. There are important links
between plan-generation activities and those involved in them and
there is a basic connection between problem formulation and solution. These dilemmas have prompted Rittel and Webber (1973) to
suggest that the search for scientific bases for confronting problems of public policy is bound to fail because

> "the problems that planners must deal with are
> wicked and incorrigible ones...(which) defy ef-
> forts to delineate their boundaries and to iden-
> tify their causes, and thus to expose their prob-
> lematic nature"

consequently

> "the formulation of a wicked problem is the prob-
> lem (and) the process of formulating the problem
> and conceiving a solution (or re-solution) are
> identical since every specification of the prob-
> lem is a specification of the direction in which
> a treatment is considered"

It is also important to note that plan-generation methods of
the kind discussed above can be seen as special cases of a more
general class of heuristic problem solving techniques in that they
involve essentially open-ended search procedures for dealing with
poorly-structured problems which offer no optimal solutions and
provide only approximations and judgemental trade-offs (Hudson,
1979:396-397). The situations in which they are used fall into the
category of non-programmed decision-making defined by Simon (1965)
which can only be resolved by special efforts involving a high element of risk because of the need to invest in research and develop-

ment. This means that, while they are suitable for tackling key
issues and trouble-shooting, they are of limited use in connection
with most of the day-to-day activities carried out by planning agencies. Consequently an entirely different class of methods must be
devised for these purposes. This can be seen as a special case of
the more general class of algorithmic problem-solving procedures
which involve set routines for dealing with known classes of problems and are capable of providing optimal solutions to these problems within specified objectives and constraints. These correspond
to the way in which standard operating procedures and precedents
drawn from that experience are used in programmed decision-making
processes.

The Emergence of Process Planning

     The development of monitoring and review activities in British
planning since the early seventies points to some of the difficulties that arise in developing algorithmic procedures to improve
programmed decision-making. Monitoring and review activities were
also given a powerful stimulus by the shift in emphasis that took
place from traditional blueprint plan-making styles towards an
explicit process style of planning which incorporated many of the
concepts of strategic choice. According to Faludi (1973:132) the
process planning approach is one

          "whereby programmes are adapted during their im-
          plementation as and when incoming information
          requires such changes...Process planning becomes
          an approach in which strategic information and
          feedback impinge directly on action, providing
          signals that lead to incremental adjustments to
          its direction and intensity."

     An early example of this approach which provided a model for
many later studies in Britain is the work of the Notts-Derbys Monitoring and Advisory Unit. This was set up in 1970 following the
completion of a major sub-regional study to undertake regular
reviews of the objectives and to monitor the implementation of the
sub-regional strategy. The basic activities of the unit have been
outlined by Harris and Scott (1974) who indicate that monitoring and
review is a cyclical and iterative process that begins with the
collection of the relevant information about strategic decision-
making and environmental change. Then, this is scrutinized and
compared with the appropriate part of the strategy to identify any
discrepancies that may have occurred between the expected levels
that are assumed in the strategy and those evident from the analysis
of the new information. The next stage of the cycle involves a
closer examination of these deviations to assess whether or not they
represent a significant departure from the strategy. A divergence
is assumed to be significant:

"if it requires some form of corrective action...
this depends not only on its size but also on its
causes and likely effects. For example, the
cause may be an unforeseen event, an inconsistency or an omission in the strategy or a failure
on the part of a local authority to implement
sub-regional policy."

Harris and Scott (1974:730)

Once the need for corrective action is accepted, the review
element of the system comes into operation to consider possible
alternative courses of action. This involves the same basic elements as the process of plan-generation described in the previous
section with its (re)formulation of objectives, development of
alternatives and their evaluation in relation to the (re)formulated
objectives.

This description emphasizes the essential complementary nature
of monitoring and review in process planning and highlights the
vital link between monitoring activities and information collection
and data management. It is essential to recognize that the process
of information collection does not only involve the creation of
files containing population, employment and other statistical information but also the storage and retrieval of qualitative information
from sources such as local authority reports, committee minutes,
government publications and even the personal views of professionals
and politicians. The experience of Harris and Scott (1974:731) suggests that

"in practice this type of information has been of
as much importance as the statistical files in
monitoring the strategy".

The vital significance of qualitative information is further confirmed by the findings of a state of the art review of subsequent
practice that has been carried out by the Institute of Operations
Research (Floyd et al, 1977). This emphasizes the central role of
intelligence in the form of verbal communications and informal
exchanges of memoranda in identifying important issues and appre-
ciating their significance. The recommendations of this study
stress the importance of facilitating information flows in this
respect and providing a more structured framework which can take
account of the intelligence derived from the personal judgement of
the individuals who have to confront changing realities in the
course of their day-to-day work. These recommendations are also
supported by the findings of a detailed investigation of information
usage in two strategic land use planning authorities (Cater, 1979).
These draw attention to the need to improve the organization of the
flow of information so that informal and qualitative information can
be of use to people other than the person directly responsible for
its collection.

In practice, statistical material is used principally to provide a series of benchmarks for comparative purposes. Generally, standard statistical sources in Great Britain are of only limited value for strategic monitoring in this respect because of their high level of aggregation or their lack of spatial detail (for example, employment statistics), and the usefulness of the comprehensive information that is collected by the population census is offset to a large extent by the long intervals that occur between censuses. As special surveys to provide data for monitoring purposes are largely ruled out on cost grounds, planning authorities have found it increasingly necessary to utilize data that is generated as a consequence of the day-to-day exercise of control over development and management of local authority activities. Information regarding planning applications occupies a particularly important position in this respect and has been used in a number of cases as the nucleus of a management information system. The cancellation of the proposed 1976 population census also stimulated efforts to make the fullest possible use of the information that is available from existing sources such as electoral registers (BURISA, 1977) for monitoring purposes. The advantages of population register systems for monitoring purposes has long been recognized in countries such as the Netherlands where a comprehensive register system has been in operation since the second world war but the usefulness of this source has been limited until recently by the restrictions on data processing. These have now been largely overcome and a special information analysis system is currently being developed to match data sources directly to user requirements (Scheurwater and Masser, 1979).

In practice, as was the case in the land use transportation studies, the task of comparing expected with observed levels during the monitoring cycle has proved much more difficult than was originally expected. Attempts have been made to represent expectations in a variety of ways by means of targets, balances and performance indicators but considerable difficulties have been experienced in specifying these precisely enough to pick out deviations with any degree of certainty by the application of algorithmic-type procedures. Even where deviations have been identified, the task of assessing whether they are significant or not can only occasionally be dealt with by mechanical means because most of the measures involved give only a partial indication of change and the unintended consequences of change have often proved the most important element. Consequently, in many cases significant deviations were not detected until they were pointed out by outsiders and the assessment of their significance proved to be very much a matter of personal or collective judgement (Wedgwood-Oppenheim et al, 1975:11).

For these reasons it has proved necessary to broaden the scope of strategic monitoring activities to check on the continuing validity of the assumptions underlying the plans themselves and to look for emerging issues and changes in emphasis and priority in respect of existing policies. With this in mind, a mixed scanning model of

strategic monitoring activities has been proposed by Wedgwood-
Oppenheim and his associates (1975:10). This approach

> "involves taking a very wide, low-resolution scan
> of all matters that could potentially affect the
> region, and "zooming in" on those matters which
> in the judgement of those involved are of region-
> al importance. For this intensive analysis,
> information will need to be collected with speci-
> fic reference to the problem and detailed quanti-
> fication will be important. In the broad scale
> we want to be able to increase our awareness of
> as wide a range of information as possible.
> Depth will give way to breadth, and in seeking to
> absorb large quantities we will look for it in
> condensed forms, such as, articles, written
> reports, etc., rather than as raw data."

British experience suggests that strategic monitoring is primarily concerned with assessing the appropriateness of policies
rather than measuring performance by predesigned indicators. For
this reason, Floyd (1978) questions whether current models based on
management control methodology are of much value in this type of
situation because of the relative absence of tangible entities such
as budgetary targets or investment programmes and argues that a more
selective approach is needed which concentrates on the identification of issues and important areas of future choice. In his view,

> "the job of monitoring is to be "a few steps ahead
> of the game" and the planners themselves will
> often be responsible for getting new issues "on
> the agenda" of the authority or of other agen-
> cies...planners must always be on the alert for
> new circumstances in the economic and social en-
> vironment of their immediate areas of concern.
> These may call for policies and actions different
> in kind, rather than degree, from those currently
> in operation."

(Floyd, 1978:480)

Even more than the preceding discussion on plan-generation
methodology, experience with monitoring and review activities highlights the difficulties of separating the methods that have been
developed for these purposes from the context in which they are
used. The process planning style gives a special significance to
monitoring and review activities which is very different from the
emphasis placed upon them by other styles of planning. Although the
nature of monitoring activities depends to some extent on the statutory duties and perceived responsibilities of the planning agency
and the style of planning that has been adopted to deal with these
tasks, their detailed implementation reflects, as the studies of

Friend, Power and Yewlett (1974) show, the personalities of those involved, the nature of existing policy guidelines and the network of internal and external contacts that have evolved in a particular planning agency.

The Environment of Planning

Experience during the last twenty years with procedural methods in the rapidly-expanding field of planning activity demonstrates the extent to which approaches to strategic land use planning are influenced by the environment in which planning takes place. The procedures that have been described above form part of what Hudson (1979) calls the dominant planning tradition which is associated with the development of rational and comprehensive planning methods in the general public interest. However, the events of the last twenty years have highlighted the limitations of this synoptic tradition and made it increasingly necessary to recognize the existence of alternative approaches to planning which place a different emphasis on procedural methods. From the middle of the sixties onwards, the traditional view of a unitary public interest has been challenged by advocacy planners (Davidoff, 1965) and the development of radical planning since the early seventies has focused attention on conflict situations whereby urban protest movements demonstrate basic structural contradictions in society (Castells, 1977).

In practice it is likely that the approach that is developed in response to a particular planning situation will involve a combination of planning styles. For this reason, it is particularly important to try and identify and classify the dimensions of different planning situations. Masser's (1980) basic distinction between planning for growth and planning for decline provides a useful starting point for further discussion. The differences between these two situations are explored in the context of an example from education related to the planning environment that condition the debate and surround the decisions that are made about the opening of new primary schools and the closure of existing primary schools in urban areas. It is argued that the school opening debate will typically revolve around technical problems regarding the expected growth in population and its impact on enrollments in existing schools. A decision to build a new school will eventually be justified by the education authority on these grounds and accepted by parents and teachers alike because of the tangible benefits that are likely to arise for them in the form of shorter distances to school and smaller classes for the parents' children and better teaching conditions and possible new job prospects for teachers. Consequently, it is in the interests of all concerned to persuade the education authority to push the matter through as quickly as possible on the basis of the technical evidence.

In the case of school closure, personal and public priorities are of paramount importance. The essential argument to justify

school closure is that it will save public money. But the savings
to the education authority must be offset against the additional
costs as perceived by teachers who may lose their jobs and parents
whose children will have to change schools, and it is in the interests of both these parties to delay the decision as long as possible
because postponement means a continuation of the preferred situation. Under these circumstances, the formulation of the problem as
a whole is likely to be challenged in the hope that other solutions
will emerge, and questions may be asked as to why this particular
school was chosen and whether there are not stronger candidates for
closure. Situations of this kind call for protracted negotiations
in which technical considerations play a relatively restricted role.

These two extreme situations demonstrate the extent to which
planning in the dominant tradition is feasible. As Hudson
(1979:393) points out,

> "where planning for the future is _feasible_ (based
> on good data and analytical skills, continuity in
> the needs being extrapolated, and effective means
> to control outcomes), then planning is unneces-
> sary — it is simply redundant to what already
> goes on. Conversely, where planning is most
> _needed_ (where there is absence of data and skills
> and controls in the presence of primitive or tur-
> bulent social conditions), planning is least
> feasible."

Some of the intermediate positions between these two extremes have
been explored by Bryson and Delbecq (1979) in a study of the way in
which experienced planners react to changing planning situations.
They define eight types of planning situation on the basis of the
extent to which they involve easy or difficult decisions regarding
goals, technical considerations and political factors. Situations
involving easy goals are those in which a single decision must be
taken in circumstances where precedents already exist and no change
in organizational structure is necessary. Conversely difficult
goals involve multiple decisions in situations where no precedents
are in existence and changes will be required in organizational
structure. Similarly, easy political situations are those involving
a limited number of groups who are substantially in agreement, while
difficult political situations imply many groups who are in conflict
over both problem formulation and the determination of priorities.
Easy technical situations are those where standardized and proven
routines can be used and there are none of the conflicts regarding
either the techniques that are to be used or the interpretation that
is to be given to their findings that arise in the difficult technical situations.

The findings of the study indicate that planning behaviour
associated with rational systems based approaches is associated most
clearly with the easiest of all situations and that the planners

become more political as the situations become increasingly complex. They also demonstrate the extent to which planners change
their tactics as the situation changes in the hope of improving
their chances of goal-achievement. The findings that behaviour is
situationally appropriate and contingent on a wide range of relevant
factors raises important questions regarding the extent to which
experience can be transferred from one situation to another and
highlight the need for classification schemes which can be used for
comparative purposes.

There is a great deal of evidence available from case study
material which supports and extends these findings concerning situationally appropriate behaviour. The studies of two Dutch town
expansion schemes that have been carried out by Masser and his colleagues (1978) provide ample evidence as to the extent to which
planners modified and changed their tactics over time in response to
both internal and external factors. The findings of these studies
suggest that a number of thresholds can be identified at the end of
each phase of negotiation which represent an increase in the degree
of commitment of those involved to the project. They also indicate
that the contribution of the agencies involved varies substantially
from one phase to another and that there are also marked variations
between phases in the tactics that are used to find an acceptable
solution. Levin's (1976) investigation of major British town expansion designation processes contains some particularly good examples
of the way in which prior commitments restrict the options that are
open to the planners. It draws attention to the way in which people
become increasingly bound by their previous decisions and become
disinclined to go back on them without a plausible excuse. Because
of this it is suggested that those involved increasingly tend to
take up a justificatory rather than a critical attitude to the
extent that the decision-making process

> "seems to fall about half-way between the "cock-
> up" theory and the "conspiracy" theory"

                                            (Levin, 1976:9)

Although case studies of this kind are valuable for the insights they give into planning procedures and decision-making behaviour, their usefulness is limited by the lack of classification
schemes which would enable their findings to be generalized and
transferred to other planning situations. Some of the main elements
that might be included in schemes of this kind were outlined more
than ten years ago by Bolan (1969) and partially tested by Bolan and
Nuttall (1975) but have yet to be fully worked out for research
design purposes. Since the mid-seventies, interest in these questions has also been stimulated by the advent of systematic studies
of procedural planning in a comparative framework, such as the
Leiden-Oxford project (Faludi et al, 1980) and the introduction of
comparative planning study programmes in some schools. Nevertheless, the development of classification schemes and analytical

frameworks for comparative research on planning methodology remains one of the major tasks for future research in this field.

CONCLUSIONS

This evaluation of experience with procedural methods in strategic land use planning over the last twenty years highlights the changes that have taken place in this field during this period. A concept of multi-faceted contingent planning strategies employing special plan-generation and process planning techniques, which forms the basis of current procedural methodology, would have been almost inconceivable to planners in 1960. With these considerations in mind, then, what are the prospects for the next twenty years? Can planners expect an equally dramatic series of changes in the methodology relating to their field? Or might they expect a slowing down in the rate of change? The answer to these questions depends to a very large extent upon the changes that may occur in the way in which planning is perceived and the effect that this has on its role in society. Events since the election of the new Conservative government in Britain in mid-1979 point to the prospect of equally dramatic changes in the role of planning during the next few years. This government has questioned much of the conventional wisdom that has been built up over the last thirty or forty years in Britain regarding the position of the public sector as a whole and there are indications that a radical rethink of past policies is underway which will have important implications for strategic land use planning methodology. If this is the case, the immediate future for planning methodology may be as turbulent as the immediate past.

REFERENCES

Abercrombie, P., 1959, Town and Country Planning, Oxford University Press, Oxford, UK.

Bolan, R.S., 1969, Community Decision Behaviour: The Culture of Planning, Journal of American Institute of Planners, 35, 301-310.

Bolan, R.S. and Nuttall, R.L., 1975, Urban Planning and Politics, Lexington Books, Lexington, Mass.

Boyce, D.E., Day, N.D. and McDonald, C., 1970, Metropolitan Plan-Making, Regional Science Research Institute, Philadephia, Penn.

Bryson, J.M. and Delbecq, A.L., 1979, A Contingent Approach to Strategy and Tactics in Project Planning, Journal of the American Planning Association, 45, 167-179.

BURISA, 1977, Living without the 1976 Census, BURISA Newsletter, 30, 3-8 and 31, 2-7.

Castells, M., 1977, The Urban Question, Edward Arnold, London.

Cater, E., 1979, Patterns of Information Use in Planning - A Study of Oxfordshire and Berkshire, Working Paper No. 39, Oxford Polytechnic Department of Town Planning, Oxford, UK.

Chadwick, G., 1971, A Systems View of Planning, Pergamon Press, Oxford.

Coventry-Solihull-Warwickshire Sub-Regional Study Team, 1971, The Report on the Subregional Planning Study, Coventry City Council, Coventry, UK.

Davidoff, P., 1965, Advocacy and Pluralism in Planning, Journal of the American Planning Association, 31, 331-338.

Dekker, F. and Mastop, H., 1979, Strategic Choice: An Application in Dutch Planning Practice, Planning Outlook, in the press.

Faludi, A., 1973, Planning Theory, Pergamon Press, Oxford, UK.

Faludi, A. et al, 1980, Leiden-Oxford: A Comparative Study of Local Planning in the Netherlands and England, Oxford Polytechnic Department of Town Planning, Oxford, UK.

Floyd, M., 1978, Structure Plan Monitoring: Looking to the Future, Town Planning Review, 49, 476-485.

Floyd, M., Sutton, A., Friend, J. and King, L., 1977, Monitoring for Development Planning, Research Report 23, Department of Environment, London.

Friend, J.K., Power, J.M. and Yewlett, C.J.L., 1974, Public Planning: The Intercorporate Dimension, Tavistock, London.

Harris, R. and Scott, D., 1974, The Role of Monitoring and Review in Planning, Journal of the Royal Town Planning Institute, 60, 729-732.

Hickling, A., 1974, Managing Decisions: The Strategic Choice Approach, Mantech, Rugby.

Hudson, B., 1979, Comparison of Current Planning Theories: Counterparts and Contradictions, Journal of the American Planning Association, 45, 387-406.

Lee, D.B., 1973, Requiem for Large Scale Models, Journal of the American Institute of Planners, 39, 163-178.

Levin, P.H., 1976, Government and the Planning Process, George Allen and Unwin, London.

Lindblom, C.E., 1965, The Intelligence of Democracy: Decision-Making through Mutual Adjustment, Free Press, New York.

McLoughlin, J.B., 1969, Urban and Regional Planning: A Systems Approach, Faber and Faber, London.

Masser, I., 1980, The Limits to Planning, Town Planning Review, 51, 39-49.

Masser, I., Van Hal, W., Post, W., and Van Schijndel, R., 1978, The Dynamics of Development Processes: Two Case Studies, Town Planning Review, 49, 127-148.

Mumford, L., 1938, The Culture of Cities, Secker and Warburg, London.

Nijkamp, P. and Van Delft, A., 1977, Multi-Criteria Analysis and Regional Decision-Making, Martinus Nijhoff, Leiden, The Netherlands.

Rittel, H.W.J. and Webber, M., 1973, Dilemmas in a General Theory of Planning, Policy Sciences, 4, 155-169.

Scheurwater, J. and Masser, I., 1979, An Information Analysis System for Monitoring Spatial Planning in The Netherlands, in Policy Analysis for Urban and Regional Planning, PTRC, London.

Simon, H.A., 1965, The Shape of Automation for Man and Management, Harper and Row, New York.

Wedgwood-Oppenheim, F., Hart, D. and Cobley, B., 1975, An Exploratory Study of Strategic Monitoring, Pergamon Press, Oxford, UK.

Wiseman, C., 1978, Selection of Major Planning Issues, Policy Sciences, 9, 71-86.

SYSTEMS ANALYSIS IN PLANNING: A CRITIQUE OF CRITIQUES

Michael Breheny

Department of Geography
University of Reading
Reading, RG6 2AB
United Kingdom

INTRODUCTION

This paper is as much concerned with a critique of the available critiques of systems analysis as with systems analysis itself. It argues that much of the assessment of systems analysis during the past decade has been misleading in the impression that it has given to the planning profession. It argues that a preoccupation with American evidence, drawn from a relatively small sample of applications, with theory rather than with practice, with substantive rather than procedural issues, with major planning issues rather than common problems, with spatial modelling rather than with methods generally and a general failure to define systems analysis, have given a distorted interpretation of the role of systems analysis in planning.

The comments made here are broached largely from a practical standpoint, a perfectly reasonable vantage from which to observe events, given Batty's (1981) plea for a multiplicity of perspectives on systems analysis. Where appropriate, evidence is drawn from the experience of using systems analysis in British Structure Planning. The conclusions drawn are not necessarily that systems analysis should be resurrected, if considered dead, or is even a reasonable basis for planning in the eighties, but to argue for more substantial critiques, for less parochialism in the drawing of conclusions. There are many lessons to be learned from the attempts to use systems analysis in planning, even for hard-line critics of the approach, but such learning can only come from an undistorted assessment. If the answer is that we should rid ourselves of all vestiges of the systems era, let's at least do so on the basis of sound evaluation and criticism.

The views expressed here are predicated on the assumption that the peak of interest in systems analysis in planning was reached in the United States in the mid-sixties, in Britain in the early seventies, and that interest has waned considerably since. Hence the use of terms like "resurrection". It is acknowledged that in some places and some sectors the approach has had greater longevity, as for instance in the energy debate (Hoos, this volume), but in general the assumption is a reasonable one. Thus, the approach adopted here is historical, but like all recent history, a clear assessment of it can have profound implications for how we act in the future. Any decent critique of systems analysis in planning should thus be looking to inform future efforts.

The next section of this paper is concerned with the definition and scope of systems analysis, and particularly with the distinction between its role in substantive and procedural issues. This discussion is useful as a foundation for the following section which highlights the partial nature of most existing critiques of systems analysis and tries to identify aspects of the field which have been neglected relatively. Having mapped out the nature and coverage of available assessments of systems analysis, the argument then focuses, in the remaining sections, on some lessons that may have been learned had these critiques been more comprehensive. The conclusion drawn is that because of these "lost lessons" our education in plan-making leaves something to be desired, but that by concentrating on more substantial historical assessments of the systems era we may find that "it is never too late to learn".

SUBSTANTIVE OR PROCEDURAL SYSTEMS ANALYSIS?

"Systems analysis" is one of those terms frequently heard in planning circles, but rarely understood or given a common meaning. Obviously, it is essential to be clear about this before any evaluation of its history or merits can be made. Sadly, very few commentators in the field bother to define the meaning they ascribe to the term, assuming, presumably, that such meaning is common knowledge. This, however, makes reading of the critical literature very difficult.

The major definitional problem seems to be whether or not systems analysis solely refers to the substantive issues that planners are concerned with, that is the nature and functioning of cities and regions, or also to the procedures or processes adopted in deriving plans, notwithstanding the argument that the two should not be divorced (Camhis, 1979). The majority of the literature carrying the term "systems analysis" is concerned with the former, with substantive issues. Under this view the problem concerns techniques of systems analysis in aiding understanding or forecasting of elements of cities. Much of the land use modelling work in the United States in the fifties and sixties, and in Britain in the sixties and early seventies, was developed in this context. Most of the damning critiques of systems analysis (eg. Lee, 1973; Sayer, 1976) concern the development and use of such models.

However, the definition used here, or at least the issues discussed in relation to systems analysis, cover procedural aspects of planning as well. In this sense, then, systems analysis is taken to mean the use of systems concepts in urban and regional planning generally, in substantive and procedural contexts. It is essential to study systems analysis in this broader sense, even if the concern is mainly with substantive issues, for many of the answers concerning the merits and demerits, successes and failures of analytical and forecasting techniques can only be found by reference to the planning process within which such techniques were used. It will be argued later that many of the critiques of systems analysis are found wanting largely because they fail to make this reference to the planning process when assessing the validity of techniques.

Another reason for stressing the procedural uses of systems concepts, as well as the analytical ones, is that these had, arguably, much the greater impact on strategic planning practice in Britain during the 1970's. Thus, the history of planning methodology during this period cannot be understood without reference to procedural uses of systems analysis. Having said this, it is necessary, however, to make the point that the adoption, even of these procedural aspects of systems analysis, may have been more apparent than real. As will be argued later, we know very little about what happened in practice in British Structure Planning during this period. What we do know about the period, suggests, in any case, that such procedures were not really followed, that the process in practice was a very pragmatic one, and any systematization came from post-rationalization of the process. Assuming, however, for the moment that systems analysis in Britain had a greater effect on procedural rather than substantive issues, the point is that this differs considerably from the American experience in which the prime influence of systems analysis was on substantive, analytical issues.

The aim of this brief discussion of the definition and scope of systems analysis has been to make two points. Firstly, that if we are to gain useful assessments of systems analysis, we must consider its procedural and substantive uses together, as separate consideration can only lead to partial and misleading conclusions. Interestingly, this viewpoint has been reached from different theoretical standpoints recently; Batty (1978) makes the case for collapsing "science and design" in planning from what may be termed an apolitical standpoint, by arguing that they must interact and feed each other, in conditions of uncertainty, if progress is to be made; Camhis (1979) argues from a political standpoint, that procedural and substantive aspects of planning cannot be divorced, because only knowledge of material substantive issues can determine the means of confronting those issues. These viewpoints are, of course, valid, but the case for relating the two uses of systems analysis is argued here on practical grounds; only by considering both can we determine the contribution that systems analysis has made. The second point made in this section, and elaborated in the next, is that the British and American experiences in the uses of systems analysis have been different, emphasizing to varying degrees substantive and procedural issues. Both points argue for broader, more comprehensive critiques.

CRITIQUES OF SYSTEMS ANALYSIS

The amount of literature available giving assessments of systems analysis in planning is considerable. We must assume that the general purpose of this criticism is positive in that it attempts to point out the weaknesses and merits of the approach as a basis for better planning theory and practice in the future. One would be forgiven, however, on a reading of the literature, for assuming that this is not the purpose, or at least not the outcome, of all this effort. It is argued here that because of the particular approaches adopted in making these assessments, the evidence used and the general willingness to transfer conclusions made in one context to another context, many of the important issues have at best become no clearer, and at worst become more confusing. One overall conclusion from looking at the literature is that the assessments are remarkably unsystematic and lacking in rigour, assuming that rigour is still a desirable characteristic of analysis. It is difficult to believe that the commentators, many of whom were the original protagonists of systems analysis, have had such a change of paradigmatic heart as to eschew such an approach entirely.

These conclusions are drawn, it must be admitted, by looking at the history of the development and criticism of systems analysis from a practical standpoint. The points to be made may be clearer if a diagram adopted from Batty (1981) is used. This diagram relates together the ideas of science, which we can assume to be concerned with substantive issues, and design, which we can equate with procedural, plan-making activities, with the notions of theory and practice. This diagram, Figure 1, provides a framework for assessing the issues with which the critiques of systems analysis have been concerned, and for showing the differences between the emphases of various approaches.

The American assessments of systems analysis, which have given the lead in these matters, and which have been responsible largely for the impression held about systems analysis, tend to have been concerned with the uses of modelling and analytical techniques in assessing substantive issues. This has been done at both a theoretical and practical level (boxes 1 and 3 in Figure 1). However, the practical critique, which has been particularly influential, has by and large been concerned with a small sample of large-scale, expensive modelling efforts (Lee, 1973; Brewer, 1973) in the US. The experience shown in this sample cast a shadow over the whole modelling effort in the sixties and early seventies, and subsequently,

	SCIENCE	DESIGN
THEORY	1. Substantive issues, knowledge of cities, models, techniques	2. Theoretical processes, rational decision model, etc.
PRACTICE	3. Applied knowledge, Practical modelling	4. Applied processes, practical procedures

Figure 1 - Adaptation of Batty's (1981) Science-Design/Theory-
          Practice Concept

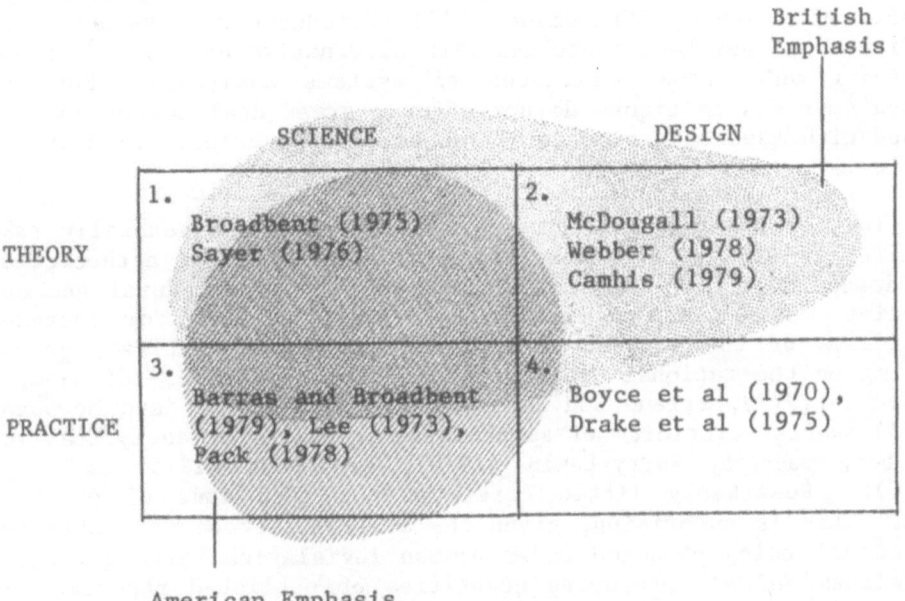

Figure 2 - Major Emphases of British and American Critiques of
           Systems Analysis

over systems analysis in general, as it revealed vast amounts of wasted resources, large cumbersome and inadequate models and planning projects undermined by heavy reliance upon modelling.

To what extent, though, is this sample representative, even in the United States? It seems that models were in widespread use in the United States in the sixties (Pack, 1978), but in how many cases were the circumstances the same as those surrounding the much-published efforts? How many planning projects were really dependent on a successful model? The British experience would suggest very few. The pragmatic nature of most planning processes, despite claims to be using explicit decision models, the prejudices of senior staff, plus the "organizational inertia" characteristic of most authorities would probably have ensured that models were not crucial elements of the plan-making process. Thus the American survey of practical applications of systems analysis may have painted a picture, a very influential picture (for instance, Wegener, (this volume) documents the profound impact these modelling critiques had on the systems movement in West Germany), not at all typical of average practice. The points that such a survey may have missed will be discussed later.

Turning back to Figure 1, the American assessments of systems analysis, with the single exception of Boyce, Day and McDonald (1970), have little of direct value to say about procedural uses of systems concepts, particularly in practice (box 4). There is some direct coverage of theoretical aspects (see for instance, Webber, 1978), but much of the American literature reviewing procedural aspects of planning (Friedmann, 1973; Friedmann and Hudson, 1974, for instance) has been concerned with alternative approaches without providing substantial critiques of systems analysis. For this reason American critiques do not offer a great deal for anyone concerned with assessing procedural issues, and practical procedures in particular.

The British critiques of systems analysis have generally taken a different line in that most of the work has been at a theoretical and academic level. The criticism covers both procedural and substantive issues at this theoretical level. Thus, for instance, criticisms of the procedural aspects of systems analysis, concentrating on the rational decision model have been made, for example, by Camhis(1979), Friend and Jessop(1969), Gutch(1970) and McDougall (1973), whilst criticism of substantive modelling or analytical work has been made by Parry-Lewis (1970), Broadbent (1975) and Sayer (1976). Remarkably little criticism has been made of practical work. This is surprising, given the availability of many Structure Plans, all being produced under common legislation, with government guidelines and all producing quantities of published reports. The major exception to this is the work of Barras and Broadbent (1979) who were concerned with analytical techniques (box 3) in the first twenty Structure Plans. As part of this work, Booth and Jaffe (1978) assessed the techniques used in certain phases of plan generation and evaluation.   There has never been a major empirical study of modelling practice of the kind carried out by Pack (1978). On the procedural side some work on early Structure Plans was carried out by Drake et al (1975), which was in part theoretical, but drew also on limited practical experience.

In summary, then, the major British and American critiques of systems analysis could be classified as follows:

1.   American work on substantive modelling and analytical techniques, at a theoretical and practical level, the latter being drawn from a small number of major planning projects.   Some theoretical work on procedural issues.

2.   British work, at a largely theoretical level, concerned with both procedural and substantive aspects of planning.   A small amount of work on the practical application of techniques and on procedures.

These contributions are superimposed on the structure from Figure 1 to give Figure 2 and the main areas of American and British influence shown.   Looking at Figure 2 the major gap would seem to be in relation to box 4, the practical aspects of procedural uses of systems analysis.   In other words, we have little evidence available upon which to draw conclusions about practical planning processes. This is a serious gap in our assessments of the value of systems analysis, both because it implies that we know little about such practical processes and because our understanding of the use of models and techniques, the substantive concerns of systems analysis, must relate to this wider process.

This gap in both the British and American literature, and the small, probably unrepresentative, sample from which the Americans have drawn conclusions about practical modelling, highlight one of the main points of this paper, namely that conclusions have been drawn from remarkably little evidence about the practice of systems analysis.   Given the immense significance of the retreat from "rational" decision-making in the seventies, it is remarkable that the critiques which undoubtedly aided this process should have been based on such little knowledge of systems analysis in practice.

A whole superstructure of critique of systems analysis is based on a tiny foundation of empirical evidence.   The point has not gone unnoticed, for Batty (this volume) accepts that the assessment of systems analysis has been too theoretical in nature.   It is interesting that the systems analysis debate is not the only one founded in this way.   A debate with a much more substantial and long-standing superstructure is that concerning the philosophy of science, in which the likes of Kuhn and Popper take part.   Yet they are castigated by Pearce-Williams (1970) for debating the philosophy of science without actually knowing what scientists do in practice:

"...but there seems to me to be a very important
gap in both theories. It is, simply, how do we
know what science is all about?... I am interested in what practitioners of mature sciences
think they are doing. To repeat, we simply do
not have this information."

To be in good company in our ignorance of practice is no reassurance.

Given that the experience of systems analysis, particularly in
Britain, is only weakly based on empirical evidence, a number of
points are made below in an effort to show some of the issues that
might have been considered in previous critiques, had they been more
concerned with practice. These observations are themselves predicated on a rather weak base, in that no comprehensive assessment of
systems analysis in practice is available. Yet they are presented
from an explicitly practical standpoint, unlike most other critiques, and they are based upon limited personal experience.

ORGANIZATIONAL INERTIA AND THE ADOPTION OF SYSTEMS ANALYSIS

Throughout the substantial critical literature, very few efforts have been made to study systems analysis in the organizational
context in which it was being used. This is, of course, partly a
function of the lack of empirical evidence on practice. There has
been a suspicion throughout the more theoretical work that organizations, consciously or unconsciously, have tended to frustrate the
efforts of those trying to implement systems analysis. Thus, Batty
(1976), for example, talks of "organizational inertia", militating
against innovation and dissipating the effects of any innovations
that are accepted. He returns to the same point in this book, but
with more force:

"...agencies seem incapable of absorbing the simplest and most obvious ideas...only capable of a
concern for survival."

The suspicions of the theoreticians are well-founded, for a
comprehensive assessment of planning practice in Britain would probably show at both the procedural and technical levels that intentions and reality are different things. Many local authority
departments, and planning departments are no exception, have traditionally operated on a pragmatic, month-by-month if not day-by-day
basis, taking care not to commit themselves to procedures or ideas
too far in advance of their usage. To the local government officer
this approach is eminently sensible, as new unforeseen issues are
continually arising and the uncertainties of the operating environment means that it is unwise to commit oneself to elaborate programmes and timetables. In a novel situation, such as producing the
first County Structure Plan submission, it also makes sense to

decide what to do in the next stage only when the current one has been completed. This method of working is common and is understandable. But it is not inevitable.

This "organizational inertia" has effects both on planning methodology or procedure and on techniques. Many Structure Planning authorities during the period 1969-1975 in Britain adopted at the beginning of their plan preparation work, carefully devised planning processes, often variants on the rational decision model. It is doubtful that many of these processes were ever followed for the organizational reasons discussed above. It is inevitable in such a context that sophisticated procedures for plan preparation, with each stage building carefully upon the previous one, and hence requiring complete adherence to the process, should flounder.

The problem with substantive techniques of systems analysis in this context is very similar. The techniques are devised to contribute to certain phases of the plan preparation procedure, be it, for example, problem identification or plan evaluation. However, if this procedure itself has been frustrated, then little wonder that techniques are not used at all or are used inappropriately. If nobody knows if or when plan alternatives are to be evaluated, it is unlikely that evaluation techniques will be adequately developed or ready for use, if and when required.

These relationships, between reality and planning methodology, and methodology and techniques, can be represented as in Figure 3 below.

This shows that techniques of systems analysis, largely concerned with substantive issues, support the plan methodology or procedural systems analysis, which should then be used in practice. It has been suggested above that organizational inertia can frustrate this arrangement both at the link between methodology and

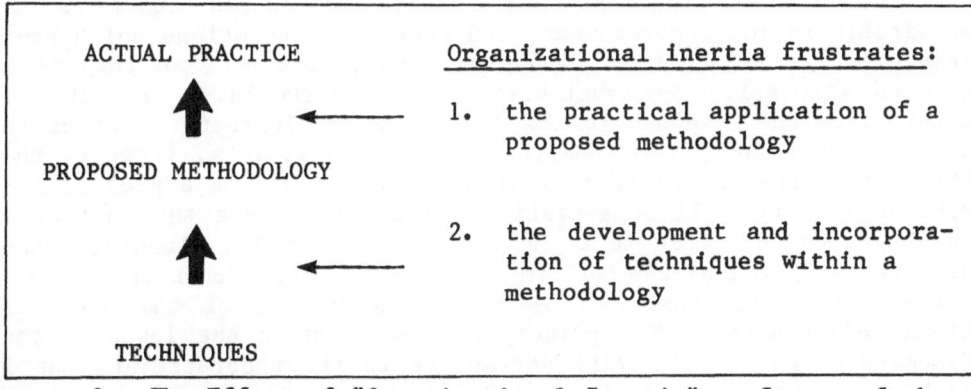

Figure 3 - The Effect of "Organizational Inertia" on Stages of the Plan-Making Process

practice, by preventing adherence to the methodology as incrementalism takes over, and at the link between technique and methodology
because the justification for using techniques, that is, the methodology, is dissipated.   The linkages in this arrangement have
usually been criticized in terms of the inadequacy of procedural
planning theory in relation to practice and the inadequacy of proposed techniques for supporting the planning procedure. A critique
in terms of organizational constraints is also important. Doubtless
some of the problems caused by these constraints have been attributed incorrectly to the systems analyses adopted.   It is interesting that in Pack's (this volume) survey of local planning agencies, such organizational constraints were not mentioned by practitioners as a reason for the non-use of models.   This is not surprising however, given the fact that the respondents are unlikely to
volunteer their own inability to implement programmes as a reason
for model failure.

Two interesting conclusions might be drawn from this practical
consideration of organizational constraints on the application of
systems analysis. Firstly, it might be said, if the above arguments
hold in practice, that the inadequacies in some British Structure
Plans could be attributed just as much to the lack of systems analysis in plan preparation as to a surfeit of it. For example, Breheny
and Roberts (1981) show that a major shortcoming of a number of
Structure Plans is the inadequacy of the methodological and technical work on forecasting. The second conclusion might be that if
systems analysis is as inadequate a basis for planning as some
critics would have it, then the "organizational inertia" evident in
British local authorities was a blessing in disguise. It prevented
the widespread application of a set of dangerous and inadequate
concepts. In this sense the natural scepticism and indifference of
the British bureaucrat was all to the good. Certainly it is difficult to see the average senior local government officer in Britain
ever receiving an innovation with the kind of "euphoria" with which
American planners greeted the advent of models in the fifties and
sixties.

Whilst the "organizational inertia" discussed above is not
inevitable in local government, and there are exceptions which prove
this, the fact that it is widespread and pernicious, in the way it
affects innovations or systematic working, should be a factor in
determining the response to the lesson we are learning. One of the
concepts used in systems analysis, requisite variety, suggests that
things that are to relate or work together should be expressed in a
similar variety. If organizational inertia reduces the variety in
local government, then it is no good having complex planning methodologies and techniques of high variety. They have to be of a
variety and scale that can be matched with that of the organizational environment. The principle of working to the lowest common
denominator may not be very attractive to the theoretician, but it
is highly practical.

It should be borne in mind when considering the value of systems analysis, and particularly when considering what might replace
it, that it was applied in public policy-making originally in order
that it might overcome some of this "organizational inertia", this
tendency towards "adhocism", and make issues clearer and decision-
making more accountable. Ironically, one of the main arguments
against systems analysis is that it has, in fact, obscured and
distorted real issues, rather than making them open and debatable
(Hoos, this volume; Camhis, 1979). Doubtless there is a lot of
truth in this, but the danger is that by discarding systems analysis, in the absence of any real alternative basis for analysis and
decision, that the "inertia" described above will dominate once
again. It has been argued here that systems analysis may have been
losing the battle against this "inertia" in many cases, but at least
battle was commenced. It may well be that rather than public policy
issues becoming more open and the decision-making process much
clearer, the ever-present fog will descend once again. Major issues
may avoid this fate, because of the attention given to them by the
politically and professionally committed, but mundane local practice
is highly vulnerable.

PASSIVE AND ACTIVE AUTHORITIES AND THE PROBLEM OF SCALE

The available critiques of systems analysis in planning,
largely American in origin, make very little reference to the degree
to which the apparent success or failure of systems analysis was
dependent upon the type of authority being investigated. The
American sample is a small one and concentrates on experience in
large cities or city regions. In Britain, the uniform Structure
Plan system means that the same planning issues can be assessed in
different types of authority. It will be argued here that the
experience of systems analysis does differ considerably between
authorities and that valuable lessons may be learned by considering
the reasons for such differences; a possibility ignored in existing
critiques.

In Britain both the problems considered and the political
responses to such problems tend to be very different between metropolitan and shire counties. The former tend to be Labour-controlled
and the latter Conservative-controlled authorities. It is difficult
to generalize, but the shire counties tend to be "passive" in that
the response to problems, which are less severe than in metropolitan
counties, is largely of an apolitical nature. Such authorities
traditionally are characterized by little political argument and a
deference by politicians to the "superior" knowledge and experience
of senior local government officers. This combination has tended to
make the operating environment in such authorities much easier for
the planner. A thorough investigation of Structure Planning in such
authorities in the early seventies might reveal some minor success
in the use of systems analysis. The effects of "organizational
inertia", which is just as prevalent in such authorities as any

others, would have taken their toll, but at least they were less likely to be compounded by a complex political environment. In such circumstances some kind of explicit planning process was more likely to be followed and the use of limited techniques more likely to have been considered valuable. In no instance, as emphasized earlier, is process or technique likely to have been fundamental to the outcome of structure planning, but in shire counties the circumstances were at least more conducive to the adoption of some kind of systems analysis.

In contrast to shire counties, metropolitan counties, dealing with inner city issues and industrial decline, tend to be highly-charged politically, with considerable debate over problems and policies. The working environment in such authorities is likely to have been very disjointed, with little opportunity for following an established process or of using sophisticated techniques. Thus the use of systems analysis in such situations was unlikely to have been very successful.

This tendency, for the working environment in metropolitan counties to militate against systematic procedures, was slightly offset by the generally greater expertise available, relative to shire counties, to develop and use such procedures. Indeed, in a number of such authorities some interesting and successful technical work was carried out, such as in Merseyside for example, where social area analysis (Webber, 1975) and input-output analysis (de Kanter and Morrison, 1978) were employed. Such technical work was, however, likely to be used entirely in problem clarification and within the context of a highly pragmatic process.

The point being made here, then, is that, in the British context at least, the experience of the use of systems analysis differed considerably between authorities. Any conclusions drawn about the practical validity of systems analysis should reflect an understanding of such differences. Existing critiques do not. This line of reasoning is not meant to imply that the highly political, pragmatic approach to planning in some authorities, particularly metropolitan ones, is inferior to a more systematic one. There are many who would suggest that it is distinctly superior. The point is simply that most existing critiques of systems analysis do not consider the type of authority as a factor in determining the adoption or value of systems analysis. Yet again we are suffering from the fact that most critiques of systems analysis have ignored everything other than major and controversial issues, and in particular have ignored a large proportion of practice.

Another related issue which has received little attention concerns the validity of substantive or technical systems analysis at different scales. Scale in this context refers mainly to the degree of importance attached to an issue or problem, but also to spatial extent. Many of the problems planners deal with, whilst being quite important in total, are individually of a relatively uncontroversial and/or localized nature. As Harris (1978) argues, this is a point often missed by those attempting to make profound critiques of the planning function. The assessment of such minor issues has often been enhanced by the use of analytical techniques, which have in turn been uncontroversial and acceptable. Such techniques play a significant role in much strategic planning work in Britain.

The major, usually American, critiques of systems analysis fail to acknowledge the effect of such techniques, developed during the peak of interest in systems analysis in the early seventies, on the planning process. Again, the focus upon experience in large-scale land use and transportation modelling, and upon major, often national issues (see Hoos, this volume, for example) is responsible for this oversight. In this way a significant legacy of the systems era in planning goes unnoticed. There is no reason to assume that the same oversight has not been made in the United States.

CONCLUSION: FLOGGING A DEAD HORSE?

Some researchers and practitioners in planning have met the criticisms of the systems era by retreating from the field of systems analysis altogether and, if possible, denying their roots. This stance is unfortunate if the arguments contained in this paper hold good, for much is still to be learned from reflection on the systems era in planning. This additional education might only strengthen the criticisms of these skeptics, but this would be no bad thing and would give added weight to their advocacy of alternatives. Others, such as the proponents of Strategic Choice (for example, Friend and Jessop, 1969, Bather, Williams and Sutton, 1976) have dealt with these criticisms in a positive way by the adoption of valuable legacies from the systems era coupled with new concepts designed to avoid the mistakes of the past.

Still others, such as Batty (this volume) have gone bravely against the general trend by going deeper into the origins of systems analysis, to tackle the subject on its original ground, in order to gain some new insights. Indeed, he argues that in order to defend the viability of an analytical paradigm, it is necessary to use the paradigm itself as the mode of analysis; to do other is to admit defeat. In this sense he could be said to be following Koestler's (1967) advice and reflecting on the value of "flogging a dead horse", and finding the experience of resurrection a valuable one, as Koestler would have predicted.

The argument presented here is that there is value in "flogging the dead horse" of systems analysis, notwithstanding the argument raised earlier that we do not really have the information on which to declare it dead. The reason is not necessarily Batty's one, that some of the original concepts may be more useful than we think if only used properly, although this argument would be subsumed, but that we know much less about theory and practice during the systems era than is usually acknowledged.

The shortcomings of many existing critiques of systems analysis have been outlined here and the kinds of issues that might be uncovered from further, more comprehensive reflection, have been discussed. What is needed is historical research into the substantive and procedural developments in planning during the systems era, drawing on wide-ranging experience of both theory and practice. The argument has been put forward here that the main benefit would derive, because of deficiencies in existing critiques, from assessments of practice, from the actual uses of systems analysis. In concluding his plea for more empirical analysis of what practicing scientists actually do, Pearce-Williams (1970) states that scientists should keep "in mind Lord Bolingbroke's remark that 'history is philosophy teaching by example'. We need a lot more examples." The same goes for systems analysis in planning.

REFERENCES

Barras, R. and Broadbent, T.A., 1979, The Analysis in English Structure Plans, Urban Studies, 16, 1-18.

Bather, N., Williams, C. and Sutton, A., 1976, Strategic Choice in Practice: The West Berkshire Structure Plan Experience, Geographical Paper 50, University of Reading, Reading, UK.

Batty, M., 1976, Models, Methods and Rationality in Urban and Regional Planning: Developments Since 1960, Area, 8, 93-97.

Batty, M., 1978, On Planning Processes, in B. Goodall, and A. Kirby, (Eds.) Resources and Planning, Pergamon Press, Oxford, UK.

Batty, M., 1981, A Perspective on Urban Systems Analysis, in D. Banister, and P. Hall, (Eds.), Transport Policy and Planning, Mansell, London.

Booth, D. and Jaffe, M., 1978, Generation and Evaluation in Structure Planning, Town Planning Review, 49, 445-458.

Boyce, D., Day, N. and McDonald, C., 1970, Metropolitan Plan-Making, Regional Science Research Institute, Philadelphia, Penn.

Breheny, M.J. and Roberts, A.J., 1981, Forecasting Methodologies in Strategic Planning: A Review, Papers of the Regional Science Association, 44, in the press.

Brewer, G.D., 1973, Politicians, Bureaucrats and the Consultant: A Critique of Urban Problem-Solving, Basic Books, New York.

Broadbent, A., 1975, The Wrong Case, The Planner, 61, 188-189.

Camhis, M., 1979, Planning Theory and Philosophy, Tavistock Publications, London.

Drake, M. et al, 1975, Aspects of Structure Planning in Britain, Research Paper 20, Centre for Environmental Studies, London.

Friedmann, J., 1973, Retracking America: A Theory of Transactive Planning, Doubleday, New York.

Friedmann, J. and Hudson, B., 1974, Knowledge and Action: A Guide to Planning Theory, Journal of the American Institute of Planners, 40, 2-16.

Friend, J. and Jessop, W.N., 1969, Local Government and Strategic Choice, Tavistock Publications, London.

Gutch, R., 1970, Planning, Philosophy and Logic, Journal of the Town Planning Institute, 56, 389-391.

Harris, B., 1978, A Note on Planning Theory, Environment and Planning A, 10, 221-224.

Kanter, J. de and Morrison, W.I., 1978, The Merseyside Input-Output Study and its Application in Structure Planning, in P.W.J. Batey, (Ed.), Theory and Method in Urban and Regional Analysis, Pion, London.

Koestler, A., 1967, The Ghost in the Machine, Hutchinson, London.

Lee, D.B., 1973, Requiem for Large Scale Models, Journal of the American Institute of Planners, 39, 163-178.

McDougall, G., 1973, The Systems Approach to Planning: A Critique, Socio-Economic Planning Sciences, 7, 79-90.

Pack, J.R., 1978, Urban Models: Diffusion and Policy Application, Regional Science Research Institute, Philadelphia, Penn.

Parry-Lewis, J., 1970, The Invasion of Planning, Journal of the Royal Town Planning Institute, 56, 100-103.

Pearce-Williams, L., 1970, Normal Science, Scientific Revolutions and the History of Science in I. Lakatos, and A. Musgrave, (Eds.), Criticism and the Growth of Knowledge, Cambridge University Press, Cambridge, UK.

Sayer, R., 1976, A Critique of Urban Modelling, Pergamon Press, Oxford, U.K.

Webber, M., 1978, A Difference Paradigm for Planning, in R.W. Burchell and G. Sternleib, (Eds.), Planning Theory in the 1980's: A Search for Future New Directions, The Center for Urban Policy Research, Rutgers University, New Brunswick, New Jersey.

Webber, R., 1975, Liverpool Social Area Study, 1971 Data: Final Report, PRAG TP 14, Centre for Environmental Studies, London.

PART 2

SYSTEMS ANALYSIS IN DIVERSE CONTEXTS

THE IMPACT OF SYSTEMS ANALYSIS ON URBAN PLANNING: THE WEST GERMAN

EXPERIENCE

Michael Wegener

Institute of Urban and Regional Planning
University of Dortmund
Postfach 500500, 4600 Dortmund 50
West Germany

INTRODUCTION[1]

What makes the experience of one particular country relevant
for others? If in that country everything is the same as elsewhere,
that is hardly noteworthy. If, however, everything is different,
other countries have no way of relating such exotic information to
their own experience. The author of a national report, therefore,
must focus on the narrow field of _variety_ in _similarity_, ie. on the
differences in otherwise similar processes that suggest new explanations or alternative conclusions.

In this case, the similarities are quickly listed. In West
Germany, like in other Western countries, systems analysis methods
in urban planning[2] had a euphoric pioneer period, a period of
criticism and decline, and a period of stagnation. In particular,
the development echoed that in the United States with considerable,
but reducing, time lag.

This information alone is not very interesting. However, there
are some peculiarities in the professional and intellectual debates
accompanying this process in West Germany, which may offer some
additional insight into the causal structure of similar processes in
other countries as well.

The peculiarities are related to the political development in
the Federal Republic and, more generally, to the intellectual and
philosophical traditions of Germany. In the controversy pro and
contra systems analysis methods in urban planning, much evidence can
be found of the old controversy between _rationalism_ and _antirational_
ideologies characteristic for German intellectual and cultural his-

tory. In this larger controversy, in Germany the antirational position always was powerful. Periods of enlightenment were always
brief, and the reaction following them was always thorough and long-
lasting.

Here, the controversy focuses on the role of rationality in
public policy-making and planning. Democratic planning has to
reconcile two conflicting objectives: to efficiently process complex
decision situations and still maintain and develop democratic norms
and procedures. If they are in conflict, which is more important?

In this paper, I will review four variations of this controversy which all are in some sense related to the application of
systems analysis methods in urban planning: the controversy between
comprehensiveness and incrementalism in the urban planning practice;
the controversy about urban planning as a science or as an art; the
controversy between social cybernetics and political economy in
planning theory (a German specialty); and the controversy on technocracy vs. advocacy in political science.

All these controversies had a common theme. They challenged
the traditional "engineering" kind of rationality which had discredited itself by its recklessness and insensibility towards human
values and natural resources in the name of economic growth and
technological "progress". In that sense, the controversies were
part of, and contributed to, the general process of reevaluation and
reformulation of societal goals going on during the last decade,
which also deeply influenced the style of public policy-making and
planning. Today in West Germany, it has become much more difficult
to carry out controversial large-scale technical projects affecting
the natural environment or existing neighbourhoods without taking
account of the reactions of a watchful and critical public.

And yet, in all the controversies, inextricably intermingled
with their progressive intention and effect, there was a common
regressive tendency to reject science and technology altogether as
tools for improving the human condition. I hope to demonstrate this
in the paper, in order to support my hypothesis that the present
aversion against systems analysis methods in urban planning is part
of a broader anti-rational tendency in society. If this hypothesis
is only partly correct, it is clear that improving the methods in
their own terms, though desirable, will probably have not much
effect on their diffusion into the planning practice. Instead, it
seems to be much more effective to concentrate on questions of
transfer, acceptance, relevance, conflict and human values.

Unfortunately, systems analysis methods have so far been undiscriminatingly associated with being technocratic, conservative and
anti-democratic. I will argue that there is nothing inherent in
these methods that would justify such classification. The point I
want to make is that the potential of such methods to support and

enlarge citizen involvement in urban planning has in the past been
ignored, and should be an area of prime concern to everyone working
in the field in the future.

THE FACTS

In this first chapter, I will give an account of the few successes and many failures of the application of systems analysis
methods in urban planning in West Germany. In order to avoid
unnecessary detail, I will keep this review as brief as possible.[3]

Arrival (1968-1971)

Systems analysis methods were introduced to West Germany in the
early sixties, mainly in economics and engineering. In the mid-
sixties, the first computer-based transportation studies demonstrated that it was possible to successfully apply these methods to
spatial planning problems. By 1967 or 1968, word had spread among
urban planners that the new methods might be the badly needed tools
to cope with the increasing complexity of urban planning problems.

Urban planners in Germany in those days were architects.
Unlike in the United States or in Great Britain, in Germany there
was no undergraduate planning education until independent planning
departments were established at the universities of Dortmund in
1968, and Berlin in 1970. At that time, the inadequacy of the planning education offered at architectural schools, with its design
orientation and its bias for physical planning, had become obvious
(Albers, 1963; 1966).

In Dortmund, a completely new beginning was made. An interdisciplinary department of Raumplanung (spatial planning) was founded,
which was to integrate all levels of spatial planning, from the
local to the national scale. The faculty was recruited from economics, law, sociology, various engineering disciplines, and mathematics. One of the major subjects of the new curriculum was systems
analysis, later called systems theory and systems engineering.
There was a broad consensus among students and the young faculty,
that fundamental changes in society were needed, and that planning
was a key instrument to bring them about. The new systems techniques were associated with being progressive, rational and objective, as opposed to the backwardness, irrationality and subjectivity
of the traditional professional practice. In brief, the systems
approach was understood as a piece of enlightenment.

This view was strongly endorsed by the Federal government. In
1969, the Staedtebaubericht (Urban Planning Report) expressly called
for the development of "models which allow insights into the dynamic
changes of spatial behaviour of people"[4], and concluded:

"This implies that mathematical techniques for
analysis and forecasting as well as the techniques of electronic data processing are of prime
importance for urban planning. The same applies
to the simulation of human behaviour relevant to
spatial processes. Also the mathematical optimization techniques which originated from operations research as well as the entire field of
systems research and decision theory should be
utilized for urban planning."

BMBau (1969)

Economists and engineers, as well as architect-planners, took
the message for granted and worked their way through A Model of
Metropolis and back issues of the Journal of the American Institute
of Planners, or specialized in linear programming, cost-benefit
analysis, critical path analysis, or the like. Some fashionable
practising architect-planners sought alliances with the new experts
for fear of being left behind by the new trend. The architectural
planning schools hastily put up courses in computer programming,
mathematical statistics and various systems techniques, usually with
the help of lecturers from outside the existing faculty.

Even research money became available. Two large research projects were launched with funds of the Federal Housing Ministry to
develop comprehensive urban simulation models.

The first of these models was the POLIS urban simulation model
developed by a group of researchers at Battelle-Frankfurt (Battelle,
1973a). POLIS had all the vices of its American predecessors: it
was large, difficult to calibrate, and costly to operate. But it
had a sound modular structure, offered a multitude of policy
options, and produced convincing results. Besides, it was one of
the first of its kind to accomplish feedback between transportation
and land use. POLIS was implemented for the city of Cologne
(Wegener and Meise, 1971), and briefly afterwards, for the city of
Vienna (Battelle, 1973b).

The second model was SIARSSY, a joint product of the universities of Mannheim, Erlangen, Munich and Stuttgart (Popp, 1974).
SIARSSY was originally based on ORL-MOD, a Lowry adaptation developed at the ETH Zurich (Stradal and Sorgo, 1971), but the authors
soon started to make it recursive and augment it by transportation,
infrastructure, ecology and budget sub-models. The housing and
service employment allocation parts of the model were calibrated for
several West German cities.

Another source of funding became the EDP promotion programmes
of the Federal Ministry of Research and Technology. They constituted a major share of the budget of DATUM, a government-funded
research organization developing computer-aided planning methods for
local, state and federal planning authorities. In a first large

project DATUM collaborated with the City of Cologne and a German computer manufacturer, Siemens, on the design and development of a local planning information system. The resulting system named KODAS contained program modules for data manipulation, aggregation, statistical analysis, diagrams and maps, as well as a package for population analysis and projection (DATUM et al, 1974). In addition, DATUM began to develop, or adapt from other sources, a variety of computer programmes for processing spatial data, such as location-allocation programmes (DATUM, 1977).

Besides these activities supported by the Federal government, many cities set out on their own to implement computerized data bases and programmes for analyzing and displaying data. One outstanding example was the work of the Stadtentwicklungsreferat (Department of Urban Development) of Munich. Founded during the preparations for the 1972 Olympic Games, its staff of about forty professionals quickly won reputation for high-level, and yet pragmatic, development and application of computer-assisted planning tools. As the only planning department in this country, it operated its own computer and developed its own interactive data management and analysis system called KOMPAS (Blum, 1973).

Decline (1972-1975)

This period of enthusiasm lasted only four or five years. It started at a time when, in the United States, the use of systems analysis techniques for public policy-making and planning was already severely criticized. When the first news about the high costs and general failure of ambitious information systems and modelling projects arrived in this country (Fehl, 1971), and translations of critical articles (Alonso, 1968; Hoos, 1968; Churchman, 1968; Hoos, 1970) were published in the Stadtbauwelt, the opinion-making journal of architect-planners, this had a disastrous effect on city administrators and funding agencies. The "Requiem for Large-Scale Models" by Lee (1973) did the rest to prevent any more funds flowing into such research.

Moreover, many of the painful experiences reported from the United States were repeated here. Almost without exception, all modelling projects took longer than expected and had to cut back their objectives, and eventually, the results appeared not as useful as the proponents had promised and the clients had hoped. However, it is also fair to say that, given the limited amount of funding and the brief time span available, these projects did not have a real chance to be successful.

Anyhow, everybody involved was disappointed. The Housing Ministry quite abruptly stopped funding research dealing with systems methodologies, the architectural schools reduced their courses in such techniques back to the barest minimum, and the ill-considered marriages between architect-planners and systems people were quickly

divorced. The KGSt, the influential advisory institute on rationalization in local government, recommended to its member cities
extreme caution with respect to the use of computers in planning
(KGSt, 1975), and dismissed its advisory committee on automation in
local planning. The cities gratefully accepted the verdict, and cut
back their plans for planning information systems down to the most
routine data manipulation and report generating functions.

Of the small number of urban planners who had seriously got
involved in systems analysis techniques, many gave up and returned
to a "normal" career in the planning administration. Others turned
to related fields where quantitative analysis and modelling continued to be an accepted practice, such as transportation or energy
planning. A third, even smaller group retreated to those few universities where it was still possible to find a niche for a sort of
research nobody seemed to be asking for.

Stagnation (1976 to date)

In fact, between about 1975 and today, only small progress in
the adoption of systems analysis methods in urban planning has been
made.

Work on the comprehensive urban models was stopped almost
altogether. The POLIS model was applied two more times, to the
cities of Karlsruhe (Ruppert and Krieger, 1976), and again, Cologne
(Ruppert and Wuerdemann, 1979), but as in the earlier applications,
none of the cities decided to adopt the model for its planning on a
regular basis. In the case of the SIARSSY model, the ambitious
extensions of the model were never completed (Popp, 1977).

The work of DATUM was gradually shifted to the regional, state
and national planning levels. There have been some DATUM projects
relevant to urban planning, in particular the GEOCODE project concerned with generating and maintaining spatial reference files
(Klitzing, 1978), and the PENTA project dealing with demographic
techniques based on the computerized population register (Blum et
al, 1977). However, in these projects the main clients of DATUM
were the surveying or statistical offices, and not the urban planning departments.

Also, there have been some notable contributions to the field
by private consultants. Perhaps the most interesting example is the
work of Volwahsen with its skillful blending of systems analysis
methods and traditional techniques (Volwahsen et al, 1975;
Volwahsen and Heide, 1978). But this kind of work has not found any
followers, be it because Volwahsen never fully disclosed the methods
he used, be it because the quality of his work rested too much on
his particular talents. Another example is a series of housing
market simulation models developed by private research institutions
for five metropolitan areas, with funds provided by the Housing

Ministry (Stahl, 1981). But, with one exception, these models, if they became operational at all, disappeared without leaving any traces in the planning practice of the client cities. The one exception is connected with the Stadtentwicklungsreferat of Munich. This department has continued work on its KOMPAS planning information system (Franke, 1978), and developed or adopted a number of computerized planning tools, such as a model of intra-urban migration, one of the housing market simulation models referred to above, an employment projection model, and a system of models for allocating public facilities (Schussmann, 1978).

Nevertheless, on the whole, the diffusion of systems analysis techniques in the planning practice of the average municipal planning department has been negligible. Although in 1977 about sixty cities claimed to operate some kind of computerized information system (Kooperationsausschuss, 1978), according to the KGSt, "only a minimal share of the capacity of municipal computers is used for planning purposes" (Ostermann, 1977). Moreover, the majority of these applications are concerned with data retrieval, sorting, selection and aggregation, and with the production of tables, diagrams and maps. Data analysis techniques are largely confined to basic statistics. Only in a few cities programs are available for transportation network analysis and accessibility calculations. Except population projections, practically no forecasting techniques are applied. The demographic models, however, are in general developed and operated by the statistical offices. In the urban planning departments, again with the exception of Munich, virtually no models are in use or development.

This description was confirmed by a survey conducted in 1976 by Hoberg (1978) in forty-two urban planning departments. Hoberg investigated the use of various methods for allocating public and private facilities. He found that only in about fifteen percent of all reported applications methods which might be called systems analysis methods, such as cost-benefit analysis, location-allocation techniques, optimization or simulation models, were employed.

These findings are well in line with surveys which try to evaluate the utility of various skills and fields of knowledge for the professional practice of planners. In a 1975 study (Kunzmann et al, 1975), planners at all planning levels were asked to rate skills and fields of knowledge in terms of relevance on a six-point scale. Systems theory and systems engineering scored an average of 1.0, compared with, for instance, 3.3 for economics and 2.7 for law. In a similar study in 1979, addressed only to urban planners, this figure dropped to 0.8 (Mengden, 1979). A third study in 1980 addressed only alumni of the department of Raumplanung of the University of Dortmund, ie. a sample of the small minority of West German planners who already in their academic education have been exposed to systems analysis methods (Nonnenmacher and Schwoerer, 1980). Of the respondents, twenty-five percent indicated that systems analysis was very important for their work; however, this high proportion was only due to the fact that <u>statistics</u> was included under that heading. Only six percent believed that <u>systems models</u> were very important, while two-thirds believed that <u>models were of</u> no importance whatsoever.

THE CAUSES

At first sight, these facts seem to speak a clear language: the efforts to establish systems analysis techniques in urban planning have failed, and probably were misconceived from the beginning.

But that story is too simple. There are other ways of telling it, and each reveals different aspects of what has happened, just as different witnesses give different accounts of an observed event. In the following sections, I will present some of these different aspects.

Urban Planning Practice: From Growth to Stagnation

At first, an attempt will be made to relate the rise and decline of systems analysis methods to the political and economic framework of urban planning. Have there been major economic or social changes, or changes in settlement policy, planning legisla- tion or organization?

In West Germany, like in other countries, the agglomeration process has, over the last thirty years, resulted in the rapid growth of a few major urban regions. In recent years, however, the agglomeration process seems to have changed its pattern. Most large cities experience a decline of population, while communities at the periphery of urban regions continue to grow at a fast rate. The consequences of this exodus from the urban centres — loss of tax income, monofunctionality of the city center, increasing spatial segregation of age and income groups, and urban sprawl at the peri- phery — make this a serious problem for many cities.

At the same time, the power of local government to control spatial development has narrowed. On the one hand, more and more local planning decisions are directly or indirectly determined by state and Federal policy due to a tightening network of government subsidies in virtually every field of local policy-making and plan- ning. On the other hand, city governments get under increasing pressure by citizen groups outside of the traditional power struc- ture of industry, commerce and the political parties. These groups, usually focused on a particular neighbourhood issue, began to organ- ize themselves in great numbers in the late sixties, and today have informally established themselves in the local decision-making pro- cess as an extremely effective stumbling block for all kinds of planning actions of the administration.

These changes were accompanied by global economic, demographic and cultural developments, which in similar form could be found in most Western countries during the seventies. They included the energy crisis of the early seventies and its reverberations through the economic and monetary systems, the new cycle of economic, ie. sectoral and technological, change connected with the breakthrough of microprocessors, the painful expansions and contractions of the educational, employment and pension systems caused by the dramatic drop of birth rates during the sixties and, last but not least, secular changes of cultural traditions and values.

This cultural revolution was only remotely related to the student protest movement of the late sixties, which deeply affected the intellectual scene as well as government and administration, but was hardly realized by the broad public. But now a general feeling pervaded all groups of society that something had gone wrong, that economic growth and prosperity for everybody had been paid for with destruction of the land, waste of natural resources and pollution of the environment. The ambiguity of progress became evident: progress towards what, for whom? In particular, younger people felt alienated by the world of their parents. Many turned to alternative, sub-cultural forms of living and working in urban or rural cooperatives. Artisan work, pre-industrial crafts, and traditional ways of farming were rediscovered, a new, more subtle relation to nature was sought. However, new energy-preserving technologies were also experimented with . A broad ecological movement developed and demonstrated through spectacular anti-nuclear or conservationist actions, and even a few successes in local and state elections, its potential power.

It is worth noting that all this happened in a relatively affluent and politically stable country, in which to see signs of a crisis would simply be to ignore the facts. During the seventies, the Federal Republic had a stable and fairly liberal government, a comparatively low level of social tension and an exceptionally cooperative relationship between unions and industry. It is true that the economy was slowing down, but it continued to grow at an average rate of three percent per year, and so did household incomes. Indeed, there was some inflation primarily caused by rising energy costs, but it never exceeded six percent, and fuel and gasoline prices were among the lowest in Western Europe. There was unemployment, but it has settled down at under four percent lately. There was a housing shortage indeed, but nevertheless, between 1950 and 1975, housing floor space per capita approximately doubled. There have been, of course, serious environmental problems, but it is also true that the pollution of most large rivers has effectively been reduced, and the sky over the Ruhr region is cleaner today than ever before in the last century.

Nevertheless, the factual and atmospheric changes had their effect on urban policy-making and planning. The sixties were the time of massive housing construction, mostly in huge new housing areas at the fringe of the urban regions. Local road networks were overlaid with urban motorways, and extensive underground systems blueprinted. The perspective of urban planning was long-range and growth-oriented. Most large cities established new administrative units for Stadtentwicklungsplanung (urban development planning), which with their strategic orientation were the ideal clients for the new systems analysis methods and models just then entering the scene.

But, with the turn of the tide of the agglomeration process, the interest in strategic planning faltered, and so did the interest in the methods and models. Now the revitalization of old neighbourhoods became the must urgent problem. The 1972 urban renewal and development act (Staedtebaufoerderungsgesetz) marked this breaking point with regulations for renewal as well as for suburbanization programmes. In addition, for the first time it institutionalized some degree of citizen participation in local planning. In the following years, the scale and time horizon of local policy-making and planning were consistently lowered. Stadtteilentwicklungsplanung (urban district development planning) was a catchword for a few years, but today the focus is on Stadtreparatur (town repair), ie. on micro-scale efforts to rehabilitate individual blocks or buildings.

With each reduction in scale and comprehensiveness, the need for sophisticated analysis or forecasting methods was reduced. That was not only a question of scale and time (ie. the strategic versus incrementalist dichotomy), but also one of clientele. The more the planner works only for a small and homogeneous section of the urban population, the less comprehensive analyses are desired which are likely to reveal conflicts with the interests of other groups, or of the community at large. In this sense, systems analysis methods are not only useless for urban planning, but are, in fact, counterproductive as they tend to impede the bargaining process. That is the situation today.

Urban Planning as a Discipline: Science or Art?

Next, I turn to urban planning as a profession and discipline, assuming that the acceptance of a new technology like systems analysis methods depends much on the training, attitudes and intellectual traditions of its potential users. And indeed, there is ample evidence that the aversion of architect-planners to innovation and change in their professional practice contributed much to the early decline of systems analysis methods in urban planning in West Germany. In this case, the controversy pro and contra such methods took the form of the old debate on science versus art in urban planning. To understand this discussion, we must look back into the 19th century where it started.

Stadtplanung (urban planning) is a relatively new word in German. The more traditional term is Staedtebau (town building), which indicates that it originally was a building discipline concerned with physical aspects of urban planning. Like in other countries, urban planning in Germany originated from two disciplines: architecture and civil engineering. For many centuries, the domain of the architect as the creator of urban form was unchallenged. With the technical and industrial revolution of the early 19th century, the construction of bridges, railways and roads, canals, water and sewerage systems required skills architects did not have. At this time, a first division of labour took place: while the architect remained responsible for the physical appearance, ie. the aesthetics of urban form, the civil engineer took responsibility for the less visible: urban structure.

In 1861, James Hobrecht, a civil engineer, prepared the first development plan for Berlin, in which fire regulations, hygiene and transportation considerations played a dominant role. In 1876, R. Baumeister published his book on town development, in which he treated urban planning strictly in engineering terms. In the following years, most German states developed building codes and zoning regulations, and laid them down in planning laws. Many countries looked to Germany as having the most advanced planning system of the time: "In Germany town planning has become a science just like the construction of machines", the Metropolitan Plan Commission of Boston admiringly wrote in a report of the year 1912 (Stuebben, 1924).

The reaction of the architects against the predominance of the engineers in urban planning was formulated by C. Sitte (1889). He laid the foundations of what was known as Stadtbaukunst (art of town building), a sort of urban design which, like the French Beaux-Art tradition, was almost totally preoccupied with the aesthetics of buildings and public spaces. In this tradition, Ludwig Hoffmann, the chief architect of Berlin, declared (Stuebben, 1924):

> "The art of town planning, like every other art,
> has no laws nor rules. It is based on exper-
> ience, sentiment, reflection and taste."

This artistic tradition was totally insensitive to the emerging social problems connected with the rapid urban growth of that time. In 1914, in Berlin, nearly 600,000 persons lived in overcrowded dwellings with more than four persons per room (Hegemann, 1930). Consequently, housing became the dominant urban problem after the war. A new generation of architect-planners like Ernst May and Martin Wagner, or Walter Gropius and Hannes Meyer of the Bauhaus attacked the Beaux-Art tradition under the sign of functionalism and modern technology. Motivated by radical economic and societal reform ideas, the architect-planners of the Neues Bauen (new building movement) created some of the most outstanding examples of mass worker housing ever produced in this country.

This brief period of <u>rationalist</u> urban planning ended in 1933.
Most of the proponents of <u>Neues Bauen</u> were denounced as communists
and lost their jobs, or had to leave the country. However, the new
government soon recognized the usefulness of rigorously centralized
spatial planning. For this, the most advanced scientific planning
methods were to be applied, after they had been purged from the
"dominance of the rationalist, causal-mechanistic principle" of the
"rational-liberal science" (Meyer, 1936). How perfectly this was
achieved, is illustrated by the sad case of Walter Christaller, one
of the fathers of central place theory (1933), who himself helped to
apply his "system" to the occupied territories of Poland under these
auspices:

> "The final domination of the <u>Generalgouvernement</u>
> will be based on the key positions of a regular
> network of central places. The central place in
> the <u>Generalgouvernement</u>, centre and leader of its
> region and focus of German culture, power, and
> economy, will contain all elements required for
> the immediate expression of German dominance."
>
>   Schepers (1942)

Three years later, most central places in and around Germany
were ruined. Of the 10.5 million dwellings existing in West Germany
before the war, five million were destroyed or severely damaged. In
addition, ten million refugees came into the country from the East
and brought the housing shortage up to five million (Power, 1976).
Consequently, the reconstruction period was largely devoted to
rebuilding the housing stock. The notion of planning had become
deeply discredited by the abuse of centralized authority by the
Nazis. So up to the sixties, about ten million dwellings were put
into place by architects and architect-planners, with almost no
planning controls in effect.

The second half of the sixties seemed to change everything.
With the Social Democrats entering the government in 1966, planning
lost its bad image. There was a broad consensus among architectural
students, architect-planners and the public that it was possible by
better planning to arrive at a better urban environment. At the
same time, the concept of urban planning was rapidly expanded to
include not only physical, but also economic, social, educational,
ecological, and various other kinds of planning. Economists, social
scientists, geographers, and many other disciplines became aware of
the city as a study object. The planning department at Dortmund
University was established as the first undergraduate planning
school, others followed. <u>Raumplanung</u> seemed to establish itself as
a new integrative, interdisciplinary science (AG.Kop, 1972).

But this period of euphoria was soon over. Somehow the inter-
disciplinarity of Raumplanung lost its appeal. Even at Dortmund,
the disciplines slowly retreated into their traditional specialized

fields. More important, however, was the fact that planning practice never really accepted the scientist-planner. Only for a brief
period, the new techniques for analysis and forecasting seemed to
point to a scientization of the field (Rautenstrauch, 1974). In the
reality of the planning department, however, the rapid expansion of
responsibilities of urban planning, and the need to respond to a
multitude of different problems under time pressure made it impossible for the architect-planners to develop a new professional identity. Instead, they felt that they were being disqualified, and
that their field of work was gradually disintegrating (Siebel,
1975).

The natural reaction to this experience was to defend, or
rather to revitalize, the old universalist position. A new discussion about the "generalist" versus the "specialist" planner arose
which ended clearly in favour of the "generalist" demonstrating that
in the daily work of the average planning department, there is no
room for scientific analysis beyond the most routine (Albers, 1979).

And this is not likely to change soon, as today the leading
positions in the planning administration are still held, and probably will be held in the future, by architect-planners. Under the
pressure of the architectural lobby, higher careers in the planning
administration continue to be reserved for candidates who "have
demonstrated their ability to apply their knowledge methodically by
several design projects, and a final thesis in urban or regional
planning of mainly conceptual character" (BMBau, 1978), ie. practically only to architects. In the light of this tendency, it is not
surprising that for the BDA, the major architects' association, the
Dortmund planning education is "a deplorable misdevelopment which
should be corrected as soon as possible" (BDA, 1979).

Planning Theory: Social Cybernetics versus Political Economy

The third kind of witness I now call on is the planning theorist, as an impartial observer of what is going on in the planning
scene. Unfortunately, there has been much disagreement in German
planning theory about the nature of planning and the role of
scientific methodology in planning. Therefore, this section again
is a description of a controversy.

Planning was discovered only recently as an object of scientific investigation and theory by political scientists in West
Germany. During the postwar and reconstruction period, the recol-
lection of the misuse of centralized control in the Nazi period, the
dominant neo-liberal economic doctrine, and the abhorrence of economic planning a la East Germany, all worked together to associate
planning with being a menace to individual freedom. However, with
the changing economic policy and the evolving Ostpolitik after 1966,
this taboo became obsolete. This meant for the political sciences
that a considerable deficit had to be compensated in a relatively
short time.

This first period of German planning theory was largely influenced by American political science, and, in particular, by authors like Parsons, Deutsch or Easton, who tried to apply systems theory concepts to societal or political organizations. Accordingly, the German planning theory of the late sixties was dominated by systems theory thinking (Senghaas, 1967; Narr, 1967; Naschold, 1968). The most influential formulation of this paradigm was presented by the sociologist N. Luhmann (1966b). For Luhmann, planning is a sophisticated kind of selection mechanism by which a social system reduces the extreme complexity of its environment. A planning decision is a choice act through which, by excluding potential actions from further choice, a planning object is fitted to a mental or internal model of itself by the planning system. Planning differs from other choice acts by its reflexivity: "Planning means to set premises for future decisions, ie. to decide on decisions" (Luhmann, 1966b).

For this school of planning theory, methodology is important. As planning is understood as a cybernetic process, the failure to adequately process problem complexity is a prime bottleneck of the process. Every possibility to increase the problem processing capacity of the planning system is appreciated as a progress towards more system rationality, ie. the "ability to predict and control the consequences of actions over as many links in the causal chain as possible" (Luhmann, 1966a). The increasing scientific character of planning methodology is accepted as a necessary correlate of the growing complexity of society, moreover, it is recognized that science for societal planning is going to be more and more technical, ie. approaching the ideals of exactness, plausibility and falsifiability associated with the natural and engineering sciences. The use of sophisticated systems methods is part of the system process itself: acting in the system requires the awareness of reality as a "network of problem-solving structures, secondary problems of such structures, solutions for such secondary problems, etc." (Luhmann, 1969).

Accordingly, most planning theorists of that period were strongly in favour of implementing the new, mostly yet unknown systems analysis methods for public planning. However, as these methods became better known, a first phase of criticism developed. This first critique challenged the methods in their own terms, ie. did not question their usefulness, but their efficiency.

Some critics generally questioned the possibility of identifying causal relationships between human acts. It was argued that human actions are rational only in a loose sense, and are determined by expectations and aspirations, roles, institutions and "latent needs" which defy quantitative analysis (Tenbruck, 1967). Other critics pointed to the "wicked" nature of societal planning problems (Rittel, 1970), and questioned the relevance of statistical and other quantitative data for the solution of such problems (Fehl, 1970), in contrast to "informal" and ad hoc (Fehl, 1971), or normative, explanatory or instrumental information (Rittel, 1973). A third group of critics challenged the claim of the methods to grasp and reproduce the complexity of human society and, of course, found severe deficits. While some of these deficits could be attributed to insufficient data or modelling techniques, at least one deficit seemed uncorrectable: the undeniable ambivalence of systems analysis methods with respect to values, in particular, democratic norms, made many critics concerned about their possible political misuse (eg. Naschold, 1968).

This last critique was the main concern of the second stream of planning theory which developed only a few years later. The polit-ical economy paradigm of planning theory is founded on the Marxist theory of fundamental conflict between the working and capitalist classes which, in this final era of capitalism, is bound to lead to perennial crises. Political planning in this context has the func-tion to overcome or avoid such crises in order to safeguard the conditions for capitalist exploitation. The ultimate goal of Marx-ist planning theory is the transformation of the political-economic structure. Two different ways to approach this are discussed: while a minority propagates to concentrate all efforts in a Marxist political party (eg. Schuon, 1971), the majority favours a long-range strategy of political consciousness-raising in "planning-oriented political base organizations" (Offe, 1969). It is hoped that by utilizing existing channels of citizen participation "move-ments of countervailing power" can be mobilized which will gradually transform the political system (Offe, 1970).

Towards the end of the decade, Marxist planning theory attrac-ted more and more followers, for whom, as for the student protest movement and the "extraparliamentary opposition" (APO) of that time, the Vietnam war, racial violence in US cities, and the pending state emergency legislation of the West German government converged into a fundamental crisis of Western civilization. This landslide carried away many earlier proponents of the systems theory paradigm, with the result that in 1970 German planning theory was more or less Marxist.

This is worth noting because it had implications for the use of systems analysis planning methods in research, education and prac-tice. For the Marxist planning theorist, "social-cybernetic" approaches are "ahistoric" and "idealistic", and therefore, cannot provide guidance for political action (Fehl et al, 1972). Moreover, their political vacuousness makes them disposable for any sort of political abuse (Ronge, 1971), while the adaptive, stabilizing mech-anisms of complex systems are associated with conservative tenden-cies in society (Kade and Hujer, 1972). This is not to say that the "heuristic value" of systems analysis methods are denied (Fehl et al, 1972), but it is believed that their "progressive aspects can unfold only in the course of a thorough transformation of the pro-duction system" (Arch+, 1972). Under the capitalist system, their application is, at best, irrelevant; in the worst case, however, an instrument to prepare "a new cycle of capital accumulation" (Kade, 1973).

These arguments were widely accepted by younger planners and planning students. The systems approach lost its progressive image, and more and more became associated with being technocratic, conservative and anti-democratic. By 1971-72, the period of innovation and optimism of the late sixties was ridiculed as the time of "planning euphoria". Unfortunately, it had been much too brief to establish any permanent tradition. Today the attitude of most planning theorists towards methodology questions is disinterested, if not hostile.

Political Theory: Technocracy versus Advocacy

The final section of this chapter, while less directly related to urban planning methodology, in fact, presents the intellectual background from where most of the ideas and arguments of the controversies reported in the preceding sections originated.

I am talking about the deep and lasting influence exerted on political life in West Germany by a series of debates in political science, social philosophy and science theory during the sixties and seventies. They all were made possible by the fortunate fact that in West Germany, unlike many other countries, the Frankfurt School has developed a rich tradition of Marxist political theory and social philosophy. The works of Horkheimer, Adorno, Marcuse and Habermas represent a most fruitful effort to unfold the political theory of Karl Marx into a meaningful tool for analyzing and criticizing modern capitalist society. The debates I am referring to were all disputes between the Critical Theory of the Frankfurt School and other, non-Marxist, theories of society. I will briefly excerpt three of them most relevant to the subject of this paper. All three deal in some way with the role of the scientist or expert in social decision-making, ie. with the old question of how scientific knowledge and human values are to be integrated into decisions or actions.

The first challenge came from the sociologist H. Schelsky (1961), who extrapolated certain tendencies of the techno-scientific development into a future where an anonymous technocracy of unaccountable experts decides about the direction of technical progress on the grounds of technical requirements, instead of human needs. The most radical opposition to this Orwellian projection was formulated by Habermas (1963a; 1968) in the form of his "pragmatistic" model, in which the division between technical requirements and human values is transcended by way of a dialogue between the scientist and the politician, ie. by "public, unrestricted and uncontrolled discussion about the suitability and desirability of action-guiding principles and norms" (Habermas, 1968).

The second challenge originated from the science theory of Karl

Popper. The controversy started from Popper's criticism of philosophical idealism and dialectical philosophy (1957; 1961), which was
attacked by the Frankfurt School as positivism, ie. a kind of
"ahistoric" empiricism unable to grasp the process of societal
development of which it is a part. In contrast, the Critical Theory
asks for a theory of society which realizes the totality of the
societal process, ie. accepts that all cognition is determined by
the emancipative interest of the scientist (Adorno, 1961; Habermas,
1963b). This position was questioned by "critical rationalists"
like H. Albert who, following Popper, insisted that even socially
progressive values can become irrational if they are set absolute
and shielded from continuous critical scrutiny (Albert, 1964).

The third challenge was caused by the diffusion of systems
theory in the social sciences, and led to an extensive debate
between Luhmann and Habermas in the early seventies. It will be
remembered that for Luhmann, society is a cybernetical system which
stabilizes its existence in a hypercomplex environment by the reduction of complexity. Reduction of complexity is thus the raison
d'etre of social systems, it is achieved by various reduction techniques, among them planning; their application is system rationality. Habermas concluded from this description of self-stabilizing
system behaviour that a systems theory of society must be "conservative" and "apologetic": inasmuch as system rationality is directed
towards system stabilization, the theory must avoid issues that
might jeopardize the existing power structures, and this makes it
"disposable for technocratic use" (Habermas, 1971).

These three debates were by no means only academic exercises.
The utopian spirit of the pragmatistic model not only contributed
much to the optimism with which in the early seventies the ex-APO
students started their "march through the institutions", it also had
great influence on the architectural and planning students of the
time, because it offered to them the attractive role of the enlightened mediator between scientific knowledge and the public. The
emancipative function of science also played an important role in
the 1967-1969 student movement, and later on, was constitutive for
the motivation and social and political involvement of younger
scientists and planners. It is the merit of the first two debates
that they, from the critique of the technocrat, developed the concept of the critical, ie. politically involved scientist or planner
who sees himself as the partisan or advocate of the emancipation of
underprivileged groups of society.

The third debate, however, made it clear that the optimistic
belief of the late sixties that advanced scientific techniques and
reform-oriented democratic planning could go together, was an illusion. The allegation that systems theory, and all methods and techniques related to it, are conservative, technocratic, and anti-
democratic persisted, was repeated over and over again, and today is
commonplace among intellectuals in this country.

CONCLUSIONS

I have attempted to show that the history of systems analysis methods in urban policy-making and planning — arrival, decline and stagnation — had remarkable parallels in other fields of the political and intellectual development of West Germany.   In the urban planning practice, long-range, strategic planning made way to incremental planning for particular client groups.   The scientization of urban planning as a discipline was brought to a halt in favour of the revitalization of the "generalist" planner.   In planning theory, systems theory thinking was replaced by the political economy paradigm.   In political theory, the critique of the technocrat led to the concept of the critical, politically involved scientist or planner.

What these four controversies have in common is the shared critique of the one-dimensional concept of rationality which has dominated public decision-making in most Western countries for a long time.   This "engineering" kind of rationality was predominantly oriented towards economic growth and technological "progress", and was completely insensitive towards aesthetic and emotional needs, environmental qualities, grown fabrics of social relations, and the concerns of minorities and underprivileged groups of society.   With this critique, these controversies are part of a larger process of reevaluation and reformulation of societal goals going on during the last decade, moreover, they contributed themselves much to it.

This larger process has also deeply influenced the style of public policy-making and planning in West Germany.   While it cannot be said that technocratic planning has completely disappeared, at least it has become much more difficult to carry out controversial large-scale technical projects affecting the natural environment, or the quality of life in existing neighbourhoods, without taking account of the reactions of a watchful and critical public.

The decline of systems analysis methods in urban planning in West Germany must be seen as primarily a consequence of these changes in the context and style of urban planning.   And there can be no doubt that the systems analysts, model builders and other proponents themselves are to be blamed in the first place.   By their irresponsible promises, their narrow-minded preoccupation with technical detail and jargon, their stubborn insistence on a type of planning process which did not exist any longer, and their failure to adapt their methods and models to the changing planning environment, they are mainly responsible for the present disrepute of the field.

And yet, if one looks closer, one can also find in all these controversies and debates, inextricably intermingled with their progressive intentions and effect, strong undertones of a general rejection of science and rationality as tools for improving the human condition.   This is obvious, not only where the architects' lobby, under the pretext of practice-orientation, tries to sabotage the new planning discipline, but also where "political" planners use science only in an opportunist fashion to support their particular cause, or return to rhetoric or other less rational techniques to produce consensus. Here is the critical point where the enlightening intention of the critique of technocratic planning is in danger of turning into its irrational counterpart, and where progressive and regressive tendencies in the present planning discussion in West Germany meet in an insidious way.

These undertones are well in line with other anti-rational tendencies of the present cultural development of West Germany. A most significant example is contemporary architecture where the achievements of half a century of socially-oriented functionalism currently are being thoughtlessly thrown overboard and replaced by the short-lived fashion of a shallow and sterile eclecticism which, ironically, calls itself the New Rationalism. Anti-rational tendencies can also be found in the contemporary theatre, in popular music, and in other fields of cultural production, as well as in certain backward-oriented changes of lifestyles, and in the nostalgic esteem for past periods and fashions and its correlate, the general aversion against our technical civilization.

It would be very surprising if these tendencies would have had no effect on urban planning. It can, therefore, be assumed that the decline of systems analysis methods in urban planning in West Germany was, at least to a certain degree, also influenced by the general turn of the Zeitgeist to the anti-rational. If this interpretation is not totally amiss, most of the discussions about technical aspects of the methods and models, eg. about model performance or model cost, have in fact missed the real issue. Because, if the hypothesis is only partly true, technical deficiencies of the methods and models were not the prime reason for their not being accepted: even if they had performed better and at less cost, they would not have been accepted anyway. Rather, it can be said that the present unsatisfactory state of the art is a consequence of the fact that society did not want these methods and models, for reasons that had not much to do with their performance or cost. If society had wanted something like these methods and models, it would have provided the conditions to improve them, regardless of cost.

What conclusions can be drawn from this analysis for the future development of systems analysis methods in urban planning?

It seems obvious that improving the methods and models in their own terms alone would probably not have much effect on their acceptance in the planning practice. Nevertheless, there is much to be said in favour of doing just that. First, nobody would disagree that the work still to be done in terms of model specification, model technology and model calibration is enormous. Second, it can realistically be expected that the near future will see even greater advances in terms of data availability and hardware performance than the past, which will make modelling concepts feasible which are still utopian today. However, those who decide to work only in this field probably will have to be prepared to work mostly in a research environment at the university, unless they are willing to offer their services to clients of questionable respectability.

Those, however, who wish to see their methods and models be put to use in the service of public policy-making and planning, must do more than that: they must make society want the methods. How can this be accomplished?

Unfortunately, systems analysis methods have so far been undiscriminatingly associated with being inseparably linked to technocratic planning. For this misunderstanding, the systems analysts and model builders themselves are to be blamed, as they have in the past failed to demonstrate that there is nothing inherent in systems analysis techniques which reserves them exclusively to one particular style of planning (Fehl, 1976). On the contrary, systems theory offers a great variety of concepts and techniques directed towards decentralization of control, system transformation, conflict resolution and learning. However, except in laboratory settings, almost nothing of this potential has been explored or demonstrated, let alone effectively introduced into urban planning.

But, in fact, this potential of systems analysis methods offers the only chance of their survival in urban planning. Various proposals have been made to exploit this potential (Wegener, 1978). All of them are based on some concept of a communicative planning process embracing all groups of urban society, in which systems analysis methods serve as channels, or intelligent communication media for conflict analysis and conflict resolution. Planning by enlightened discussion is an old dream of planning theorists in the United States (Etzioni, 1968; Friedmann, 1973), as well as in West Germany (Senghaas, 1967; Naschold, 1968; Offe, 1969; Fester, 1970; Fehl, 1971; Habermas, 1973). However, none of them offer any advice how in the face of the "unalterable low attention potential of human experience" (Luhmann, 1967), it can be brought about.

Of course, there is no guarantee that systems analysis techniques can. Too many problems have to be solved, eg. how to make systems analysis techniques available to a large public, how to overcome the enormous didactic difficulties, how to handle the privacy issue, how to channel the information explosion, how to structure the communication process, and how to prevent its abuse. And yet, two-way TV communication, home computing and remote access to computer networks are a technical potential too powerful to be rejected without careful scrutiny.

To choose this strategy would mean to shift the emphasis away from model refinement to questions of transfer, acceptance, man-model and man-machine interfaces, and, of course, questions of relevance, conflict and human values. It would force the field to undergo a fundamental transformation of goals and standards, but would, at least, promise the chance of a modest revival.

NOTES

1. I am grateful to Ekkehard Brunn, Klaus R. Kunzmann, Claus Schoenebeck and Hans-Georg Tillmann for their helpful comments on a draft of this paper.

2. The term "systems analysis methods" is used throughout the paper in a loose fashion to summarize a variety of methods from the fields of mathematical statistics, decision analysis, and operations research directed towards the organized or systematic processing of complex information for policy-making and planning. These methods, sometimes also called "systems engineering methods", have in common that they attempt to analyze, explain, forecast and evaluate observed phenomena and processes, including societal and economic ones, in terms of quantitative dimensions in the fashion of the natural and engineering sciences. Another common feature of these methods is the fact that their application usually requires the use of electronic computers.

3. More details on computer applications, urban models, and software development for urban planning in West Germany (with references) are contained in an earlier paper (Wegener, 1979), which is, in a way, a companion paper to the present one.

4. This, and all following quotations throughout the paper, are my own translations.

REFERENCES

The first date in any reference shows the original date in which the item was written, the second date shows the date of the publication refered to.

Adorno, T.W., 1961, Zur Logik der Sozialwissenschaften, in T.W. Adorno, et al, (Eds.), 1969, Der Positivismusstreit in der deutschen Soziologie, Luchterhand, Neuwied/Berlin.

AG.Kop - Arbeitsgruppe Kommunale Planung, 1972, Thesen zu einer Theorie der Raumplanung, in E. Brunn and W. Pannitschka, (Eds.), 1978, Raumplanung und Planerausbildung, Dortmunder Beitraege zur Raumplanung 9, Universitaet Dortmund, Dortmund.

Albers, G., 1963, Hochschulausbildung und Kommunalplanung, Archiv fuer Kommunalwissenschaften, 2, 33-43.

Albers, G., 1966, 1969, Ueber das Wesen der raeumlichen Planung, Stadtbauwelt, 21, 10-14.

Albers, G., 1979, Zur Rolle der Planung und des Planers, Stadtbauwelt, 62, 196-198.

Albert, H., 1964, Der Mythos der totalen Vernunft, in T.W. Adorno, et al, (Eds.), 1969, Der Positivismusstreit in der deutschen Soziologie, Luchterhand, Neuwied/Berlin.

Alonso, W., 1968, 1969, Bestmoegliche Voraussagen mit unzulaenglichen Daten (Predicting Best with Imperfect Data), Stadtbauwelt, 21, 30-34.

Arch+ (editorial), 1972, Was heisst "fortschrittlich" in Bezug auf Planungstheorie und Planungsmethode? Arch+, 15, 2.

Battelle-Institut e.V., 1973a, Simulationsmodell POLIS: Benutzerhandbuch, Staedtebauliche Forschung 03.012, BMBau, Bonn.

Battelle-Institut e.V., 1973b, POLIS Wien: Anwendung des Simulationsmodells POLIS fuer die Stadtentwicklungsplanung Wiens, Battelle-Institut e.V., Frankfurt.

Baumeister, R., 1876, Stadt-Erweiterungen in technischer, baupolizeilicher und wirtschaftlicher Beziehung, Ernst & Korn, Berlin.

BDA - Bund Deutscher Architekten, 1979, Bedauerliche Fehlentwicklung, Vorgaenge, Informationen des BDA Nordrhein-Westfalen, 2.

Blum, H., 1973, Das Kommunale Planungsinformations- und Analyse-System (KOMPAS) der Landeshauptstadt Muenchen, OEVD (Oeffentliche Verwaltung und Datenverarbeitung), 11, 495-507.

Blum, H., Ruhland, S. and Tuellmann, H., 1977, Projektschwerpunkte undLeistungsbeschreibung fuer die PENTA-Methoden, DATUM e.V., Muenchen.

BMBau - Bundesministerium fuer Raumordnung, Bauwesen und Staedtebau, 1969, Staedtebaubericht 1969, Stadtbau-Verlag, Bonn.

BMBau - Bundesministerium fuer Raumplanung, Bauwesen und Staedtebau, 1978, Leitfaden fuer den Vorbereitungsdienst der Baureferendare in der Fachrichtung Staedtebau, BMBau, Bonn.

Christaller, W., 1933, Die Zentralen Orte in Sueddeutschland, Wissenschaftliche Buchgesellschaft, Darmstadt.

Churchman, C.W., 1968, 1972, Informationssysteme und die Informierte Gesellschaft (Real Time Systems and Public Information), Stadtbauwelt, 36, 339-340.

DATUM e.V., 1977, Daten ueber DATUM, DATUM e.V., Bonn.

DATUM e.V., Siemens AG and Stadt Koeln, 1974, Datenverarbeitung fuer die kommunale Planung, Research Report DV 74-03, BMFT, Karlsruhe.

Etzioni, A., 1968, The Active Society: A Theory of Society and Political Processes, The Free Press, New York.

Fehl, G., 1970, 1971, Informations-Systeme in der Stadt- und Regionalplanung, Kraemer, Stuttgart.

Fehl, G., 1971, Informations-Systeme, Verwaltungsrationalisierung und die Stadtplaner, Stadtbau-Verlag, Bonn.

Fehl, G., 1976, Systemmuedigkeit und Systemoptimismus, in G. Fehl and E. Brunn, (Eds.), Systemtheorie und Systemtechnik in der Raumplanung, ISR 21 (P1), Birkhaeuser, Basel/Stuttgart.

Fehl, G., Fester, M., and Kuhnert, N., (Eds.), 1972, Planung und Information, Bertelsmann, Guetersloh.

Fester, M., 1970, 1971, Vorstudien zu einer Theorie kommunikativer Planung, Arch+, 12, 42-72.

Franke, D., 1978, Das Kommunale Planungsinformations- und Analyse-System (KOMPAS) der Landeshauptstadt Muenchen, OEVD (Oeffentliche Verwaltung und Datenverarbeitung), 3, 11-14.

Friedmann, J., 1973, Retracking America: A Theory of Transactive Planning, Doubleday, Garden City, New York.

Habermas, J., 1963a, Verwissenschaftlichte Politik und oeffentliche Meinung, in J. Habermas, 1968, Technik und Wissenschaft als "Ideologie", Suhrkamp, Frankfurt.

Habermas, J., 1963b, Analytische Wissenschaftstheorie und Dialektik, in T.W. Adorno, et al, (Eds.), 1969, Der Positivismusstreit in der deutschen Soziologie, Luchterhand, Neuwied/Berlin.

Habermas, J., 1968, Technik und Wissenschaft als "Ideologie", in J. Habermas, 1968, Technik und Wissenschaft als "Ideologie", Suhrkamp, Frankfurt.

Habermas, J., 1971, Vorbereitende Bemerkungen zu einer Theorie der kommunikativen Kompetenz, in J. Habermas and N. Luhmann, 1974, Theorie der Gesellschaft oder Sozialtechnologie: Was Leistet die Systemforschung?, Suhrkamp, Frankfurt.

Habermas, J., 1973, Legitimationsprobleme im Spaetkapitalismus, Suhrkamp, Frankfurt.

Hegemann, W., 1930, Das Steinerne Berlin: Geschichte der Groessten Mietskasernenstadt der Welt, Ullstein, Berlin/Frankfurt/Wien.

Hoberg, R., 1978, Methodenanwendung in der kommunalen Planung, Schriftenreihe des IfR 13, Universitaet Karlsruhe, Karlsruhe.

Hoos, I.R., 1968, 1970, Rumpelstilzchen oder: eine Kritik der Anwendungder Systemanalyse auf gesellschaftliche Probleme, Stadtbauwelt, 25, 21-27.

Hoos, I.R., 1970, 1972, Sozialplanung und die kontrollierte Gesellschaft (Information Systems and Public Planning), Stadtbauwelt, 35, 221-228.

Kade, G., 1973, Die Grenzen des Wachstums - Das Elend der buergerlichen Oekonomie, in C. Freeman and M. Jahoda, et al, (Eds.), Die Zukunft aus dem Computer? Eine Antwort auf "Die Grenzen des Wachstums", Luchterhand, Neuwied/Berlin.

Kade, G. and Hujer, R., 1972, Planung der kleinen Schritte und Politikdes "Status quo", in G. Fehl, M. Fester, and N. Kuhnert, (Eds.), Planung und Information, Bertelsmann, Guetersloh.

KGSt - Kommunale Gemeinschaftsstelle fuer Verwaltungsvereinfachung, (Ed.), 1975, Automation und Planung: Bestandsanalyse I, KGSt-Bericht 13/1975, KGSt, Koeln.

Kooperationsausschuss Bund/Laender/kommunaler Bereich, Arbeitsgruppe Planungsinformationssysteme, 1978, Umfrage zum Stand der wichtigsten mit automatisierten Datenverarbeitungsanlagen unterstuetzten Informationssysteme fuer Statistik und Planung (PLIS) bei Bund, Laendern und im kommunalen Bereich am 1.1.1977, KGSt, Koeln.

Kunzmann, K.R., Mueller, S., Warlitzer, V., 1975, 1980, Der Bedarf an Raumplanern in der Bundesrepublik Deutschland, Universitaet Dortmund, Dortmund.

Lee, D.B., Jr., 1973, Requiem for Large-Scale Models, Journal of the American Institute of Planners, 39, 163-178.

Luhmann, N., 1966a, Reflexive Mechanismen, in N. Luhmann, 1970, Soziologische Aufklaerung 1, Westdeutscher Verlag, Opladen.

Luhmann, N., 1966b, Politische Planung, in N. Luhmann, 1971, Politische Planung, Westdeutscher Verlag, Opladen.

Luhmann, N., 1967, Soziologische Aufklaerung, in N. Luhmann, 1970, Soziologische Aufklaerung 1, Westdeutscher Verlag, Opladen.

Luhmann, N., 1969, Komplexitaet und Demokratie, in N. Luhmann, 1971, Politische Planung, Westdeutscher Verlag, Opladen.

Mengden, J., 1979, Bestimmung des quantitativen Bedarfs fuer Ausbildungsgaenge mit planerischer Orientierung, Universitaet Dortmund, Dortmund.

Meyer, K., 1936, Raumforschung, eine Pflich wissenschaftlicher Gemeinschaftsarbeit!, Neumann, Neudamm/Berlin.

Narr, W.D., 1967, Systemzwang als neue Kategorie in Wissenschaft und Politik, in C. Koch and D. Senghaas, (Eds.), 1970, Texte zur Technokratiediskussion, Europaeische Verlagsanstalt, Frankfurt.

Naschold, F., 1968, Demokratie und Komplexitaet, in C. Koch and D. Senghaas, (Eds.), 1970, Texte zur Technokratiediskussion, Europaeische Verlagsanstalt, Frankfurt.

Nonnenmacher, W. and Schwoerer, I., 1980, Der berufliche Werdegang Dortmunder Raumplaner, Universitaet Dortmund, Dortmund.

Offe, C., 1969, Sachzwang und Entscheidungsspielraum, Stadtbauwelt, 23, 187-191.

Offe, C., 1970, Das Politische Dilemma der Technokratie, in C. Koch and D. Senghaas, (Eds.), Texte zur Technokratiediskussion, Europaeische Verlagsanstalt, Frankfurt.

Ostermann, J., 1977, Das Konzept der Gemeinsamen Kommunalen Datenverarbeitung - Gemeinsamkeit als Chance und Risiko, KGSt-Mitteilungen, 6, 1-10.

Popp, W., (Ed.), 1974, Entwicklung des Planungsmodelles SIARSSY, Staedtebauliche Forschung 03.018, BMBau, Bonn.

Popp, W., (Ed.), 1977, SIARSSY: Ein Modell zur Simulation von staedtischen und regionalen Systemen, Haupt, Bern/Stuttgart.

Popper, K., 1957, 1965, Das Elend des Historizismus (The Poverty of Historicism), Mohr, Tuebingen.

Popper, K., 1961, Die Logik der Sozialwissenschaften, in T.W. Adorno, et al, (Eds.), 1969, Der Positivismusstreit in der Deutschen Soziologie, Luchterhand, Neuwied/Berlin.

Power, A., 1976, France, Holland, Belgium and Germany: A Look at their Housing Problems and Policies, Habitat, 1, 81-103.

Rautenstrauch, L., 1974, Berufsbild von Planern in Stadtplanungsaemtern, Staedtebauliche Forschung 03.026, BMBau, Bonn.

Rittel, H., 1970, Der Planungsprozess als Iterativer Vorgang von Varietaetserzeugung und Varietaetseinschraenkung, in J. Joedicke, et al, (Eds.), Entwurfsmethoden in der Bauplanung, Kraemer, Stuttgart.

Rittel, H., 1973, Informationswissenschaften: Ihr Beitrag fuer die Planung, in G. Fehl, (Ed.), Planungsinformationssysteme fuer die Raumplanung, Staedtebauliche Beitraege 2/1973, Institut fuer Staedtebau und Wohnungswesen, Muenchen.

Ronge, V., 1971, Politoekonomische Planungsforschung, in V. Ronge and G. Schmieg, (Eds.), 1971, Politische Planung in Theorie und Praxis, Piper, Muenchen.

Ruppert, W.-R. and Krieger, E., 1976, Stadtentwicklungssimulation, Battelle-Institut e.V., Frankfurt.

Ruppert, W.-R. and Wuerdemann, G., 1979, Anwendung des Simulations-modells POLIS fuer die Stadtentwicklungsplanung Koelns, Staedte-bauliche Forschung 03.072, BMBau, Bonn:

Schelsky, H., 1961, Der Mensch in der Wissenschaftlichen Zivilisa-tion, Westdeutscher Verlag, Koeln/Opladen.

Schepers, H.J., 1942, Raumordnung im Generalgouvernement, in A. Teut, (Ed.), 1967, Architektur im Dritten Reich 1933-1945, Ullstein, Berlin/Frankfurt/Wien.

Schuon, K.T., 1971, Wissenschaft, Politik und wissenschaftliche Politik, Pahl-Rugenstein, Koeln.

Schussmann, K., 1978, Relevanz von Stadtentwicklungsmodellen, in M. Pfaff and W. Asam, (Eds.), 1980, Integrierte Infrastruktur-planung zur Verbesserung der Lebensbedingungen in Staedten und Gemeinden, Duncker & Humblot, Berlin.

Senghaas, D., 1967, Sozialkybernetik und Herrschaft, in C. Koch and D. Senghaas, (Eds.), 1970, Texte zur Technokratiediskussion, Euro-paeische Verlagsanstalt, Frankfurt.

Siebel, W., 1975, Wandlungen Kommunaler Planung und die Qualifika-tion von Stadtplanern, in R.-R. Grauhan, (Ed.), Lokale Politik-forschung, Campus, Frankfurt.

Sitte, C., 1889, Der Staedtebau nach seinen kuenstlerischen Grund-saetzen, Prachner, Wien.

Stahl, K., (Ed.), 1981, Quantitative Wohnungsmarktmodelle, Wohnungs-markt und Wohnungspolitik, BMBau, Bonn.

Stradal, O. and Sorgo, K., 1971, ORL-MOD-1: Ein Modell zur region-alen Allokation von Aktivitaeten, Arbeitsbericht 24.1, ETH Zuerich, Zuerich.

Stuebben, J., 1924, Der Staedtebau, Gebhardt, Leipzig.

Tenbruck, F.H., 1967, Zu einer Theorie der Planung, in V. Ronge and G. Schmieg, (Eds.), 1971, Politische Planung in Theorie und Praxis, Piper, Muenchen.

Volwahsen, A. and Heide, R., 1978, Stadtteilentwicklungsplanung am Beispiel Wuppertal-Barmen, TH Darmstadt, Darmstadt.

Volwahsen, A., Meise, J., Feldmann, H., Breeding, D. and Siemering, W., 1975, Siedlungsstruktur im Ruhrgebiet, Stadtbauplan GmbH, Essen/Darmstadt.

v. Klitzing, F., 1978, Raumbezug fuer Kommunale Planung und Statistik - GEOCODE, Vermessungswesen und Raumordnung, 7, 346-366.

Wegener, M., 1978, Mensch-Maschine-Systeme fuer die Stadtplanung, ISR 61 (P5), Birkhaeuser, Basel/Stuttgart.

Wegener, M., 1979, The Use of Computers for Urban and Regional Planning: A Review, in Proceedings of PArC 79, International Conference on the Application of Computers in Architecture, Building Design and Urban Planning, AMK, Berlin.

Wegener, M. and J. Meise, J., 1971, Stadtentwicklungssimulation, Stadtbauwelt, 29, 26-31.

ON THE USE OF STRATEGIC PLANNING MODELS IN IBERIAN CITIES

Pedro Geraldes

Departamento de Vias de Comunicacao
Laboratorio Nacional de Engenharia Civil
Avenida do Brasil, 1799 Lisbon Codex
Portugal

INTRODUCTION

The basic motivation behind the elaboration of the present paper relates to the considerable scepticism which has until now prevented a wider acceptance of urban models as strategic policy-making tools in Iberian cities. The objective of the paper is to contribute to a more general evaluation of past experience on the use of urban models, towards possible improvements in the use of systems analysis in urban planning. The organization of the paper reflects three main factors: availability of information, time constraints and personal judgement.

As a matter of fact, some references are available with respect to certain of the modelling aspects, although almost no references were found relating to the corresponding planning impact. Time availability prevented us from a more exhaustive search, which necessarily would have involved a good amount of direct contacts. Finally, in the absence of more information, the assessment of the experiments is likely to be biased by our own subjective judgement, both through our modelling background and through the experiments in which we have been directly involved or that we indirectly accompanied.

Given the above constraints, the paper is structured as follows. Firstly, a broad characterization of the environment where the modelling experiments took place is summarily attempted, by means of general indicators capable of reflecting the existing similitudes and differences between Iberian countries on their own, and between them and their industrialized and middle-income counterparts.

153

Next, some experiments in the field of urban systems modelling in Portuguese and Spanish cities are reviewed and, whenever possible, characterized in terms of the context where they took place. This review does not pretend to be exhaustive, and its coverage is uniquely a function of the available references. Consequently, some of the case studies referred to are analysed in greater detail than others. At this stage, the likely planning impact of those experiments is also addressed.

Finally, some proposals are advanced in terms of what urban systems analysts can do, in the short range, to improve their contribution to strategic urban policy-making.

Throughout this paper, the emphasis is mainly on those aspects relating to the planning framework where the decision to implement a certain type of model takes place, rather than on the aspects concerned with the details of the structure of the models used.

GENERAL CHARACTERIZATION

Portugal and Spain share some common broad characteristics, such as parallel cultures and history, and similar geographical location. Both countries experienced major socio-economic transformations in the last decade, and both are in the process of negotiating their admission to the EEC, for which they have traditionally constituted important sources of labour supply. International classifications, such as the World Bank's, currently assign both countries to the category of middle-income developing countries, which corresponds to developing countries with per capita gross national product above US$300 (World Bank, 1979).

Nevertheless, the stage of development is unlikely to be identical in both countries. As a matter of fact, considering as an indicator the GNP per capita, the corresponding values exhibit a wide difference. Using the same source, in US$1977 the value for Portugal was 1,890, the corresponding Spanish value being 3,190. With respect to this indicator, Portugal can indeed be considered a middle-income country, although Spain is in the boundary between middle-income and industrialized countries.

Still taking World Bank figures, the values of total population and of indicators relating to urbanization in Portugal, Spain and the groups of middle-income and industrialized countries are presented in the following Table. Despite the fact that these figures reflect the criteria involved in the classification of urban population, which are likely to vary from country to country, we will assume that they are able to provide an overall pattern for the process of urbanization.

Considering first the values for total population, their average annual growth in the periods 1960-70 and 1970-77 for Portugal

TABLE 1

INDICATORS CONCERNING GLOBAL URBANIZATION TRENDS

Population Countries	TOTAL			URBAN				% OF URBAN				No. of cities >500,000 inh.	
	Mid-1977 (M)	Average annual growth (%)		% of total population		Average annual growth (%)		In largest city		In cities >500,000 inh.			
		1960/70	1970/77	1960	1975	1960/70	1970/75	1960	1975	1960	1975	1960	1975
Portugal	10	0.0	0.8	23	28	1.5	2.3	47	44	47	44	1	1
Spain	36	1.1	1.0	57	71	2.6	2.4	13	16	37	43	5	6
Middle-income countries	na	2.5	2.6	37	47	3.7	4.2	23	25	35	44	na	na
Industrialised countries	na	1.0	0.8	67	74	1.8	1.4	18	17	48	54	na	na

na: not available

and Spain looks closer to the industrialized countries value than to the developing countries. In both countries, the registered tendency is likely to have been influenced by emigration, although to varying degrees. Taking, for example, the case of Portugal in which this effect is likely to have had a more important impact, it is worthwhile mentioning that the country had in 1977, a work force of 3.786 M; in 1973, the expatriate labour force in Western Europe alone was estimated at 0.745 M (Kavalsky and Agarwal, 1978). That effect was, however, counterbalanced in the period 1970-77, because of the stoppage of new emigration to Western Europe, and because of the end of the war in Africa. This last factor, in particular, in addition to demobilization, implied the return from Africa of some 0.5 M people. The above facts probably help to explain why the percentage of Portuguese urban population is considerably lower than that of the group of middle-income countries, the same tendency being exhibited by the average annual growth values.

The Spanish 1975 figure for the urban population as a percentage of total population looks rather close to the one for the industrialized countries, although the process of urbanization progressed at a higher rate than the one verified for that group of countries. Nevertheless, the same tendency towards a decrease in the average annual growth of the percentage of urban population in the periods 1960-70 and 1970-75 can be detected both for Spain and the industrialized countries. Finally, Spain had in 1975, six cities exceeding 500,000 inhabitants, although Portugal had one, perhaps two.

But, if the process of urban concentration in the Iberian countries has not been so dramatic as in most developing countries, the pattern in the rise of car ownership, not accompanied by the corresponding investment in transport infrastructures, looks rather similar. For the middle-income developing countries, Willoughby (1978) estimates that the average investment in transport systems for every new entrant in the labour force each year of the mid-1970's has been US$3,000, this value being roughly one-fifteenth of the corresponding value for the industrialized countries. In the meantime, considering the Spanish and Portuguese case, the number of passenger cars per 1,000 population in the period 1960-70 multiplied almost eight-fold in the former case, and more than three-fold in the latter (World Bank, 1976). A good illustration of this process is the case of the Portuguese city of Oporto where, it is claimed, no new road infrastructure has been provided in the period 1963-74, although in the same period, traffic is likely to have doubled (DGTT/GEPTRP, 1979).

After this brief characterisation of the global pattern of urbanization in Portugal and Spain, it is now useful to address summarily the problem of urban legislation in both countries.

Urban legislation is regulated in Spain by the commonly called "Ley del Suelo" (Legislacion del Suelo, 1976), according to which the main levels of planning are: national, provincial, metropolitan and municipal. Metropolitan plans concern a set of municipalities agreeing on a general territorial plan involving the set of municipalities. Once this plan is approved, the municipalities are co-responsible for its implementation, as well as for the elaboration of intra-municipal plans in a way consistent with the metropolitan plan. The overall objectives of a metropolitan plan are: zoning land for specific uses; general definition of the transport systems and right-of-way zones; overall specification of public service requirements (parks, schools, hospitals, etc.); environmental protection; programming the execution of the plan through time and guaranteeing coordination among the various sectoral initiatives of investment (private and public) in a way consistent with the relevant Central Government Departments.

In particular, with respect to urban land, a metropolitan plan may specify: detailed land zoning at the level of each municipality; reserve of land for the provision of public services; design (alignments and remaining characteristics) of the road network and provision of parking facilities; overall design of sewerage and water and energy supply infrastructures; specification of the average intensity of use of the land zoned; evaluation of costs of implementation. The control of floorspace demolition and reconstruction and the constitution of land reserves are also allowed by the legislation.

Portuguese urban legislation does not appear to be, for the purpose of our analysis, substantially different from the Spanish. As a matter of fact, the existence of regional plans for the two big metropolitan areas is stipulated since 1970, as is the existence of territorial plans involving a set of municipalities. Both levels of plans are complemented by intra-municipal plans.

The objectives of the elaboration of these plans are rather similar to their Spanish counterparts. Nevertheless, at least two important differences must be stressed at this stage. The first one relates to the initiative of developing the Portuguese regional and territorial plans, which belongs to the central urban administration, through the Ministry of Public Works. The second one relates to the coordination between various bodies of the central administration and between them and the local administration, the responsibilities of which are not so clearly specified at the spatial level, as they appear to be in the Spanish legislation.

From the above general characterization, it now seems useful to synthesize the following points.

The overall situation of the Portuguese economy is, accepting Paelinck's hypothesis, "...still a relatively supply-oriented economy, and has to be brought up to the stage of a demand-oriented and later on, a demand-creating economy" (Paelinck, 1979). Spain exhibits a higher degree of economic development and, with the exception of some peripheral regions, probably lies somewhere between the two last stages referred to by Paelinck. The stage of economic development is naturally correlated with the performance of the planning machinery, particularly if seen from the point of view of overall efficiency, coordination and speed of adaptation to evolving situations.

In both Iberian countries the process of urban concentration has not manifested itself with the same dramatic intensity as in most developing countries. However, spatially well defined nuclei of urban poverty are still apparent, and their elimination became a top priority for the town planners. Simultaneously, transport planners had to face a very fast rise in car ownership rates in the decade 1960-70, which was not followed up by the corresponding investment in road infrastructure. Traffic congestion on the one hand, and the rising costs of operation of the private car on the other, in the context of an almost total dependency on imported fuel and of a growing scarcity of capital resources, inexorably forced the public transport authorities to join their efforts with operators, in order to improve management and to restructure the corresponding networks. These efforts have been facilitated by the considerable capacity of intervention of the administrations in the sector, arising either as a consequence of their being an important shareholder of some of the operators, or of the latter's dependence on public subsidies.

It is in the preceding framework that most of the experiments in urban systems analysis which will be analyzed below took place.

SOME EXPERIMENTS IN STRATEGIC URBAN SYSTEMS PLANNING

After an overall characterization of the urban processes in both Iberian countries, some experiments on the use of systems analysis in urban planning will now be summarily reviewed. Only the urban scale is considered and, hence, interesting exercises at the regional scale, such as the use of a Forrester-type model in the Viscaya Region (Sener-Preyser, 1977) or the models proposed for Andalucia by Eyser and the Netherlands Economics Institute (Ancot and Paelinck, 1979) are out of the scope of the present review.

In the late sixties, an urban transport model was implemented in the Lisbon municipality, in order to help design the future transport network, given an assumed land use plan. This model was initially developed by the French group Metra/Sema, and it is included in Wilson's 1972 review (Wilson, 1972). The model was further applied in Paris, Brussels, Dijon, Jerusalem and Abidjan. The main particularity of such a model is the use of a special type of trip distribution technique (the so-called preferential equilibrium) for the estimation of home-based trips to work (Le Boulanger, 1971). The model has been formally used in policy-making and concrete planning proposals were made concerning the structure of the main transport network in the municipality of Lisbon. Nevertheless, the plan proposed by the municipality in 1967 was approved by the central administration only ten years later, and no further use of the model has been reported.

Another study has been made in the Lisbon region for the Portuguese National Highway Administration, by the German Dorsch Consultants, under the sponsorship of the OECD. This study involved a comprehensive household survey (75,000 households surveyed), the implementation of a standard four-step travel demand forecasting model and a detailed economic evaluation sub-model. The study was commissioned in 1972 and finished in 1976, thus including the period in which the political situation in the country changed. The model was used in order to prepare a transport plan for the Lisbon region, focusing in particular on the private transport network.

A detailed feasibility study justified the selection of the chosen alternatives. The consultants (Dorsch Consultants/JAE, 1976) reported difficulties in carrying out their tasks, mainly as a consequence of the lack of definition concerning economic structure and land use intentions (constituting the basic input for their model), of changing of government priorities from private to public transport and of lack of accuracy in the National 1970 Census. No further use of the model is reported and it is unlikely that the optimal years recommended for the implementation of the selected projects will be respected.

Another study involving the public transport subsystem in the Lisbon region was initiated in 1973 by the Ministry of Transport, with the collaboration of the Swiss University of Lausanne, under a protocol of technical assistance with the Swiss Federal Government. This study has as its main objective a short and medium-range operational planning of the public transport network, and particular emphasis is assigned to continuous planning and to adequate coordination between the different interventions involved in the public transport sector — central government, municipalities, operators, population, etc. (DGTT/ITEP, 1979). This study makes use of the model NOPTS, developed by the University of Lausanne in collaboration with W. & J. Rapp AG. Basically, the model is a non-capacity restraint multipath assignment technique, complemented by other procedures such as time-table optimization and operations management (staff, rolling-stock, etc.). The model allows an operational evaluation, particularly in terms of graphic outputs, representing desire-lines, transfers, running costs, intensity of use and other indices of exploitation. The model has also been implemented in the Portuguese city of Oporto, and in the Swiss cities of Basle, Lausanne and Zurich. A detailed account of these experiments, the logic of the approach and some examples of the graphic capabilities of the system, involving, namely, the case of Lisbon and Oporto, is provided by Rapp and Mattenberger (1978). The overall interest of the Portuguese public transport administration in these experiments can probably be assessed by two indices: the implementation of the

model in the Northern city of Oporto, following the pioneer experience in Lisbon, and the present installation of the model software
in Portugal. Although the implementation of some of the proposals
of the planning group which is now using the model has been carried
out (such as bus lanes), the report containing more formal proposals
on the development of the public transport system in the Lisbon
region, generated and selected with the help of the model, is still
under discussion (DGTT/ITEP, 1979). Consequently, it is as yet too
early to evaluate properly the corresponding planning impact.

The Valencia planning study, in the early 1970's, has been the
first of a series of major land use/transport planning studies commissioned by the Spanish Planning Authorities. This study involved
a land use model of the Lowry type, combined with a standard transport model. The implementation was carried out by the Spanish consultants Eyser, together with an international team including Martin
and Voorhees Associates (Martin and Voorhees Associates, 1974). One
of the objectives of the study was the elaboration of a
comprehensive land use/transport strategic plan for the region,
involving a detailed analysis of the public transport subsystem. In
order to carry out this task, Martin and Voorhees used the GREYS
analysis, a technique in which, for a given pattern of demand, and
through assignments to an idealized network, links are successively
removed in order to maximize the intensity of use in the remainder
of the network. This, complemented by a pre-feasibility study
including detailed operational considerations, allowed the selection
of a reduced number of alternatives for the development of the
network.   These alternatives were then simulated using the
comprehensive transport model, and subsequently subjected to a
cost-benefit evaluation.

The GREYS analysis has also been used in the context of
policy-generation in the study realised for the "Comision de
Comunicaciones de Viscaya" by the Spanish consultants Sener and
Eyser (Comision de Comunicaciones de Viscaya, 1974). This study
involved three "Comarcas" (groups of municipalities) centred around
the main city of Bilbao. The fundamental objective of the study was
the development of a comprehensive public transport plan for the
region. The alternatives pre-selected using GREYS were evaluated
through simulation in a standard four-step transport model, by means
of cost-benefit appraisal. The study proposals contemplate the
comprehensive integration of the existing heavy public transport
infrastructure with a new underground railway. This proposal has
been accepted to the extent that a more detailed engineering analysis of the underground system has been executed, commissioned by the
central transport administration.

Also, in the "Comarca" of Bilbao (a spatial subset of the
region considered in the transport study referred to above) a comprehensive integrated land use and transport model is at present
being used as a simulation tool by the metropolitan planning authority. The development of this model was commissioned from Applied
Research of Cambridge and Marcial Echenique and Partners, and the

main purpose of its implementation was to help in the revision of the metropolitan plan concerning the group of nineteen municipalities headed by the city of Bilbao. Basically, the model was designed to cope with detailed land use policies, such as a major increase in the existing supply of urban land, provision of public services, zoning and plot-ratios, taxation and subsidies in housing and land. The model also has the capability of providing an impact analysis at the urban scale of the existing projects in the public transport sector (such as the underground system referred to above), as well as the ability to simulate the proposed projects concerning the highway network (such as the construction of a tunnel and a bridge).

The simulation system has three main modules: land use, transport and economic evaluation. The land use module combines an input/output framework where nine types of activities are considered, with a floorspace capacity-restrained locational utility formulation, utility being mainly a function of space and trip consumption. The transport module includes a price-elastic transport demand sub-model and a standard modal-split and capacity-restrained multipath assignment sub-model. The details of the mathematical and iterative formulation of the model are provided in Geraldes et al (1978). It has been possible to calibrate the model using available information contained in the household survey tapes of the previous transport study, complemented by data provided by several sectoral studies elaborated by different consultants for the planning authority (Geraldes, 1980).

Finally, in a somewhat different context, it is worthwhile mentioning the model implemented for the region of Sines, in Portugal, in the early 1970's. Basically, the model has been used in order to help design the New Town of Santo Andre, part of a huge industrial complex the Portuguese Government was strongly committed to develop. Considerable multi-sectoral administrative powers were delegated to the Board responsible for the development, reporting directly to the Prime Minister. The model implemented was of a mathematical programming type, and was initially used by the planning team (see also Coelho, 1980). Once more, the changing of the political context in Portugal, associated with major alterations in the international economic scene, introduced, from 1974 on, a considerable amount of uncertainty about the feasibility of the project, including the proposed urban development.

It is now useful to synthesize some comments that the analysis of the previous experiments in urban systems modelling suggests.

The environment where the implementation and use of the models referred to above took place, can hardly be claimed to be stable. As a matter of fact, Iberian countries have experienced considerable political change in the last decade, and they seem to be on their way from one equilibrium point to another. The fact that structural changes occurred in a period of world economic recession introduced

still greater uncertainty in the process of adjustment, particularly
given the dependencies of both economies.

Under the circumstances, it is natural that the greater the
uncertainty, the more politicians and bureaucrats express a greater
concern with short-range decision-making at the expense of long-term
planning. Hence, the first reason for an attitude of scepticism
towards strategic planning modelling.

Nevertheless, strategic models have been widely used, triggered
mainly by stringent requirements of reorganization of the public
transport networks, given the cumulative effects of congestion, the
scarcity of capital and the rising cost of operation of the private
car fleet.

The type of model selected in each case, and the way in which
it has been used, seems mainly to be dictated by the institutional
competence of the body commissioning the study. For example, the
Portuguese central public transport authority expresses its interest
in a sub-modal-split multipath assignment model, considering solely
the public transport mode. In a similar context, with respect to
the objectives of the study, the Spanish metropolitan authorities
used GREYS, but within the framework of a comprehensive land use and
transport model in Valencia and in Bilbao, although, in the latter
case, within an interval of two years.

The conflict between institutional competence whose sectoral
decisions affect the same system is patent in the case of Lisbon,
where the National Highway Authority, pressed (at least until 1974)
by rising car ownership and growing intensity of car use, had to
initiate a major comprehensive modelling study affecting the whole
of the metropolitan area, its sole interest being the improvement of
suburban highway facilities. In the absence of a comprehensive
decision-making framework, some claim that there is no role for any
comprehensive model. Hence, the second reason for an attitude of
scepticism towards strategic planning modelling.

The previous comments, however, although capable of providing
some insight, are not able to explain the case of the remaining
metropolitan areas included in Table 1 above, where, to our knowledge, no strategic models have been used at all. A multitude of
plausible unstated reasons probably lies behind that anti-modelling
approach. Nevertheless, we feel tempted to advance two likely
interrelated reasons. The first one concerns the apprehension of
the responsible planning officers about becoming engaged in exercises relying massively on computer support, and of which, they
feel, they cannot control the details. The second reason relates to
a more general attitude, that a Portuguese journalist present at an
urban modelling seminar held in Lisbon spontaneously synthesized for
his readers as follows:

"Does it seem believable, to our average reader, that,
in order to plan the construction of a city - starting with the houses where you are going to live and
finishing with the supply of milk to the nurseries -
everything will have to be translated into 'complicated' mathematical equations, only instantaneously
'digestible' by a frigid computer?

And, if so, would not this hypothesis be - because of
its inaccessibility - a danger to the society, the
solution of whose problems is being approached in
such a way?"

(Diario de Noticias, 8th September 1976;
my translation)

THE WAY AHEAD

Past experience suggests that, in the turmoil introduced by
conflict and uncertainties in the environment where decision-making
occurs, simulation models involved in a particular policy-making
exercise often tend to be abandoned once they have been employed in
the selection of alternative courses of development, hence compromising their continuous use, and the possibility of validation
tests.

Under such circumstances, the non-validity of the model as a
predictive tool can hardly be claimed. Assuming the inner contradiction in the social sciences between positivism and normativism,
we will first concentrate on both assessments individually, before
turning to their interface.

The predictive capability of any planning simulation model can
be considered a function of three interrelated factors: the hypotheses underlying the theory behind its formulation, the calibration
process and the model's sensitivity to the independent variables
included in its formulation.

With respect to the theoretical foundations, a fair amount of
progress has been reported recently, and it can easily be realized
that most of the available theory, particularly with respect to
consumer behaviour, is now included in model formulation. Nevertheless, it can hardly be claimed that major increases in existing
knowledge have been achieved. Conventional theory lacks convincing
explanatory power, particularly with respect to the supply-side and
to technological improvements. We are probably now in a position to
foresee that, in the period we are entering, considerable change is
going to be induced by technological innovation, namely where the
impact of the silicon chip is concerned. Here lies a proposed point
for reflection.

Within the technicalities of the calibration process, considerable improvements have also been taking place. The price to achieve
a satisfactory representation of reality generally passes, nevertheless, through the collection and processing of massive quantities of
data. Here reality and models appear as relative concepts to one
another, the feedback between them being performed by data flows.
In fact, it seems that there cannot exist an information system
without a framework for the interpretation of reality (even if just
conceptual), and that there can exist no model without reality, as
perceived through data.

Model predictions are conditional upon the exogenously estimated independent variables included in their formulation. Some of
them are clearly stipulated by legislation to be the urban policy-
maker's attributes. This is the case of land zoning, floorspace
controls and design of the road networks. However, others are just
partially controlled, such as transport costs, floorspace construction, taxation and subsidies, or not controlled at all, such as
income distribution. The influence of the last set of variables on
city structures is an important determinant, hence the necessity of
its consideration in model design. But this number of degrees of
freedom introduces additional uncertainty in model prediction. Or,
in other words, even if our theoretical capability for explaining is
correct, the predictive power is conditional upon the capacity for
intervention.

The assessment of the validity of the use of systems analysis
in urban policy-making is therefore inseparable from the normative
content of decision-making as related to what ought to be done,
given a set of value judgements. And this set of values backing
distinct concepts of evaluation, often associated with different
disciplines, introduces, in turn, conflict among bureaucrats and
politicians.

Long ago, John Neville Keynes called for the necessity of distinction between "a positive science..., a body of systematized
knowledge concerning what is; a normative or regulative science...,
a body of systematized knowledge discussing criteria of what ought
to be...; an art...a system of rules for the attainment of a given
end" (quoted by Friedman, 1979). In such a context, it seems that
comprehensive strategic urban planning models have provided very
detailed frameworks for the explanation of reality, capable of simulating the likely impact of alternative policies, for the best
available predictions of the variables out of the control of urban
planners, thus allowing for the selection of the more desirable
policy, given certain evaluation criteria. Nevertheless, it can
hardly be claimed that they have had a great impact in terms of the
highly creative process of policy-generation.

It seems that the process of policy-generation is more easily
developed within the context of each sector, which currently is the
object of a discipline. The development of the latter has, at least

partially, been determined by the practical necessity of filling the gap between positivism and normativism, by means of the elaboration of specific techniques, strongly motivated by the 'ends'. These techniques are, in most cases, associated with specific operational criteria of evaluation, with a varied economic content. OR based models, CAD techniques or detailed simulation models are currently used at this stage, the main emphasis being on finding some optimal solution, or simply reducing the spectrum of alternative sectoral projects. However, inter-sectoral consistency among these projects has still to be achieved. Sketch-planning models, here assumed to be of an embryonic systemic nature and thus characterized by a low degree of spatial resolution, small amount of data requirements and reduced computer support, can play an important role at this stage. This is the case of the portable models developed for the transport sub-system by Manheim and his associates (quoted by Schofer and Stopher, 1979) or for residential location, floorspace and trip demand by the author (Geraldes, 1980), using programmable calculators.

Nevertheless, it does not seem that these reduced models can be considered as valid substitutes for the more comprehensive general strategic models, where the relationships among the components of the system tend to be fully considered. It is our belief that only in the framework of a simultaneous urban land use and transport analysis, is it possible to achieve full consistency within each policy-package to be considered, and to assess the corresponding relative advantages. If this is the case, general urban megamodels are necessary in order to provide the comprehensive outputs which are required, if a consensus for intervention is to be achieved among the conflicting sectoral selection criteria backing the actors intervening in urban policy-making. However, as the analysis of the Iberian experiences strongly suggests, it is important to stress that the success of the planning use of any comprehensive model is definitely a function of the comprehensiveness of the decision-making framework.

The interface between decision-makers and systems analysts consequently becomes critical, particularly in developing countries where most of the know-how in modelling techniques still has a strong imported component. A possible way of improving that interface seems to lie in a coordinated research effort between universities and national research institutes, the latter providing adequate environment for contacts with policy-makers, both through their institutional links and medium-term applied research concerns. That coordination could, for example, include the design and the execution of a detailed survey of the planning officials' assessment of their own experience with the practical use of models in the member countries, the realization of international courses and seminars focussing on the methods, but with a special emphasis on case-studies, or even the elaboration of a multi-lingual handbook in urban systems analysis which could provide uniform guide-lines for technical vocabulary, mathematical symbols and flow chart techniques at different levels of presentation, thus improving communication.

Finally, a last suggestion for debate. It is a well recognized fact that the development of urban systems models has been strictly interdependent with the development of hardware facilities. Trying to assess the likely impact of the increasing capabilities of microcomputing and computer networks, with their (probably attractive) decentralization characteristics, as related to the urban decision-making process and the philosophy behind systemic modelling design, seems to us to be a challenging proposal.

REFERENCES

Ancot, J.P. and Paelinck, J.H.P., 1979, Modeles et Sous-Modeles pour L'Amenagement de L'Andalousie, Vth Meeting of Regional Studies, Zaragoza, Spain.

Coelho, J.D., 1980, Combinatorial Optimisation Methods in Public Facility Location: An Experience, Unpublished paper, Department of Mathematics, University of Lisbon, Portugal.

Comision de Comunicaciones de Viscaya, 1974, Estudio Coordinado de Transportes Urbanos Colectivos – Bilbao y Zona de Influencia, Bilbao, Spain.

DGTT/ITEP, 1979, Estudo dos Transportes da Regiao de Lisboa – Relatorio de Progresso, Julho, Portugal.

DGTT – Grupo de Estudo do Plano de Transportes da Regiao do Porto, Politica de Gestao da Circulacao – Cidade do Porto, 1979, Simposio Sobre Transportes Urbanos e Suburbanos, Ordem dos Engenheiros, Lisbon, Portugal.

Dorsch Consultants/JAE, 1976, Transportation Study of the Lisbon Region, Volume I, Lisbon, Portugal.

Friedman, M., 1979, The Methodology of Positive Economics. F. Hahn and M. Hollis, (Eds.), Philosophy and Economic Theory, Oxford University Press, Oxford, UK.

Geraldes, P., 1980, Integrating Land Use and Transport Components in Urban Systems Modelling, PhD Thesis, University of Cambridge, Cambridge, UK.

Geraldes, P., Echenique, M. and Williams, I., 1978, A Spatial-Economic Model for Bilbao. Proceedings of the PTRC Summer Annual Meeting, Warwick, UK.

Kavalsky, B. and Agarwal, S., 1978, Portugal – Current and Prospective Economic Trends, The World Bank, Washington, DC.

Le Boulanger, H., 1971, Research into the Urban Traveller's Behaviour, Transportation Research, 5, 113-125.

LEGISLACION DEL SUELO, 1976, Editorial Civitas.

Martin and Voorhees Associates, 1974, The Valencia Planning Study, Martin and Voorhees, London.

Paelinck, J.H.P., 1979, Portugal as a Peripheral Country in Western Europe - A Systematic View of the Problem, International Seminar on "Portuguese Economic Development in a Changing International Environment", Lisbon, Portugal.

Rapp, M.H. and Mattenberger, P., 1978, Planification Operationelle des Transports Urbains en Commun: Approches et Applications, in Proceedings of the World Conference on Transport Research - Transport Decisions in an Age of Uncertainty, Martin Nijhoff, Leiden, The Netherlands.

Schofer, J.L. and Stopher, P.R., 1979, Specifications for a New Long-Range Urban Transportation Planning Process. Transportation, 8, 199-218.

Sener-Preyser, 1977, Modelo de Simulacion de Viscaya, Informe No. 1, Corporacion Administrativa "Gran Bilbao", Spain.

Willoughby, C.R., 1978, A Development Banker's View, in Proceedings of the World Conference on Transport Research - Transport Decisions in an Age of Uncertainty, Martinus Nijhoff, Leiden, The Netherlands.

Wilson, A., 1972, Travel Demand Forecasting: Achievements and Problems. Urban Travel Demand Forecasting, Special Report 143, Highway Research Board, Washington, DC.

World Bank, 1976, World Tables 1976, The John Hopkins University Press, Baltimore, MD.

World Bank, 1979, World Development Report, 1979, Oxford University Press, New York.

ACKNOWLEDGEMENTS

I am grateful to colleagues and friends who helped to keep me informed of their experiences in urban systems modelling. I am also grateful to Michael Batty, for his suggestions on the overall structure of the paper. Marcial Echenique and Robert Laurini stimulated me in long discussions concerning some of the points which are analyzed here. The Portuguese "Instituto Nacional de Investigacao Cientifica" sponsored the present work.

SYSTEMS ANALYSIS IN A DEVELOPING COUNTRY: THE CASE IN TURKEY

Gunduz Ulusoy

Department of Industrial Engineering
Bogazici University
PKZ Bebek, Istanbul
Turkey

INTRODUCTION

In this short paper I will first try to summarize the environment in which systems analysis is applied to policy-making and planning in Turkey. Then I will elaborate on how this environment might
be improved, and what strategy to adopt for that purpose. I hope
that the suggestions made will be of a general enough nature so as
to be useful to systems analysts in other developing countries.

In order to put the observations and suggestions made into a
larger framework, I have not restricted my analysis to the urban
setting. In the following I will restrict my observations to the
academic field and the public sector, since the problems addressed
by the systems analysts in the private sector in Turkey are of a
rather limited scope, being of interest only to the particular firm
or to a set of firms. This omission is certainly justified if one
recalls the dominant role played by the public sector in the
development of Turkey.

PROBLEMS IN APPLYING SYSTEMS ANALYSIS IN TURKEY

The application of systems analysis for solving problems is
rather new in Turkey — almost a decade and a half. And it is only
recently that it has started to diffuse through the planning departments of different organizations as a worthwhile approach. It has
yet to prove itself. Let me briefly mention some of the reasons for
this slow diffusion process.

Education

One of the reasons has been education in systems analysis. The systems analysis field in education and practice has been heavily dominated by operations research. It is operations research which preaches the systems approach to problems and provides mathematical modelling techniques to facilitate it. But, unfortunately, it has penetrated into only a very limited number of educational programmes. There is only one Operations Research Department, and that is in the Middle East Technical University in Ankara offering programmes both in undergraduate and graduate levels. The operations research techniques are also taught in Industrial Engineering and Business Administration Schools. Lately, other engineering departments and departments involved in teaching planning practice (such as Public Administration Departments) have adopted courses in operations research. A program based on a good balance between qualitative and quantitative aspects of systems analysis is yet to develop.

As a result of the limited progress in systems analysis education currently, the number of people who can be called systems analysts is relatively small. Positions normally in need of people with a good background in systems analysis are mostly filled with people with little, if any, idea about systems analysis. This statement becomes increasingly true as one moves to higher levels in the hierarchy. There is presently no mentionable effort being made either by the universities to educate these people in the basics of systems analysis, or by the organizations employing these people to motivate them to seek and acquire such knowledge. Much seems to have been left to the good will of these people.

Missing Services and Support

One example of this category is the lack of reliable data in most areas. Available data is, in general, unreliable, insufficient and outdated. When a systems analyst starts with a project he recognizes in almost all cases the need for field work to collect data because data is non-existent, or because existing data has not been updated. For example, data for the origin-destination matrix for Istanbul was collected by the Greater Istanbul Metropolitan Council in 1971. It has not been updated since then, and when an urban transportation study was initiated by the Municipal Bus Company of Istanbul in 1978, the data for the origin-destination matrix had to be collected anew. All these result in a tremendous waste of time and other resources. Especially the updating and monitoring of urban data has become very important, due to rapid urbanization and the dynamic nature of cities.

It is interesting to note that some projects are even requested with the hidden intention of data collection. But once data has been collected and used for the purposes of that particular project, usually the client is no longer interested in updating it regularly to ensure that it is an integral part of the decision-making process. The use of data, not as a justification of preconceived ideas and already-made decisions, but as an essential tool in decision-making has yet to be demonstrated to the decision-makers.

It is true that lack of data bases hampers systems analysis studies, and lack of data bases can mainly be attributed to the attitude taken by the organizations involved. But another difficulty is encountered in the attempts to remove this one. It is the proper and efficient use of the available computer facilities, and the availability of dependable computer personnel. This, of course, is a problem with effects beyond the data problem. In most cases, the support of the systems analyst by computer programmers and analysts is missing. The computational aspects of the project work has to be overtaken, or closely overseen by the systems analyst himself -- another source of the drain on time and energy. Any suggestions made by systems analysts, as a result of a study which requires additional computer facilities and work, meet difficulties in implementation due to the reasons mentioned previously. Thus, the lack of support and services not only restricts the systems analyst during his work, but also limits the set of feasible solutions open to him.

Inefficiency and Ineffectiveness of Government Agencies

The inefficiency and ineffectiveness of most government agencies is well-known to the systems analyst in Turkey. A clear-cut jurisdictional and functional organization among different government agencies is missing. On any particular issue, several government agencies have a say, and since there is a total lack of coordination among them, finding solutions to problems in a reasonable time becomes almost impossible. It is hard to conceive, but these agencies have developed a certain rivalry among themselves which leads to conflicting views, and sometimes even to conflicting data. This puts the systems analyst working on a particular issue in the unique position of arbiter between these government agencies. Although difficult, it is imperative for the systems analyst to try to bring the views of the different agencies into his analysis of the issues involved. This is imperative if the solution proposed should have a chance of implementation. For example, there are eleven agencies involved in the decision-making process about issues of traffic in Istanbul. With so many parties present, any fundamental suggestion usually has to go through a long compromising process, taking the problem beyond the systems analyst.

THE STATUS OF SYSTEMS ANALYSTS IN TURKEY

The Systems Analyst in the Public Sector

The problems with the systems analyst in the public sector are twofold: (1) public sector jobs are not well-paid jobs, and thus attract only a few able systems analysts, and (2) they might be subject to pressure to conform with the views of the organization they are in, and be requested to work on certain selected problems to produce results not opposing those views.

These lead to a high turnover ratio among systems analysts in the public sector which prevents the accumulation of information and experience in these areas in the public sector. The fact that it took more than a decade for the Greater Istanbul Metropolitan Council to prepare the land use plan for Istanbul is an example of the above observations.

The Systems Analyst in the University

The systems analyst who is also a faculty member is not subject to the pressures mentioned above for his in-house research. There are no restrictions imposed on him to make public the results he has obtained, which a regular government employee can only do at the expense of losing his job. The faculty member is quite free in the selection of topics for research, and in the execution of the research as long as no outside funding is needed. If needed, the project is controlled by the sponsoring institution, but this control is usually exercised only to check conformance of the scope of the research to that indicated in the research proposal.

It is striking to observe how few sponsored systems analysis projects are conducted in universities. Almost all the sponsored research consists of research into a particular problem of the sponsoring organization. Both the faculty members, and potential sponsoring organizations contribute to this situation. People in a position to extend such a support might not sincerely believe that research as delivered by faculty members is worth such support. They seem to have come to this conclusion since they feel that university-based research usually takes more time and is more expensive, and the faculty members are accused of not devoting enough time and effort in identifying the real problem. Consequently, they are not of much help to the decision-maker. On the other hand, the faculty member might be reluctant to seek such sponsored research for the following reasons. The academic reward system puts more emphasis on excellence in teaching than in research. The faculty member might disagree with the sponsoring organization on which issues have a research value, and those which are straightforward project-type work. He might claim that much precious time is wasted due to the procedures and regulations of the sponsoring organization, and in the efforts to convince all parties involved. As a result of his dealings with the sponsoring organization, he might reject cooperation because he has concluded that the particular organization is after a pre-determined result, and wants only the ratification of it as the result of his research.

The first part of the paper has summarized the difficulties encountered in trying to apply systems analysis and the status of systems analysts in the public sector and universities. It is now useful to explore possible future directions.

FUTURE DEVELOPMENTS

The observation has been made previously that, in general, the decision-maker is sceptical of systems analysis, and that this is partly due to the unsuccessful work completed previously, and partly due to the fact that they are not familiar with this approach. Starting from this observation, we can distinguish between three main areas where efforts have to be concentrated, and these are (1) education, (2) selection and execution of systems analysis projects, and (3) implementation.

Education

Educating both the system analyst and the decision-maker is essential for the acceptance of systems analysis as a tool for problem-solving. The diffusion of systems analysis concepts and formal course work into more branches than is currently the case is needed. When preparing the student to become a systems analyst, one should not only emphasize quantitative techniques, but also stress the qualitative aspects of systems analysis. A good preparation in social sciences is of considerable help in forming the systems analyst's "Weltanschauung".

An effort in formal teaching at university level is not enough. Attention should also be given to education beyond the formal university level education, such as short courses, workshop, etc. One can distinguish here between the content of the programme for the systems analyst who would go through such a programme to update his knowledge, and that of the layman who should be exposed to such an educational programme with the purpose of becoming familiar with the means and capabilities of systems analysis.

We can go one more step and suggest that in order to incorporate systems thinking as part of the culture, it should be introduced at earlier stages of education, ie. during secondary education.

Selection and Execution of Systems Analysis Projects

It is true that nothing can convince decision-makers more as to the effectiveness of systems analysis than successful systems analysis studies. Successful systems analysis studies, on the other hand, start with correct selection and proper execution of these studies. The observations made earlier about the lack of qualified manpower and the lack of support services, sceptical decision-makers ready to blame systems analysis in general when faced by a failure, and other observations mentioned, all point to the adoption of the following strategy: never oversell, and attempt to solve only problems of a relatively small scale where the chances of successful completion of the project and its implementation are high. This strategy conforms also with the experiences of the systems analysts in developed countries.

Implementation

Successful project implementation is the ultimate test of a project, and therefore, it is important that systems analysis studies are successfully implemented. It is true, though, that in some cases one has a hard time defining a measure for success, and in some cases it might take a relatively long time for the results of implementation to appear, and by that time, due to changes in the environment, it might be hard to identify which of the results really originated from a particular study.

Implementation chances are increased if attention is paid to the following two points.

The systems analyst should not spare any effort in attempting to understand the real problem. It might turn out that the real problem is beyond the control of the decision-maker, and hence the systems analyst might not be of any direct help to him. In such a case, one might locate the actual "owner" of the problem, or stop the study preventing further waste of effort.

The systems analyst should be in close contact with those who execute power. Wherever possible, there should be a direct link between the systems analyst and those who execute power. This does not need to imply a compromise on the "right" as he has established the "right" by independent and objective research, to the best of his knowledge and abilities. It is the recognition of the fact that political realities are the limiting factors to the implementation of the results obtained. Furthermore, the flow of information from the decision-maker representing the organization to the systems analyst can be useful to the systems analyst in shaping his studies. Ideally, the decision-maker should also take part in the evolution phase of the project, which might make him identify himself with the results obtained, thus increasing the chance for implementation.

Need for Information Flow

Gathering and dissemination of information generated in the most effective way is essential for the smooth operation of systems. If we consider the systems analyst and the decision-maker (representing the organization) as sub-systems forming a system, we can talk about information flow within the sub-systems and between them. In the paragraph above, we have mentioned the information flow between the two sub-systems. Let us also look briefly at the flow of information within the sub-systems.

Flow of information within an organization implies the installation of a data-gathering and updating mechanism as an integral part of the organization. The decision-makers should be convinced, if not yet, that reliable and current data constitute an essential tool in decision-making, and thus the cost involved in carrying an information system is well-justified.

Dealing with everybody else's information systems, the systems analyst should not neglect to form the network of communications covering systems analysts and related institutions in Turkey and abroad. The operation of such a network is a major task in itself, and appears to be only realizable by extensive government funding. The lack of qualified manpower in the systems analysis area in Turkey was mentioned above. Such a network would help to overcome this difficulty, even if only partially.

Problem Areas with Urgency

There appear to be certain areas of application which deserve more attention than others, and these are mentioned briefly below.

Institutional studies are needed which are directed at organizational changes. A clear picture in terms of jurisdictional and functional aspects of these institutions is required to reduce inefficiency and ineffectiveness.

Problems of equity which constitute a rather broad spectrum of problems, may be classified into two groups: equity in income distribution, and equity in service distribution. Most urban problems can be considered as equity problems in service distribution.

State enterprises form an important component of the Turkish economy, and a set of measures of effectiveness are badly needed to evaluate these enterprises. Realistic long-range plans of these enterprises constitute another challenge for systems analysis. Long-range plans should be emphasized in general since they are the means for restating and reevaluating the purpose of a particular organization.

A WORD OF HOPE

I have indicated at the beginning of the paper that systems analysis has yet to prove itself in Turkey, and later in the paper I have mentioned that some critics are ready to present the failure of a particular study as the failure of systems analysis in general. Systems analysis is a valuable scientific approach for solving problems of mankind, and there should be no reason to doubt that it will be successful eventually, if applied as a scientific approach.

One recalls the poem "Ulm 1592" by Bertolt Brecht, in which he tells about a tailor who boasts to the bishop that he has prepared a set of wings and can fly, and is then killed during his first attempt. Brecht finished his poem with the following statement of the bishop to the people surrounding:

> "The bells should ring,
> It was nothing but all lies,
> Man is no bird,
> Man will never fly."

Let us hope that we will not be as clumsy and naive in our approach to problems as the tailor, and that it won't take that long for systems analysis to take-off.

SYSTEMS ANALYSIS IN URBAN POLICY-MAKING AND PLANNING: THE MUNICH

EXPERIENCE

Klaus Schussmann

Referat fuer Stadtplanung und Planordnung
Stadt Muenchen, Muelterstrasse 34
Munich 2, D-8000
West Germany

INTRODUCTION

More than many or even most other cities in Germany, Munich has developed and adopted methods of systems analysis (in the broad sense of the term) in urban planning (Wegener, this volume). The former "Stadtentwicklungsreferat", emerging (1970) from a planning body which was initiated for the planning of investments of the City in general, and of the public aspects of the 1972 Olympic Games in particular, undertook first steps in the direction of systems analysis in the early years of its existence. An urban information and planning system (KOMPAS, see below) was developed. Although there was some work and reflection on large scale urban models of Forrester's "Urban Dynamics" type, and some effort to develop a municipal housing model, it took until about 1975 to develop and introduce computerized models as well as other fairly formalized methods of systems analysis for different parts of the urban system, covering most urban sub-systems, at least to some degree.

The Munich efforts and experiences are described in this paper, beginning with a purely descriptive and rather abstract model of the urban system as a point of reference. This reference model has the function of demonstrating the logical inter-relationships between the models of the various urban sub-systems, and of illustrating the level of aspiration of integrated urban planning; integrated urban planning has been, although considered as difficult to undertake, a fundamental claim of urban planning in Munich since the early seventies.

Following the discussion of the central models used in Munich (omitting transportation models for reasons of spatial and temporal

177

limitations of this contribution), I will turn to the main questions raised by the conference, such as the decision-making environment, the philosophic basis and use of systems analysis in planning
and the question of future directions in this field.

A MODEL OF REFERENCE FOR URBAN SYSTEMS ANALYSIS

Figure 1 gives an idea of how the urban system can be understood on a rather high level of aggregation, showing the main sectors of political interest; the places of employment and of population, the labour market, land use, infrastructure (in the very broad
sense of the term, including the housing sector and private facilities open to the public) as well as the urban financial sector.
The main linkages between these sectors are shown including the
influences being exerted from the goal system (comprising the political and administrative system and the decisions of both sub-
systems). For each sector of political interest a distinction is
made between the demand and supply side with the possibility of a
surplus or a deficit which is considered as being reported back to
the goal system (political and administrative sector) and calling
for adjustments in an iterative way.

One may consider the demands of the population and of the
places of employment, both either considered as given in their
development or influenced by political action, as a starting point
for the demands of employment (labour force), land (spatially disaggregated and distinguished with respect to different types of land
use) and infrastructure, as defined previously. Likewise, the
places of employment and population are considered as the origin of
the financial consequences for the municipal budget. All of these
demands and consequences are influenced in some way by the goal
system (political and administrative sector), which in turn receives
feedback from the different sectors of political interest within the
urban context; ie. whenever conflicts become apparent between the
actual goals development and goals at a particular point in time.

This rather descriptive model serves as a framework for the
actual models applied in Munich. Firstly, the lines starting from
the population side are described, and then those coming from the
lines of the working place-branch are described. Finally, questions
of interaction between the various models and their applications are
discussed.

SYSTEMS ANALYSIS IN MUNICH URBAN PLANNING

KOMPAS - Urban Information and Planning System

KOMPAS provides demographic, residential, place of employment
data, and the details about infrastructure, building structures,
etc., with all of these data being spatially, and in other respects,

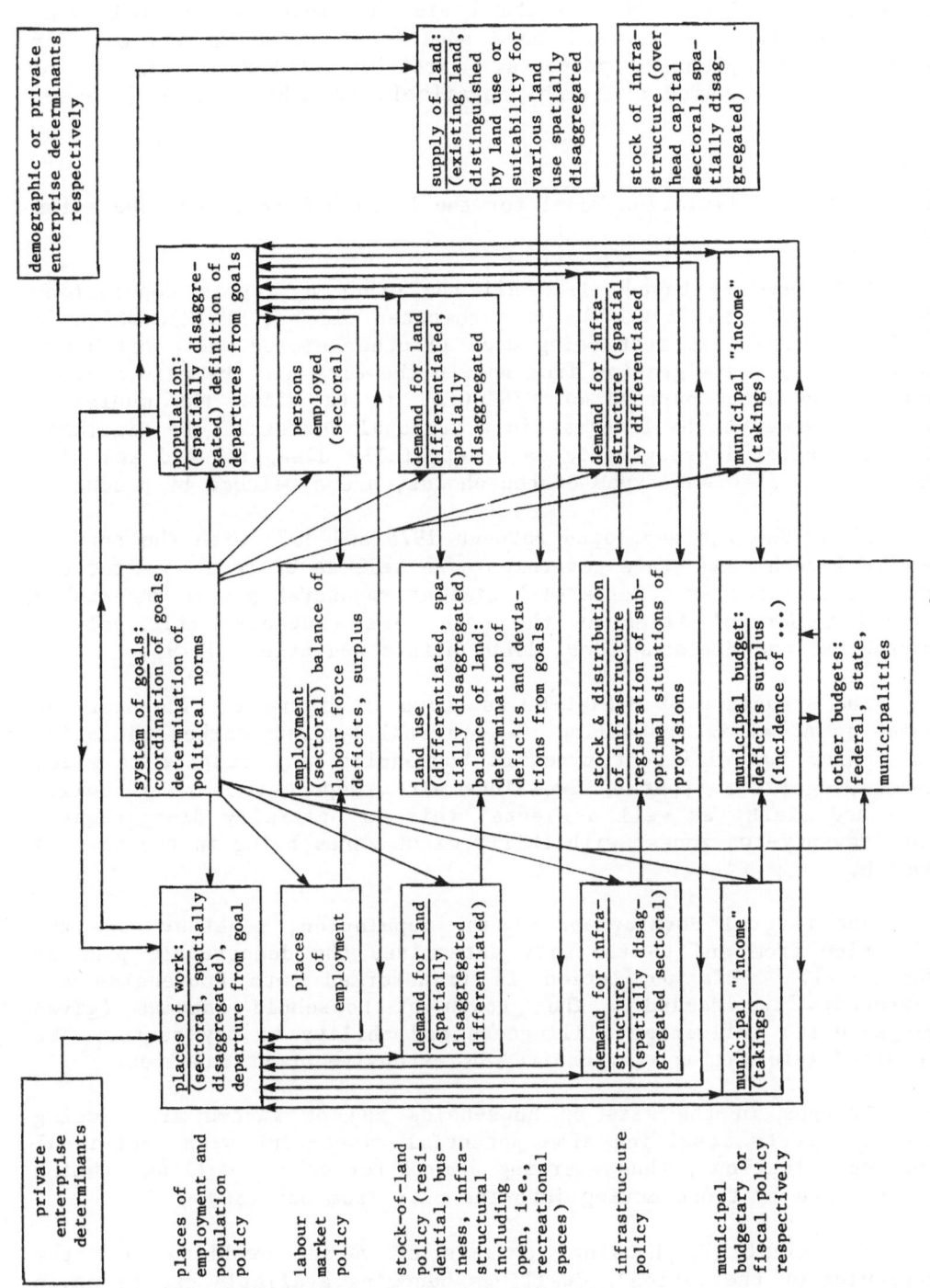

Figure 1 - An Outline of a Simulation-Model of Integrated Urban Planning

highly disaggregated. All of these data can easily be combined, and
therefore, KOMPAS serves as the basis for most of the following
models and has, of course, some additional value on its own for
analyzing and planning purposes. Probably KOMPAS is similar to
planning tools which are widely available in other cities (Franke,
1978).

MINIWOPRO - A Simulation Model for the Housing Market and Population
Development

MINIWOPRO is based, like all the other aggregate population-
related models in Figure 1, on a computer model (BEVSIM), which is
probably similar to that being used in other cities where birth and
death rates, and migration from and to the city are the main components. The main purpose of MINIWOPRO is to allow the municipal
housing market to be forecast in a regional context. The simultaneously predicted demographic data, spatially disaggregated and with
respect to different types of households, are a welcome by-product.

MINIWOPRO was developed between 1975 and 1977 with the support
of the federal and state governments as well as by local contractors
and banks. Similar efforts with similar sponsorship were undertaken
in other German cities at the same time (Wegener, this volume;
Bartholmai and Bretschneider, 1980; Meintz and Stahl, 1979).

The structure of MINIWOPRO is shown in Figure 2. The starting
point is a matrix resembling the original housing market situation
containing two different types of households with respect to size,
income, age; two types of dwellings with respect to size of buildings and flats, as well as rents: this is spatially disaggregated
into seventy-two zones, with thirty-eight zones being in the City of
Munich.

The natural development of the population, together with the
migration from and to the city determines the demographic part of
the model. This population is transformed into households and
represents the demand. The growth of household incomes (given
exogenously) influences willingness and ability to pay rents. The
required information was acquired by extensive questionnaires.

Changes in the size of households and of income are causing
demand effects resulting from potential discontent with individual
housing situations, thus exerting demand for other dwellings (which
is enlarged by those moving into the city from outside).

The supply of housing consists of vacant dwellings (at the
beginning of the period), dwellings becoming available due to moves
within the city and finally, newly-built housing units. This
housing supply is reduced by destruction of (obsolete) housing,
conversion of dwellings into business-use (offices), etc. Housing
construction is provided by external supply forecasts with two

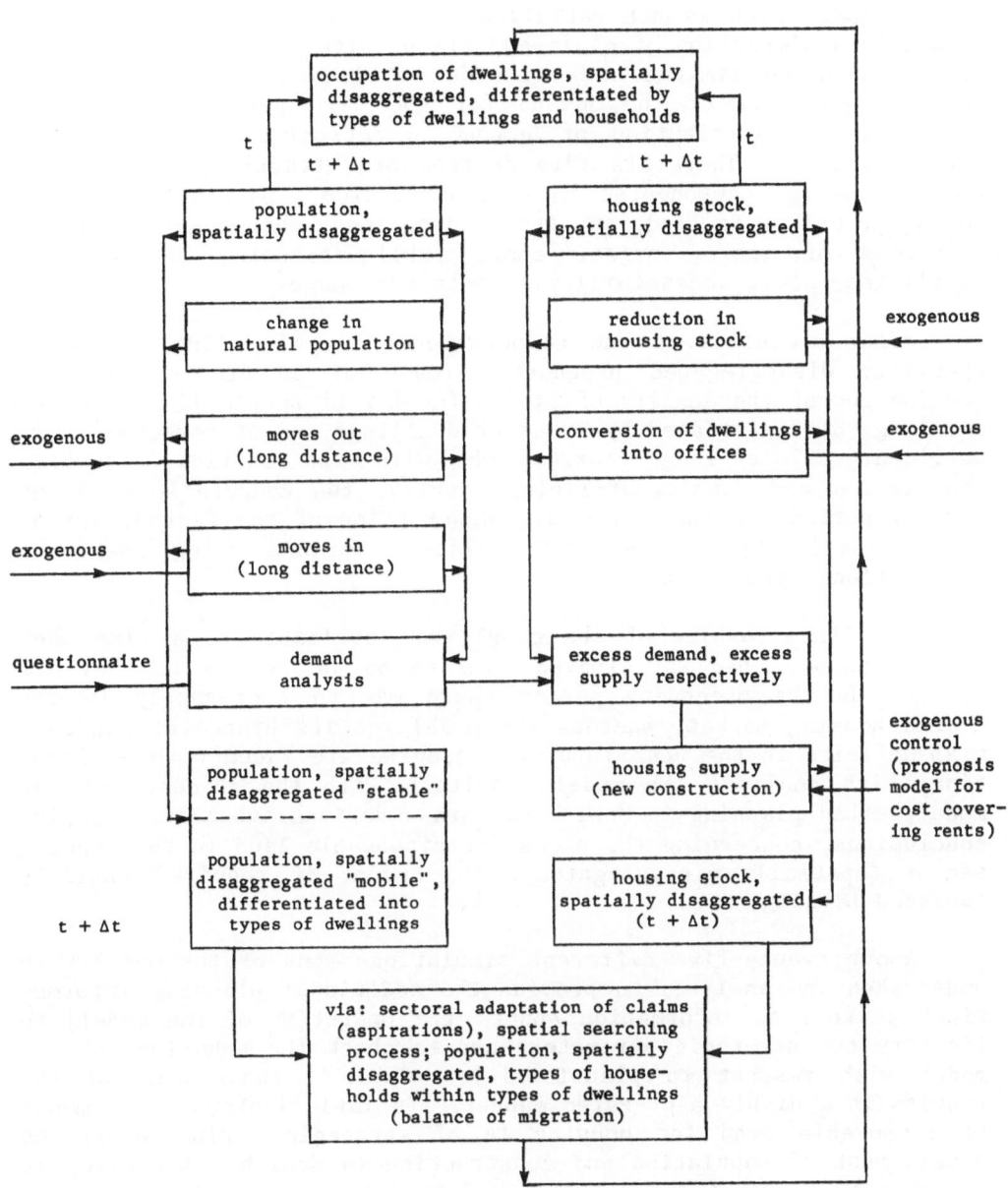

Figure 2 - The Structure of MINIWOPRO

built-in filters: one is a special forecast of cost-covering rents,
and the other one consists of existing legal restrictions with
respect to land use, etc.

Demand, which is not satisfied by its "ideal offer" is transformed by a simulation of claim-adjustment with the number of iteration runs being limited.   Households not having found a suitable
dwelling are part of the demand for the next period.   Spatially
disaggregated distribution of demand is determined by indices-of-
attractiveness.   There are five degrees of attractiveness based on
the following components: new construction, possibilities for
building, prices of land, quality of dwellings, shopping-facilities,
emissions of smoke, noise, etc., public schools, recreational
facilities, etc., accessibility aspects and image.

After execution of the computational steps outlined above, a
spatially disaggregated prognostic image of an area's stock of
housing and of the density of living (number of people living in one
dwelling, differentiated by types of dwellings and of households) is
outlined.   This image corresponds with the initial situation
(population and density-of-living matrix), the computational image
for any period (in the sense of a description of the facts), serves
as the basis for subsequent simulation periods using identical
computational processes.

The first results of the model were available at a time when
almost "nobody" had anticipated them to be as obvious as they are
today.   In the preceding period there was an over-supply in the
Munich housing market, whereas the model results hinted at substantial deficits in the near future.   Today we are faced with a situation which confirms the model results.   With the guidance of the
model, urban planning in Munich has drawn various substantial policy
conclusions, concerning the needs for disposable land in the housing
sector (spatially disaggregated), the amount of required publicly
assisted housing, etc.

About twenty-five different simulations runs of the model were
undertaken and analyzed to provide the additional planning information, as well as information about the operation of the model, to
identify the strategic parameters and to check the behaviour of the
model with respect to plausibility.   Table 1 shows some of the
results in a highly aggregated manner.   As Table 1 shows, the amount
of disposable land for housing is of strategic influence on the
development of population and construction in Munich.   Likewise, it
is demonstrated that the provision of land for housing in the region
surrounding Munich has astrong influence on what happens in the
city.   It is also shown that a competitive situation with respect to
the population split between the city and the suburbs, a topic of
regional policy at that time, is rather disadvantageous for the
central city.

TABLE 1

SIMULATION RUNS ON THE DEVELOPMENT OF THE HOUSING MARKET,
VARYING ASSUMPTIONS ABOUT THE DEVELOPMENT OF DISPOSABLE
LAND FOR HOUSING IN THE PLANNING REGION OF MUNICH.

Simula-tion Runs	Population (Inhabitants) in 1990			Housing Construction (Housing Units) 1978-90		
	Munich (1)	Suburban (2)	Planning Region (1)+(2)	Munich (1)	Suburban (2)	Planning Region (1)+(2)
variant 1 a)	1,286,604	971,174	2,257,778	130,221	78,333	208,554
1st run b)	1,268,803	989,128	2,257,932	130,087	90,055	220,142
2nd run c)	1,274,515	983,380	2,257,895	135,755	90,022	225,777
3rd run d)	1,224,544	1,033,605	2,258,148	136,460	141,715	378,183

Explanation:

a) Variant 1 is the reference standard introduced by PROGNOS AG, Consultants, Basel 1976.

b) Assumption: as opposed to Variant 1, additional disposable land for housing in the suburban region for approximately 12,000 additional housing units.

c) Assumption: as opposed to Variant 1, additional disposable land for housing in both Munich and the suburban region for approximately 12,000 additional housing units each.

d) Assumption: as opposed to Variant 1, triplication of the development of disposable land for housing in the suburban region.

_____

Other simulation runs were not as helpful in illuminating the particular problem in question, but finally for obvious reasons (for instance, the impact of intensified modernization of the housing stock was not clarified with the above model) almost no influence of different modernization policies could be analyzed by the model. The reason for this, we think, is that the model does not contain housing characteristics in sufficient detail for that question, as the model simply moves modernized flats to the categories of higher rents without indicating much about the improvements. But, on the other hand, if there were an effort to include all the relevant characteristics of the housing units in the model, this would inevitably lead to the model's "explosion", not to speak of the data requirements. So this is just one further illustration of the fact that even large-scale models cannot solve all questions in the field for which they were designed.

Some other simulation runs have shown the results to be rather insensitive to other policy variables, for instance, those which alter the attractiveness of parts of the city. Within the range of actions, reasonably considered as realistic, not much influence on migration processes was detected, which does not necessarily mean that an increase in the population's welfare is impossible by that kind of policy. The result, however, that migration is not influenced substantially by these policy variables, does not seem implausible.

With the help of the model, it was last, but not least, possible to derive quantitative goals for the development of the housing sector, although for this task, guidance by some other models (see MIPROS and infrastructure models below) was considered to be helpful. MINIWOPRO is viewed to be a useful planning tool, not only within the urban planning department, but also seems to be accepted so far among other parts of the urban administration and among the members of the city council.

MIPROS - A Highly Disaggregated Population Simulation Model

Apart from the information obtainable from MINIWOPRO concerning the development of the population, there is a much more detailed need for information, estimates or even guesses about the structure and local distribution of the city population. Particularly for planning the size and location of physical infrastructure, eg. schools, kindergartens, service facilities for old people, facilities for guest workers and so on, very detailed data requirements are expressed by many parts of the municipal administrative and political body. Corresponding to this, a large-scale computer model was developed for five types of households by household size, 429 spatial units of the city (average about 3,000 inhabitants) further disaggregated with respect to age (0-99 years), sex and German population versus other nationalities. Of course, such an undertaking may seem rather hybrid and therefore, the results may be considered as highly uncertain. Nevertheless, as the planning needs in this field are vital, the community itself has put substantial efforts into the design of MIPROS (Kubatzky, 1980). The fundamental philosophy of the model is, recognizing its inherent uncertainties, to keep it open for dialogue with experts and politicians as well, providing a tool for deriving the implications of fundamental assumptions in a highly disaggregated manner in a very short period of time.

For MIPROS, the city population as a whole can be considered as given. As statistical indications in the past have shown, an average of 90 percent of the change of population in the above-mentioned 429 spatial units is caused by migration, with only 10 percent by births and deaths. That is why the understanding of migration is central to this model. As is known from questionnaires, the motives for migration to and from the Munich region are dominated by labour market aspects and within the Munich region by housing market conditions. Consequently, detailed data on the housing market play a major role in the MIPROS model. Among others, a distinction is made between the existing housing stock and new buildings, because statistics show that about 90-95 percent of migration takes place within the existing housing stock, and only 5-10 percent is due to new construction.

As a whole, the forecast data are generated by considering the population/housing distribution in some original period, deriving from the natural development on an annual basis and mapping migration with respect to the existing housing stock and new construction, as, for instance, is obtainable from MINIWOPRO. As this model is more aggregative, some further methods for the disaggregation of new construction and the change of the existing housing stock are required.

MIPROS, meanwhile, has found various applications, mainly in the field of school planning, and although the results from the model have invariably been combined with hints about uncertainties, they were nonetheless received favourably by the authorities in question. The consequences, in fact, are considerable because the amount of schools still to be built, and their location, is heavily affected by the model results, so are the plans for closing down schools in areas of anticipated substantial out-migration, and last, but not least, the planning of the municipal budget.

Models for Infrastructure Planning

For infrastructure planning, among other tools, two computerized models are in use in Munich. One is called the accessibility-model and the second, the allocation-model. The accessibility-model is able to accumulate the people expected to use a particular type of infrastructure (eg. school or library) within a certain range of accessibility (eg. up to fifteen minutes travel time by public transport). With this planning tool, it is possible to map the provision to the public of public services with agreed-upon ranges of accessibility, and therefore, also mapping those who are not adequately serviced by the stock of infrastructure at a particular point in time. Likewise, with the help of this model, it may be demonstrated when there is an over- or under-supply of particular services, given a pre-specified range of accessibility. Obviously, by iterative steps of the model application, optimization of infrastructure can be achieved.

The allocation-model identifies the location of new infrastructure for planning purposes. In any planning case, a location is to be found which minimizes the average travel time of users, whereby an upper limit of tolerable travel time is defined according to the philosophy of the accessibility-model. One important aspect of optimization consists in taking into account the capacity of the existing unit of infrastructure. Optimization is approached iteratively, changing the location of the unit of infrastructure to be planned. With the aid of the allocation-model the number of users being mapped by the accessibility-model with overlapping zones (zones, representing maximum travel time considered as given) should be attributed to either one of the infrastructure units in question, in order to minimize average travel time.

Both models together are a main part of optimization in the Munich infrastructure planning system (MIPLAS) as a whole (Schussmann, 1980), whereas the level of infrastructure supply to be obtained is derived by population data on the one hand (see MIPROS above), and the defined aspiration-levels in the form of quantities and qualities to be confirmed in the political process on the other.

Systems Analysis of Employment Location and the Labour Market

Considering the urban system, places of employment are of interest, at least in respect to their demands and needs (infrastructure, land, etc.) in the municipal realm, as well as under aspects of supply (jobs, facilities, municipal income). Compared with the population-side of Figure 1, the employment identified there actually has much less relevant data available, and, were it only for that reason, there is less formal modelling of the systems analysis type. Perhaps interest in these questions was also lacking until recently due to the fact that most cities were growth centres for places of employment, and that more abundant land (less density in land use) in cities seemed to allow for some flexible possibilities of "muddling-through".

Meanwhile, most cities are losing employment, and are facing labour market problems (at least more so than in the past). Additionally, the competition and conflicts between different kinds of land use have been aggravated, requiring better tools for planning to cope with these problems. For these and other reasons Munich has relied heavily on modelling the employment location in the city. As a result, forecast figures of places of employment in a regional context (sectorally disaggregated) became influential in urban planning about 1975 (Prognos AG, 1975) after some preceding efforts in the early 1970's. The forecast data were derived from a model (essentially a computer-type model, though availability of data finally did not allow it to be run in that way), designed for subregions of Germany with simultaneous solution for all of them (Schroeder, 1968). The underlying model was in use until recently in the context of federal regional planning, although the necessary adjustments were undertaken using an insufficient data base (in terms of the model) and in spite of widespread methodological critique (Flore, 1977). Nevertheless, the model delivered fairly good results on a whole. Very recently, new efforts have been made, sponsored by federal funds, to develop a better and more realistic model for the sub-regions of Germany in the regional planning context, which promises to be of some use for cities like Munich.

At present, traditional methods of shift-analysis are used in Munich to estimate data on the places of employment. These data serve for labour-market considerations (together with data concerning the labour force, which are derived from the population models discussed above), as well as a starting point for estimating the demands for (or needs of) land, of potentials of certain facilities and places of employment, which may support the development of sub-centres in the city, etc. Also, spatially disaggregated forecast data are derived from the aggregated data as follows. Firstly, a rather mechanical procedure is applied (eg. shift-analysis methods) to sub-regions of the city; secondly, all branches of urban planning are confronted with these data and asked to comment on the desirability of such an outcome, followed by several rounds of discussion among experts, until a compromise is found, under certain analytically derived constraints (eg. a certain trend-spectrum of outward migration cannot be reversed).

The underlying philosophy of this procedure is that recognizing the analytical givens and the demands of consistency, desirable results are obtained which may serve for orientation in planning rather than formal forecasts (which are uncertain) or their formal derivatives, including some optimization claims. Obviously, employment location in Munich's urban planning (Figure 1) is much less formalized and computer-assisted than the population side. Nevertheless, we think that this line of analysis is also pursued in the spirit of systems analysis, and probably is not necessarily less useful than more formal and morecomputerized approaches. Anyhow, we see little sense in formalizing and computerizing this part of the urban system any further, as long as the data situation and the available knowledge about the most relevant and interesting interrelationships cannot be improved substantially.

On the Interaction of Models in Munich's Urban Planning

Although there is no formal mechanism provided for the integration of the models discussed above, there would not be much sense in trying this with the present state of the art anyway. Of course, the results of different models have to be used and reconciled against each other in the context of certain planning tasks. A few examples are provided in the following paragraphs to illustrate this point.

Firstly, there is a rather imminent need to reconcile the results of MINIWOPRO and MIPROS, when population numbers are com-

pared at different aggregation levels on a uniform basis of aggregation (38 spatial units for MINIWOPRO, comprising the 429 MIPROS spatial units), as a planning body should obviously not use inconsistent figures. Since the figures produced by the two models are somewhat different at the uniform level of aggregation, we do not undertake so many simulation runs of both models, until the figures are
exactly the same. We just do not believe, this being possible under
the constraint, that then there will still be plausible sets of
parameters in either of the models. A second possibility would be
to use MINIWOPRO for questions of population alone down to 38 zones
and then to apply MIPROS for further disaggregation; but this would
require special information contained in MIPROS at higher aggregation levels. So we decided to undertake some further plausibility
checks and to average the results in those cases (due to a common
criterion for decisions under uncertainty) where no further evidence
is available concerning the plausibility of either of the figures.
So finally with some correction factors to provide for overall
consistency, the above-mentioned problem may be solved.

We think that similar problems very probably arise whenever the
results of overlapping models of (urban) sub-systems are brought
together. Most apparently, interaction of the above models seems to
be required whenever a comprehensive plan of the city is worked out,
or when it is adjusted to new developments, as will be done in
Munich in the near future. So the different parts of the plan have
to be worked out by different specialists in a process very much
as trial and error, which I think is well-known wherever urban planning takes place. Finally, some consistent and realistic plan
should be the outcome.

At other levels, planning tasks could be set up like those in
Munich at the present, eg. developing land use plans (drawing
heavily on the above models and also being guided by cost-benefit
techniques) for the city as a whole, or by more detailed analysis
for particular zones.

CONCLUSIONS

Although Munich's urban planning in general has taken a
positive view of systems analysis techniques, not much progress in
this field is expected in the near future. This does not exclude
some further initiatives such as developing a model for planning
technical infrastructure with respect to cost minimizing goals, the
results being in principle of major importance both for planning
technical infrastructure, and for almost all other branches of urban
planning.

Firstly, the data situation will probably not improve much;
there is a serious shortage of census data, and recent regulations
to prevent misuse of data have become major obstacles for much
analytic work to be done; questionnaires on municipal accounts are

not only becoming increasingly costly, their value is also doubtful in some respects, with an important reason being that the public is becoming increasingly resistant to additional surveys. Of course, there are other sources of information as far as experts are concerned, but this is hardly sufficient for the operation of large-scale models. Secondly, the further development of existing models is a burdensome task, eg. whenever we might wish to develop MINI-WOPRO further with respect to some points of criticism. This probably is common to any extension of almost all existing large-scale models, since it may cost almost as much as the initial development. Although the model was set up by ample sponsorship outside the community budget, does there now exist a similar possibility whenever necessary or desirable? Thirdly, though more and more theoretical insight into the nature of the inter-relationships among elements of the urban system seems to be achieved through progress in the social sciences, the empirically founded knowledge about those inter-relationships does not keep pace with the progress of theory.

The administrative and political authorities are not demanding additional results which are derived solely from large-scale or other highly sophisticated models, which apparently is also true of the public. On the contrary, there is some aversion against complicated reasoning of this kind. Yet when the reasoning is clear-cut and not too complicated, we have experienced no adverse reaction to the underlying models, however complicated they may have been. Therefore, while there are not too many disincentives to use models of the systems-analysis type as described above, there are not too many incentives from outside the planning body either.

Last, but not least, it is our profession itself which provides us with definite insights, prone to lead to nihilism of some kind: was it not welfare theory and political science which taught us that there is almost no (more or less) unambiguous way to define social improvements — except from the very restrictive and, as a whole, almost irrelevant cases — not to speak of (attainable) social welfare maximization? Turning back to the more positive side of economics, is there not reason to believe that progress in the discovery of well-established economic laws is very reluctantly, if ever, achieved? Just remember, to leave systems analysis for a moment, the Keynesian-Monetarist-Controversy and what we can really learn from it — for instance, on how to cope with stagflation!

Nevertheless in principle, no alternative to systems analysis seems to exist. Firstly, one may take some comfort from the belief that as a decision rule, the adoption of rational methods in general, whenever decisions are to be made, is widely acceptable. Moreover, probably many planners find themselves in a situation where they have a clear idea about the problem to be solved, in one way or another. Taking this for granted, systems analysis seems in principle an adequate tool to cope with the ever-increasingly perceived complexity of the world in general and urban systems in particular (Nowak, 1973). Also, the insight that systems (however defined) are always embedded in still larger systems (Goldstone, 1971) and that ignoring this may lead to severe problems (Mesarovic, 1971; Popp, 1977) is a call for urban planning to use systems analysis techniques. This seems to hold, even when it creates the paradox of adding to the complexity of the world, although systems analysis is intended to make complexity manageable (Batty, 1980); perhaps we have to face another example of the prisoner's dilemma.

Bearing this in mind, and in order to avoid the failures and mistakes of the past (Lee, 1973), it seems sensible to proceed in this field with caution. That is, to seek a balance with respect to available theory, empirical foundation and available data, all to be delivered at reasonable cost. It seems that a system of sub-models of the urban system will be the most adequate approach given the present state of the art. Formal connections between these sub-models seems neither necessary nor advisable at the present, although some effort should be made to gain relevant information from the inter-related sub-models.

Therefore, the preferred course of action is modular development of models which are part of the urban system as a whole, the setting of rather restricted goals for the sub-models (as to their complexity), concentration on more empirical foundations, and a dialogue with the interested public (widely defined) rather than an approach favouring further formalization and the probable development of empty concepts.

REFERENCES

Bartholmai, B. and Bretschneider, M., 1980, Querschnittsanalyse regionaler Wohnungsmarktanalysen, Deutsches Institut fuer Wirtschaftsforschung, Berlin.

Batty, M., 1980, A Perspective on Urban Systems Analysis, in D. Banister and P. Hall, (Eds.), Transport Policy and Planning, Mansell, London.

Flore, G., 1977, Zur Regionalisierung der Arbeitsplaetze, in der Faumordnung, (Ed.), Informationen zur Raumentwicklung, 1/2, 69 ff.

Franke, D., 1978, Das Kommunale Planungsinformations- und Analysesystem (KOMPAS) der Landeshauptstadt Muenchen, OEVD (Oeffentliche Verwaltung und Datenverarbeitung), 3, 11-14.

Goldstone, S., 1971, Dilemmas for Systems Analysis of Urban Programs, in: M.D. Mesarovic and A. Reisman, (Eds.), The Systems Approach and the City, North Holland, Amsterdam.

Kubatzky, U., 1980, Kleinraeumiges Bevoelkerungsprognosemodell MIPROS, Planungsreferat, Muenchen, West Germany.

Lee, D.B., 1973, Requiem for Large-Scale Models, Journal of the American Institute of Planners, 39, 163-178.

Meintz, Th. and Stahl, K., 1979, Modelle zur Simulation der Regionalen Wohungsmarktentwicklung, Vergleich und Bewertung, University of Dortmund, Arbeitspapier Nr. 7903, Dortmund, West Germany.

Mesarovic, M.D., (Ed.), 1971, The Systems Approach and the City, North Holland, Amsterdam.

Nowak, J., 1973, Simulation and Stadtentwicklungsplanung, Stuttgart, West Germany.

PROGNOS AG, 1975, Ueberarbeitung der Sektoralen Beschaeftigtenprognose in der Stadt und im Raum Muenchen bis 1985, Basel, Switzerland.

Popp, W., (Ed.), 1977, SIARSSY, Ein Modell zur Simulation on Staedtischen und Regionalen Systemen, Bern, Switzerland.

Schroeder, D., 1968, Strukturwandel, Standortwahl und Regionales Wachstum, Stuttgart, West Germany.

Schussmann, K., 1980, Relevanz von Stadtentwicklungsmodellen, in: M. Pfaff and W. Asam, (Eds.), Integrierte Infrastrukturplanung zur Verbesserung der Lebensbedingungen in Staedten und Gemeinden, Berlin, West Germany.

FRENCH LOCAL PLANNING PRACTICE

Robert Laurini

Institut National des Sciences Appliquees de Lyon
69621 Villeurbanne
Lyon
France

Two years ago with Philippe Doat (Doat and Laurini, 1979; Doat, 1978), INSA conducted a survey on the working methods of local planning offices for city-size strategic plan-making, in order to examine the following points:

1. To estimate the divergence between official recommendations and actual practices.

2. To assess how well the systems approach is understood and used.

In this paper, after having rapidly presented the legislative context of French town planning, we exhibit and emphasize the most promising practices by reconsidering conventional systems analysis in a context of conflict.

Currently, French planning is organized according to the "Loi d'Orientation Fonciere" approved on 31/12/67 by the parliament. This law distinguishes two sorts of plans and these are (i) medium- and long-term prospective plans (SDAU and PAR), and (ii) special short-term action plans (POS), (Laurini, 1980). Table 1 lists the main differences between SDAU and POS.

SDAU (SCHEMA DIRECTEUR D'AMENAGEMENT ET D'URBANISME)

This consists of a map and written statements providing a framework for key issues such as housing, employment, green spaces and so on. As a first approximation, it resembles British Structure Plans or US Land Use Plans. SDAU's are concerned with urban planning and PAR's deal with countryside planning.

TABLE 1

MAIN DIFFERENCES BETWEEN SDAU AND POS

	SDAU "Schema directeur d'amenagement et d'urbanisme"	POS "Plan d'occupation des sols"
equivalent	British Structure Plans US land use plans	British Local Plans
definition	Long-term indicative outline	Legislatively detailed prescription
scale range	1/20,000 1/50,000	1/2,000 1/10,000
scopes	- indicative land use assignment - transportation sketch - location of activities and services - indicative zoning of renewal or sprawl	- legislative prescription for built forms - protection of natural spaces - preparation for public services - accurate land use assignment - definition of every land use change
planners	- local state representatives - local authorities	- local authorities - local state representatives - approved local citizen's interest groups
geographical area	- multi-parish level - metropolitan level	- generally one parish
presentation	- maps - written statements (livre blanc)	- maps - written statements (15 articles)
approval	- by the prefect	- by the prefect
citizen involvement	- no	- variable according to local councils

POS (PLAN D'OCCUPATION DES SOLS)

Resembling British Local Plans, POS's focus more on planning details giving an extremely itemized framework for built forms. They are based on the concept of COS which is the prescriptive floorspace ratio not to be exceeded for every plot of land. For example, for a given area of 1000 m^2, if the COS is 1.5 it is only possible to construct a building having less than 1500 m^2 floorspace. In this paper, only the POS process will be examined and the broad flow of activities is illustrated in Figure 1.

About Joint Plan-making (Elaboration Conjointe)

In France, "elaboration conjointe" means both corporate and participative planning: that is to say that all French plans must be designed both by citizens and planning authorities. But, in fact, there is a contradiction between (i) the official language which advises more public participation in the planning process and (ii) the effective behaviour of some planning authorities who neglect to take into account the citizens' opinions about their own environment (Dahan et al, 1979).

THE STORY OF A POS

In France, a POS is mandatory for every city with more than 10,000 people, and for several smaller cities with particular conditions. Moreover, if the council of a smaller town desires to get a POS, it may be designated.

The first step of a POS process is the initial decree given by the departmental prefect and the constitution of a planning committee which consists of (i) the local prefect, (ii) several politicians of the city (town councillors, MP's), (iii) some civil servants of local government offices such as environment, agriculture, commerce, industry, transport and so on, (iv) some representatives of commerce and industry chambers and (v) some approved local associations.

All of these people provide directions to the local planning agency in order to design a plan. A type of detailed plan is designed, but if the prefect dislikes it the plan must be modified accordingly. Afterwards, the corporate planning takes place and all the local governmental offices give their opinions. Eventually, only the approved local citizen groups may be asked their advice.

Following the above activities, the first version is designed and submitted to the local council which may ask for some modifications. At this point of the process, every building licence must match the POS plan. Finally, there is a public inquiry which generally lasts two months. From the formal point of view, a copy of the

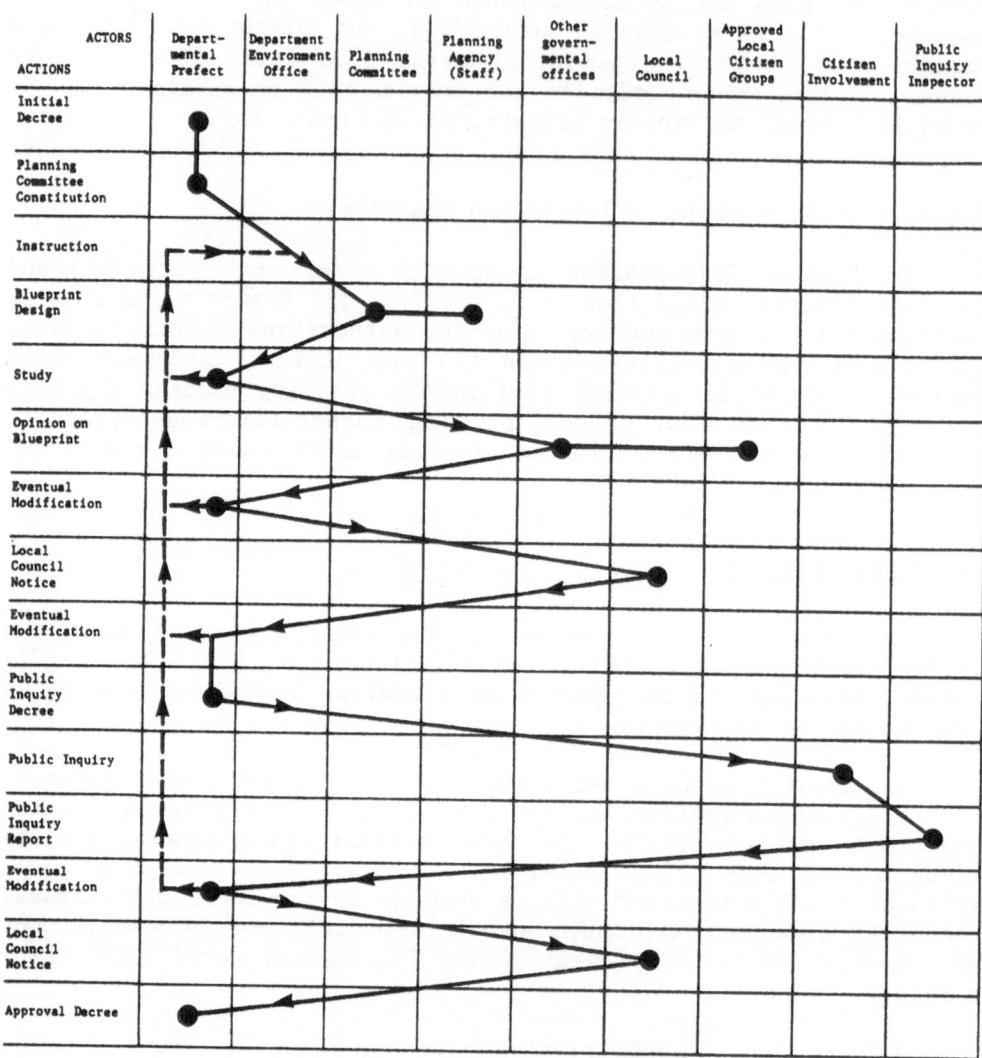

Figure 1 - The French Legislative Planning Process of
the "Plan d'Occupation des Sols"

POS dossier and an observation book is provided at the Town Hall in which all the people may write their opinions about it.   Then, an inspector, called "commissaire enqueteur" is asked to make a report on the plan.   Theoretically, afterwards, the POS is modified according to this report, but generally speaking, only minor modifications are accepted.   Finally, the local council approves and the prefect gives the approval decree for implementation.

It is very important to notice that the POS process is like every French legal procedure within the Napoleonic state in which people are considered as secondary to the laws.   This viewpoint may be highlighted by two legislative remarks: (i) at every step of the POS procedure, the prefect, as a representative of the state accepts any modification and moreover does approve the POS and (ii) legislatively speaking, the citizens are not involved, except as landowners seeing their property rights modified in the public inquiry step.

From a conceptual point of view, the POS design process is based on a succession of parallel arbitration procedures, as suggested by Figure 2:

So, the planner's attitude is to anticipate the other actors' roles in order to propose something interesting.   Bearing in mind this consideration and knowing the effectiveness of urban systems analysis, it seemed very useful to us to do a survey to improve our knowledge of the actual practices.

The results of this survey were very interesting at several levels, but the main difficulty was to ask planners to present their working practices.   This was very difficult for several reasons. The first one was that nobody was using exactly the official recommendation and so the survey was considered as inquisitorial for many planners especially for civil servant planning officers.   The second reason is due to the fact that several town planners considered themselves as artists: that is to say, people with no methods but only a strong feeling how best to evolve in this context.

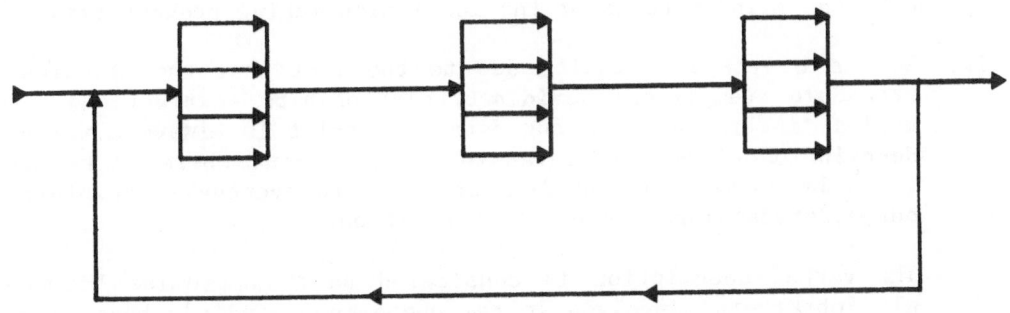

Figure 2 - Structure of the POS Design Process

But, some other planners did answer and give a diagrammatic model of their practices, and some of the most interesting and promising are presented:

1. An official recommendation given by an important Department of Environment civil servant, as illustrated in Figure 3.

2. The plan-making process of an Alsacian planner, as illustrated in Figure 4.

3. One based on stake evaluation, as illustrated in Figure 5.

4. An approach based on residents' associations, as illustrated in Figure 6.

The proposed official process (Figure 3) is essentially based on a balance between previous projects and new objectives, in order to give floorspace prescription and to locate public facilities accordingly. It is very interesting to see that the official recommendation does not point out conflicts between issues or people.

The second process (Figure 4), designed by an Alsacian planner, is more or less based on politicians' requests and pressure group arbitration in order to seek a balance between conflicting interests, but bearing in mind the necessity of a consistent plan. So, in this process, the planner's role is more considered as an arbitrator between people.

The third process (Figure 5) is very similar to the previous one, but emphasizes stakes and the manoeuvring of margins.

The fourth process (Figure 6) is dissimilar. Instead of evaluating all stakes and actors, this process is based on ward committees. That is to say that the planner asks them to make their ward plan, and the city-level plan represents the amalgamation of all these plans. In this process, the planner's role is an advocate to ward committees especially concerning the legislative consistency of their proposals.

In comparison with the conventional (or Anglo-Saxon) systems approach, the main features of the local plan-making process are:

1. No use of computer tools, due to the fact that the classical procedure (empirical evidence/relationships extraction) is useless in France since our departure point is always a verbal description of our understanding of phenomena where facts are taken by themselves and data are used to eventually reinforce our understanding, but not to base it on.

2. The verbal description is considered as a compromise between all "observers" involved in the phenomenon studied, based more on qualitative data rather than quantitative.

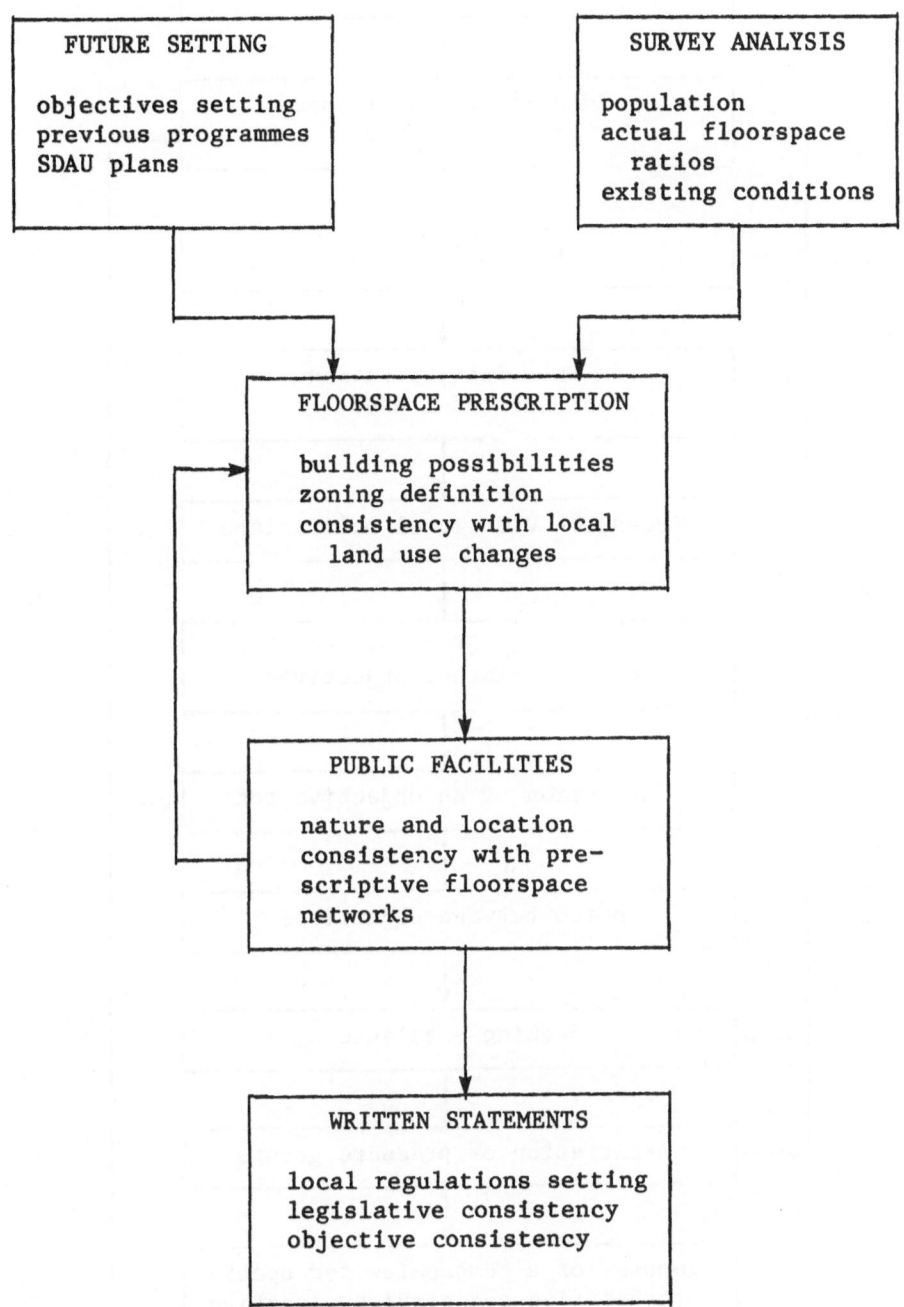

Figure 3 - Methodology for POS Design Proposed by the French
Department of Environment (Givaudan, 1973)

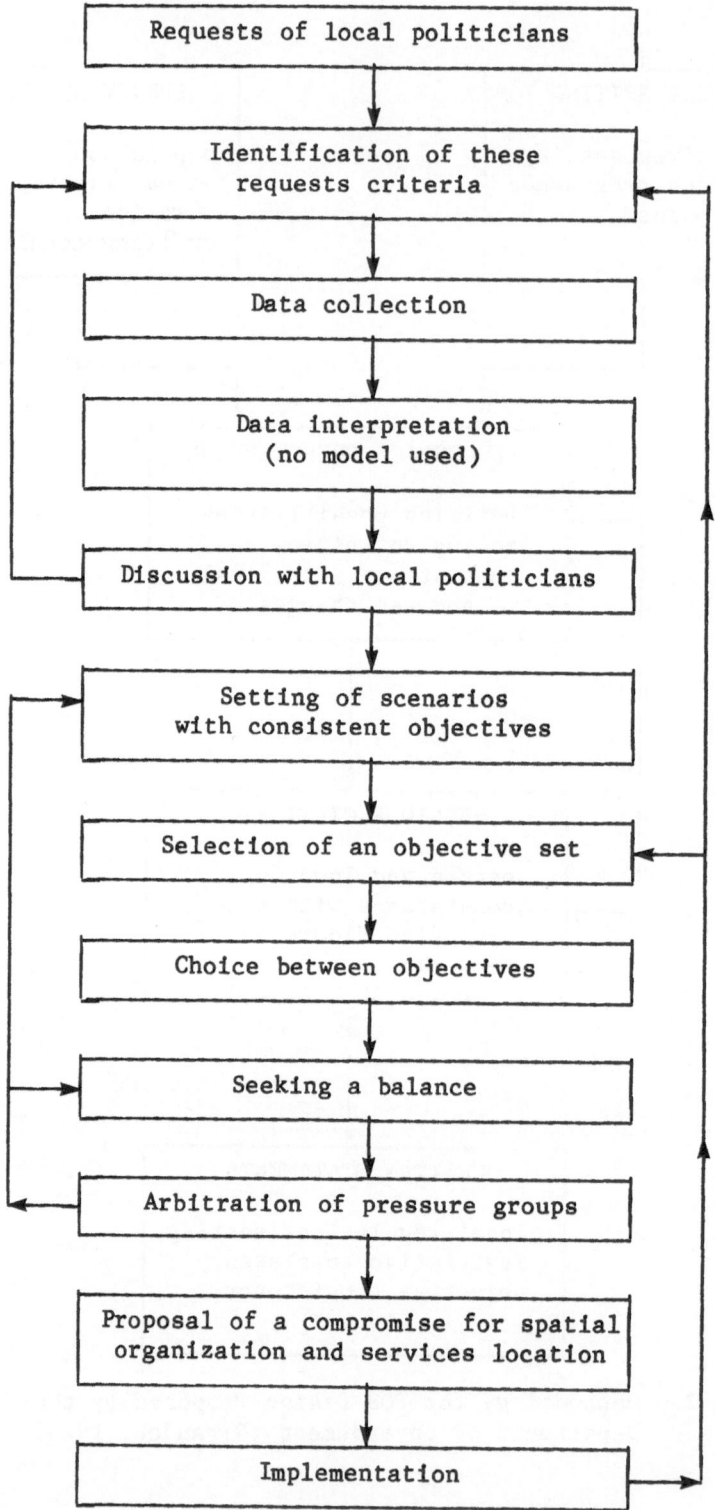

Figure 4 – Plan–making Process of an Alsacian Planner

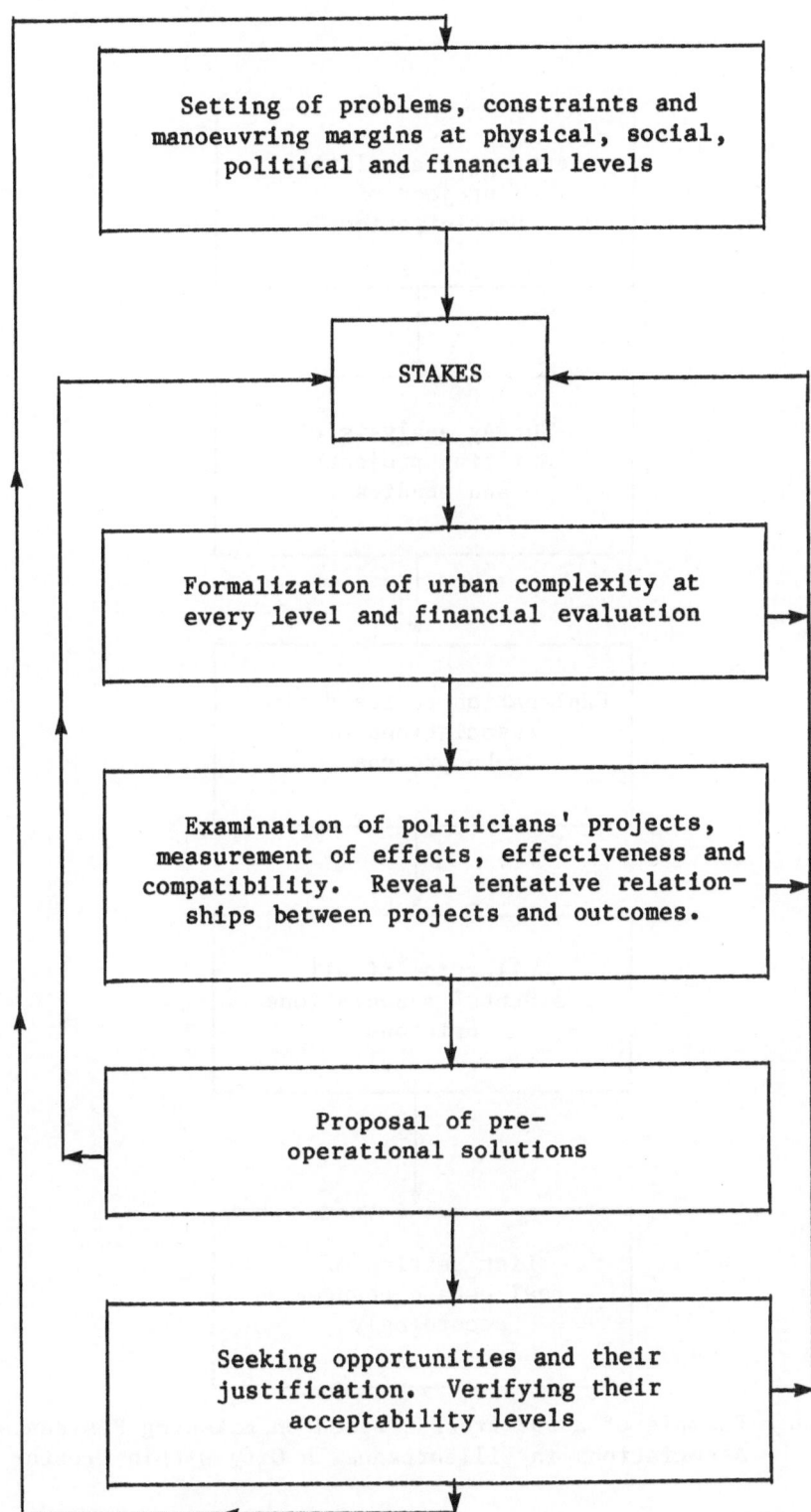

Figure 5 - Plan-making Process of another Local Planner

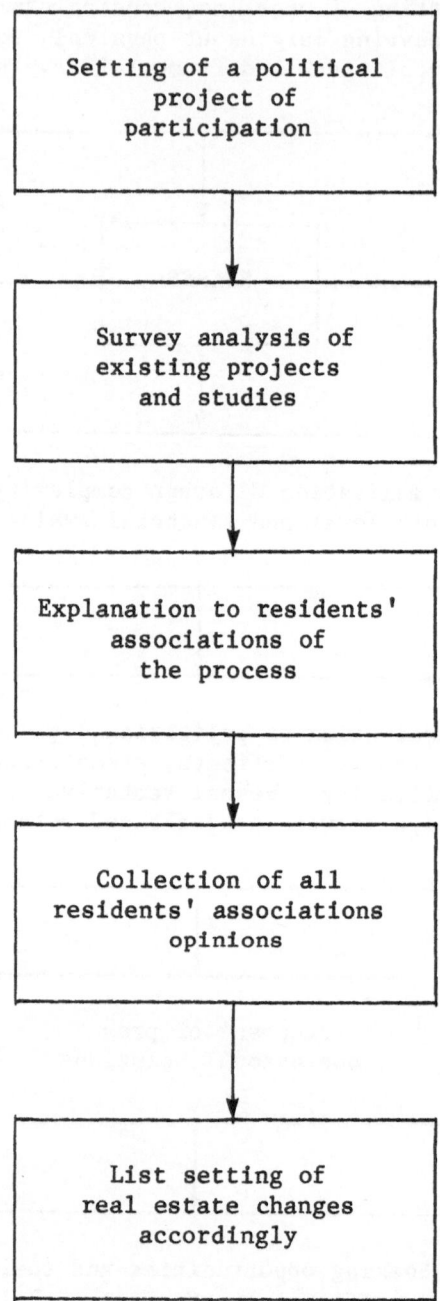

Figure 6 - Example of a POS Process Based on Existing Residents'
Associations in Villeurbanne, a City within Greater Lyon.

3.   As a consequence, the study is based more on stakes and actors (soft data) rather than issue measurements (hard data).

4.   French local planning process is more a conflict-resolving procedure, rather than a problem-solving one.

REFERENCES

Dahan, Y.M., Forget, J.P., Morel, J.P., and Vidal, J., 1979, PROCESSOR: Procedures et Institutions d'Amenagement et d'Urbanisme, Ministere de l'Environnement et du Cadre de Vie, Paris.

Doat, P., 1978, Processus d'Elaboration des Plans d'Urbanisme, Unpublished DEA dissertation INSA, Universite Lyon, Lyon, France.

Doat, P., and Laurini, R., 1979, An Outline of the French Planning System, PTRC Warwick Conference, Preliminary Session, Warwick, UK.

Givaudan, A., 1973, Elements pour une Methodologie des POS, Urbanisme, 138, 17-23.

Laurini, R., 1980, Contributions Systemiques et Informatiques au Multipilotage des Villes, These d'Etat, Universite Lyon, Lyon, France.

PART 3

APPLICATIONS OF SYSTEMS MODELS

EARLY WARNING SYSTEMS FOR URBAN POLICY-MAKING AND PLANNING

John Dickey

Division of Environmental and Urban Systems
Virginia Polytechnic Institute and State University
Blacksburg, VA, 24061
United States of America

BACKGROUND

Aside from their educational value, models are supposed to be
beneficial to the decision-making process. There are, as we well
know, many reasons why this cannot be so. Both technical and political decision-makers often are guided by factors other than those
embodied in the model. These might include, for instance, past
decisions made by these people or agreements made to trade votes in
exchange for each other's favourite projects. Under these circumstances there is relatively little contribution that a model can
make to influence the "proper" decision.

In more suitable situations we still have to answer the question of what is "proper". Models, which are by definition simplifications of reality, rarely can address the broad spectrum of actual-
ity in which decisions are made -- that is, the number of variables
and relationships is limited. Further, the variables employed may
not be accepted as the most appropriate measures of decision or
impact criteria, if in fact such can be "measured".

Another reason models may not be very useful in decision-making
is that they might not provide the best information at the time it
is needed. This has been a particular hazard recently. Urban
decision-makers (as well as those on other levels) have been faced
with a seemingly endless procession of events, most of which have
never been experienced before. Rapidly rising oil prices and
general inflation; calls for increased rights for women, children,
gays, the handicapped, and small businesses; and propositions to
reduce tax revenues are but a few examples. Many decision-makers
are overwhelmed with the complexity and uncertainty of change. A

common complaint is that they cannot forecast what will happen one
month from now, much less several years ahead. Most of the models
that could be employed under these conditions are severely (and
quickly) outdated.

Under these circumstances, it appears somewhat ludicrous, and
certainly trite, to suggest that models, to be of use, should be
developed and updated rapidly. The real question is "how"?

I will not pretend to know the complete answer to this, but it
appears to me to centre on the timing of the decision. Faced with
an uncertain situation, those charged with making choices have a
crucial decision — (1) act now, and risk over- or under-responding
to forecasted situations, if and when they actually occur, or (2)
act later, and risk being unable to avert a crisis (or build on a
large potential). A useful model should help to point out the
benefits/disbenefits of these two options.

For a model to do this, it must be up-to-date. Since this
usually is not possible, a "sacrifice" in accuracy would appear to
be in order. We must be willing, as the expression goes, "to be
roughly right, not precisely wrong". When little time exists for
field data collection and calibration, the best that can be done is
to insure that the variables in the model are at least related
logically and in about the proper magnitude with respect to each
other. In other words, we should be able to say, based on past
studies and the best experience available, if A increases 1 unit, B
will decrease about 2 units, which will in turn lead C to drop 1/2.

A willingness to accept "roughness" does not mean, however, we
should forsake the most recent information. This implies monitoring. This, in turn, means searching the literature and as much of
the latest personal experiences (and, yes, opinions) as possible.

In reverse order, then, the elements needed to provide decision-makers with useful models to help forecast the implication of
the act/do not act now decision are:

1.   A monitoring system;

2.   A "roughly right" logical model.

These two are not separate. We cannot monitor everything, so
the model should give clues as to the factors most important in the
search. Conversely, monitoring should turn up new factors which
should be included in the model. Together they provide an early
warning system that should be of assistance to decision-makers in
determining whether and how to act.

AN EXAMPLE

Early warning systems are common in defence (eg. the DEW —

distant early warning -- line), in anticipating natural disasters
(eg. hurricane tracking), in gauging national economic conditions,
and in guiding some private industries. To my knowledge none has
been developed for urban administration purposes. For this reason
the following example will be hypothetical, lacking the honing of
actual experience.

The Study Area

     The geographic area selected for illustration of the proposed
early warning system coincides with the boundaries of (what used to
be) the Miami-Dade County Model City Area. It also has been called
Liberty City or the Dr. Martin Luther King Boulevard area, in honour
of the late civil rights advocate. The section is a relatively low-
income, predominately black (part Jamaican) neighbourhood in north-
western Miami, occupying an area of approximately five square miles,
and bounded by major expressways running east-west and north-south.
During the summer of 1968 this area underwent social unrest and
riots (during the Republican National Convention in Miami). Indications of social disorganization in the study area include unusually
high population densities, a high number of female heads of households, high crime and delinquency rates, and almost total non-resi-
dent ownership. Interestingly, the area just was subject to another
riot due (at least according to the papers) to the declining economic standing of blacks versus the fast-growing Latin community.
Fourteen people were killed and about $200 million in damage done.

     In 1973-74 I participated in developing a systems dynamics-type
model to help determine and forecast those factors which could
influence the education, health, and income characteristics of the
residents (Steiss et al, 1975). Since it has been over six years
since this effort, we now have the opportunity here to review the
model and its results critically, particularly in terms of its value
for early warning.

The Original Model

     The three main components of the original model -- education,
health and income -- and the factors affecting them, are portrayed
in Figures 1 to 3 respectively. Educational success (Figure 1) is
represented (negatively) by the dropout rate. This is seen as
affected by factors such as the number of school facilities, teacher
years of education, and "student background" (with income used as a
surrogate).[1] This latter indicates the interconnectivity of the
components since "income" is estimated in the third sub-model. The
influence of each factor in the dropout rate is calculated in the
model via "table functions" of the kind somewhat unique to DYNAMO
(Forrester, 1968). Each influence in turn is scaled by the relative
importance and direction of each factor as indicated by the numbers
on the corresponding arrow in Figure 1.

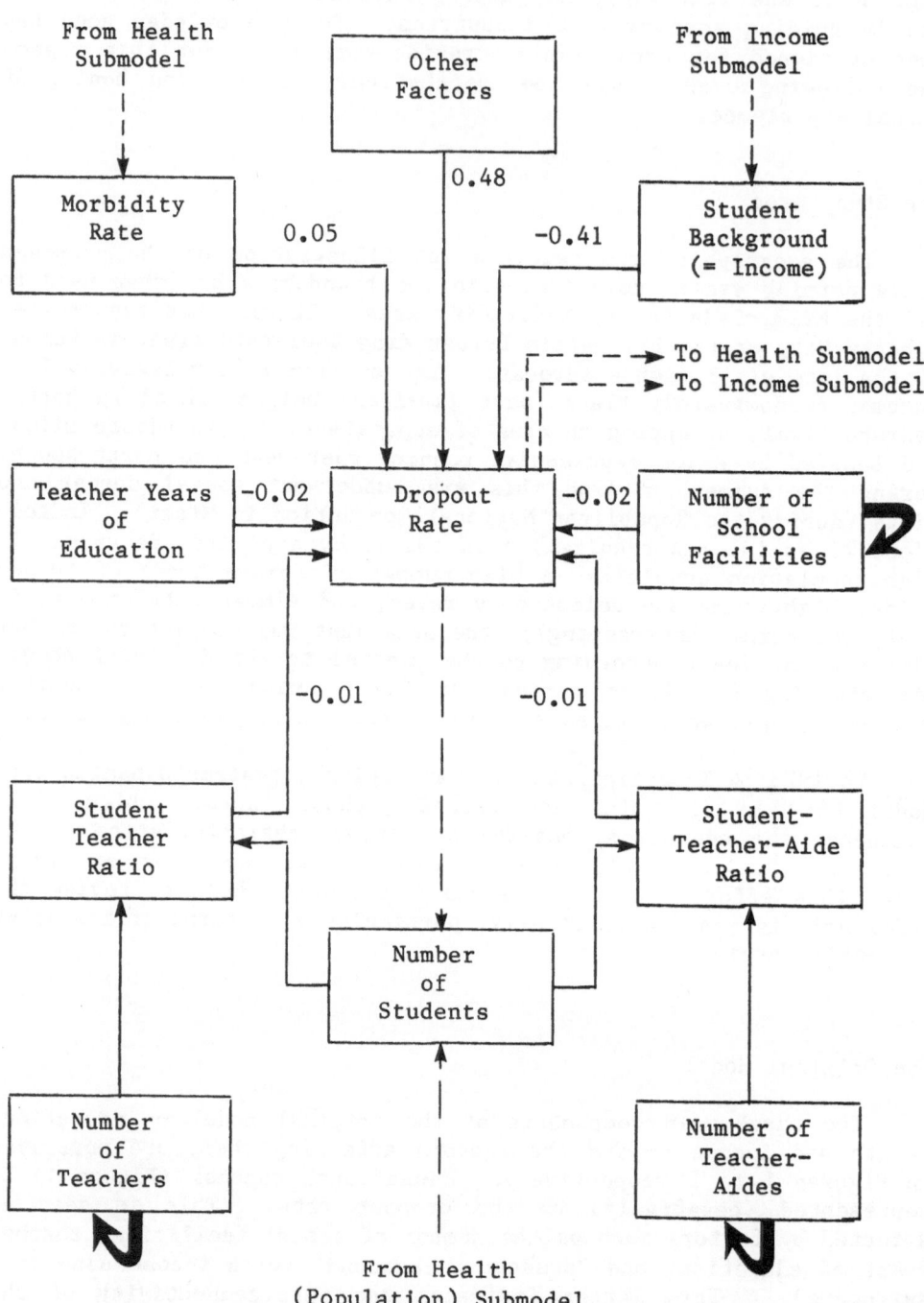

Figure 1 - Educational Component of the Model

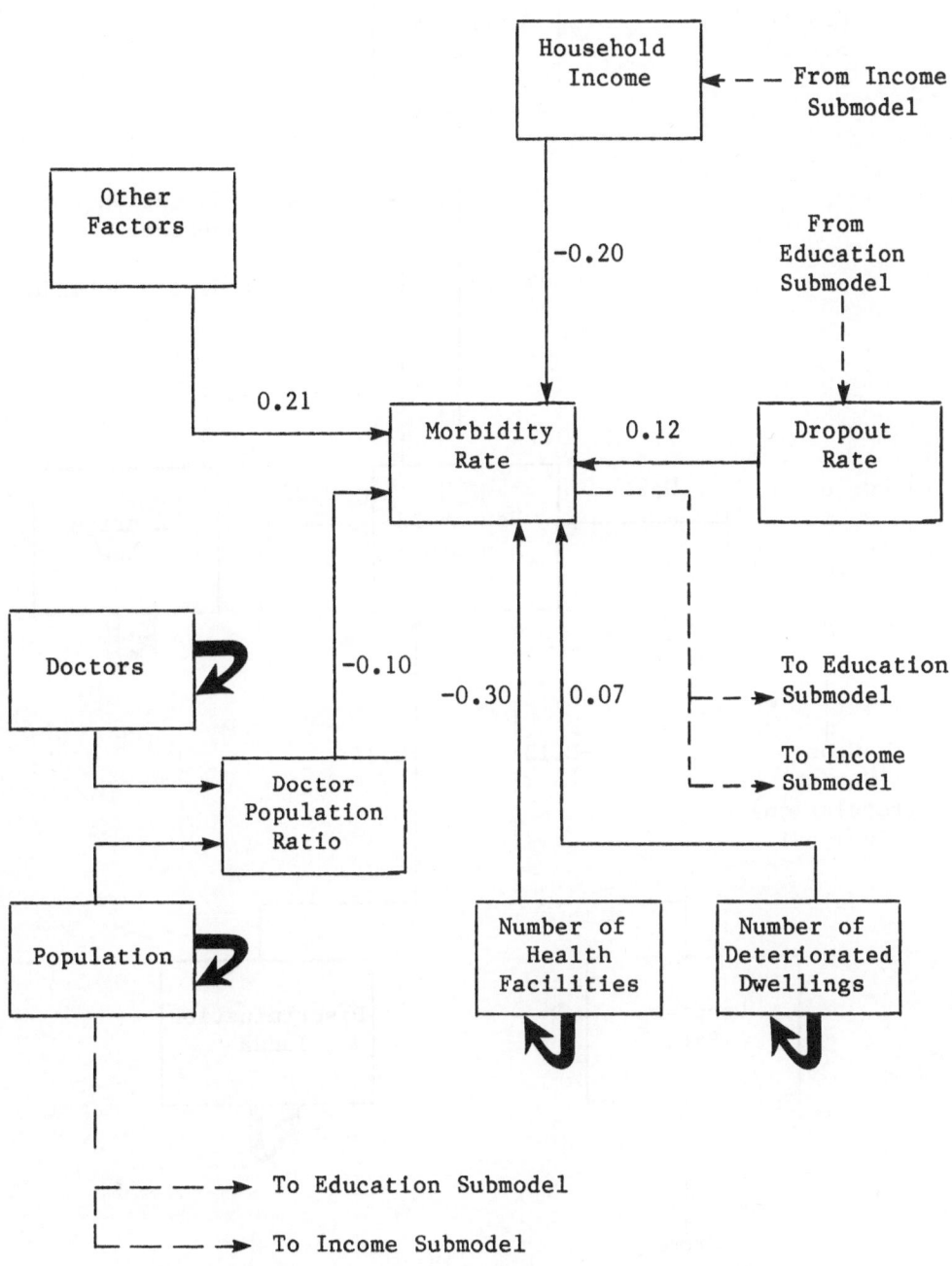

Figure 2 - Health Component of the Model

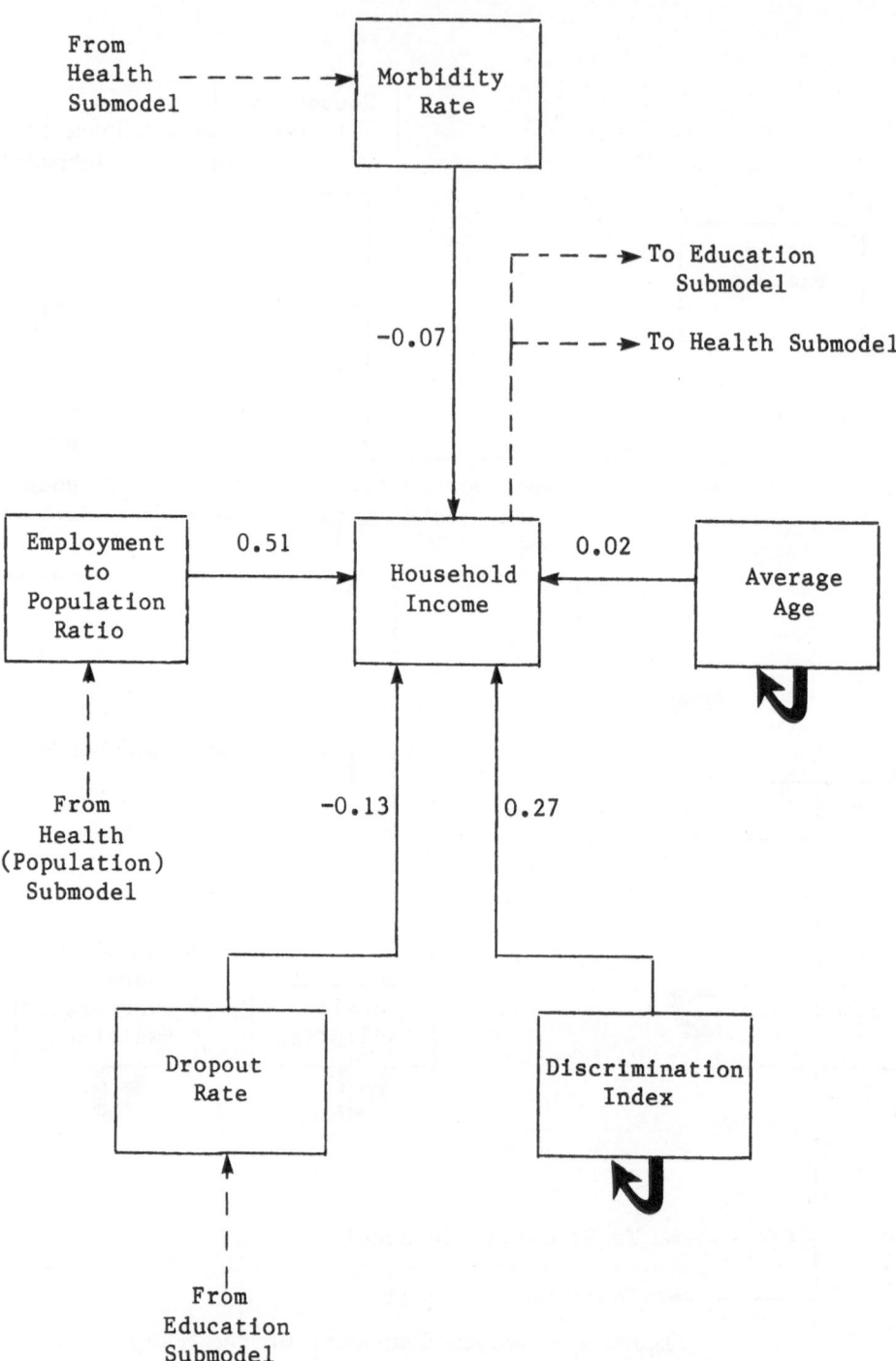

Figure 3 - Household Income Component of the Model

A "Roughly-Right" Model

The first step in the development of the early warning system was to "correct" and update the original model. As an example of the former, hindsight showed that the number of students never was related to the population level or to the number of dropouts itself. So this obviously was in need of alteration. The latter type of change is exemplified in part in Figure 4, where the employment-to-population ratio has been hypothesized to be impacted by local economic conditions, which in turn have been influenced by the national economic situation. For instance, severe inflation in 1979-80, especially in gasoline prices (lower left corner) has reduced Miami's attractiveness to tourists, and led to reductions in the need for black employees who find their livelihood in tourist-serving industries. Since this occurrence was never anticipated in 1973, the model obviously had to be altered to take it into account now. The other major changes suggested for the 1973-80 period are listed in Table 1.

The updating procedure illustrated above highlights the before-mentioned need for a simplified, "logical" model. When new affecting factors emerge over time, we have to relate them to local conditions. In most cases, precise data are missing, as are identifications of the strength of the relationships. Still, it is important, particularly for discussion purposes, to lay out the likely connections and try to gauge, at least roughly, the magnitude of their impacts. Such estimation obviously is crude, but still is better than disregarding the factors completely.

TABLE 1

NEW CONDITIONS ADDED TO THE OLD MODEL

---

1. Starting in 1976, because of increased medical school graduates and because of Medicaid/Medicare programs, the number of doctors increases 4% per year instead of 2.95%.

2. For similar reasons, starting in 1976, medical facilities increase at 2% per year instead of 1%.

3. Starting in 1981, because of a tightened school budget, the number of teacher aides is held constant (instead of rising at 5% per year).

4. Starting in 1979, because of national and local "stagflation", employment as a proportion of population drops from 29% to 25%.

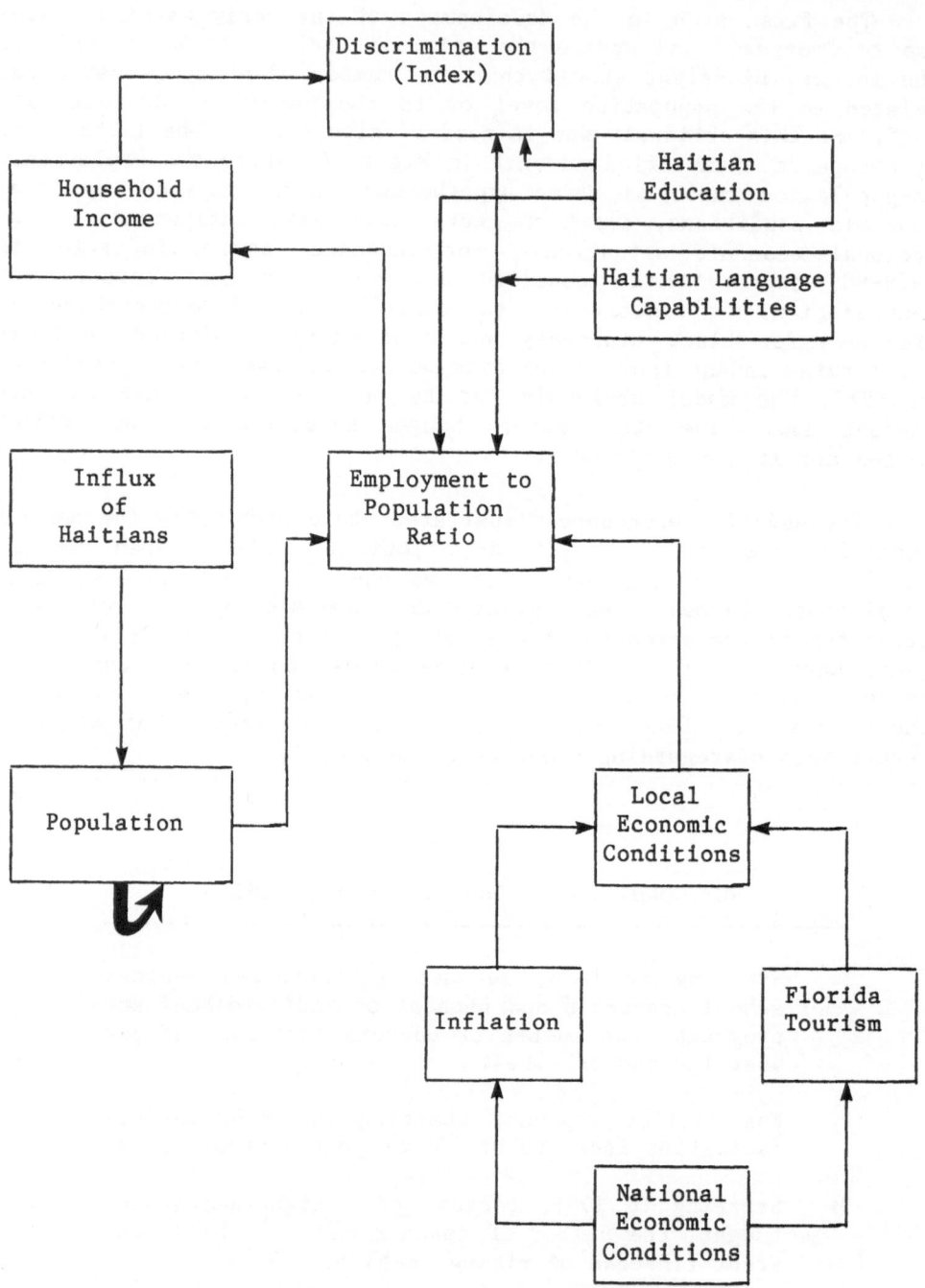

Figure 4 – Example Restructuring of Model

The Monitoring Process

The monitoring process as part of an EWS involves two basic steps:

1.   Collection of data;

2.   Screening and evaluation of the data.

The model provides the background for both of these efforts. Initially a list of the model variables is made (Table 2) and an indication given of the priority of importance of each. Such would be established on the basis of factors like the elasticity of the variable (eg. the numbers in Figures 1 to 3) with respect to health, education and income levels, the likely amount of change in the variable, and the extent to which it relates to public decision-making. For instance, employment levels (Figure 3) are highly related to income, are subject to substantial change, and can be the subject of many Federal-level programs, so that variable has been given the highest priority (a "1" in Table 2) for monitoring purposes.

Collection of data now can take place, with emphasis given to the first priority items. There are three main sources of data relevant to a monitoring system:

1.   Discussions with people knowledgeable in particular fields;

2.   Retrieval from available and continually-updated information systems;

3.   Direct search of the literature.

Assuming that knowledgeable people can be found, the first source probably is most useful because the information is as up-to-date and exhaustive as desired. Studies reported in the literautre usually are at least two years old and may not describe all the kinds of results and implications of importance to the early warning process.

Discussions with Knowledgeable People - People with experience and expertise related to each main variable in Table 2 would be identified. They would be contacted by telephone or in person, say, once every two months. They would be asked leading questions about recent changes, likely short- and long-term trends, and the implications for future policy decisions. As an example, the head of a community development agency in the area may be a good person to contact about changes in employment opportunities and instances of new discrimination.

The data would be recorded on a form such as that displayed in Figure 5. In some cases, the experts themselves could fill out most

of the form.   In other cases a specially developed information
centre could do it.   Key words, like those used in a library search,
would be entered in block 13, and then translated to index terms
standard to the particular EWS.[2]

Retrieval From Available Information Systems - The second possibility for collecting data is through readily available information
retrieval systems.   These provide regularly updated and transmitted
tests of abstracts in general areas of requested search.   Some
possible sources for our example situation are:

> NTISearch (US Department of Commerce)

> Social Sciences Citation Index (Institute for Scientific
> Information)

> Domestic Statistics (Predicasts, Inc.)

> SHARE (US Department of Health, Education and Welfare)

TABLE 2
MAIN FACTORS IN MODEL

Number	Name of Factor	Priority
1	Dropout Rate	1
2	Student Background	1
3	Number of School Facilities	3
4	Number of Teacher Aides	3
5	Number of Students	1
6	Number of Teachers	3
7	Teacher Years of Education	3
8	Morbidity Rate	1
9	"Other" Factors Affecting the Dropout Rate	1
10	Number of Doctors	2
11	Population	1
12	Number of Health Facilities	1
13	Number of Deteriorated Dwellings	3
14	"Other" Factors Affecting Health	1
15	Household Income	1
16	Employment-Population Ratio	1
17	Discrimination	1
18	Average Age	3
19	National Economic Conditions	1
20	Florida Tourism	2
21	Inflation	2
22	Local Economic Conditions	1
23	"Other" Factors Affecting Income	3

The resultant information again would be recorded on forms like that in Figure 5.

Direct Search of the Literature - The third possibility is direct reading of journals, trade magazines, newsletters and the like. It has been found (Davis, 1973) that trade magazines and newsletters are more informative for this purpose than "academic" journals. Moreover, those directed to the particular geographic area under study naturally are most relevant. During the process of the search it is possible to reduce the number of sources and add new ones focusing on additional significant factors (eg. in Figure 4).

MONITORING SYSTEM	Summary Sheet	
1. Title (or Citation)	3. Class Category:	
	4. Division:	
	5. Department:	
	6. Report Date:	
2. Submittor	7. Report File Number:	
	8. Number of Pages:	
9. Summary:		
10. Supplementary Notes:	12. Key Words: (Suggested by Submittor)	13. Index Terms:
11. Distribution:		

Figure 5 - Sample Summary Sheet for Monitoring System

Screening and Evaluation of the Data

The screening and evaluation phase is that in which the model and the data are brought together to highlight implications for decision-making. There are two obvious questions by those empowered to make choices -- what is to be decided, and when should the response take place?[3] Obviously the data should relate to these, or else be purged from the system as inconsequential.

The first step in this phase is the generation of "events" based on the trends seen in the data. These events are in effect scenarios to be employed in the model. To illustrate, over the last few years there has been a steadily increasing migration of Haitians into the south Florida area. With the recent massive inflow of Cubans, we might imagine that even more Haitians would see how relatively easy it is to slip in (by boat usually) and escape their poverty-stricken island in much greater numbers. How many might be induced to do so is open to question, of course. Further is the question of how many would move into the Liberty City area, again difficult to answer with any certainty.

Still, there are some clues. One comes from the fact that the Liberty City area has a preponderant black population, which should appeal to the Haitians. Second, there already are blacks from the Caribbean (Jamaica) in that area. Third, and alternately, there is a rough limit on how many people can move into Liberty City, given the existing amounts of housing. These and other "logical" considerations, in addition to the data trends, help in the generation of events like that shown in Table 3. Basically, it has been assumed in that scenario that the number of Haitians will increase fairly drastically in 1981 and 1982, then taper off to a more moderate rate. In addition, because of their inability to speak either English or Spanish, their low skill level and their colour, discrimination against them will be high and unemployment and the school dropout rate might be expected to increase.

Assume now that the year is 1977 and decision-makers have been presented with the forecasted event shown in Table 3. Assume further (for simplicity) that the best type of response is known and agreed upon. This consists of the actions and the corresponding times to implementation listed in Table 4, and involves mainly more teachers, school facilities and employment programmes. The question then is, "Should the response be undertaken now so that the programmes are "in place" when the event occurs, or should action be withheld until the event actually occurs (if it does), perhaps even until the impacts on health, education and income make themselves apparent?"

The tradeoffs are clear. If action is taken in 1977 (particularly for the construction of school facilities, which takes three years), the Haitians coming in 1980 will be greeted with available programmes and their status improved fairly quickly. On the other

TABLE 3

CHARACTERISTICS OF THE "EVENT"

---

1. Population (mostly French-speaking Haitians) jumps to 23,850 in 1981 (from 21,673); to 26,850 in 1982; and increases by only 1% per year thereafter (instead of 3%).

2. Being unaccustomed to "high quality" housing, the Haitians cause the number of deteriorated structures to increase by 4% per year (instead of decreasing by 3.1%).

3. Because of the poor English (and Spanish) language skills and generally poor education of the Haitians, the discrimination index goes to 7 (out of 12) in 1981 and 1982. It decreases at 3% per year thereafter.

4. For the same reasons as in #3, employment drops to 20% of the population (instead of 25%).

5. Again for the same reasons as in #3, the dropout rate as a function of income increases for most income levels.

TABLE 4

CHARACTERISTICS OF THE RESPONSE

---

1. The number of teachers is set and maintained at 1 per 30 students (1 year to implement).

2. Teachers are trained in French and Creole language skills and Haitian culture (2 years to implement).

3. The number of teacher-aides is set and maintained at 1 per 60 students (1 year to implement).

4. The number of school facilities is set and maintained at 1 per 500 students (3 years to implement).

5. Employment programs are generated, bringing the employment level to 28% of the population (2 years to implement).

---

hand, it is highly uncertain they will come at all, in which case, the response expenditures in 1977-1980 will create relatively few benefits. Yet, if the decision is made to wait until 1980 to see if the event does occur, it will be three years before, say, the needed schools are built, and people will suffer in the interim. And, if the decision is made to wait even longer to determine the seriousness of the situation, the potential suffering will be prolonged even further.

Results from use of the model to help address these questions are displayed in Figures 6 to 8. The first of these shows that the dropout rate is affected almost imperceptibly if the response is made (whether or not the event occurs). This tells decision-makers they certainly can afford to wait to take action, and perhaps not even do so at all (at least insofar as the dropout rate is concerned). It also tells those concerned with the EWS that much less attention need be given data collection on educational matters.

The results in terms of household income (Figure 7) show a different story, however. Whether or not the event occurs, if the response is made, income will increase in the range of $300 to $400 (about 10 percent) per household per year. Decision-makers thus would be likely to give more attention to the timing of the response, and those people concerned with the EWS should focus more heavily on data related to factors affecting income.

The question of the timing of the response and its influence on income is addressed in Figure 8 (in which it is assumed that the event occurs). If the response is made in 1977, three years beforehand, households are forecasted to gain about $29.5 million in personal income over the 1980-1990 period, compared to the situation where no response is made. If the response is delayed to 1980 to see if the event actually occurs, the gain is only $24.0 million. And if the response is held off another year to see the impacts of the event, the gain is reduced to approximately $21.0 million. Where economic and financial considerations rate highly with decision-makers, the above income gains could be weighed against the cost of the response detailed in Table 4. Where this is not the case, reference could be made to the (small) changes in health and education levels. In either situation the decision-makers have some rough estimates of a relatively broad variety of impacts, especially as they correspond to the act/do not act now choice.

OBSERVATIONS

It was stated initially the main purpose of a model, outside of education, is to aid decision-making. In what many modellers would view as the most desirable situation, a model would give highly accurate forecasts of all relevant variables under different alternatives, leaving decision-makers the single task of assigning their values to the appropriate impacts and then automatically selecting

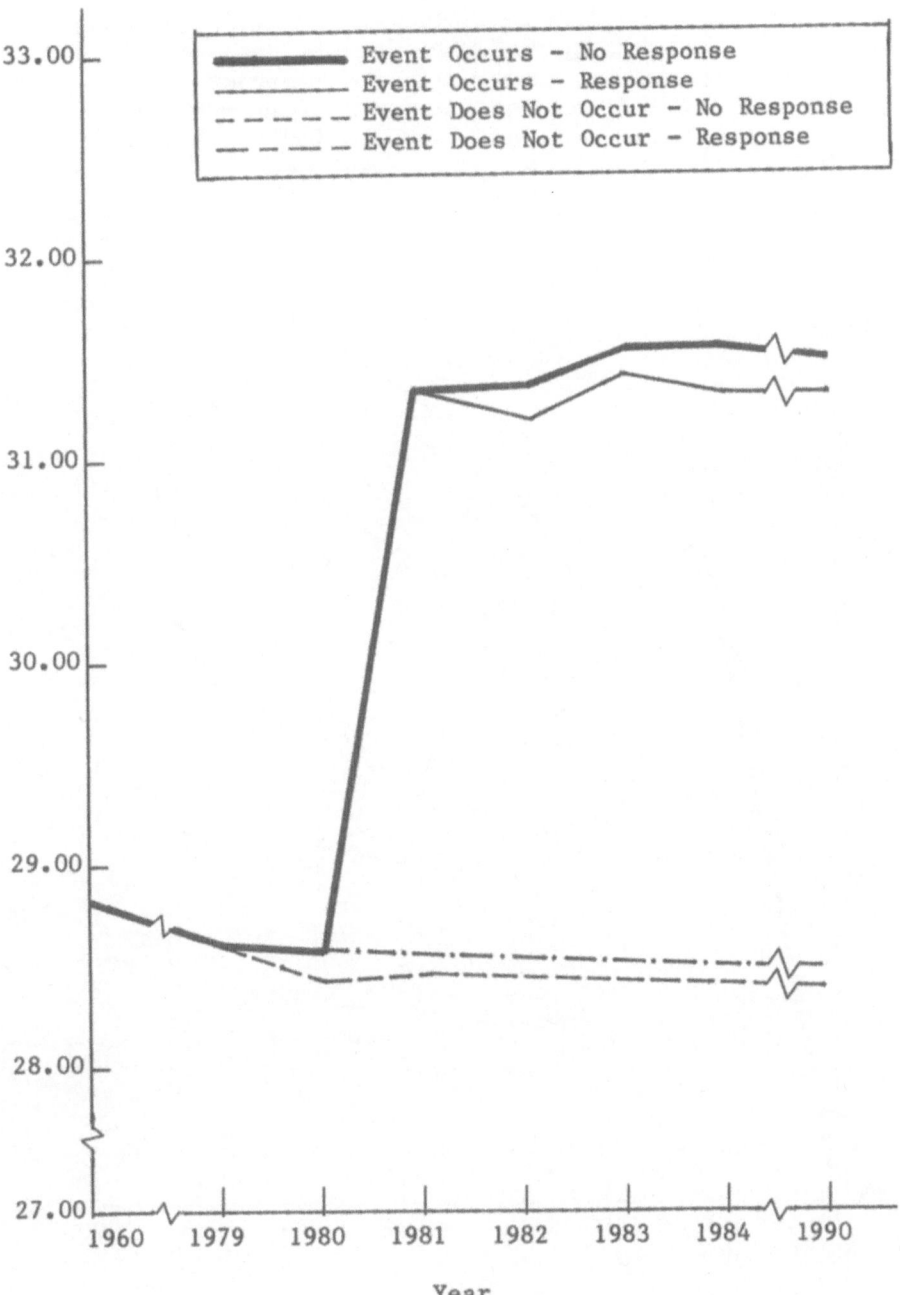

Figure 6 – Forecast Dropout Rate If the Event Does/Not Occur and the Response Is/Not Made (In 1977)

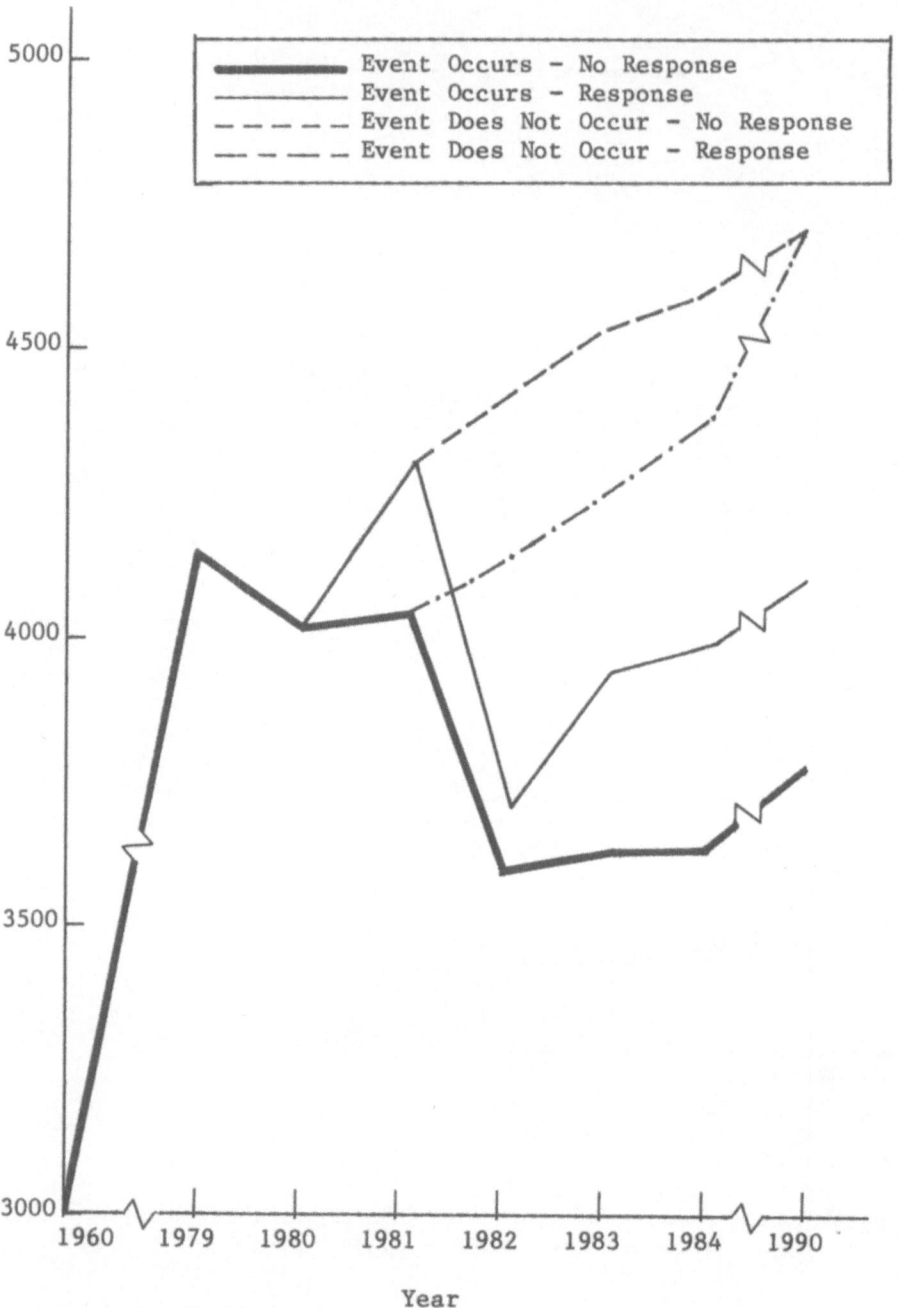

Figure 7 – Forecast Household Income If the Event Does/Not Occur and
          the Response Is/Not Made (In 1977)

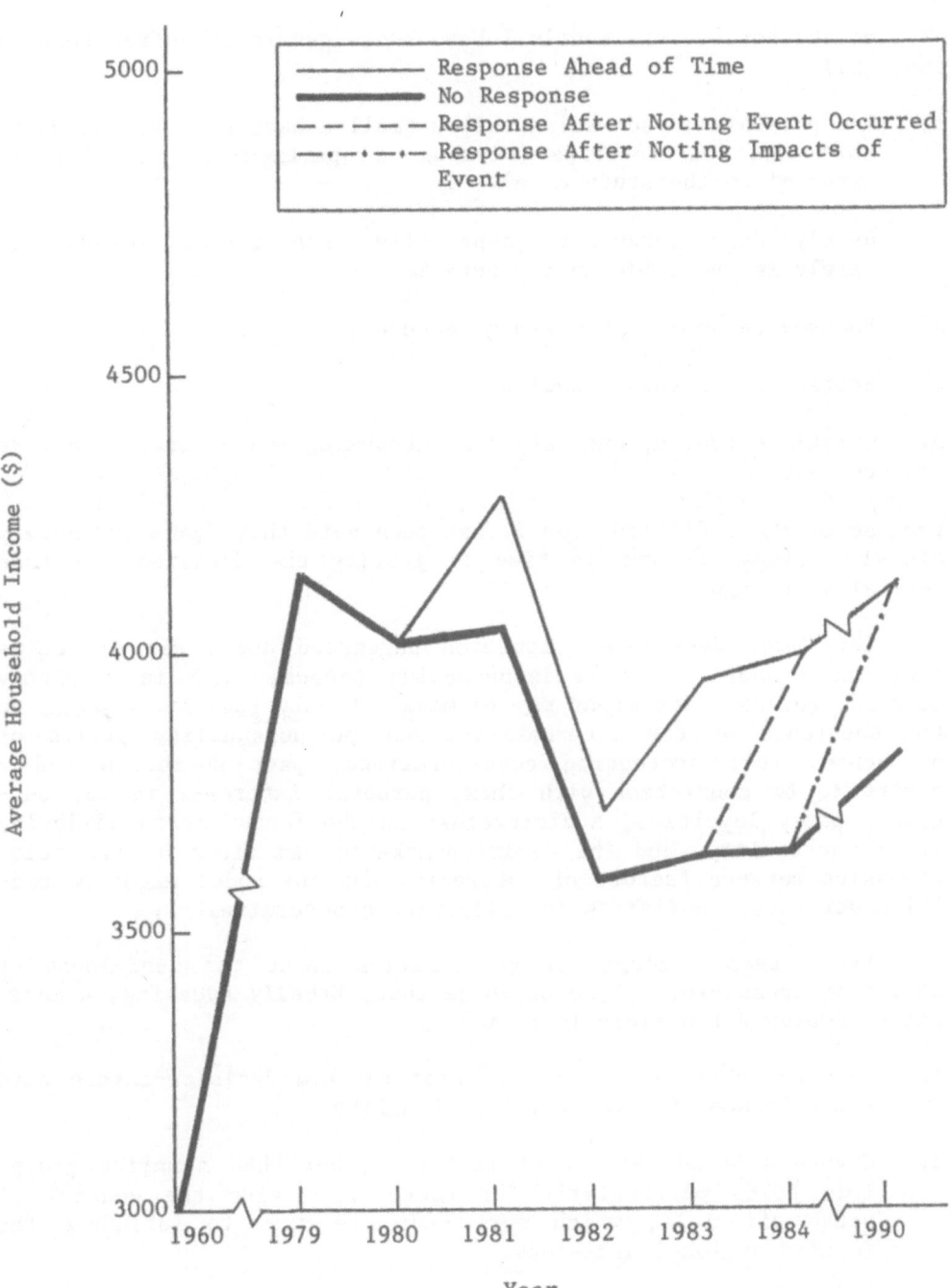

Figure 8 - Household Income Depending on Timing of Response (If
          Event Occurs)

the best alternative. This, of course, is naive for reasons related both to modelling and to the decision-making process.

On the former, the models I have known generally suffer because they are:

1. Too parochial, addressing only a small subset of relevant variables and relationships (missing, in particular, many factors external to the study area).

2. Overly data intensive, especially when recent information rarely is available in any detail.

3. Focused on some of the wrong measures.

4. Static rather than dynamic.

5. Quickly outdated, and very time-consuming and expensive to make current.

Because of these difficulties it has been said that "data and models are the things you get in time to justify the decision you made several years ago".

The "most desirable" situation suggested above also is naive from the standpoint of decision-making because such is a highly tangled process. Decisions may be made not only (and not always) on the substance of the alternatives, but on personality traits of opponents, turf protection considerations, past decisions and a desire to be consistent with them, personal interests in the outcomes, party loyalties, a disinterest in the facts, confused logic, great uncertainty, and the decision-maker's own views of the relationships between factors of interest. How any model might be useful under these conditions is difficult to determine.

Yet I feel a couple of generalities about the usefulness of models are possible. It seems to me that, broadly speaking, a model can be employed beneficially if it:

1. Focuses on the near future. Most elected decision-makers have a high discount rate on future benefits.

2. Covers a broad variety of factors rather than a narrow group. This holds particularly for factors, outside the control of urban officials, which more and more seem to influence the impacts of local decisions.

3. Has a transparent structure. Few decision-makers will trust a black box against their own intuition.

4. Has an "agreeable" structure. If there are obvious fallacies or inconsistencies, or relationships which are foreign to most decision-makers, then the model will not be trusted.

5. Makes use of the most up-to-date information. If such is not included, most decision-makers will feel the model is incomplete, and therefore, unusable.

6. Shows the value of acting now versus sometime in the future.

7. Highlights the likely amount of uncertainty.

An early warning system, exemplified in the semi-hypothetical case in this paper, is one attempt to help put models in a context more amenable to the decision-making process. It does this by concentrating on getting the most up-to-date information (possibly using decision-makers to provide this); highlighting basic structure and newly-formed relationships; tracing the logical connections between factors; and providing information on the likely advantages of acting now (perhaps even in advance of probable events) versus waiting.

Development of an EWS is only one step in bringing decision-makers in closer contact with analysts. Yet it might be an easily achieved one. Almost any person concerned with public activities spends some time scanning newspapers, professional publications and the like to keep "up-to-the-minute" on events. These people, and their knowledge, can be readily tapped with a monitoring programme. Similarly, with personal computers becoming more prevalent[4], many trends and scenarios can be analyzed readily with simple models. The result I foresee is more informed decisions, taken more on the merits of the alternatives than on personal political considerations.

NOTES

1. Most of these numbers come from regression analyses of census data and from relevant studies, such as the early Coleman report (Coleman, 1966).

2. An example index system is Urbandoc (Sessions and Sloan, 1971a,b).

3. Concern in this paper is more for the latter question, since it is often overlooked.

4. In fact, I performed the calculations in this exercise on my own Apple II computer.

REFERENCES

American Council of Life Insurance, Trend Analysis Program, 1980, Health Care: Three Reports from 2030 AD, TAP #19, Washington, DC.

* Bright, J.R., 1973, Forecasting by Monitoring Signals of Techno-
   logical Change, in J.R. Bright and M.E.F. Schoeman, (Eds.), A
   Guide to Practical Technological Forecasting, Prentice-Hall,
   Englewood Cliffs, New Jersey.

Coleman, J., 1966, Equality of Educational Opportunity, US Government Printing Office, Washington, DC.

Davis, R.C., 1973, Organizing and Conducting Technological Fore-
   casting in a Consumer Goods Firm, in J.R. Bright and M.E.F.
   Schoeman, (Eds.), A Guide to Practical Technological Forecasting,
   Prentice-Hall, Englewood Cliffs, New Jersey.

* Dickey, J.W., 1976, A Proposed Early Warning System for the Division
   of Motor Vehicles, Interim Technical Report to the Virginia Division of Motor Vehicles, Virginia Polytechnic Institute, Blacksburg, Virginia.

Forrester, J.W., 1968, Principles of Systems, Wright-Allen Press,
   Cambridge, Mass.

Sessions, V.S. and Sloan, L.W., 1971a, Urbandoc/A Bibliographic
   Information System: Technical Supplement 1/General Manual, Report
   Urbandoc-71-2, The Graduate Division, The City University of New
   York, New York.

Sessions, V.S. and Sloan, L.W., 1971b, Urbandoc/A Bibliographic
   Information System: Technical Supplement 2/Operations Manual,
   Report Urbandoc-71-3, The Graduate Division, The City University
   of New York, New York.

Steiss, A.W., Dickey, J.W., Phelps, B. and Harvey, M., 1975, Dynamic
   Change and the Urban Ghetto, Lexington Books, D.C. Heath, Lexington, Mass.

* US Department of Commerce, Office of Technology Assessment and Forecast, 1973, Early Warning Report, COM 74-10150, NTIS, Springfield,
   Virginia.

*References not referred to in the text.

MULTIREGIONAL POPULATION ANALYSIS FOR URBAN AND REGIONAL PLANNING

Frans Willekens

Netherlands Interuniversity Demographic Institute
Postbus 955, Prinses Beatrixlaan 428
2270 AZ Voorburg
The Netherlands

INTRODUCTION

During the past few years, we have witnessed the appearance of a new generation of models for urban and regional analysis and planning. Following the systems idea that everything depends on everything, analytically-oriented demographers, economists and planners have expressed a growing dissatisfaction with "one-at-a-time" regional models. As an alternative, a multiregional approach has been advocated, in which regions are represented as components of a larger, interconnected system. This new model structure is proposed as an improved tool for the assessment of spatially differentiated impacts of national and/or regional policies. Multiregional models not only recognize the relationships that exist between regions, and are able to trace both the direct and the indirect interaction effects, but they also yield consistent regional and national statistical measures. The consistency of results with national totals has always been a difficult issue in regional modelling; it was never adequately resolved until a model structure was developed in which all regions of a nation were considered simultaneously.

In trying to construct multiregional model structures, economists and demographers went through conceptually similar designs, which may be classified into four groups: (i) the top-down approach, (ii) the bottom-up approach, (iii) the hybrid approach, and (iv) the multiregional approach. The approaches are reviewed in the first section of this paper. Many of the models actually used for regional forecasting and policy analysis are still of types (i) to (iii). However, as scientists simplify the mathematics of multiregional models, gather experience in using them, facilitate the application by providing packages of computer programmes, and, last

227

but not least, develop accurate estimation techniques to fulfil data requirements, the multiregional perspective is within reach of more agencies, and it is being adopted rapidly.

The selection of a particular model-type for regional fore-casting and analysis is frequently the outcome of an evaluation pro-cess in which a more or less standard set of questions is asked. The second part of this paper reviews some of the important ques-tions and attempts to provide an answer. The questions relate to model characteristics (degree of complexity to adequately represent reality, suitability of the model for impact assessment, consistency and reliability of the numerical results), and to the feasibility and even desirability of its application in view of the limited availability of statistical data and the lack of manpower. These are two of the major factors that Pack (this volume) puts forward as hffecting the actual use and institutionalization of models in regional planning agencies.

APPROACHES TO MODELLING A SYSTEM OF REGIONS

The rationale for modelling a system of regions instead of a set of independent regions is threefold. First, the awareness of the regional identity: regions behave differently and models should be able to fully represent these differences. Second, interregional dependence: regions do not exist in a vacuum; they are intercon-nected with other regions, which they influence and by which they are influenced. Third, the need for consistency in regional projec-tions and impact assessments: for each variable, the value for the nation must equal the sum of regional values (stock variables), or must be a weighted average (rate variables). Several ad hoc proce-dures, that exist to impose consistency, are no longer acceptable on scientific grounds. The approaches to modelling systems of regions reflect the response to these three requirements.

The Top-down Approach

The top-down approach is a common way to project regional econ-omic and demographic variables. It is attractive because it ensures consistency. The procedure is simple: national variables are esti-mated first and then distributed among regions on the basis of a predefined allocation procedure. The allocation procedure may be extremely simple, such as the fixed-ratio technique, in which the regional variable is a fixed share of the national total. Some procedures, particularly in economic models, are quite complex and apply mathematical programming to determine the regional distribu-tion that optimizes a welfare, cost or information gain index, and at the same time satisfy the national control totals. Regional growth models, if used at all, are satellites of the national model.

Most demographic models are of the top-down type. Ter Heide (1976) and Pittenger (1976) provide an extensive discussion of distribution functions. National totals are generally taken from projections of a country's population, that are prepared by statistical offices. The official or semi-official nature of national statistical offices and of the numbers they produce explains the reluctance of analysts to produce deviating projections, and hence the popularity of the top-down approach. As in many other countries, this approach is being followed in the Netherlands to produce regional population forecasts. The Dutch model has, however, several features that show the form of a hybrid model. By intensive use of standardization techniques, the model can account for autonomous changes in regional population compositions (Eichperger, Hamel and Nieuwenhuis, 1979).

In a review of so-called multiregional economic models, Bolton (1980) classified most of them as top-down models. An illustration is MULTIREGION, developed by Olson and associates (1978) at Oak Ridge National Laboratories; it is an input-output model, driven by Almon's INFORUM model of the national economy. MULTIREGION predicts the region's share of national employment in each industry on the basis of changes in key economic indicators, of which the most important is the region's accessibility to the national market.

The top-down approach is convenient from a practical point of view since the consistency problem does not exist. It may also be defended on a theoretical basis. What happens in a region is very much determined by what happens in the nation. This statement underlies, for instance, the economic base theory which claims that regional growth is driven by a "basic" sector producing for the national, or even, international market. The economic base multiplier denotes the impact on total regional employment of changes in basic sectors. The employment describes, therefore, the effects on the region of changes originating at the national level. The validity of the economic base theory has been questioned recently, mainly because of its pure demand orientation and ignorance of the supply constraints of production factors. A focus on supply conditions would logically call for a more region-oriented perspective, and hence, for a different modelling approach.

The selection of a top-down model of regional change implies the adoption of a particular theoretical perspective which may be questioned. The main drawback of the approach is, however, that at the outset regional differences are disregarded. National projections or indicators are produced by assuming "average" values of model parameter and initial conditions. Regional differences are only introduced at a secondary stage, in which national projections are distributed.

The Bottom-up Approach

The second approach is to analyze or project regional variables completely independent of national control totals. Each region is treated as a separate sub-system without any explicit links with any other region. This "uniregional" perspective gives maximum weight to regional characteristics and is therefore attractive when the emphasis is on regional differences rather than on consistency or interdependence. This is the case, for instance, when attention is limited to a single region, or when supply conditions largely determine regional change.

The bottom-up approach leads to inconsistent results. The value of a national variable is the sum of regional variables, and nothing assures that this sum equals the national control total. Although the uniregional perspective implies that each region is independent from the others, it does not require the regions to be closed. Interactions with other regions of the same system may be accounted for through variables representing net exchange. A typical illustration is the net migration variable. Usually the regional population is projected by a cohort-survival model, a technique originally developed for projecting national populations. The assumption of closedness may be realistic at the national level, but it does not hold for sub-national units. The cohort-survival model is therefore adapted to the new situation by adding a net migration variable. The use of net migration, and of net exchange in general, is a simple procedure, but has a number of important theoretical and methodological shortcomings, as will be shown in the second part of this paper.

The Hybrid Approach

The hybrid approach is a first attempt to combine the advantages of the top-down approach (consistency) and the bottom-up approach (explicit consideration of regional differentials). In it, some regional variables are sum-constrained, ie. constrained by values determined by a national model. The predetermined national totals are regionally-independent, while the other totals, which are obtained by simply summing the regional variables, are regionally-dependent. The movement away from the pure top-down approach is a response to the need for more realism in models of systems of regions. Regions do not always follow national developments: they have their own internal dynamics, and may also generate national change. There is, therefore, a need for a double linkage: national-regional (top-down) and regional-national (bottom-up).

The hybrid approach is used in several "multiregional" economic models developed in Europe. The REGINA model for France, developed by Courbis (1975), is the best illustration. Courbis's ideas have recently been adopted by Ballard and Associates (1980a, 1980b) of the US Bureau for Economic Analysis in NRIES.

The hybrid approach combines a national model and several regional models. Employment and labour supply are the most important variables generated by the regional models, reflecting the fact that labour markets are regionally bound. Population level, interest rates, exports and imports are among the variables determined by the national model, and are subsequently used as control totals for the regional models.

For population projection, a hybrid model is being used in the US, Canada and Sweden, and the hybrid nature relates to the way migration is treated. Migration has always caused major problems for sub-national population projections. Application of a net migration variable violates an important consistency requirement: without international migration, the net migration rate for the nation should be zero. To ensure consistency, the concept of "migrant pool" was introduced. For each region, the number of outmigrants is calculated independently by applying fixed region-specific outmigration rates. The national total of outmigrants is obtained by simple summation. To get the number of inmigrants in each region, the migrants in the "pool" are allocated to the regions by using a distribution function. The choice of this procedure rests on theoretical and empirical bases. Theoretically, inmigration rates should not be calculated since the base population is not the population at risk of migrating. A "propensity to inmigrate" has no theoretical meaning. In addition, there is the empirical observation that the outmigration rate of a region is much less volatile than its inmigration rate, and hence, easier to predict.

In some instances, it may be acceptable to calculate the inmigrants first, to generate an "inmigrant pool" and to obtain outmigrants by a distribution function. For instance, Gordijn and Heida (1979) use an "inmigrant pool" since in the Netherlands migration is very much determined by housing opportunities. Because of the relation between housing construction and inmigration, inmigration is easier to predict.

The Multiregional Approach

In the first three approaches to modelling systems of regions, the emphasis was on consistency with national totals, and on regional differences. In the multiregional approach, the concern for consistency and regional differences is augmented by a concern for correctly representing and projecting interregional dependencies. Regions interact with one another by exchanging people, goods and services. Modelling migration, commuter and trade flows by region of origin and of destination has generated substantial research in geography and regional science, which is currently referred to as spatial interaction analysis, and which is a growing sub-field in both disciplines. Spatial interaction analysis may formally be defined as the study of observed interregional (origin-destination) flows, and the estimation of flows from incomplete data. A critical

ingredient of multiregional models is a spatial interaction model.
The development of the spatial interaction analysis is appropriate,
particularly if not all the required statistical data are available. The next section of the paper looks at data availability as a
limiting factor in multiregional analysis, and at estimation procedures to alleviate the data problem.

The dominant feature of multiregional models is that they study
and project all regions of a multiregional system simultaneously.
In the previous approaches, the regional variables and/or national
variables are solved sequentially. Consistency is assured by a
negative feedback mechanism. The simultaneous solution of all
regional variables not only assures internal consistency, but, at
the same time, enables the introduction of regional differences and
the representation and projection of linkages that exist among
regions. It is an appropriate approach to study the indirect, as
well as the direct, effects on the system of changes in a particular
region. The multiregional perspective is therefore advocated for
regional and urban impact assessment of policies at national and
sub-national levels (Glickman, 1980). As Glickman indicates, an
essential aspect of sound impact studies is that the impact of
policy is analyzed in terms of both indirect and direct effect. In
addition to the scientific and policy relevances of the multiregional approach, there is a practical reason for adopting this
approach. The level of detail of multiregional models forces the
analyst or user to devote more attention to the input, and to spell
out the assumptions with greater care than is needed in other types
of models of multiregional systems. This demand for increased
attention to assumption specification may prove to be a very useful
fringe benefit of multiregional models.

This approach to modelling systems of regions is not widely
applied yet. A prototype of a multiregional model of the economic
system is Polenske's (1975) input-output model. Regional tables are
linked by industry-specific interregional trade coefficients.
Although in the original version of the model the trade pattern is
assumed to remain constant, there is no technical barrier to introducing changes. More important is the economic theory imbedded in
the models, which is Keynesian. Input-output models are demand-
oriented and assume a perfectly elastic supply. This paradigm of
the economy is transposed to interregional trade. Trade flows are
completely determined by input requirements, and are independent of
production constraints. The trade coefficients are <u>admission</u> rates.

In demographic analysis, the multiregional approach was introduced by Rogers (1975), who generalized conventional demographic
concepts and models to multiregional systems. The multiregional
perspective is also adopted by Rees in his work on demographic
accounting (see, for instance, Rees and Wilson, 1977). Regions are
linked through age-specific directional migration flows. The migration coefficients are derived from base-year data. They are <u>transi-
tion</u> rates, ie. outmigration rates, and applied to the population in

the region of origin. If flow data are not available, they may be estimated from whatever relevant information exists (Willekens, Por and Raquillet, 1979). Regional differences are easily accounted for: the fertility and mortality parameters are region-specific. This leads to the question of consistency: are the aggregates of multiregional population analysis and projections consistent with national characteristics and projections? The consistency issue is investigated in the next part. Whether consistency is achieved depends primarily on the relative differences in regional fertility and mortality levels, and not on the migration pattern. The consistency condition is satisfied if, and only if, the multiregional model imbeds the same set of assumptions as the national model. This sounds promising, but it is not. One could defend the thesis that the aggregates of multiregional analysis should deviate from the totals obtained by a national model, and that the national totals obtained by a multiregional model are closer to reality.

ISSUES IN MODELLING A SYSTEM OF REGIONS

In order to be a useful tool for forecasting and policy analysis, a model should satisfy a few conditions. It should be theoretically sound, represent reality, have a high internal validity (internally consistent), yield reliable and accurate results of an acceptable level of geographical detail, and it should be simple, transparent, easy to implement and should respond to "what if" questions. Some of these requirements are associated with some fundamental modelling issues. I would like to focus on four issues: (i) (i) consistency, (ii) level of disaggregation, (iii) ability to simulate, ie. to answer "what if" questions, and (iv) implementation (application).

Consistency

The rationale of the top-down approach is that the sum of regional projections should be equal to national control totals. Although, from a practical point of view, this consistency requirement may be necessary, it is theoretically unacceptable. Control totals are obtained by applying nationally average rates of growth. Control totals may therefore be thought of as the sum of regional projections with identical growth rates. In other words, in adopting control totals, one assumes the system to be homogeneous, and to completely ignore regional differences. The introduction of heterogeneity in the system by allowing regional growth rates to differ would generally lead to other totals. This can easily be demonstrated. Assume a system consisting of two regions. The first region has 2000 people, and the second, 8000. The regional growth rates are -0.01 and 0.015 respectively. The average growth rate is 0.01. Applying the average growth rate to the population of 10,000 yields a population of 10,510 after five periods. Applying the regional growth rates to the regions separately would give a population of 10,520. That the sum of separate projections is larger than the projection of the combined population was already shown by Keyfitz (1972). Now assume that people can migrate between the regions. If both regions grow at the same rate, then migration does not affect the total population, but only its distribution over space. If the regional rates of natural growth differ, then migration neutralizes some of the effects of these differences: the total population is closer to the population figure obtained by applying the national rate. The effect of migration is, therefore, to level off the impact of regional differences. This phenomenon has been denoted as the regression-towards-the-mean effect of migration in multiregional analysis.

In demographic and economic analysis, the consistency issue has been dealt with extensively; for a recent review of the state of discussion, see Gibberd (1980). Gibberd, building on Rogers (1976) shows that the necessary and sufficient conditions for perfect aggregation in multiregional models is that the Markov property holds for both the aggregated and the disaggregated process; in other words, that the Markov property is not destroyed by the aggregation. The reader is referred to the literature for a detailed discussion.

How Disaggregated Should a Model Be?

Do we need a highly disaggregated and complex model if the only interest is on some aggregate indicators, such as the total population? Isn't a simple exponential growth model sufficient? The optimal level and type of disaggregation of a population system for projection and analysis purposes depends on three considerations: use of projection or analysis, demographic theory and statistical theory (Willekens, 1979).

Use of Projection: The elements of the population structure may be selected because predictions of sub-categories are of value in their own right. If age-specific population data are required, an age-specific model should be used. An alternative, which is still used in some instances, is to project the total population and to apply a distribution function to decompose the population into age groups. A problem is that the "user" of projections is generally not known or is highly diverse, in particular, if projections are made by statistical offices. It shows a lack of realism if one wants to please all potential users by a single projection output. Flexibility in output format is, instead, a more realistic strategy.

The interest in a projection model may not be limited to its output. In "what if" analysis, the input is as important as the output. A highly disaggregated regional model with a detailed input structure offers more entry points for simulation use. This increased ability to simulate may justify the additional effort of disaggregation.

Demographic Theory: How large should a demographic projection model
be?   In other words, how complex should the projection system be?
This issue is often the subject of debate between model developers
and users.   Complexity and forecasting accuracy are not proportional.   A simple projection model, such as the scalar exponential
growth model in which the projection system consists of a single
element, may yield highly accurate results, in particular, in the
short run.   Whether this could be a result of simple dynamics, or of
a complex interaction of neutralizing forces is always uncertain.
To eliminate this type of uncertainty, the choice of the appropriate
projection system and its constituent elements must also be determined by demographic insight.   An aim of demography as a science is
the identification and explanation of regularities in demographic
processes.   A review of the demographer's search for regularity and
for the fundamental components of demographic change is presented by
Brass (1974).   This knowledge provides a theoretical basis for the
design of the projection system.   "Demographic forecasting is seen
as the search for functions of population that are constant through
time, or about which fluctuations are random and small." (Keyfitz,
1972:347).   A first step to this ideal situation is the detection of
homogeneous sub-groups of the population.   This statement has an
important implication for the definition of the spatial system: the
spatial units of analysis, selected for projection purposes, should
be as homogeneous as possible with regard to the demographic characteristics (Ter Heide, 1973:460).   This is contrary to the geogra-
pher's search for natural spatial units (settlement clusters) which
are both territorial and functional (see, for instance, Dziewonski,
1978:40).   The search for homogeneous categories at the outset of
the projection exercise, facilitates the scenario-analysis at a
later stage, and the interpretation of the results.

Statistical Theory: The demand for disaggregation to achieve
greater homogeneity within categories is constrained by the need for
statistical data. Disaggregation becomes excessive if the number of
observations (sample size) in a certain category is too small to
allow a calculation of representative indicators, eg. transition
probabilities. Although this requirement seems trivial, it is often
not met, particularly in large simulation models in which size and
empirics prevail over quality and substance.

     The optimal internal structure of the projection system is a
compromise.   Demographic detail and statistical significance are
conflicting objectives.   The statistical requirement of adequate
sample size in each category can be dealt with by using auxiliary
models to generate the values of the demographic model parameters
(Willekens, 1980a).   In highly disaggregate systems, the observations in each category (cell frequencies or probabilities) may not
be independent.   The pattern of association may therefore be described by relatively simple models.   These models are intermediate
between the data and the parameters of the demographic models, and
are tools for information reduction.   The demographic parameters are
not directly calculated from the observed data, but are derived from

models fitted to the observed data. Examples of auxiliary models
that may be used when the observations are too small in number, or
even missing completely, are the so-called model schedules. The
observed age-specificity of fertility, mortality and migration has
induced demographers to describe the curves of age-specific rates by
simple models (Coale and Demeny, 1966; Coale and Trussel, 1974;
Rogers and Castro, 1979). The results of this research may fruitfully be applied in simulations of highly disaggregated population
systems.

"What if" Analysis

     To be useful for planning and policy-making, models should
assist in answering relevant questions. They can do so if they
incorporate an "ability to simulate". Forecasting, policy impact
assessments and sensitivity analysis are particular illustrations of
"what if" analysis or simulation. In forecasting, the interest is
on the evolution of the system if the external conditions (environment) and the model parameters take on their most likely values. In
policy analysis, the interest is in knowing how the system responds
to anticipated changes in policy variables. To perform "what if"
analysis, the mathematical model must be linked to a procedure to
translate general judgments about prospective changes into specific
assumption statements directly related to model variables and parameters. Such a procedure is scenario-writing. The scenario-writing
process consists of two steps: identification of scenario-variables
and assignment of specific values to these variables.

     Identification of scenario-variables: scenario-variables are a
subset of the variables and parameters of the mathematical model.
Parameters or variables not in the model can, therefore, not be part
of the scenario. As already mentioned above, the model structure
should reflect the desired scenario analysis. A demographic model,
for instance, is unable to trace the impact of the energy crisis on
population growth and distribution, since it does not contain energy
variables. The choice of the appropriate scenario-variables has
generated already considerable discussion. A remark, made
frequently with regard to purely demographic models of regional
population growth, is that migration levels and directions are
closely tied to variations in socio-economic conditions, and that,
therefore, migration should be endogenized by linking the
demographic model to one or a series of models of socio-economic
change. Scenario-variables should not directly relate to migration
levels and directions, but to the factors explaining migration.
Although very realistic at first glance, this claim is only justified if three conditions are met (Brass, 1974:565): (i) the relation
between migration and the explanatory (socio-economic) variables
must be a close one, (ii) the relation must persist over time, and
(iii) the explanatory variables must be predictable, with a greater
accuracy than the demographic measures.

If one of these conditions is not met, then the endogenization of the migration variable cannot improve the reliability of the results. In an era of economic uncertainty, the potential of demographic-economic models for improving population forecasts may be limited. Instead, an intermediate approach may be fruitful. It consists of two steps. First, migration rates by migrant categories are treated as scenario-variables. It presents the impact on population distribution of changes in migration pattern. The impact can completely be attributed to the features of the demographic system. Second, an explanatory model of migration is added, and the impact on migration of changes in some of these explanatory variables is studied. Assuming independence of effects, the total impact of changes in socio-economic conditions on population distribution is the product of the two impact measures.

Assignment of values to scenario-variables: Unlike in policy impact assessment, the most difficult part in forecasting is the assignment of values to scenario-variables. The selection of values, the scenario-variables are most likely to take on in the future, requires courage, demographic insight, some vision and a good deal of luck. Most forecasters solve this problem by providing the user with a set of alternatives. In many instances, this strategy creates more problems than it solves. Since forecasts frequently have their own independent lives, it becomes very difficult, if not impossible, to select between them without reference to the details of their preparation. The proper debate on the population forecasting issue should centre on the process of assigning values to scenario-variables, since it is in this process that judgments are translated into specific assumption statements. The debate may focus on a few questions: (i) who should assign values to scenario-variables (scientists, user, policy-maker, panel of experts, a socio-economic model)? In the Netherlands, scenario-variable values are selected by a Commission, including representatives of the Central Bureau of Statistics, planning agencies, and government agencies; (ii) how should values be assigned (common practice, Delphi technique, random number generation, educated guess, crystal ball)?; (iii) how often should values be adjusted (ie. how often should forecasts be updated?); the answer lies between no adjustment at all, to a highly-developed monitoring system); (iv) which values should be assigned (extrapolation, most likely values, desired values)?

Model Implementation

In her contribution to this conference, Pack (this volume)lists a few conditions that must be satisfied for a smooth implementation of a policy analysis or forecasting model. Similar conditions are discussed by Brooking and Hake (1979) in their paper on the adoption of models for economic policy analysis. A major factor is the environment of the client agency and the model. Pack found that most important for the decision to adopt a model was the presence on

the agency staff of an energetic model enthusiast, who would
advocate the model within his organization. She places the main
determinant of acceptance or rejection outside of the model
characteristics. Some scientists, eager to do socially relevant
research, and to see their intellectual product be integrated in
planning practice, may be frustrated by this fact of life. There
are, however, some limiting conditions that must be satisfied before
model adoption can even be considered. One of them is data
availability.

The lack of adequate statistical data is frequently given as a
reason for not adopting advanced techniques for policy analysis and
forecasting. Data limitations are severe for multiregional models
which require knowledge of interregional flows by origin and destination. The choice may be between less advanced, ie. less realistic
models and the estimation of missing data. The estimation of
spatial interactions has interested scientists for many years (for a
review, see Nijkamp, 1979). The gravity model, entropy and biproportional adjustment techniques have been studied and applied extensively. Recently, a new dimension has been added to this estimation
problem that may well lead to a complete change in our perception of
the estimation problem of spatial interaction data. Spatial interaction data constitute a contingency table. The estimation of missing elements of the table from available (frequently aggregate)
information implies that certain assumptions are made on the statistical dependence between the cross-tabulated variables. Recent
findings in discrete multivariate analysis or categorical data
analysis may fruitfully be applied to describe in explicit and simple terms, the interaction pattern imposed on the estimates by the
known information. The quality of the estimates may be increased if
one can find a simple expression that relates the interaction patterns in the estimates (and therefore, the values of the estimates
themselves) to each element in the set of prior information. The
log-linear model of contingency table analysis, developed by
Goodman, is such an expression. Bishop, Fienberg and Holland
(1975), and Goodman (1978) provide good discussions of the analysis
of contingency tables. This observation could have even a greater
impact on spatial interaction modelling than Friedlander's (1961)
paper on the estimation of contingency tables when marginals and
some supplementary data are known. Willekens (1980b) shows how the
gravity model, the entropy model and the biproportional adjustment
model relate to the log-linear model. The parameters of each of
these models are also related. The functional form of this relationship is a valuable aid for the statistical interpretation of the
parameters.

The relationship between conventional spatial interaction
models and Goodman's log-linear model also enables the analyst to
answer the question whether additional prior information may be
expected to increase the accuracy of the estimates substantially.
This is of particular relevance for the decision process in terms of

which data generation technique to adopt (survey or non-survey), since the contribution of each piece of information to the final quality of the estimates (goodness-of-fit) can be quantified. For instance, Willekens, Por and Raquillet (1979), in a report on the estimation of age-specific migration flows from aggregate data show how additional prior information affects the estimation accuracy. The log-linear model helps to explain why. Data limitations will always exist, and data collection will always be costly. If we focus on gaining substantial insight in estimation procedures instead of on developing more complex techniques, we may find an efficient way to determine a minimum of prior information that is necessary to obtain estimates of an acceptable level of accuracy.

CONCLUSION

The systems approach in urban and regional planning and policy-making is evolving rapidly. New modelling perspectives are being developed in response to the demand for realism, diversity and consistency in describing complex phenomena and processes. This paper compares the approaches to modelling systems of regions and elaborates on a few fundamental issues. The paper discusses theoretical, methodological and organizational aspects of the selection of models for planning purposes. The multiregional approach is being proposed as a strategy to deal appropriately with several "hot" issues in modelling systems of regions. There is, however, a warning. Models may help to clarify issues and problems, and may pave the way for their solution by decomposing a problem into its components, and by focusing a discussion on key items. Models will not, however, solve all problems; they will never be able to replace completely human intuition and common sense.

REFERENCES

Ballard, K., N. Glickman and R. Gusteley, 1980, A Bottom-up Approach to Multiregional Modeling: NRIES, in G. Adams and N. Glickman, (Eds.), Modeling the Multiregional Economic System: Perspectives of the Eighties, D.C. Heath, Lexington, Mass.

Ballard, K., R. Gusteley and R. Wendling, 1980b, NRIES: Structure, Performance, and Application of a Bottom-up Interregional Econometric Model, US Department of Commerce, Washington, DC.

Bishop, Y., S. Fienberg and P. Holland, 1975, Discrete Multivariate Analysis. Theory and Practice, MIT Press, Cambridge, Mass.

Bolton, R., 1980, Multiregional Models in Policy Analysis, in G. Adams and N. Glickman, (Eds.), Modeling the Multiregional Economic System: Perspectives for the Eighties, D.C. Heath, Lexington, Mass.

Brass, W., 1974, Perspectives in Population Prediction: Illustrated by the Statistics of England and Wales. Journal of the Royal Statistical Society A, 137, 532-570.

Brooking, C.G. and D.A. Hake, 1980, Impacting the Decision Process with a Regional Econometric Model: Some Practical Considerations, in G. Adams and N. Glickman, (Eds.), Modeling the Multiregional Economic System: Perspectives for the Eighties, D.C. Heath and Co., Lexington, Mass.

Coale, A. and P. Demeny, 1966, Regional Model Life Tables and Stable Populations, Princeton University Press, Princeton, New Jersey.

Coale, A. and T.J. Trussell, 1974, Model Fertility Schedules, Population Index, 40, 185-258.

Courbis, R., 1975, Le Modele Regina, Modele du Developpment National, Regional et Urbain de l'Economie Francaise, Economie Appliquee, 28, 569-600.

Dziewonski, K., 1978, Analysis of Settlement Systems: the State of the Art, Papers of the Regional Science Association, 40, 39-49.

Eichperger, Ch.L., B.A. van Hamel and W.P. Nieuwenhuis, 1979, A Methodology for Regional Population Forecasting, Research Report 15, Physical Planning Agency, The Hague, The Netherlands.

Friedlander, D., 1961, A Technique for Estimating a Contingency Table, given the Marginal Totals and Some Supplementary Data, Journal of the Royal Statistical Society A, 124, 412-420.

Gibberd, R., 1980, Aggregation of Population Projection Models, Working Paper WP-80-00, IIASA, Laxenburg, Austria.

Glickman, N. (Ed.), 1980, The Urban Impacts of Federal Policies, The John Hopkins University Press, Baltimore.

Goodman, L., 1978, Analyzing Qualitative/Categorical Data, Abt Associates, Abt Books, Cambridge, Mass.

Gordijn, H. and H. Heida, 1979, Multiregional Demographic Model and Development of a Monitoring System. Research Report 79/PS/231, Research Centre for Physical Planning, Delft, The Netherlands.

Keyfitz, N., 1972, On Future Population, Journal of the American Statistical Association, 67, 347-363.

Nijkamp, P., 1979, Gravity and Entropy Models: the State of the Art, in G.R.M. Jansen et al, (Eds.), New Developments in Modelling Travel Demand and Urban Systems, Saxon House, Westmead, Eng.

Olson, R.J., 1978, MULTIREGION: A Socioeconomic Computer Model for Labour Market Forecasting in: N. Hansen, (Ed.), Human Settlement Systems: International Perspectives on Structure, Change and Public Policy, Ballinger Publishing Company, Cambridge, Mass.

Pittenger, D., 1976, Projecting State and Local Populations, Ballinger Publishing Company, Cambridge, Mass.

Polenske, K., 1975, The United States Multiregional Input-Output Model, D.C. Heath, Lexington, Mass.

Rees, P. and A. Wilson, 1977, Spatial Population Analysis, Edward Arnold, London.

Rogers, A., 1975, Introduction to Multiregional Mathematical Demography, John Wiley, New York.

Rogers, A., 1976, Shrinking Large-Scale Population Projection Models by Aggregation and Decomposition, Environment and Planning A, 8, 515-541.

Rogers, A. and L. Castro, 1979, Migration Age Patterns: II. Cause-Specific Profiles, Working Paper WP-79-65, IIASA, Laxenburg, Austria.

Ter Heide, H., 1973, Approaches and Concepts for Regional Population Projections, Bevolking en Gezin, (1973/3) 445-470.

Ter Heide, H., 1976, Demographic Distribution Functions, Bevolking en Gezin, (1976/1) 77-96.

Willekens, F., 1979, Regional Population Prospects, Paper presented at the Seminar on the Impact of Current Population Trends on Europe's Cities and Regions, Council of Europe, Strasbourg, France.

Willekens, F., 1980a, A Simulation Procedure for Multiregional Population Analysis and Forecasting, Paper presented at the First World Regional Science Congress, Cambridge, Mass.

Willekens, F., 1980b, "Entropy Maximization, Multiproportional Adjustment and the Analysis of Contingency Tables", Systemi Urbani, 2/3, 171-201.

Willekens, F., A. Por and R. Raquillet, 1979, Entropy, Multiproportional and Quadratic Techniques for Inferring Detailed Migration Patterns from Aggregate Data. Mathematical Theories, Algorithms, Applications and Computer Programs, Working Paper WP-79-88, IIASA, Laxenburg, Austria.

THE SAO PAULO METROPOLITAN STUDY: A CASE STUDY OF THE EFFECTIVENESS

OF URBAN SYSTEMS ANALYSIS

Marcial Echenique

Department of Architecture
University of Cambridge
Cambridge, CB2 1PX
United Kingdom

INTRODUCTION

In 1974 the Local Authorities of Sao Paulo were faced with the difficult decision of either stopping the construction of the under-ground or stopping the construction of a system of urban motorways. They could no longer afford to finance the construction of both facilities. Also some doubts were expressed at that time about the advisability of building car-oriented facilities with the emerging energy crisis. The mayor, at that time Mr. Colasuono, suspended the construction of an important sunken motorway (Paulista), and issued an order to bury the foundations and to pave a normal arterial high-way on top. The decision to build the underground was taken in the late 1960's and the plan is illustrated in Figure 1. By 1974, the cost of building it had escalated considerably, with an estimated cost of over US $1,000 millions for the first line (North-South line), which was to open in 1975. The decision to go ahead with the second line (East-West line), to be open by the late 1970's, was already taken, but the construction of the line was running into difficulties due to serious money shortages. During the late 1960's and early 1970's a number of important urban motorways had been constructed as part of a plan of building over 500 km of expressways as illustrated in Figure 2. By 1974 the cost of sustaining the motorway building programme was prohibitive and thus the authorities decided to commission a study to evaluate the two alternatives. SMT (municipal transport office) and EMPLASA (metropolitan planning office) commissioned the study SISTRAN with the objectives of "form-ulating a coherent programme for urban passenger transport for the metropolitan area of Sao Paulo and the creation of technical and operational capabilities for the continuous planning of Sao Paulo", SISTRAN (1975a).

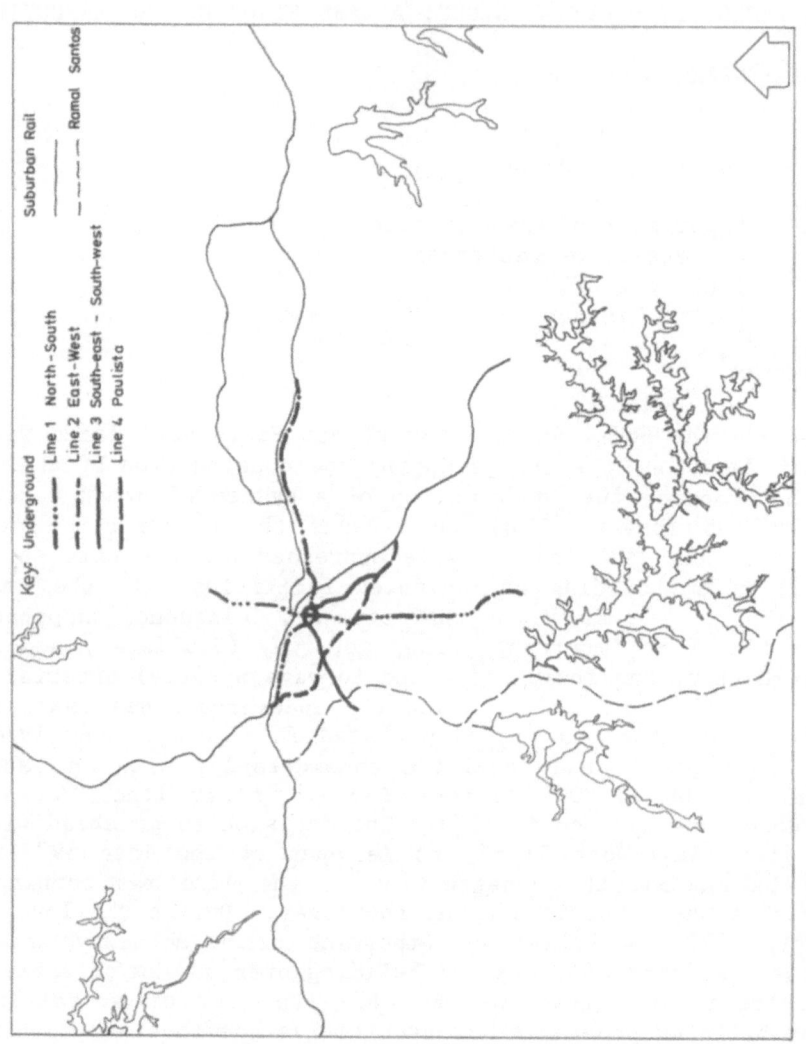

Figure 1 – Proposed Underground System

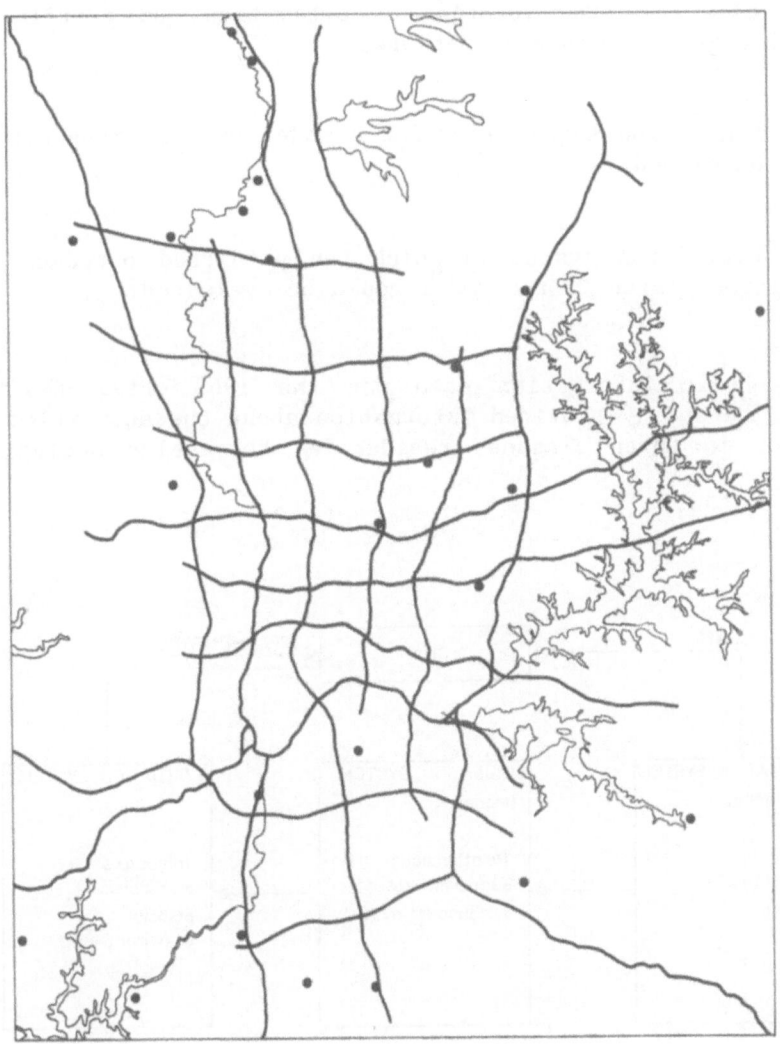

Figure 2 - Proposed Expressway System

METHODOLOGY

In order to evaluate these alternative investment policies a planning methodology was developed for the study. The methodology contains three interrelated systems as illustrated in Figure 3:

1. The information system which coordinated the information about the city, ie. the location of activities and traffic flows, buildings and transport networks.

2. The simulation system in which a model of the transport market was developed.

3. The evaluation system in which the predicted outcomes of the policies tested in the model could be evaluated.

The basic information used was the 1968 origin-destination survey. The survey provided information about the activities of the population in three classes· residential households owning a car;

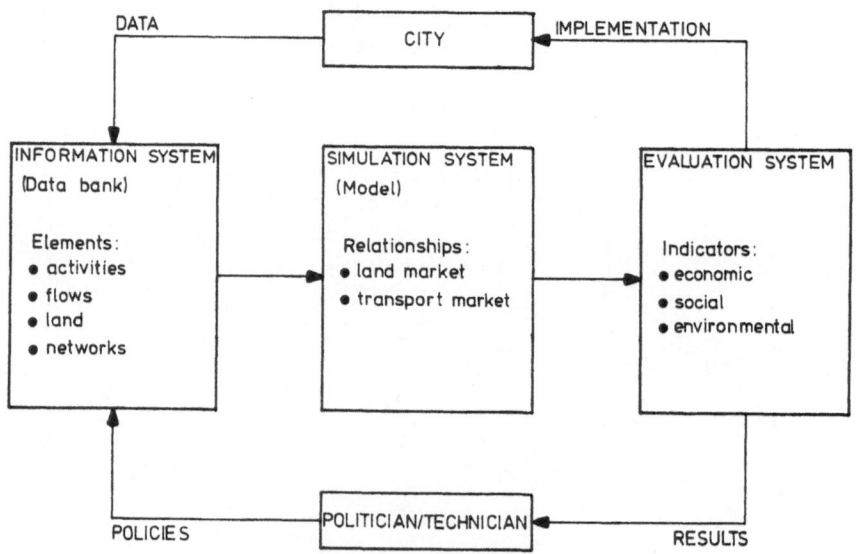

Figure 3 - The Planning System

residential households without a car; and non-residential activities (eg. commercial, industrial and services). The metropolitan area was divided into seventy-eight zones (as shown in Figure 4 with each zone containing information about the three classes of activities (the measures employed for each class were the number of trip origins by residences with and without cars, and the number of trip destinations in non-residential areas). The traffic flows from origin zone to destination zone were also provided by the survey, including the time of day and the mode of travel. Information about the building stock in each zone was provided for 1968 and 1975 by the municipal rating files (cadastre), and the network information was provided by the public transport operators and the road engineers. The network was aggregated into "spiders" (a notional link containing all the characteristics of the secondary road system which connects adjacent zones) together with a primary network: motorways, arterial roads of importance, rail and metro as shown in Figure 5.

Due to the paucity of data, a simple model was developed to simulate the operations of the transport market. The model essentially allocated trips by person type (car owning households and non car owning households) from zones of origin to destination zones, by mode of transport (car, bus, train, metro, trolley bus and combinations) and to routes within a mode. The allocation to alternative destinations, modes and routes was a function of two factors: (i) the price of transport (the price included the time as well) and (ii) the location of activities or land uses (as measured by the number of origins and destinations). The transport model reached an equilibrium between the supply of transport given by the capacity, speed and cost in the networks and the allocated demand for transport. The model was iterated until the supply and demand were in equilibrium in each link of the network. Once the equilibrium price was obtained, the model calculated changes in the land uses and activities, by reallocating the trip origins and destinations in the next time period. This reallocation included the growth of trips for the whole area due to increases in population and income as shown in Figure 6. The allocation of activities (or trips) to each zone was a function of the transport price calculated by the transport model, the availability of land and the availability of infrastructure in each zone (eg. water and sewerage). Figure 7 shows diagrammatically the operation of the model. The Appendix describes the mathematical description of the model.

The Evaluation system made a comparison between two or more outputs of the simulation system. In essence the evaluation system built a set of indicators comparing one policy against another. These indicators reflected the economic efficiency of the policy implemented (measured by the relationship between the benefits obtained and the cost associated with the implementation), the

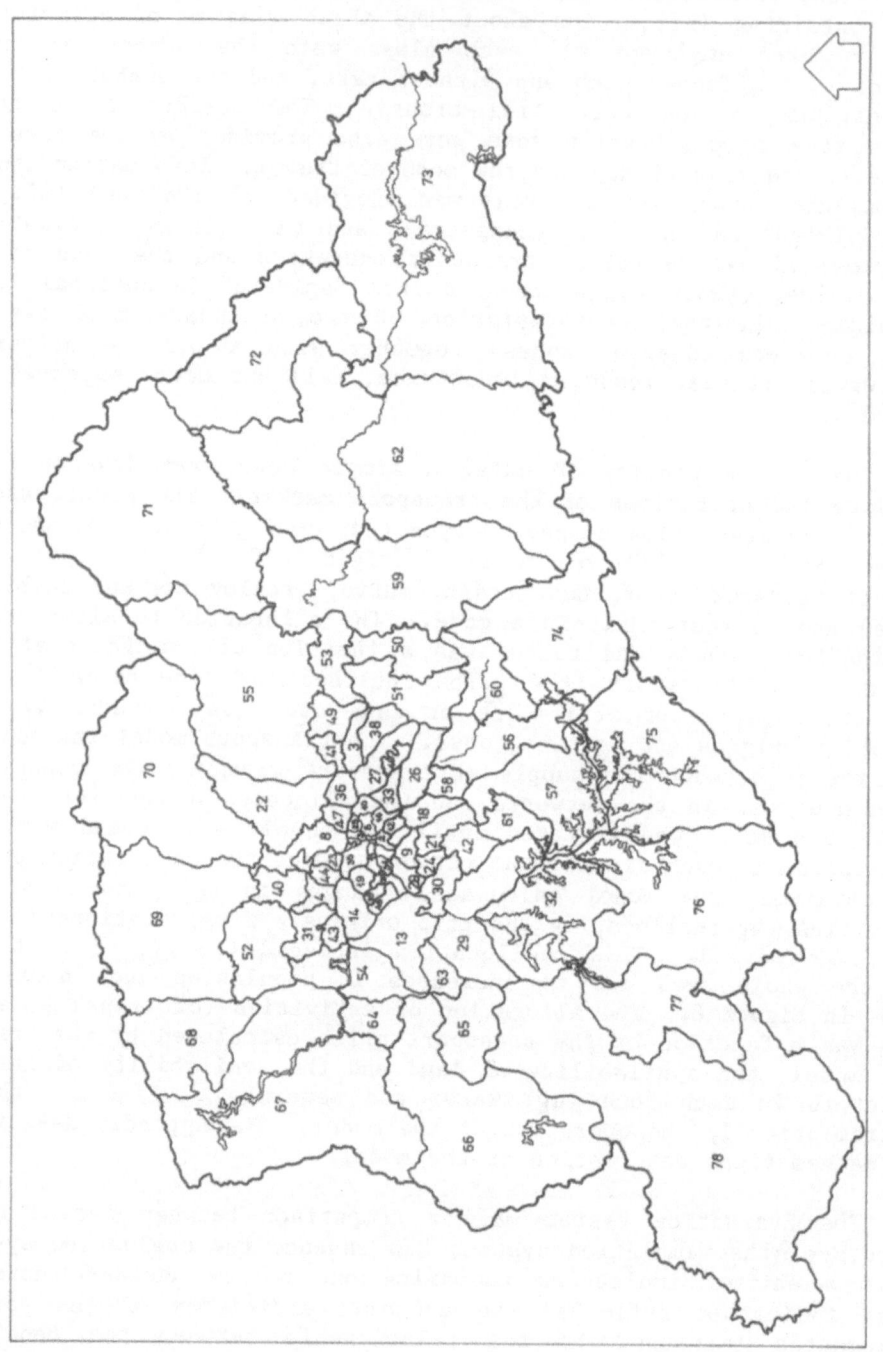

Figure 4 - Zoning System

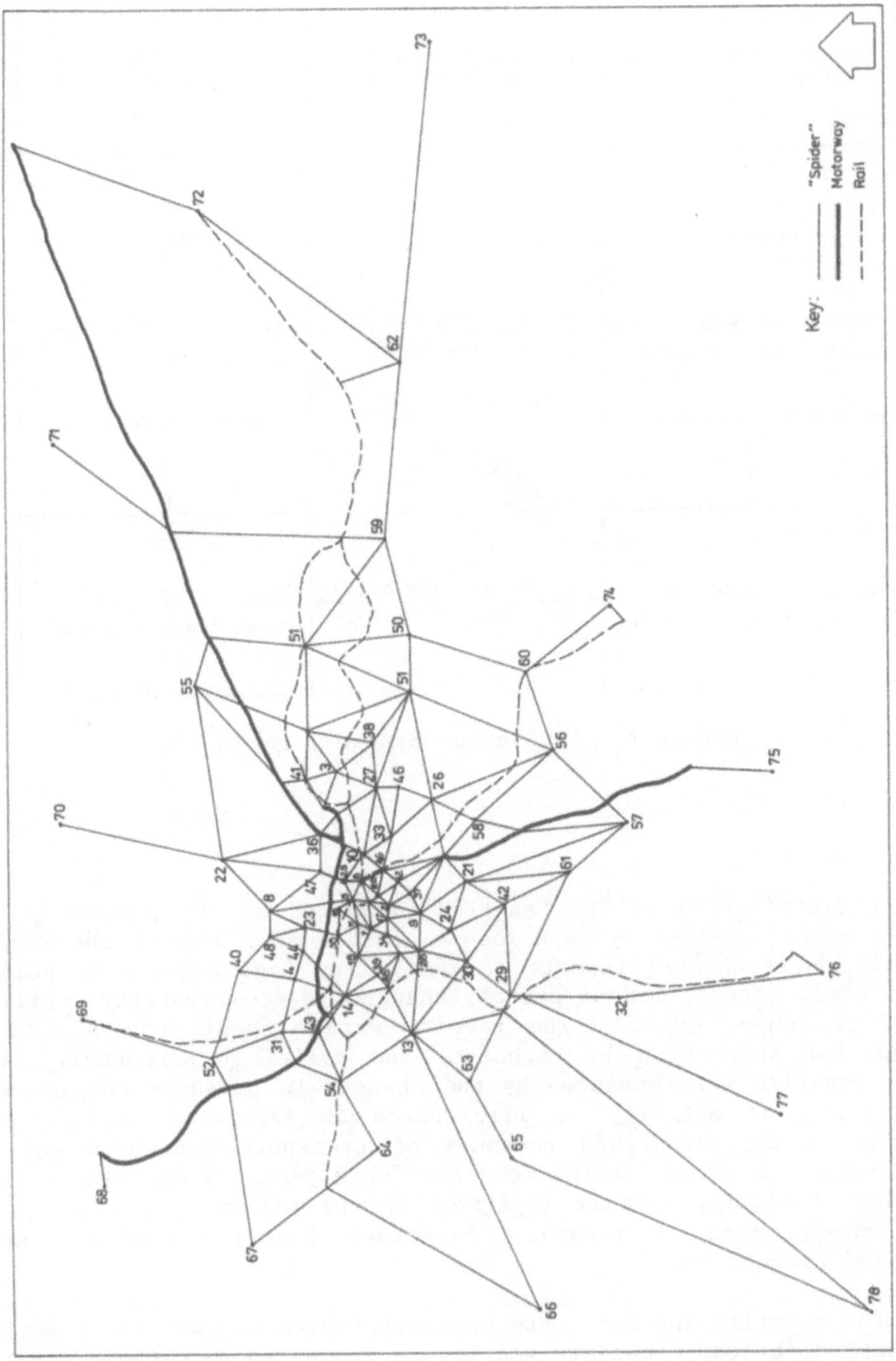

Figure 5 – Network 1968

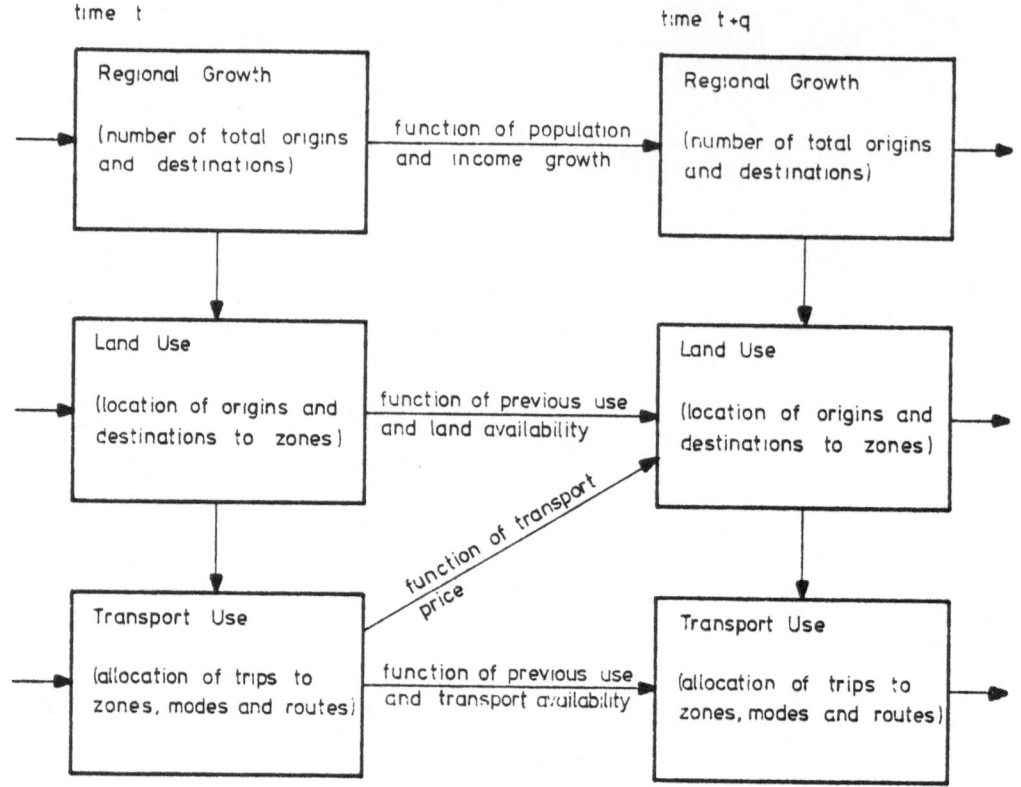

Figure 6 - Simulation System Through Time

social distribution of the benefits (measured by the percentage of the benefits received by each income group and by area of the city), and the environmental impacts of the policy (eg. changes in pollution level, consumption of petrol, changes in accessibility, utilization of zones, etc.). The results were presented in a set of tables for inspection by technical and political personnel. The total benefits were measured by the changes in consumer surplus for each class of activity as illustrated in Figure 8, and by the changes in the costs and revenues of transport operators and of government. Summing all the benefits for a year and dividing by the discounted capital cost of implementing the policy gives the first year annual rate of return. Flowerdew (1977) provides a more detailed discussion.

The benefits and the costs were attributed to each group of the population (by car ownership and income group) so that their distribution could be ascertained. Also the spatial distribution was

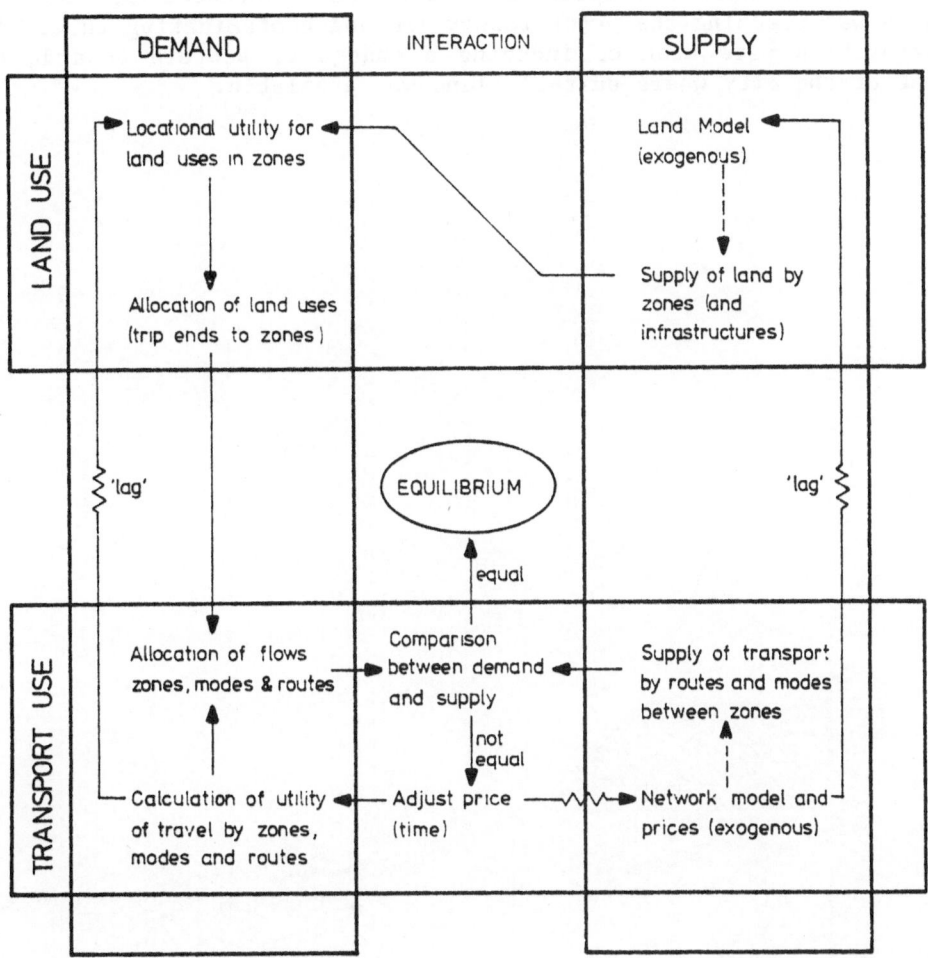

Figure 7 - Simulation System - Model of the Transport Market

calculated for areas of the city such as the periphery or the core
area (where most rich people live). With these indices it was possible to estimate who was paying and who was benefiting from the
policies. Finally a set of environmental indices was constructed to
highlight particular planning issues such as changes in pollution
levels, in the consumption of petrol, in accessibility of various
groups of people, and changes in the pattern of urbanization. The
political authorities were particularly concerned with the continuous suburbanization towards the South and Southwest of the city
which was reaching the water reservoirs and contaminating them. The
authorities were keen on inducing a change of pattern towards the
east of the city where suitable land was available.

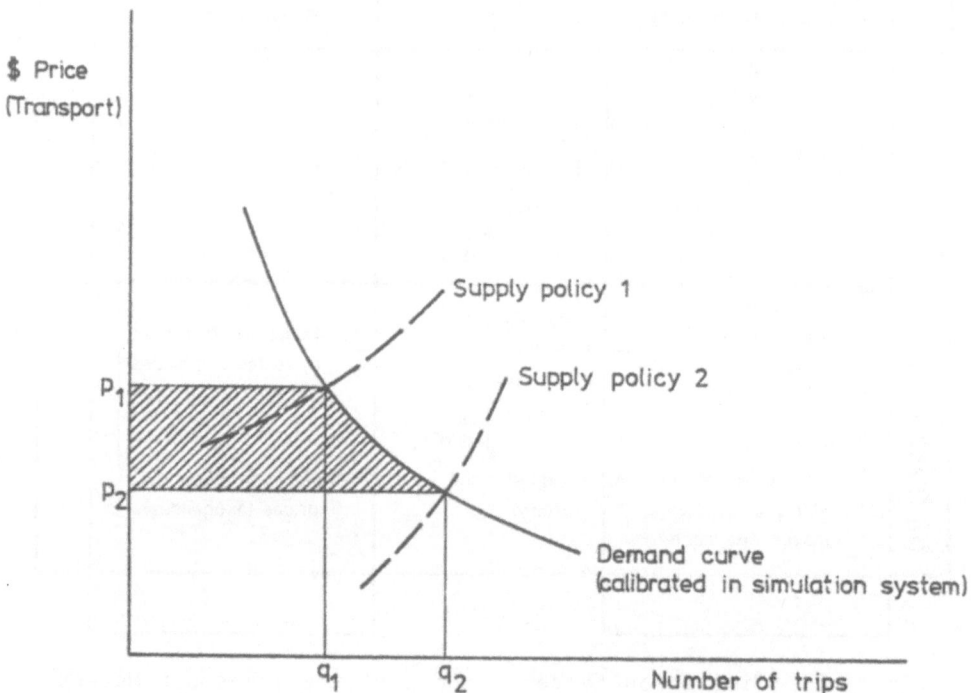

The predicted trips (q_1 and q_2) for each policy and the corresponding equilibrium price (p_1 and p_2) are outputs of the simulation
system.

Figure 8 - Consumer Surplus (shaded area) as a Result of Implement-
          ing Policy 2.

THE EVOLUTION OF SAO PAULO

Sao Paulo is one of the fastest growing metropolitan areas of the world. The rate of growth during the decade of the 1960's averaged 5.5 percent per annum. By 1975 the population was estimated to be 9,855,000 inhabitants. The extrapolation of the trend up to 1990 gives over 22 million inhabitants. The SISTRAN study took a lower rate (3.08 percent) which gives 12 million by 1980 and 14.5 million by 1985. Due to the growth of services and industries the growth of incomes has gone up. In 1970 the industrial sector of Sao Paulo represented 40 percent of the Brazilian industrial product and 35 percent of the work force. The rate of growth of average household incomes has gone up to 7 percent in the 1970's. It was estimated that in 1975 the average monthly income was cr. $3,148. The SISTRAN study took a more conservative rate of growth of incomes at 5 percent per annum. The car ownership figure was closely related to household incomes. In 1968 25 percent of the households had one or more cars, by 1975 44 percent of households were estimated to have cars, and by 1980 it was forecast that 58 percent of the households could potentially afford a car.

The effect of the combined growth of population and incomes on the demand for travel is formidable. The number of daily trips in 1968 was close to 7 million, by 1975 it was estimated to be 12 million trips per day, rising to 17 million by 1980 and 23 million by 1985. The increase in trips was partially due to the increase in population (45 percent) and partially to the increase in incomes (55 percent). A household with a car available in 1968 generated 7 trips per day as compared to 3.54 trips per day for a household without a car. The increase in trips and the relative importance of the use of a private car, which had gone up from 29 percent of all trips in 1968 to 36 percent by 1975, has decreased the overall accessibility of Sao Paulo despite the enormous investment in transport during the period. The average cost of trips has gone up by 22 percent for car users and by 19 percent for public transport users from car owning households. Through the use of the model it was predicted that further deterioration would take place, despite the committed new transport investment (the North-South and East-West line of the metro, suburban railway upgrading and committed highway construction). The values predicted for 1985 were: for car users a 24.3 percent deterioration; for public transport users from car owning households one of 16.2 percent; for non car owning households a 15.5 percent deterioration. It is interesting to point out that due to the relatively greater deterioration in the accessibility by car an increasing number of trips by public transport from car owning households was taking place: from 5 percent in 1968 to 31 percent in 1975. It was predicted that by 1985 34 percent of the trips made from car owning households would be by public transport with the committed transport investment. This could happen despite the relative decline of public transport usage from 71 percent of all trips in 1968 to 64 percent in 1975 to 51 percent in 1985.

The average peak hour congestion in the main radials to the
expanded centre (which excludes the CBD and New Centre -- Paulista)
as measured by the relationship between flows and capacity has gone
up from .63 in 1968 to an estimated .76 in 1975. It was predicted
that by 1985 a saturation level of .97 would be reached if the presently committed investment was all that would be introduced. However, according to the model the CBD was declining due to increased
congestion (shift of trip destinations) and the whole of the expanded centre would reduce its relative share of the total constructed area from 24.5 percent in 1968 to 16.8 percent in 1985,
with an increased participation of the Southwest area of the city
going from 25.8 percent to 33.2 percent by 1985. It is interesting
to note that the shift in destinations as a result of congestion
cost is not usually considered by conventional transport analysis
(ie. lack of "feedback" between transport and land use.

THE EVALUATION OF ALTERNATIVE POLICIES

It was clear from the outset of the study that the simplistic
notions of either investing in the road building programme or in the
public transport system were untenable. It was necessary to invest
in both as they are mutually complementary. Naturally, the particular forms of the proposed programmes needed drastic revisions.

The urban motorway system could not be built within the projected financial resources of Sao Paulo. It was also essential that
a number of improvements in important links in the system needed to
take place as well as road maintenance, signals, etc. Thus a more
realistic reduced motorway system was designed based on the projects
already in existence. The urban motorway programme was divided into
two sets: internal area programme, roughly corresponding to the
municipality of Sao Paulo and external area programme roughly
corresponding to the rest of the metropolitan area. The first was
costed at 16,384 million cruzeiros by 1985 and the second at 11,684
million cruzeiros by 1985. In contrast to these, a cheaper alternative was devised by substituting the motorways by arterial highways
(with traffic light controlled intersections and a reduced capacity
from 2,000 vehicles per hour to 1,200 vehicles per hour) this
option, illustrated in Figure 9, was costed at 6,284 million
cruzeiros by 1985.

The underground project could cope only with a small fraction
of the public transport trips in the best of the cases. Thus a
complementary basic network of public transport was studied. The
basic network would be initially operated by conventional buses in
corridors of bus-only lanes and streets. This was developed after
some tests of the model with the underground network introduced,
revealed a rather poor performance of the first metro line (North-
South). At best, 25 percent of the capacity would be utilized. The
only way to increase its utilization was to suppress all competing
bus lines and install an integrated bus-feeder service to bring
passengers to the metro stations. (Actually after the line was

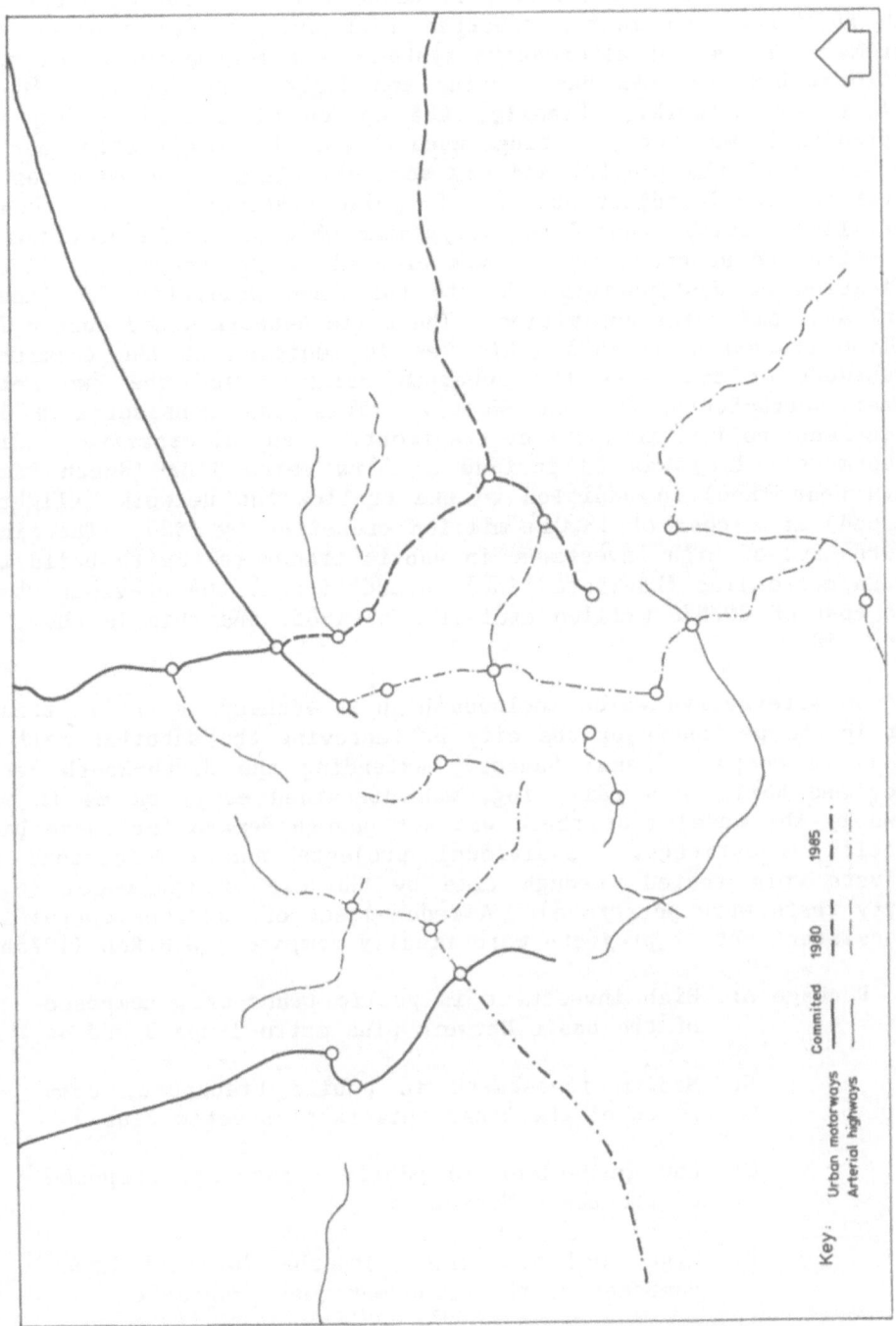

Figure 9 – Low Investment Alternatives in Highways

Key: Urban motorways   Committed   1980   1985
        Arterial highways

opened in late 1975, the predictions of the model were confirmed and since then this type of feeder-system has been implemented for the metro.) The Basic Public Transport Network would eventually operate with an integrated metro, suburban rail and special trolley bus system. A number of alternative systems were tested including conventional bus, express bus, trains and light rail for the 258 km trolley bus network. Finally, the option of a trolley bus was adopted as it was non-polluting, used cheaper hydro-electric energy, was easy to build locally and was more efficient. The need for an integrated complementary network of public transport, rather than a centralized radial trunk line type system which would be provided by the metro and suburban trains, was created by the increasing diversification of destinations, due to the decentralization of industrial and commercial activities. The Basic Network would cost 6,786 million cruzeiros by 1985 (this was in addition to the committed investment of improving the suburban railways and the two metro lines: North-South and East-West). This was considered a low investment policy in public transport. An alternative medium investment policy was to include a third metro line (South East-South West line) in addition to the trolley bus network (slightly reduced) at a cost of 13,014 million cruzeiros by 1985. The final alternative of high investment in public transport was to build the fourth metro line (Paulista line) in addition to the previous items at a cost of 20,836 million cruzeiros by 1985, and this is shown in Figure 10.

An alternative which included high investment in public transport in the periphery of the city by improving the suburban rail of the state company (Ramal Santos), extending the North-South metro line, and building a rail ring, was discarded early on as it was shown by the model that there was not enough demand for these high capacity investments. Individual projects and combinations of projects were tested through time by the use of the model (over thirty tests were performed). A reduced set of packages containing a consistent set of projects were finally compared, SISTRAN (1975b).

Package A: High investment in public transport, composed of the Basic Network plus metro lines 3 and 4.

B: Medium investment in public transport, composed of the Basic Network plus metro line 3.

C: Low investment in public transport, composed of the Basic Network only.

D: High road investment in the internal area, composed of the urban motorway programme.

E: Low road investment in the internal and external area by building arterial highways.

F: High road investment in the external area, composed of urban motorways.

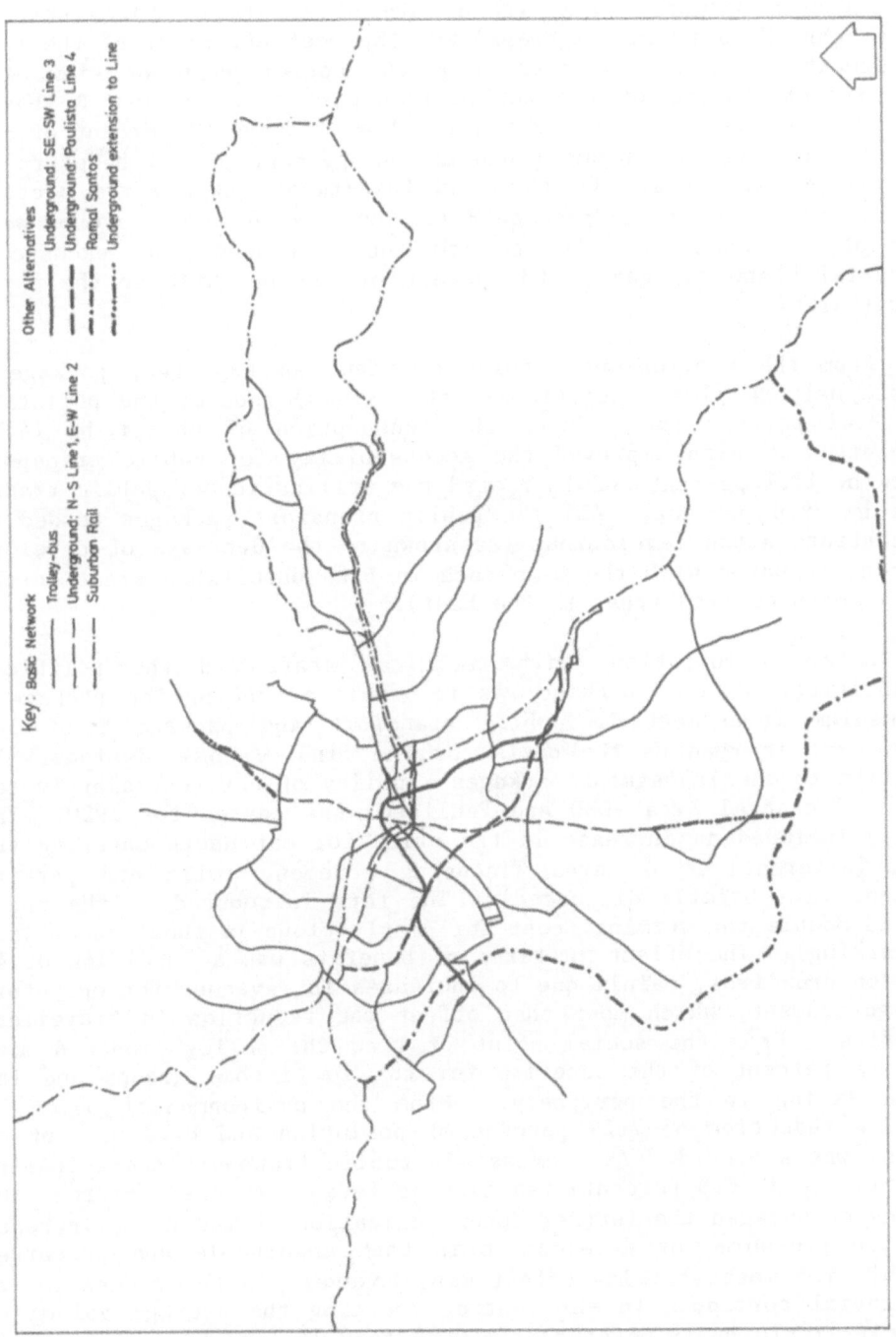

Key: Basic Network

———— Trolley-bus
— — — Underground: N-S Line 1, E-W Line 2
— · · — Suburban Rail

Other Alternatives

———— Underground: SE-SW Line 3
— — — Underground: Paulista Line 4
— · — Ramal Santos
···—···— Underground extension to Line

Figure 10 - Public Transport Alternatives

Table 1 shows the main indicators summarized for each policy compared with the Base Run (only committed investment by 1985). As it can be seen the most efficient of the public investment packages was C (basic network only) with a rate of return of 13.75 percent and package E (arterial highways) was the most efficient of the road investment packages. However, from the social point of view, the most redistributive of the public transport packages was B (basic network plus one additional metro line) as it gave 57 percent of all benefits to the low income group and 43 percent of all benefits to the peripheral areas. On the road investment program the picture was not so clear-cut as package F (external motorways) gave 50 percent of the total benefits to low income groups, but package E (arterial highways) gave a 63 percent of all benefits to the peripheral area.

From the environmental point of view, as expected, package A (basic network plus 2 additional metro lines) reduced the pollution by 19.47 percent and reduced the consumption of petrol by 14.57 percent. It also improved the accessibility for public transport users by 13.1 percent and increased the utilization of public transport by 29.6 percent. All the public transport packages tended to concentrate urban development (as shown by the decrease of development as compared with the base case in both unsuitable areas to the South and preferred areas to the East).

After consultations with technical staff and the political authorities, a final package was reformulated taking the package B of medium investment in public transport and package E of low investment in road as the basis for the final recommendations. In addition to the investment packages a policy of car restraint in the expanded central area (CBD and Paulista) was tested for 1980. The policy included an increase in the price for car users entering the area (alternatives of area licensing schemes, tolls and parking charges were briefly discussed). The form introduced in the model was to double the terminal cost for destinations in those zones (tax on parking). The effect in terms of benefits was an addition of 48 million cruzeiros, mainly due to increases in revenues for operators and government, which more than offset the reduction in travellers benefits. From the social point of view the policy produced more than 64 percent of the benefits for the low income groups and for those living in the periphery. From the environmental point of view, a reduction of 7.29 percent of pollution and 8.70 percent of petrol was achieved. An increase in public transport accessibility and ridership (6.5 percent) was also achieved. As was expected, the policy encouraged the further decentralization of the city, increasing the pressure for development in both unsuitable and preferred areas. The most striking effect was, however, in the congestion of the radial corridors to the centre, reducing the average volume to capacity ratio by 19 percent from 0.88 to 0.74, and this is shown in Figure 11. To achieve something remotely comparable by public transport investment the cost would be 14,000 million cruzeiros.

TABLE 1

EXAMPLE OF EVALUATION OF PACKAGES AGAINST THE BASIC RUN – 1985
(ie. committed investment only)

Evaluation Criteria		PACKAGES					
		Public Transport Investment			Road Investment		
		A. High: Basic Network & two Metro lines	B. Medium: Basic Network & one Metro line	C. Low: Basic Network only (Trolley)	D. High Internal Motorway	E. Low Arterials	F. High External Motorways
	Cost (Discounted)	20,836	13,014	6,786	16,384	6,284	11,684
I Economic Efficiency (millions of Cruz)	Annual Benefits:						
	Consumer (Traveller)	1,828	1,370	673	1,553	940	601
	Operators and Government	- 184	77	259	333	355	112
	Total	1,644	1,447	932	1,886	1,295	713
	Rate of Return	7.89	11.12	13.75	11.51	20.60	6.10
II Social Distribution (%)	Low Income % of Total Benefits	52	57	44	44	48	50
	High Income % of Total Benefits	48	43	56	56	52	50
	Peripheral Area % of Total Benefits	34	43	31	42	63	60
	Core Area % of Total Benefits	66	57	69	58	37	40
III Other Criteria (Environmental) (% Change over base)	Pollution % Change	-19.47	-15.96	-10.73	3.89	7.80	-1.90
	Petrol. Consum. Change %	-14.57	-12.31	- 8.59	8.76	7.82	2.10
	Accessibility for:						
	Car Users	- 1.9	- 1.7	- 1.8	13.8	10.4	5.4
	Public Users	13.1	11.1	3.4	4.3	2.8	1.9
	Development In						
	Preferred Areas (East)	- 4.15	- 3.39	- 3.19	3.75	-0.68	5.50
	Unsuitable Areas (South)	- 5.74	- 5.45	- 6.63	3.33	2.10	5.33
	Use of Public Transport	29.6	24.9	17.5	-8	-7.6	0

Source: (SISTRAN 1975b) Alternativas Estudadas (Vol. II, 1975 – Secretaria do Estado dos Negocios Metropolitanos – EMPLASA and Prefeitura do Sao Paulo)

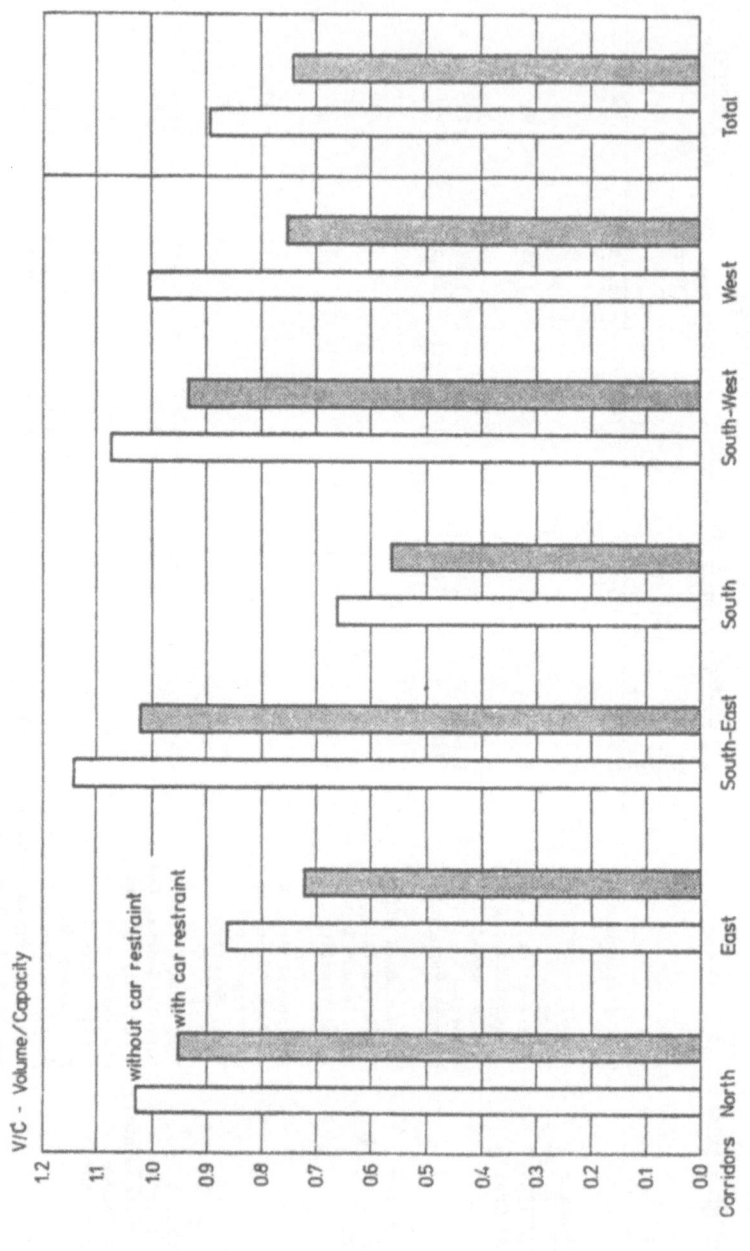

Figure 11 — Effect on Congestion on the Main Radials With and Without Car Restraint Policies

THE FINAL RECOMMENDATIONS

The recommendation for immediate action was the institution of a metropolitan authority for urban transport (EMTU) to operate the Basic Public Transport Network (metro, suburban rail, trolley buses and municipal buses). This authority would have also the responsibility for licensing private operators (bus) and for the establishment of fare prices. Another recommendation was to implement the basic public transport network by establishing a system of bus only lanes and streets and to operate the network initially with conventional buses, and later with trolley buses as they became operational. Also recommended were (i) a road improvement programme of arterial highways; (ii) the improvement of the East-West suburban rail (state and federal rail); and (iii) the implementation of the car restraint policy in the central area.

For the medium term the recommendation was the expansion of the Basic Public Network through the construction of the third metro line and the extension of the trolley bus system. A number of road investment programmes were recommended including some external motorways and arterial highways. A complete list of projects and financial resources was established, and the participation of the state in the total metropolitan transport budget was increased, SISTRAN (1975a).

THE RESULTS

Most of the recommendations for immediate action have been implemented since 1975. The new Metropolitan Transport Authority (EMTU) is now in operation and the public transport corridors have been developed. The municipal traffic department (DSV-CET) has implemented the bus-only ways with some interesting innovations in the use of conventional buses: this includes the "convoying" of buses in some corridors, increasing substantially the carrying capacity of each lane. The development of the trolley bus network is underway, including the construction of new special trolley buses. The increases in parking costs have been implemented, but are still well below the level recommended for the car restraint policy.

The medium term recommendations are going ahead with some difficulties in regard to the metro lines and the only major discrepancy from the original recommendations is that a high cost investment in the suburban rail of the state rail (Ramal Santos) is being developed.

Although it cannot be claimed that the SISTRAN study was absolutely essential in generating the programme actually developed, it is clear however that the municipal administration led by Mayor Setubal during the period of 1975-79 followed the recommended policies. It seems that the study contributed to establish a consensus of opinion amongst the different agencies as to the most realistic and effective combination of policies.

Unfortunately the other objective of the study, the establishment of a group for continuous planning and monitoring of the transport system, was not achieved. As soon as the project was ended, most of the technical group disbanded. Fortunately, however, the collection of data carried on, finishing with a new origin-destination survey completed in 1978, EMPLASA (1977).

The new information and the relative success of the policies implemented created a renewed interest by the same agencies for updating the model. The new Metropolitan Transport Authority (EMTU) joined COGEP (municipal planning) and EMPLASA (metropolitan planning) have commissioned a new study, MUT, with the expressed intention of updating the transport model and expanding the land use model. The new study has developed an explicit model of the property market of the metropolitan area as shown in Figure 12. The data base has been considerably expanded and a set of more detailed policies are in the course of being evaluated as illustrated in

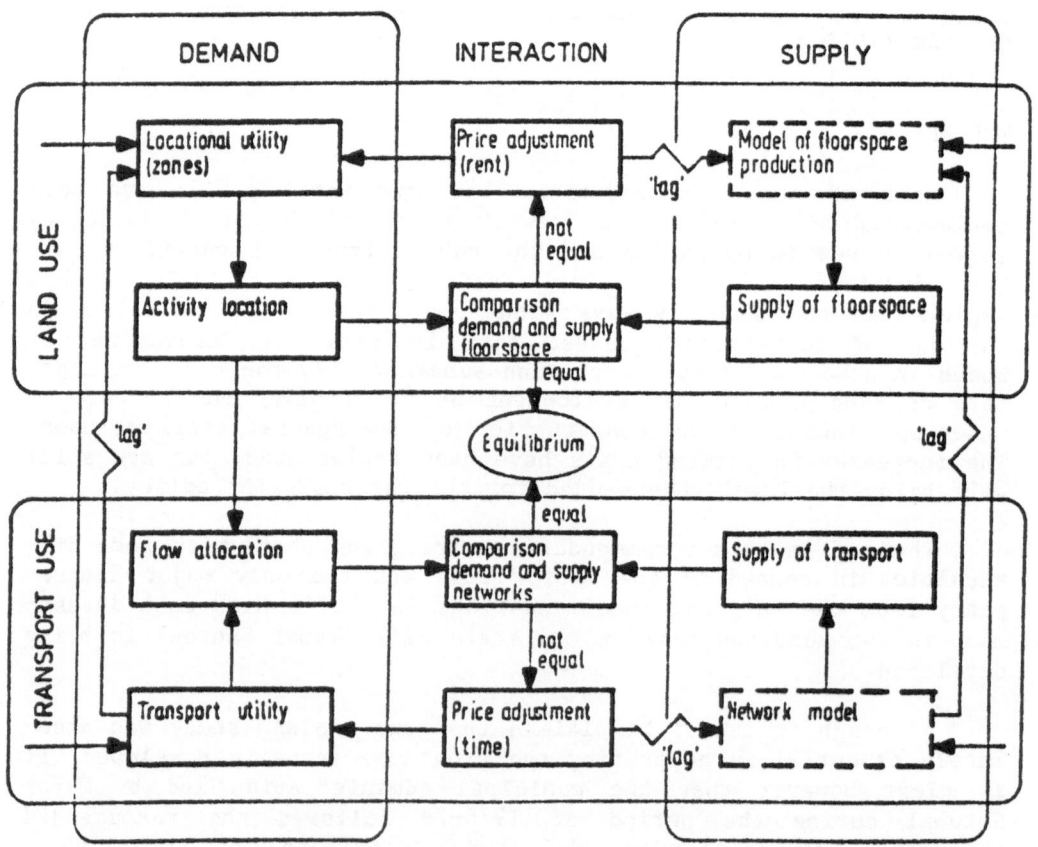

Figure 12 - The New Model MUT with Explicit Land Market

Figure 13.   The policies explored include betterment taxes for development, subsidies for operators of public transport and developers etc. (CET, 1980).   Finally, it is interesting to compare the predictions for 1977 made in 1974-75 by the SISTRAN study, based on the 1968 data and the actual values obtained by the 1977 surveys (CET, 1978).

The comparison between the prediction of SISTRAN and actual values obtained from the Survey are summarized in Table 2.   The values of SISTRAN were obtained by interpolating results between the 1975 run and the 1980 run.   The values obtained are from run No. 80-12 with car restrictions in the centre.   The results in aggregate are remarkably close.   The only major discrepancies are in the number of car-owning households which SISTRAN over-predicted by 13.6 percent.   In fact the predictions of SISTRAN are the number of potential car-owning households due to income and population growth.   The actual number could be less if improvements in public transport can be achieved.

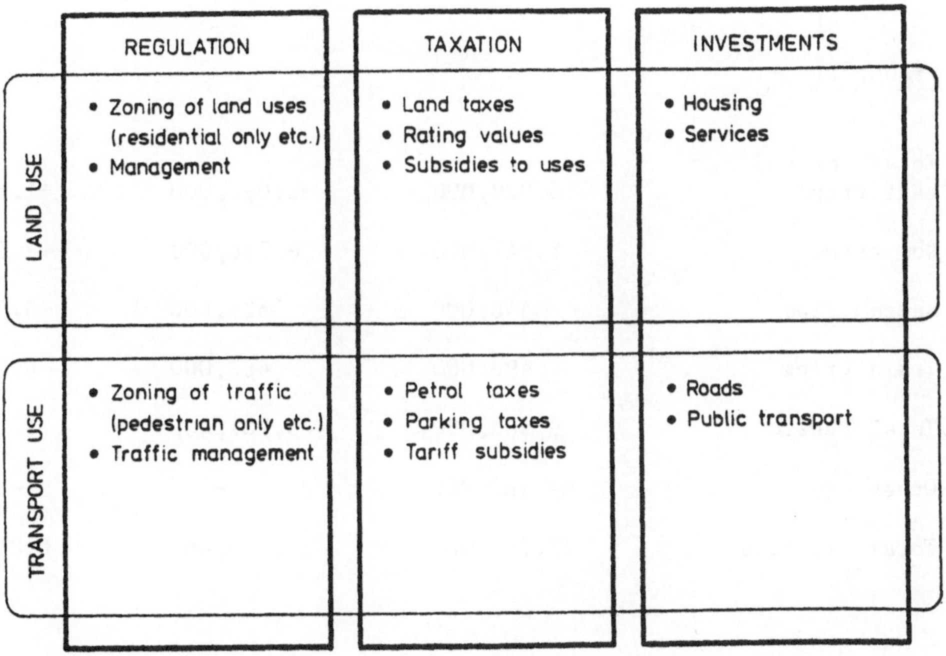

	REGULATION	TAXATION	INVESTMENTS
LAND USE	• Zoning of land uses (residential only etc.) • Management	• Land taxes • Rating values • Subsidies to uses	• Housing • Services
TRANSPORT USE	• Zoning of traffic (pedestrian only etc.) • Traffic management	• Petrol taxes • Parking taxes • Tariff subsidies	• Roads • Public transport

Figure 13 - Types of Policies being Tested with the MUT Model

TABLE 2

COMPARISON BETWEEN PREDICTIONS OF SISTRAN MODEL AND SURVEY

	Survey	SISTRAN	Percent Difference
Socio-Economic			
Total population	10,382,000	10,650,000	+2.6%
Number of households	2,443,000	2,500,000	+2.3%
Average monthly income	$7,747 cr	$7,634 cr	−1.5%
Car owning households	1,091,000	1,240,000	+13.6%
Non-car owning households	1,352,000	1,260,000	−7.0%
Transport			
Total car and taxi trips	6,029,000	6,098,000	+1.1%
Bus trips	8,542,000	8,710,000	+2.0%
Metro trips	548,000	529,000	−3.5%
Train trips	493,000	462,000	−6.3%
Total Public	9,583,000	9,701,000	+1.2%
Other	180,000	−	−
Total all modes	15,792,000	15,799,000	0.04%

The spatial comparisons are not as accurate, but are still quite good. Figure 14 shows the comparison of average income by aggregated zone (18 macrozones) made by SISTRAN and the Survey. The main discrepancy is zone 1, the centre, where SISTRAN overestimated the income of the population. In reality there has been a considerable decentralization of high income households from the centre. Figure 15 shows the comparison between the number of destinations in peak hour predicted by SISTRAN and the Survey. Again SISTRAN overestimated the concentration of destinations to macro-zone 1. The trend of decentralization in reality has been higher than that predicted by SISTRAN. Finally, Figure 16 shows the distribution of peak hour car flows in the main corridors.

As can be seen, the model, despite its simplicity, has been able to capture most of the pattern of location and passenger traffic occurring in Sao Paulo for ten years. It is remarkable that the model, calibrated with data for 1968, predicted so closely the actual behaviour of people after such major changes in policies as the opening of the underground line, changes in parking charges, increases in petrol costs, the opening of new public transport corridors, etc. This must give some reassurance to people using analytical techniques for forecasting the impact of transport policies in metropolitan regions.

APPENDIX: MATHEMATICAL FORMULATION OF THE MODEL

The Sao Paulo area was divided into 78 zones $(1,2,..i,j,..78)$. The total number of activities to be allocated in a time period was determined by the rate of growth of population and incomes. The residential activities were defined as the number of trip origins with car available, $O^{n=1}$, and without car available, $O^{n=2}$. The non-residential activities were defined as trip destinations, D, equal to the sum of trip origins. The origins and destinations were allocated to zones $\underline{i}$ of the city as a function of the availability of land in the zone for the activity type (L_i^n), the availability of public utilities: water and sewerage (U_i), and the aggregated price of transport for the activity in the zone (C_i^n), i.e:

$$O_i^n \sim L_i^{n\alpha_1^n} U_i^{\alpha_2^n} C_i^{n\alpha_3^n}$$ (1)

where $\alpha_{...}^n$ were elasticities calibrated as follows:

		α_1^n (Land)	α_2^n (Utilities)	α_3^n (Transport)	R^2
origins with car available	$O^{n=1}$	.14	1.22	-.24	.86
origins without car available	$O^{n=2}$	.34	not significant	-.84	.85
destinations	D	.58	1.31	-.32	.95

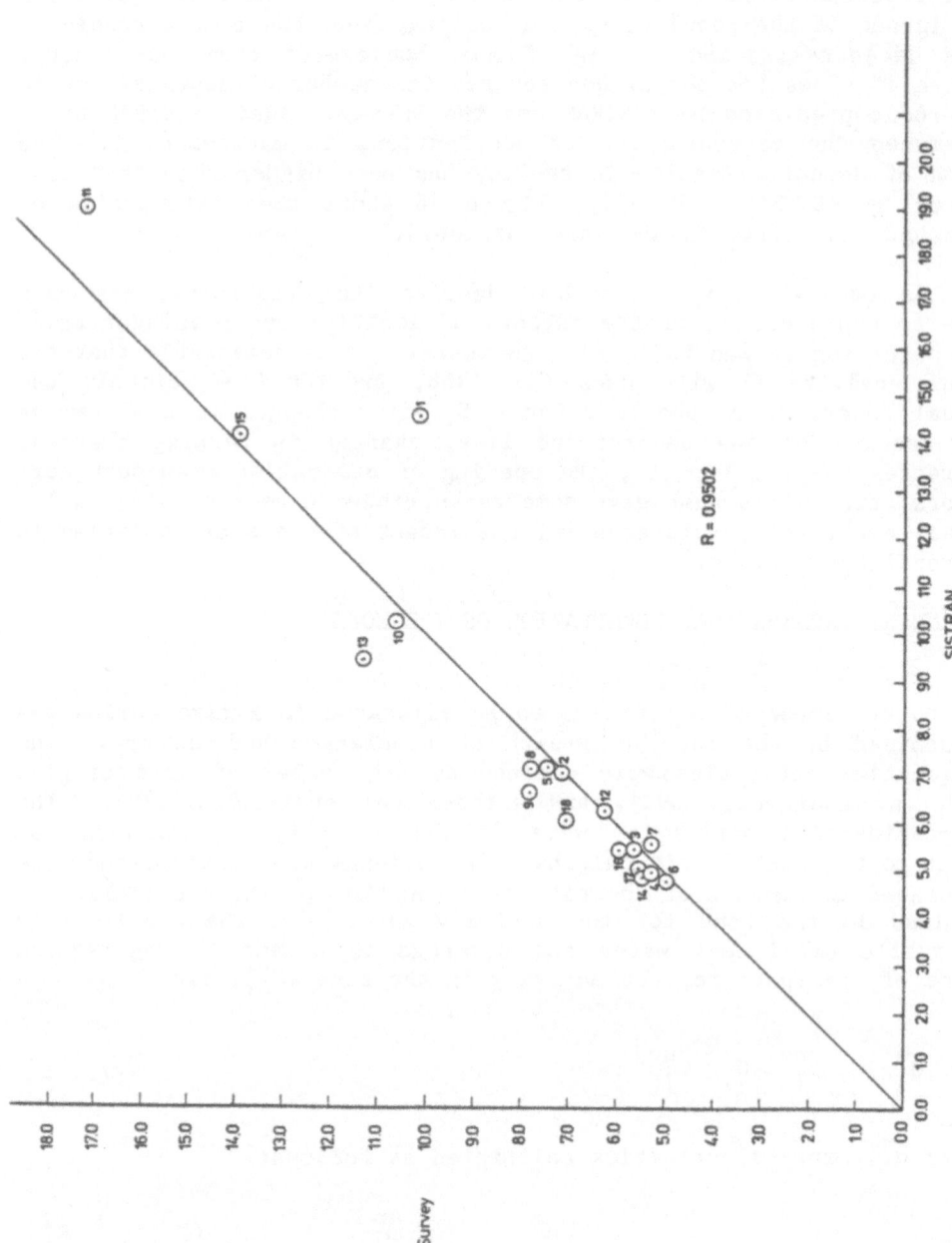

Figure 14 – Comparison of Average Zonal Income

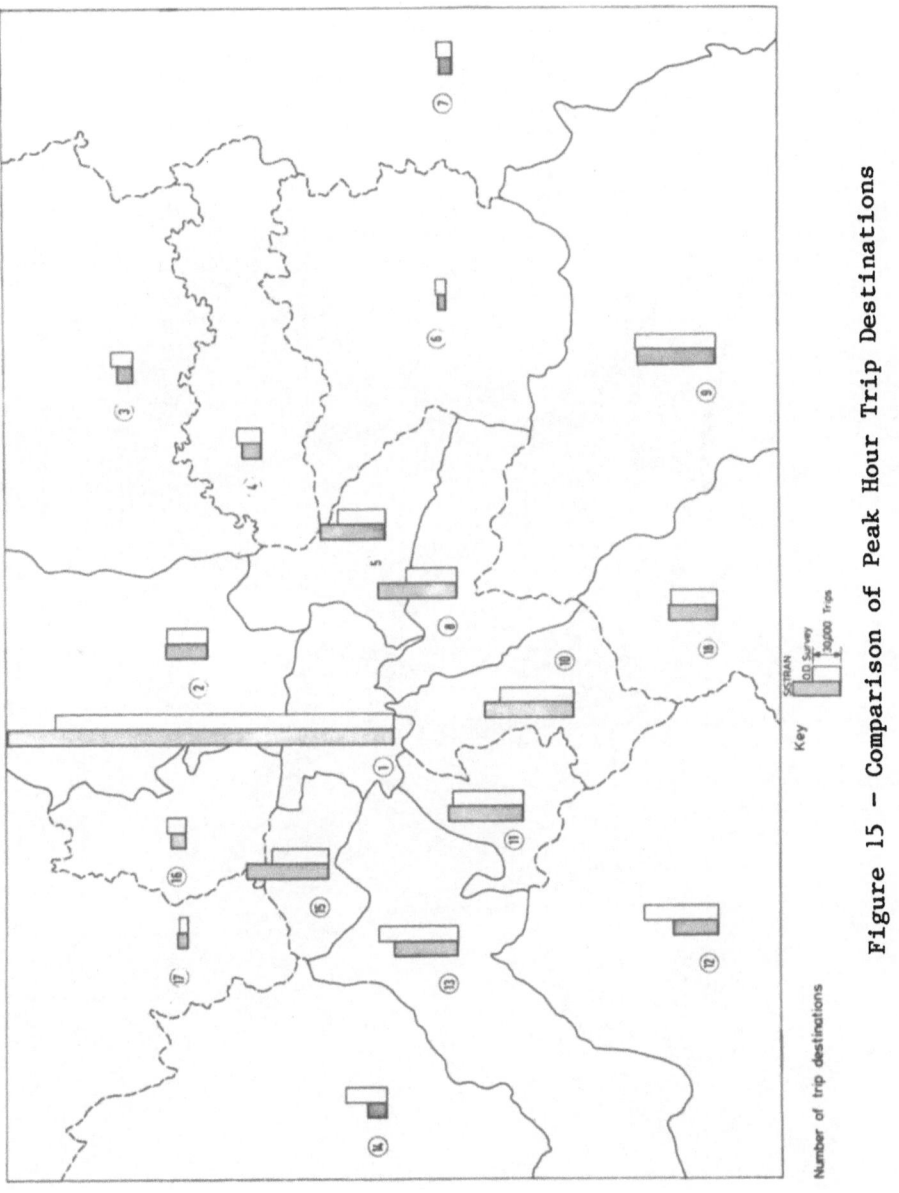

Number of trip destinations

Figure 15 — Comparison of Peak Hour Trip Destinations

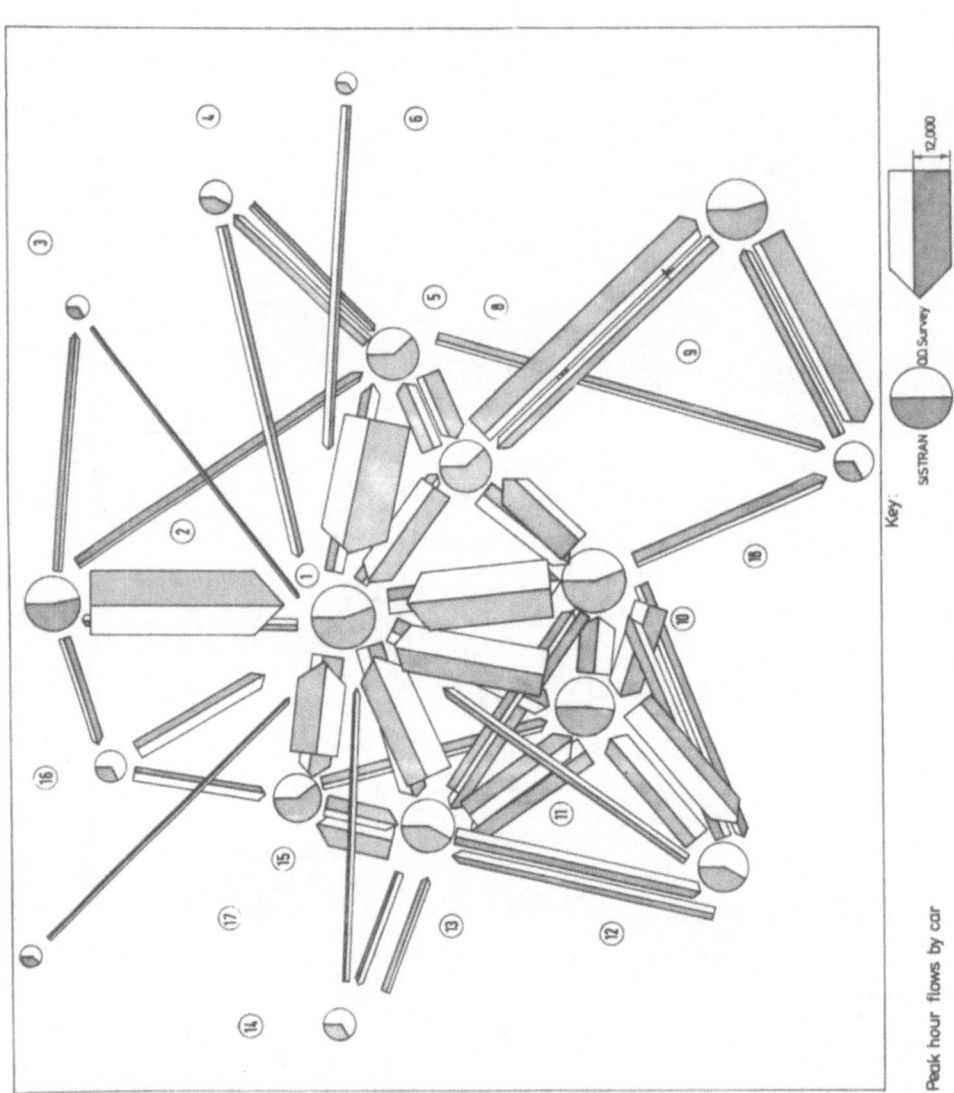

Peak hour flows by car

Key:

Figure 16 – Comparison of Peak Hour Car Flows

The procedure allocated first the destinations considering all land available (zoned land), then the origins with a car available and finally the non car available origins, reflecting the purchasing power of each activity (for more details see Echenique and de La Barra, 1977). The transport model allocated the flows, F_{ij}^{nkr}, from origins i to destinations j, by person type n, using mode k available for person n (ie. car was not available for n=2) and routes r. The flows, F_{ij}^{nkr}, were estimated as a function of the number of origins, O_i, destinations, D_j, and the price of transport between i and j by mode k and route r, C_{ij}^{kr}, i.e.

$$F_{ij}^{nkr} \sim O_i^n \, D_j \, \exp(-\beta^n \, C_{ij}^{kr}) \qquad (2)$$

where β^n were elasticities calibrated as follows:

		β^n
with car available	n=1	.172
without car available	n=2	.086

Once the flows were allocated to routes, the recalculation of the price (C_{ij}^{kr}) was achieved as a function of the volume of flows in each link (F_{ij}^{nkr}), the capacity of link r (Q_{ij}^{kr}), and the length of each link r (d^r).

$$C_{ij}^{kr} \sim f\{\Sigma_n \, F_{ij}^{nkr}/Q_{ij}^{kr}, \, d^r\} \qquad (3)$$

Once the price was recalculated, the equation (2) was re-evaluated until no more changes occurred. At that stage the aggregated price of transport by origins and destination was recalculated for each zone and introduced in equation (1) for the next time period.

REFERENCES

Companhia de Engenharia de Trafego (CET), 1978, Planejamiento de Transportes e Uso de Solo Urbano: Estudo Comparativo, OD-77/ SISTRAN, Sao Paulo, Brazil.

Companhia de Engenharia de Trafego (CET), 1979, COMONOR, Sao Paulo, Brazil.

Companhia de Engenharia de Trafego (CET), 1980, MUT: Modelo de Uso de Solo y Transporte, Sao Paulo, Brazil.

Echenique, M. and De La Barra, T., 1977, Compact Land Use/Transport Models, in P. Bonsall, Q. Dalvi and P.H. Hills, (Eds.), Urban Transportation Planning, Abacus Press, London.

Empresa Metropolitana de Planejamiento da Grande Sao Paulo (EMPLASA), 1977, Pesquisa Origem-Destino /77, Sao Paulo, Brazil.

Flowerdew, A.D.J., 1977, An Evaluation Package for a Strategic Land Use/Transportation Plan, in P. Bonsall, Q. Dalvi and P.H. Hills, (Eds.), Urban Transportation Planning, Abacus Press, London.

SISTRAN 1975a, Programa de Transporte Recomendado, Secretaria de Estado dos Negocios Metropolitanos, Prefeitura de Sao Paulo, Brazil.

SISTRAN, 1975b, Alternativas Estudadas, Secretaria de Estado dos Negocios Metropolitanos, Prefeitura de Sao Paulo, Brazil.

SISTRAN, 1975c, Metodologia de Trabalho, Secretaria de Estado dos Negocios Metropolitanos, Prefeitura de Sao Paulo, Brazil.

ACKNOWLEDGEMENTS

The SISTRAN study was developed by the consortium of Brazilian firms MONTREAL-SONDOTECNICA under the direction of Mario Larangeiras de Mendoca. Special mention is given to Tony Flowerdew who developed the evaluation methodology, to Richard Stibbs who developed the computer programmes and to Ian Williams who calibrated the model parameters. The clients for the work were the municipality of Sao Paulo's transport department (SMT) and the state of Sao Paulo's metropolitan planning agency (EMPLASA).

AGENCY POLICY REQUIREMENTS AND SYSTEM DESIGN

William Goldner

Institute of Governmental Studies
University of California at Berkeley
Berkeley, CA, 94720
United States of America

INTRODUCTION

The conventional wisdom about urban modelling systems suggests that their design is directed toward providing constructive guidance to decision-makers in evaluating policy proposals. But this judgement, loaded with linked conditions and qualifying adjectives, needs clarification if it is to serve as a launching pad to the theme of this paper.

First of all, the "conventional wisdom" regarding urban modelling systems is changing and therefore losing its conventionality. Increasing advocacy for the use of simpler models, of disaggregate demand models, of interactive computer programmes, of Delphi processes, of sophisticated calibration algorithms, and of many other developments are muddying the waters and diverting the thrust of the conventional into new and different directions.

Second, "design of urban modelling systems" also requires clarification. Here it means the conceptual framework or structural relationships among a set of interfacing economic, demographic, data bases to input and provide outputs of relevant and interrelated variables. This usage of "design", contrasts with that used by Batty (1979), where he states "in urban planning the organized action from the viewpoint of the planner has traditionally been called design...although it is important to note that this word is rapidly being replaced by the term "policy"." In this paper, "policy" will be reserved for use in just this sense.

Third, "providing constructive guidance to decision-makers" means that we are concerned with practical everyday uses of the

modelling system rather than experimental or developmental exercises, as important as the latter may be. "Guidance" is meant to
emphasize that the outputs of urban modelling systems are neither
optimal nor definitive, for in most cases, decision-makers and
decision-making bodies have their own objective functions, mostly
unspecified and only partially articulated, but nevertheless real
and operational when the decision has to be made.

Fourth, by "evaluating policy proposals" I want to place emphasis on the process of considering conditional alternative projections, embodying variations in inputs and model parameters but also
incorporating errors of measurement, specification, and sampling.
The comforting but spurious illusions of certainty that most decision-makers, many planners, and even a few model designers conjure
up are finally confronted by the bottom line -- and even here an
important proviso has to be recognized: the distinction between
policy and the instruments of policy has to be made. For the
changes from the present state to a future desired state take place
through policy instruments that are meant to be but sometimes are
not consistent with policies.

Buried in the preceding paragraphs are some definitions, clar-
ifications, assumptions, and reflections of an implicit point of
view that has been absorbed by this writer from others or that has
been personally experienced over a period of fifteen years. From
the original developmental phases, through experimental trials
associated with conceptual changes, to operational uses in varied
environments and continuing into a period of benign oversight, my
association with the Projective Land Use Model (PLUM) and the
supporting elements of its surrounding modelling system has been
continuous. The purpose of this paper is to document one selective
theme gleaned from that durable relationship: namely, how the
conceptual framework of PLUM has been modified and adapted to the
changing requirements of the planning environments in which the
modelling system was used.

The context within which this paper is written has to be
described in order to contrast the theme selected here from complementary theses elaborated in at least three other papers of interest (Harris, 1975; Hoffman et al, 1978; Batty, 1978). The domain of
concern is the process whereby technical support is used in the
planning process to achieve some change in existing conditions. In
these three discourses, the causal sequence moves from a conceptual
approach to a modelling system to the planning process. However,
Goldman (1975) and Hoffman et al (1978) point out the changing
framework of planning as it moves from optimal end-state planning,
characteristic of the 1960's, to the more contemporary form -- the
policy analytic approach. Using this approach, greater attention is
paid to the incremental planning perspective and to policy management; ie. anticipating the future consequence of present planning
actions or policies. To quote Goldman:

> "The intent here is to place the development and
> use of projections not only within the technical
> planning process, but also within the political
> decision-making process."

Of great importance to this shift in focus is the emphasis on the
distinction between "policies" and "policy instruments". Goldman
continues:

> "Roughly, a policy is a guiding principle while a
> policy instrument -- either legal, financial, or
> administrative -- is a specific action or device
> by means of which a policy is implemented ...
> through such investments as zoning ordinances,
> public capital investments, land reserves,
> etc... Extensive structuring and analysis of pol-
> icy information was necessary both prior to model
> applications and after the model outputs were
> available. This analysis and judgment outside
> the modeling system required certain changes in
> the subregional models to make them responsive to
> the policy information used as input and to make
> the output useful for policy analysis. This
> moved the models substantially from their theor-
> etical structure to be more sensitive to the
> policy analysis."

As will soon be apparent, the identification of policy instruments is hampered by the several levels of governmental participation in the planning process. Policy instruments that are directed
toward the ultimately controlling decision are supplemented by
related instruments that are more indirect in their effect, but are
linked to the decision-making process. An example of this is the
Federal project review programme. Under this programme, applications for Federal grants-in-aid are reviewed for the extent to which
a project contributes to the effective development of an area and
how it fits in with overall priorities. Reviewing authority under
this programme has been delegated to the regional level organization
in the San Francisco Bay Area, the Association of Bay Area Governments (ABAG). Implementing this responsibility involves review of
applications for Federal grants involving open space, water development, land conservation projects, hospital, airport, library, water
supply and distribution, and sewerage facilities (ABAG, 1970).

In the final analysis, the policy instrument is implemented at
the Federal level where the grant award is made. However, a handle
to the instrument is at the delegated level and thus the instrument
is segmented, with some levels exerting necessary but not sufficient
power to decide. The implications of this change, in effect, turns
the causal sequence mentioned earlier on its head. Instead of input
data, assumptions, and model concepts, leading to model structure,
it is output useful for policy analysis that calls the tune for the
structure and design of the modelling system.

Although these modifications of the modelling system are described as recent adaptations to the changed character of the planning process, adaptations in model structure and design have been
going on continuously. The purpose of this paper is to document
both these earlier adaptations as well as the more recent ones.

OPERATIONAL MODES OF THE MODELLING SYSTEM

A shorthand reference to the manner in which modelling systems
are used will conveniently serve to identify several characteristics
of the system's outputs. In addition, the nature of projections
made to accommodate the policy analytic process previously described
will be identified and made clear.

A further purpose for these definitions is to enable a correspondence to be made between the use of these descriptors here and in
other works where the meanings are not identical. (See Batty,
1978). The several system applications that are to be described
fall into three modes: (1) projective modelling, (2) policy modelling, and (3) prescriptive modelling. (Lembke, 1977).

Projective modelling is the use of a modelling system to forecast the likely state of a region, implementing the most plausible
values of parameters, and future scenarios of public policies.
Judgement is brought to bear, preferably of informed and studied
quality, and in the light of this uncertainty, ranges for quantitative inputs and outputs should be integrated into this mode of operation. This mode characteristically generates positive as opposed
to normative outputs.

Policy modelling is the implementation of policy alternatives
to compare differences in outputs, measure impacts, and provide a
basis for decision-makers to evaluate trade-offs and select the
"best" alternative. Built into this mode of operation is the necessity for the model to be able to incorporate the variables that
exactly reflect the policies to be tested, and the further requirement that the modelling system be sensitive to the differences among
policy candidates. This involves using the modelling system with
positive capabilities for a process that requires a normative outcome.

Prescriptive modelling is the process whereby mid-range or
target values are externally prescribed for selected variables at
the zone or city-wide level. Operating in this mode, the modelling
system maintains the conceptual relationships and constrained limits
with other variables and for those portions of the study area not
subject to the exogenous target values. Typically, data and adjust-
able parameters are entered into the system as overrides or substitutes for existing values after having been processed carefully by
hand from externally provided sources. It is in this mode that the
policy analytic process takes form. Hoffman et al (1978:22) iden-

tify over thirty land development policy instruments in effect in the San Francisco Bay Area in 1975. The transformation of this large number of instruments into variables operable in the modelling system is the justification for the large amount of handicraft work necessary when this mode of operation is used.

THE LIFE CYCLE OF A MODELLING SYSTEM

In reviewing the sequence of design responses to the external requirements of policy guidance, the treatment below will consider these responses in a broad framework. The whole of the system will be considered fair game, and modifications will comprehend data base revision and modification, parameter adjustments and overrides, and model concept revisions. From the continuum of the system's life cycle, selected points of transition of use and practice will be identified. The associated policy and policy instruments will be characterized, (see Goldman, 1975), and the system changes described.

The Development Phase: 1964-1969

For summary background, a short detour from the highway of discourse will serve. In 1962, a small group of scholars at the Center for Real Estate and Urban Economics, University of California, Berkeley, decided to implement the urban allocation model developed by Dr. Ira S. Lowry (Wendt et al, 1968:13). This model was documented at that early date in a preliminary mimeographed report subsequently published by the RAND Corporation (Lowry, 1964). The model, along with an appropriate data base, was used to generate some preliminary test runs for Santa Clara County (Goldner and Graybeal, 1965). In 1966, further development work was undertaken at the Bay Area Transportation Study (BATSC) and this phase culminated in the development of an area-wide operational system known as Projective Land Use Model: PLUM (Goldner, 1968).

Further development work was carried out under the aegis of the Institute of Transportation and Traffic Engineering, University of California, Berkeley, where the model was separately adapted to the policy planning needs of several regional agencies (Goldner et al, 1972). These applications of PLUM are the starting point of the life-cycle which constitutes the focus of this paper.

The Regional Airport System Study (RASSC): 1971

In 1970, the administrative planning apparatus at ABAG recognized the possibilities of contravening interests among the three major airports in the San Francisco Bay Area as they developed plans to expand capacity in the immediate future. As a policy, it was decided to develop an airport element for the ABAG regional plan.

The specific policy instruments involved developing criteria for the granting of support funds and the establishment of priorities among existing and potential major airport facilities.

Providing the projections for the evaluations required by this study became the first planning application of the newly-developed PLUM model (Goldner et al, 1971).   The study design encompassed policy modelling, with sensitivity analysis carried out among eleven alternative configurations of input assumptions regarding airport sizes and locations.

For this study, the system framework was augmented by a series of cross section regression equations involving (1) the level and distribution of zonal household incomes, and (2) the generation of income, sales, and residential property taxes.   In addition, the PLUM model design was reprogrammed to generate a growth increment directly, in contrast to its previous equilibrium form where growth increments were obtained using comparative statics, ie. subtracting an output at time t from output at t+1.

The incremental form was particularly useful in this study to identify the impacts of the input alternatives at a magnitude that was significant.   In the equilibrium form, the output changes were less sensitive because their magnitudes were almost swamped by the base data.

Special programming was required to create a series of reports that differed from the normal output format of PLUM.   Zonal values of selected variables were aggregated to cities and counties within the region and presented in a columnar frame that facilitated comparisons across the eleven alternatives.

The San Diego Comprehensive Planning Organization (CPO) Version of PLUM: 1972

Concurrent with the modifications for the RASSC Study, the San Diego CPO decided to install a regional modelling system, and use the PLUM model as the prototype (Goldner et al, 1972). As a newly-developed installation of the agency, the modelling system was adapted to provide quantitative output for a range of planning policies.   As the installation was being effected, the CPO planning programme was confronted by several important issues which became the focus of modifications to the existing PLUM modelling framework and pointed toward policy instruments further downstream.

First of all, the agency was deeply committed at the time to a regional programme of traffic management and transportation planning.   Consequently, their traffic models needed specific inputs (which were outputs of PLUM) for school populations, shopping and trade centres, government buildings, and construction activities. This set of requirements led to a disaggregation of the local-serving algorithm into eight industry categories. Supplementing this disaggregation was the development of variations in the allocation function for each disaggregated type.

A second requirement, also focused at the transportation needs of CPO, was the expansion of the computer programme so that it could handle 663 traffic zones. This requirement was responsive to the transportation planning requirements for explicitly-defined locations for the activities being modelled. As a corollary, the modelling process was established on the basis of a nested set of zones with the 663 traffic zones grouping into a smaller set of 312 census tracts and a still smaller set of 85 regional traffic zones. The PLUM programme was designed to operate at each of these levels, a more economical method than running at the 663-zone level and then aggregating the results to the coarser net of zones.

A third requirement was to modify the PLUM programme so that trip tables aggregating the allocations of the journey from workplace to place of residence could be quantified and presented separately by two transportation modes, auto and bus/rail transit.

Finally, in anticipation of the projections of infrastructural developments that would be required in the future, the CPO requested that land uses be distinguished through time between developable areas served by infrastructure (ie. streets, utilities) and those not served by these correlates of urban development.

The modelling system was used to provide the inputs to the traffic and transportation models available to CPO from the then-named State Division of Highways (presently CALTRANS). In addition, several route alternatives were tested for a potential light rail system. Also airport locations were simulated in a manner similar to that used for the RASSC study mentioned above. Finally, this version of the model was used for forecasts and sensitivity tests. These uses involved projective and policy modes of operation of the system.

State Water Resources Control Board (ABAG Series 1): 1973

The first exercise of the modelling system by ABAG was embodied in a set of regional projections made in 1973 for the State Water Resources Control Board (ABAG, 1973). The particular requirement that this set of projections had to accommodate involved (1) quantified reflection of the spatial distribution of activities in ABAG's Regional Plan: 1970-1990, (2) adherence to regional population projections supplied by the Population Research Unit of the State Department of Finance (DOF), and (3) the adaptation of these projections to a differing set of geographical units than the traffic analysis zones that were embedded in the modelling system. These requirements involved primarily the projective mode of modelling system usage.

    This effort represented ABAG's tooling up for continuing use of
a regional modelling system. Consequently, substantial effort was
allocated to a series of input requirements of the system. These
included: (1) shifting the data base from 1965 to 1970; (2) implementing an intensive enumeration of land use data to reflect the
Regional Plan and to distinguish between developable land served by
infrastructure and that which was outside the served area; (3)
developing a set of regional employment projections that were consistent with the DOF population projections; (4) establishing a
regional structure of industries that was able to be disaggregated
into the basic/local serving dichotomy; (5) devising and programming
a method of transforming the allocations from 290 PLUM traffic zones
to 50 Hydrographic Subareas, and 5 Basin or Sub-Basin Subtotals.

    These activities were implemented by a team of technical spe-
cialists from the ABAG and Metropolitan Transportation Commission
(MTC) staffs. The activities were executed concurrently by both
staffs to meet urgent deadlines, hampered by changing specifications
which were brought to bear in midstream as the several tasks were
underway.

The Regional Transit Travel Projections Project (RTTPP): 1973
(ABAG, 1973; ABAG, 1974)

    These midstream redefinitions of goals and purposes were
extremely significant and this importance justified the interrup-
tions to the continuity of work that was underway. At that point of
time in 1972, during the period of construction of the Bay Area
Rapid Transit (BART) System, pressures for and against the further
extension of the still uncompleted system were being exerted. A
comprehensive study to evaluate these extensions in a consistent
regional transportation planning framework was initiated: The
Regional Transit Travel Projections Project (RTTPP) (Smith, W. and
Associates, 1973).

    The SWRCB effort, already underway, could provide the popula-
tion, employment, and land use inputs at the traffic zone level that
were required for the transportation models. Beneficial as these
symbiotic developments were, they also generated frustrations growing out of the enlarged group of organizational interactions.
Coordination among three agencies become enlarged to accommodation
among at least seven agencies and consulting firms.

    This organizational complexity was not without benefit. As the
drafts of the several reports were circulated a host of inconsistencies popped out. Many of these were data glitches: the most serious
involved the categorization and measurements of land uses, particularly in relation to the limits of developable land, the distinctions among infrastructural service areas, and the zoned versus
actual residential densities.

Two major developments had emerged from the Series 1 projections: (1) the direct modelling of the work-to-home trip table as an adjunct to the model's allocation process, and (2) wholesale revision of the land use inventory data.

The provision of a trip table directly from the model's allocation of workers to their place of residence seemed to be a most useful input to conventional transportation analysis. That process, consisting of trip generation, distribution, modal split, and assignment, is partially duplicated by the allocation process in PLUM. Therefore, in anticipation of the system's continuing use in transportation planning by MTC, the model was enlarged to generate trip tables for auto and transit modes. The programming of these modifications for the computer were complex, incorporating bold assumptions, difficult data manipulation, and the design of a new set of reports summarizing the output matrices. Ironically, this complex modification to the PLUM programme, although tested for operationality, has never been used.

The land use data became a target for scrutiny as the actual quantities of the land use inventory were made apparent and their significance for projections into the future was recognized. This significance translated into concern whether the initial data were correct; then that the assumptions regarding the projections were properly expressed; and finally, that the outputs corresponded with the policy expectations implicit in local zoning, density, and land use instruments. In effect, this sequence strongly reflected the use of the modelling system first in a descriptive and projective mode, and finally in a prescriptive mode. In order to carry out the theme of this paper, the saga of land use measurement and its implications for model design will be placed in a separate category which follows.

The Local Development Policy Survey and Land Use Measurement: 1976-1977

Among the criticisms of the statistical data used in Series 1, one particularly piercing cry was the virtual oversight of local policy provisions, especially as they were reflected in zoning, densities, and potential land uses. To allay this criticism, a prestigious national consulting organization was commissioned to prepare an exhaustive set of regional maps comprehending virtually every quantitative variable that could be represented in the spatial dimension. Each of the maps was presented on a uniform regional base map at the 1:125,000 scale. Among these were maps reflecting five categories of potential expansionary growth. These were:

1.   Areas within a water district.

2.   Areas within a sewer district.

3.   Incorporated areas.

4.   Areas not incorporated and not restrictively zoned.

5.   Areas that do not fall in permit areas of the Coastal Commis-
     sion or in priority use areas of BCDC's San Francisco Bay Plan
     designated as water-related recreation and wildlife areas.

Because these definitions are not mutually exclusive, by over-
laying mapped transparencies, the planning consultant designated
areas with three or more of these categories as reflecting "high"
potential for development in terms of "policy". Areas with only one
or two of these characteristics were considered to have "low" potential for development. These two categories were measured by plani-
meter for the zones in the study area with the intention of using
them in the modelling system's data base.

Testing these data with the records and maps of several local
jurisdictions again revealed serious inadequacies. The scale of the
base map and even the width of the boundary lines defining the use
categories were sources of inaccuracy in measurement. More important, the concept of high and low development potential based on the
three or more categories of "policy" was considered to reflect local
policy most inadequately. The outcome of this mapping exercise for
the provision of "policy" land use data proved to be visually
appealing but quantitatively unacceptable.

The importance of the policy-oriented land use data to the
users at the local jurisdictional level and the sensitivity of the
system to the accurate measurement of developable lands and their
associated policy densities decreed that a unified large-scale
effort be undertaken to supply the required data. A comprehensive
survey of vacant land categorized by relevant zoning and development
policy instruments as administered by the several levels of local
government was planned, designed, reviewed, revised, and finally
completed over a period of two years. Much of the time was consumed
in solidifying hazy concepts and inaccurate measurements through a
continuing interchange of comments and review with local jurisdictions. Comparable time was spent translating the complex structure
of local policy instruments into the more aggregated form that was
confirmed with the structure of the modelling system.

Table 1 presents the details of the aggregation process. Examination of the table reveals how the zoning, development, and infrastructural instruments established, primarily but not exclusively,
by local governmental units were translated into categories that
were usable in the modelling system. Also of great significance was
the imputation of a time sequence to the availability of developable
lands.

The modelling system changes that matched the Local Development
Policy Survey were extensive, and also represented a shift in model

TABLE 1

LAND USE DEVELOPMENT ASSUMPTIONS, ACREAGE, AND ASSOCIATED
POLICY INSTRUMENTS: SAN FRANCISCO BAY AREA, 1975

Development Assumptions and Staging	Acreage (1,000's)	Policy Instruments
Prime Land Developable in 1975	171	Committed Developments Redevelopment Projects Service capacity exists for sewer/water/roads
Developable in 1980 or 1985		Planned for Development Service committed in 1975 capital programmes for sewer/water/roads
Secondary Land Developable in 1975	141	Sensitive Development with Special Problems Urban zoning but special regulations for steep slopes, flooding, etc. Service capacity exists or committed in 1975 capital programmes
Developable in 1980 or 1985		Ready for Development but with Service Constraints Urban zoning Services not yet committed for sewer/water/roads
Industrial Land Developable in 1975	62	Urban Industrial Reserves Industrial Parks Zoned Industrial Districts
Developable in 1975 or 1980		Industrial Redevelopment Projects Committed projects zones for industry in 1975 Rural Industrial Reserves Zoned for industry Roads, sewers, and water existing or planned
Vacant Land Not Developable	3,662	Precluded from Development by Policy Action Public owned Public acquisition committed Agricultural preserves Open space zoning Industrial reserves not yet zoned for industry Remote from existing roads

Total Vacant Land	4,036	
Urbanized Land	445	
Streets, highways	118	
Basic industries	55	
Local-serving ind.	51	
Residential	221	
Regional Total	4,481	

Sources: ABAG, 1978, A12; ABAG, 1980, 111-117.

use from the projective to prescriptive mode. As important as the forecasts into the future were, greater concern was directed to the consistent accounting of land acreages among use categories, of correspondence among population and housing units, of balances between jobs and employed residents.

The modelling system changes that were responsive to the Local Development Policy Survey (LDPS) outputs included the following:

1. The LDPS provided an accurate and exhaustive quantification of acreage zoned for "basic" industry. In the earlier versions of the Base Employment Model (BEMOD) and PLUM, these two models operated interactively, passing vectors of zonal employment and land uses back and forth across the interface between them. With the more complete quantification of industrial vacant land, the interface programme was abandoned, BEMOD was run independently through all iterations from base to target year, absorbing land exclusively from industrial vacant land. BEMOD outputs were then passed to the PLUM input files for an independent and uninterrupted sequence of iterations to the target year. This modification in the system generated substantial economies in computer time and costs, particularly reflecting the ability to establish an unchanged pattern of basic employment once and for all in BEMOD. This allowed data and parameter changes to be made in PLUM runs without concurrent runs in BEMOD. Comparisons of outputs run interactively and independently displayed differences that were insignificant.

2. The availability of LDPS data incorporating fine-grain differentiation of development and zoning concepts allowed these categories to be identified with the maturing of land from raw to developable and then usable status. This process was incorporated into the PLUM programme as "staging". This process required the programme to override the identification of vacant land categories in selected zones, for a prescribed number of acres, setting this land in a deferred category for a specified number of years, and then retrieving the deferred land, emplacing it into its original category. Staging plays an important part in smoothing the trajectory of development in a zone, and also modifies the spatial pattern of allocation significantly. The deferred land category, in effect a temporary land reserve, also has implications for use as a direct policy instrument.

3. Another modification involved the capability to provide modified data, parameters, and coefficients which overrode the standard operation of the programme at any present or future iteration. Most important among these was the treatment of residential densities. These coefficients could be adjusted from their base year relationships by a predetermined factor across all zones, or for selected sequences or groups of zones. In addition, densities in individually targeted zones could be emplaced. Further, a redevelopment sequence was simu-

lated in zones where a specified built-up acreage was transferred to prime developable status and built out at a predetermined density.

4.   The linkage between occupied housing units and resident popula-
      tion is household size. Although the PLUM model was modified
      in the time dimension to adapt to regional demographic trends
      (declining household sizes and increasing labour force partici-
      pation rates), individual zones required adjustment to reflect
      prescribed populations or housing units. An algorithm was
      developed to express zonal household size and also the propor-
      tion of single and multiple unit dwellings as a function of
      residential density.

5.   Many local jurisdictions and special districts have aggregate
      policy target values for employment, housing units, and popula-
      tion, but have no fine locational differentiation to the zonal
      or census tract scale. Aggregating those variables to conform
      to the geographical limits of the decision-making unit ran up
      against two problems of geographical definition. First, unlike
      the fixed boundaries of the zones established in the modelling
      system, local administrative boundaries were modifiable as the
      community enlarged its jurisdiction through time. Second, even
      existing municipal and special district boundaries did not
      conform or, in some cases, even approximate the zonal boun-
      daries of the modelling system's data base. Fortunately, there
      was a solution to the first problem that had the full flavour
      of a policy instrument. In California, within each county,
      there exists a Local Agency Formation Commission (LAFCO) whose
      duties are to approve the annexation or consolidation of local
      districts. In administering these powers delegated under state
      law, they define community "spheres of influence" outside of
      which new subdivisions and developments are essentially
      banned. These LAFCO spheres represent rational future bounds
      within which scatteration and leap-frogging of local jurisdic-
      tions is proscribed. Therefore, these limits as they existed
      in 1976 were used to define local jurisdictions instead of
      using the community's actual administrative boundaries.

     With regard to the problem of partitioning of variables among
contiguous LAFCO spheres, intrazonal allocations were made in proportion to mapped land uses from the LDPS. Base period partitioning
of population, housing units, and jobs was proportionalized to
existing land acreages for residential use and a corresponding estimation was made between jobs and industrial/commercial uses. Future
period allocations were made in terms of the categories of vacant
available land including the consideration of projected or prescribed densities and land absorption coefficients.

     A separate but comparable treatment of output data was made for
sewer service areas in the region. Boundaries for these areas constitute yet another semi-independent set used for sewage treatment

facility planning. Procedures for partitioning zonal data followed
the sequence described above.

The richly detailed mixture of policy concepts that were made
available by the LDPS mandated a comparable package of system modifications. Fortunately, there was time for these to be designed and
emplaced as the survey was carried out. This episode constitutes
the most substantial set of revisions to the ABAG modelling system
since its inception. It illustrates the theme of this paper by
showing the inter-relation between users' concerns, their data, and
adaptations of the system to these policy-focused inputs.

Sensitivity Tests for the BART Impact Study: 1977

As a mechanism for documenting the lessons to be learned from
the planning, construction, and operation of the BART system, the
US Department of Transportation commissioned a comprehensive
research study to assess the multi-faceted impacts of the system on
the social, economic, and physical characteristics of the San
Francisco Bay Area. The lead agency in this effort was MTC under
whose direction a wide range of in-house and consulting efforts were
coordinated.

In connection with the BART Impact Study's concern with the
relationship of the system to the patterns of development and land
use, an experiment was designed in the policy mode to answer several
questions. The most significant was whether the presence and geographical layout of the BART system made any difference to the
development patterns of the region and its component parts. To
accomplish this, the modelling system was run with transit networks
excluding BART (the no-BART alternative) and the resulting outputs
were compared with those which included the BART system.

A further aspect of these simulations was to test the scale by
which development patterns were influenced by varying degrees of
accessibility. This was to be accomplished by applying across-the-
board adjustments of 25 and 50 percent exclusively to the time-
distance variables in the modelling system. All other variables
were unchanged.

The projections based on these runs of the modelling system
were substantially inconclusive. These runs confirmed previously
determined conclusions regarding the insensitivity of the allocations to the distance variable taken alone. However, the assessment
of these differences was flawed by the use of region-wide aggregate
measures that were swamped by the effects of zones where no development was possible, rather than measuring the changes only in zones
which were subject to some degree of potential development.

Although these simulations, performed in the manner described,
were not conclusive, they do point to the importance of the policy

instrument that was being tested. In effect, the simulations asked whether rail transit speeds had any relationship to the patterns of urban development. In this set of experiments, the modelling system was utilized without any design modifications except the input adjustments described above.

The experiment's outcome highlights the difference between policy and prescriptive modes in the operation of the modelling system. Under the prescriptive approach, expected development under the differing assumptions would be simulated and the input modifications to achieve those expected states would provide the focus of evaluation. Still another approach would involve modifying the model to incorporate constraints on the commuting capacities of the highway and transit networks (Putman, 1973). Given existing knowledge of the peak-hour congestion on bottleneck segments of the highway network, the transit network (including BART) may have been favoured in such comparison.

ABAG's Projections 1979 (Series 3): 1980

The implementation of the modelling system using the LDPS data described previously was the basis for a new set of ABAG projections, Series 3. In addition to the policy concepts that were reflected in the newly-collected land use data, this set of projections elicited a substantial amount of technical review before being considered complete. The original set of projections termed "provisional" was published in March, 1977 to test whether the modelling system reflected the intent and quantities that the local jurisdictions had reported in the LDPS (ABAG, 1977). In effect, this phase was a step in validating the data that was input to the system.

Substantial feedback in fine detail was received and a succession of projection runs reflected these revisions to the input data. This review phase was brought to completion with the publication of Series 3 "Revised" Projections (ABAG, 1978). These projections were used in a number of ABAG's planning programmes including housing, environmental management, and joint transportation studies with MTC. A variant of the "Revised" projections was developed for the Air Quality Maintenance Plan, emphasizing higher densities of development and compacting of growth more centrally within the region's cities.

The technical review process generated increasing concern with the outputs of the projections and the sequence of development to the target years. The modelling mode shifted from projective to prescriptive as base year data, coefficients, and assumptions reflecting local policies were increasingly brought to bear upon the system. Major effort was applied to have the models reflect infilling opportunities, rebuilding, the timing of development, the zoned residential densities, and household sizes. Still another round of review and feedback from local jurisdictions was elicited and incorporated into the system. The final version of Series 3 was published in January 1980 (ABAG, 1980).

The sequence of projections that constituted Series 3 Provisional, Revised, and Final, were not implemented in a policy vacuum, which goes a long way in explaining the careful attention paid to their modifications. The careful honing of policy instruments at several governmental levels was being carried on in parallel with the technical process of developing and reviewing the projections.

At the Federal level, the US Environmental Protection Agency (EPA) required that grants for publicly-owned facilities (wastewater facilities, for example) be restricted to scales consistent with these projections. In addition, at the state level, the Governor's Office of Planning and Research (OPR) administered a regulation that State projects be based on OPR-approved regional agency population projections. At the regional level, ABAG's Environmental Plan (EMP) identifies the list of wastewater projects and provides for its review and modification annually. This cascading authority suggests how the policy instruments at each level of government are linked together.

San Francisco Bay Area Regional Energy Plan: 1980

As this is being written (June, 1980), the ABAG modelling system is undergoing another set of modifications to adapt it to the requirements of the Regional Energy Plan. At the regional level, an economic model incorporating an input-output framework is designed to operate interactively with a demographic model. These models provide regional outputs to a set of energy demand models for residential, commercial, industrial and transportation energy uses. The residential and commercial models are usable at subregional as well as regional levels, and use outputs of the existing modelling system to "drive" the energy models. The early projections from this system will incorporate the projection mode, developing magnitudes of energy demand for the region and for counties.

CONCLUDING REMARKS

As heroic as it seems to generalize on the basis of a sample of one, several propositions emerge from this discourse.

First and perhaps most important, there is no fixed end-state to the structure of an urban modelling system. It is virtually always in a stage of modification as it is adapted to the changing form of the planning process and to the policy issues that emerge through time. These modifications have three major sources:

1. Computer hardware is undergoing such rapid technological improvement that programmatic changes are required merely to keep up with the new hardware as it is put into use.

2. The software is also being revamped to accommodate more efficient algorithms, to incorporate changed configurations of the computer system, to test minor conceptual changes, and to correct unanticipated and unplanned outcomes.

3. Changes in the model's usage, as exemplified in the earlier sections of this paper, are building up in the work queue, as the varied and changing objectives of an actively operating agency come into view.

A corollary of this constantly changing structure is the difficulty that it raises regarding the stability of calibrated parameters in the system. Many of the changes require a newly fitted set of coefficients generated by extremely difficult and abstract processes. If these parameters are estimated by optimization procedures that produce very slight changes, if any, the value of sophisticated calibration procedures may be questioned, and even abandoned in favour of "quick and dirty" pragmatic estimation.

Under these circumstances, it is virtually impossible to test a modelling system objectively and conclusively. To remove a system from its computing environment, its technical personnel, its knowledgable users, and its continuing use makes the precise replication of the system's outputs a difficult, challenging and often impossible task. The pejoration of large-scale modelling systems as "black boxes" is an outcome of this set of characteristics.

Second, not only are the urban modelling systems continuously changing but so are the problems, issues, and policies with which they are confronted. This stormy environment is representative of two major sets of influences; one natural and the other governmental.

Within the life cycle of urban modelling systems, several pervasive "natural" phenomena have occurred. There has been a significant drop in birth rates and in household size (ABAG, 1980:111-112). Further, there has been a worldwide energy crisis which has provided economic and institutional shocks that have widespread spatial repercussions. Along with these changes, there have been continuous shifts in national and regional industrial structures as the demands for services have overtaken the demands for goods. This shift has manifested itself in the enlargement of office activities by both private industry and the public sector, and also by the expansion of health and educational activities (ABAG, 1979). Finally, the furtive awareness of the deterioration of the environment has been transformed to overt governmental programmes.

Matching these "natural" phenomena has been the broad range of governmental responses to the ongoing changes that have been mentioned above. Most significant for urban modelling systems has been the encouragement given to regional decision-making by a stream of federal programmes, many carrying grants-in-aid to governmental units at state and local levels. Collectively, these programmes have been wide-ranging, although individually they have been targeted at specific problem areas. They encompass urban transportation, resource conservation, water and sewer facilities, medical programmes, solid waste planning, air pollution, airport systems, and many other related programmes.

Third, this proliferation of programmes impacting regional development concerns has required coordination and balanced and interrelated application. Here is where the plethora of policy instruments confronted their surrogates in the modelling systems. This confrontation, of expansive force as new programmes were added, and with contractive offsets as model outputs were related to several policy instruments at the same time, was the source of the continuing adaptations that are the subject of this paper.

Fourth and last, the less-than-perfect match between model outputs, policies, and policy instruments points to the uncertainty that is inherent in these urban modelling systems. Their practical use in policy analysis and decision-making leaves a substantial place for a critical kind of knowledge -- that of judgement, good practice, and experience -- that must be brought to bear along with the programmed quantitative model outputs. This is reinforced by the awareness that data used in these systems is imperfect, that model specification errors generate error ranges, and that interpretation and translation of the output's meaning is often hampered by inadequate documentation.

REFERENCES

ABAG (Association of Bay Area Governments), 1970, Regional Plan: 1970-1990, San Francisco Bay Region, The Association, Berkeley, California.

ABAG, 1973, Population, Employment, and Land Use Projections, San Francisco Bay Region, 1970-2000, by Basin, Sub-basin, County, and Hydrographic Subunit for the Comprehensive Water Quality Management Plan, The Association, Berkeley, California.

ABAG, 1974, Projections of the Region's Future: Population, Employment and Land Use Alternatives in the San Francisco Bay Region: 1970-2000, The Association, Berkeley, California.

ABAG, 1977, Provisional Series 3 Projections, Population, Housing, Employment, and Land Uses: San Francisco Bay Region, The Association, Berkeley, California.

ABAG, 1978, Working Paper, Revised Series 3 Projections, Population, Housing, Employment, and Land Uses: San Francisco Bay Region, The Association, Berkeley, California.

ABAG, 1979, San Francisco Bay Area Economic Profile, The Association, Berkeley, California.

ABAG, 1980, Projections, 1980–2000, Population, Employment,
Housing, The Association, Berkeley, California.

Batty, M., 1978, Logical Problems Concerning the Use of Models in
Strategic Planning, a Paper presented to the Conference: Inte-
grierte Infrastrukturplanung, Universitat Augsburg, Memmingen,
West Germany.

Batty, M., 1979, Progress, Success, and Failure in Urban Modelling,
Environment and Planning A, 11, 863–878.

Goldman, R.E., 1975, The Role of Urban Development Models in Plan-
ning, in J.R. Pack, (Ed.), Proceedings of the Conference on the
Use of Models by Planning Agencies, Fels Center of Government,
University of Pennsylvania, Philadelphia, Penn.

Goldner, W., and Graybeal, R.S., 1965, The Bay Area Simulation
Study: Pilot Model of Santa Clara County and Some Applications,
Center for Real Estate and Urban Economics, University of Cali-
fornia, Berkeley, California.

Goldner, W., 1968, Projective Land Use Model (PLUM), Bay Area Transportation Study Commission, Berkeley, California.

Goldner, W., et al, 1971, Economic and Spatial Impacts of Alternative Airport Locations, Vol. 1: Report; Vol. 2: Appendices; Institute of Transportation and Traffic Engineering, University of
California, Berkeley, California.

Goldner, W., Reynolds, M.M., Rosenthal, S., and Meredith, J.R.,
1972, Vol. I: Plan Making with a Computer Model: Projective Land
Use Model; Vol. II: PLUM/SD, the Urban Development Model; Vol.
III: PLUM/SD, Computer Systems Guide, Institute of Transportation
and Traffic Engineering, University of California, Berkeley,
California.

Harris, B., 1975, Bridging the Gap between Theory and Practice in
Urban Planning, in J.R. Pack, (Ed.), Proceedings of the Conference
on the Use of Models by Planning Agencies, The Fels Center,
University of Pennsylvania, Philadelphia, Penn.

Hoffman, S.R., Carlson, R., and Goldman, R.E., 1978, The Development
and Applications of Regional Projections: Balancing Policy Concerns and Technical Factors, Paper presented at the 61st Annual
Conference, American Institute of Planners, New Orleans,
Louisiana.

Lembke, R.G., 1977, Operational Modes of the Modelling System, Basic
Forecasting Memorandum 80, Delaware Valley Regional Planning
Commission, Philadelphia, Penn.

Lowry, I.S., 1964, A Model of Metropolis, The RAND Corporation, Santa Monica, California.

Putman, S.H., 1973, The Interrelationship of Transportation Development and Land Development, Vol. 1 (Main Report); Vol. 2 (Appendices), Department of City and Regional Planning, University of Pennsylvania, Philadelphia, Penn.

Smith, W. and Associates, 1973, Regional Transit/Travel Projections Project, San Francisco, California.

Wendt, P.F., et al, 1968, Jobs, People and Land: Bay Area Simulation Study (BASS), Special Report No. 6, Center for Real Estate and Urban Economics, University of California, Berkeley, California.

PART 4

POLICY ANALYSIS

THE FAILURE OF MODEL USE FOR POLICY ANALYSIS IN REGIONAL PLANNING

Janet Rothenberg Pack

School of Public and Urban Policy
University of Pennsylvania
Philadelphia, PA 19104
United States of America

In this paper we bring up-to-date the results of an earlier inquiry into the adoption and use of urban models by planning agencies in the United States.[1] In particular, we are concerned here with the continued application of the models once adopted. Are they incorporated into the routine analytic apparatus of the agencies? For what purposes are they used? The reasons for concern with such questions are many. The undertakings are expensive, using considerable amounts of skilled professional as well as computer time, and often they supplant alternative modes of analysis. We need to ask whether the adoption of models and their subsequent use justify the expense.

In our attempt to begin the investigation of these questions, seven of the eleven model users interviewed in 1975 were successfully contacted in 1980 to determine whether or not the models were still being used; if so, in what ways and if not, why not? The earlier case studies involved a site visit for about one week by two persons, with interviews conducted throughout the regional agency and its constituent jurisdictions, as well as massive reviews of formal and informal documents. The brief updating reported here is based only upon telephone conversations with one or two persons in the regional planning agency itself. Nevertheless, the findings of this more limited inquiry are quite consistent with the earlier results. Where the models are still being used, they are generally activated only to update regional comprehensive plans and regional activity allocations, that is population and employment forecasts for sub-areas of the region. Policy analysis, the simulation of specific real or hypothetical proposals, is almost entirely absent.

THE EARLIER STUDY

In the 1975 study, the extent of model adoption and subsequent use in regional planning agencies was investigated. The limited evidence available when that study began was largely impressionistic and idiosyncratic, e.g., the documentation of a particular success or failure, usually by a major participant. The stories of failure were generally failure in model development, the stories of success generally were told in terms of potential, rather than realized, model development and application. The overwhelming impression at the time was that there was very little active model development and application by planning agencies.

Through a comprehensive survey conducted by mail, of 1500 planning agencies as well as extensive site-visits with eighteen planning agencies, we were able to determine:

1. That model adoption was far more common than had been previously assumed.

2. That model development, that is the calibration of a model, was generally successful (the availability for local adaptation of the PLUM and EMPIRIC models was an important part of the reason).

3. The needed data were generally available or had been developed as part of the overall process of model development.

4. Agency personnel were better prepared for the inevitable difficulties to be encountered and had far more realistic expectations since they had learned fairly well the lessons of the earlier pioneering "failures".

But what of actual model use? Expectations were generally more modest than the extravagant claims of the early enthusiasts but still included the full panoply of purposes. Planners sought models for impact analysis, that is for the determination of the effect of policy alternatives (or of a single policy or plan) on a wide range of variables, e.g., on the distribution of population and employment, or on land use densities. It was also assumed that the models could assist in policy design, that is to determine the appropriate policy or set of policies (mixes of land use constraints, transportation investment, sewer placement, for example) required to achieve desired outcomes.

The case studies showed little variation in model application. Simulations were carried out almost exclusively for the analysis of regional comprehensive plan alternatives. Often these were executed only to derive the local activity forecasts (population and employment) required by federal and state governments and by the regional agency itself for various planning purposes, eg. to justify the construction of a new hospital, the placement of a highway link, or the extension of sewer lines. Specific individual policy simulations, whether of hypothetical or real proposals, were undertaken only rarely. Indeed, the major impact on the planning process of model simulations was the formal adoption and widespread dissemination and use of the model generated activity allocations.

Despite the clearly more modest expectations, it was still true in 1975 that the model adopters' expectations exceeded the state-of-the-art with respect to policy analysis.[2] Many of the anticipated uses, e.g., the calculation of the effect on the tax base of alternative policies, required additional models or the incorporation of price variables (missing from all the models being used). Paradoxically, although expectations generally exceeded current model capabilities, actual model application fell short of what was possible (at least insofar as the modellers and planners understood the capabilities of the models). More analysis could have been carried out than was observed. This is explained largely by the emphasis in the regional planning agencies on producing a comprehensive plan and a single agreed upon set of activity and land use forecasts. Moreover, the range of policies considered, even within this restricted framework, was further constrained by policy differences among the political jurisdictions making up the region. Thus, the plan simulations were often restricted to those which deviated relatively little from existing preferences and practices. Finally, the model's characteritics often became more important in the discussions accompanying the exercise of the model for comprehensive plan simulation -- from preparation of model inputs expressing the policies to evaluation of the resulting outputs. Only at this late point did the shortcomings of the model structure, the adequacy with which policy changes could be incorporated, and the empirical relationships receive intensive scrutiny, often resulting in greater scepticism.

Nonetheless, despite this restricted application, of the eleven agencies visited in 1975 which were using land use/activity allocation models (i) ten of the eleven had carried out at least one regional plan simulation and seven of eleven had simulated multiple alternatives (as few as two and as many as six), (ii) five of the eleven had done no other analyses with their models, that is they had carried out no simulations of specific policy alternatives, (iii) of the six which had carried out impact analyses of individual proposals, four had done special highway impact studies, three had analyzed the impacts of airport location, two had simulated public transit proposals, and two had simulated local land use proposals.

Our overall conclusions in 1975 were that many more regional planning agencies would adopt models in the future (this conclusion was based largely upon the evidence from the mail surveys) and that model use would also probably increase, but that sustained use would be far less common than model adoption.

THE FAILURE OF MODEL APPLICATION

In the summer of 1980, seven of the eleven model using agencies visited in 1975 were contacted. Table 1 summarizes the evidence obtained on model applications. It is clear in 1980, as in 1975, that comprehensive plan simulation is the principal and in most cases the only application of the activity allocation model. Indeed in three of the seven agencies only one plan has been simulated.

Between 1975 and 1980 only two agencies carried out any additional simulations. Only one of the two had done a specific policy impact analysis. In two other agencies revised comprehensive plan simulations were planned for 1980. It is striking that in three of the seven agencies nothing had been done with the models since the initial runs and nothing was planned.

PUGET SOUND COUNCIL OF GOVERNMENTS (PSCOG)

In June of 1972 it had been decided to develop the EMPIRIC model and by May of 1973, three initial runs of the model had been completed. These first three simulations combined alternative assumptions about regional development, their immediate purpose being to form the basis for studies of water quality, water resources, urban drainage, and solid waste management. One of the alternatives was the previously articulated Interim Regional Development Plan, a second was a trends alternative, and the third was a concentrated corridor policy, which confined growth along major transportation routes.

Puget Sound was unusual, in 1975, for its actual and planned uses of the model. By 1975, EMPIRIC had also been used to analyze the impacts of a proposed extension of a major transportation corridor compared with a "no-build" alternative. Also underway then was a study of the regional and local impacts of a revised county land use plan (for the major county in the region), and alternative simulations for a revised Regional Development Plan.

By 1980, the Regional Development Plan had been revised using EMPIRIC and the county analysis had been completed. In addition the model had been rerun with new regional population and employment control totals in 1979. Two alternative futures for the year 1990 were considered in revising the population and employment forecasts (ie. the allocations to sub-regions of the aggregate forecast totals), based upon "differing sets of urban development activities which could be implemented during the next decade".[3] The first alternative, "Trends", assumed a continuation of existing development forces, subject to the normal land use constraints and continued, although limited, infrastructure development. The second alternative, "Policy", was based upon the enactment of "major restrictions ... to slow the conversion of rural land to urban uses". These were manifested in the model inputs by severe limits

TABLE 1

MODEL USING AGENCIES 1975 AND 1980

EMPIRIC	Year Agency Formed (a)	Year Model Adopted	Model Uses as of 1975 Comp. Plan (c)	Other (c)	Model Uses Since 1975 Comp. Plan	Other
Puget Sound Council of Governments (formerly Puget Sound Govt. Conf.)	1957	1972	1 alternative	Hway – 2 Water – 3(d) City Land Use-1	1979-2	Reg. Shop. Centres-2 Airport Develop.-2 Pick. Farm – 2(x2)
Atlanta Regional Comm.	1971	1972	6 alternatives	none	Planned 1980	none
Washington DC Council of Governments	1957-66	1969	3 alternatives	none	none	none
Minneapolis – St. Paul Metro Council	1965	1970	none	subregional forecast – 1	none	none
PLUM						
San Diego Comprehensive Planning Organization	1966-72	1970	4+ alternatives	airport location	1977 (base line) 1978 1980 (planned)	none
Miami Valley Regional Planning Commission (b)	1964-67	1973	1 alternative	none	none	none
Other						
Eastgate Development and Trans. Agency	1967-72	1970	1 alternative	none	planned	none

(a) Where two dates shown, the second refers to the date the agency was formally constituted in its present form; the earlier date refers to the formation of a predecessor, but substantially overlapping agency.

(b) Operates jointly with the Transportation Coordinating Committee.

(c) The figures refer to the number of alternatives simulated.

(d) One of these was the comprehensive plan simulation.

on the amount of land available for development and on sewer service extensions.

These two sets of assumptions yielded quite different sub-regional development forecasts, with "Trends" showing continued residential spread "outward from existing urban centres". "Jobs also spread outwards ... although not as much as ... population". In contrast, "the 'Policy' forecasts indicated that the growth ... would be more concentrated in existing urban areas....".

At PSCOG, the EMPIRIC model was used not only for regional plan simulation and the updating of forecasts, but more unusually, for specific policy impact analyses. Three sets of policy analyses were carried out between 1975 and 1980 and these were: (i) the comparison of two sets of employment changes around the airport, (ii) the comparison of concentrated shopping centre development with no shopping centres, (iii) the development of a 139-acre commercial, light industrial and retail space centre.[4]

Bellevue/Issaquah Airport Area: Two alternative development scenarios were analyzed for their broader impacts on the region. In each, 6,000 jobs were pre-located around the airport: in the first there were 3,000 service, 2,000 manufacturing, and 1,000 wholesale, transportation, communication and utilities jobs and in the second, 4,000 manufacturing and 2,000 wholesale, communications, transportation and utilities jobs. Overall regional employment was held constant. The greatest change occurred in CBD employment, but the impact on household location was negligible in both simulations.

Shopping Centres Impact Test: Two alternatives were compared: (i) four major regional shopping centres versus (ii) no regional shopping centres. Here, too, there were some straightforward employment contrasts, but almost no differences in household location.

Pickering Farms: Four comparisons were made. In the first two, the impacts on the region of the Pickering Farms development of 139 acres of commercial, light industrial, and retail space, plus multi-family residential housing and a park were compared with both the "Trends" and "Policy" alternatives described above. The third and fourth contrasts combined the Pickering Farms development with the Bellevue/Issaquah airport area proposal and compared their joint impacts with the "Trends" and "Policy" simulations. In none of these was there any substantial effect on household location.

In 1975, it was clear "... that the model was being developed in order to assist in the analysis of alternative plans and policies ... (not for) the development of a single set of forecasts. The emphasis was always on the capability which the model would provide for the analysis of specific proposals or alternative plans and proposals".[5] Although EMPIRIC is still in use in 1980, and has been used in the ways anticipated, the relative lack of sensitivity of activity allocations to what are considered major development policy changes has led the agency to consider the development of an alternative model with greater policy relevance.

It should be clear from Table 1 that PSCOG stood alone in 1975 in its application of its land use model to the analysis of specific policy proposals -- outside of the context of regional plan simulation -- and remains unique in this respect in 1980. Moreover, Puget Sound's continued use of EMPIRIC withstood the loss of key personnel involved in the decision to adopt the model initially and in its development and early application. (However, at least one person technically capable of working with the model has been present throughout.)

ATLANTA REGIONAL COUNCIL (ARC)

At the time of the 1975 visits, the ARC had carried out two cycles of three alternatives each for their Regional Development Plan using the EMPIRIC model. In the first cycle, existing land use controls in the region were combined with three different sets of transportation systems. In the second cycle, both land use and transportation policy alternatives were combined. Moreover, considerable effort was devoted to developing an evaluation system by which these alternatives could be compared to determine how well they helped to achieve regional objectives. On the basis of these evaluations a final Regional Development Plan was adopted in September, 1975.

Attitudes of staff members in the planning agency toward the model and its contribution to both the analysis of comprehensive plans and the forecasting of local activity distributions were very favourable in 1975 and appear not to have changed. In 1980, plans were underway for revising the regional development plan and the land use inventory and transportation system data were being updated for that purpose. Although other models are being examined, it is assumed that the agency will continue to use EMPIRIC: it was thought to be about as good as other available models and therefore that a change would not be worthwhile given the heavy investment in the EMPIRIC data base and software.

Several things should be noted: (i) no simulations had been carried out since the 1975 Regional Development Plan analyses, despite the interest expressed earlier in more specific policy evaluations, (ii) the model would be used again, only for updating the regional development plan, (iii) a model (EMPIRIC or some other) was considered necessary for regional plan analysis.

Here, as at Puget Sound, the model has survived the loss of key personnel involved in its development and initial application. This is in marked contrast to earlier reports of enormous investment in modelling efforts being totally lost with personnel turnover.

SAN DIEGO COMPREHENSIVE PLANNING ORGANIZATION (CPO)

In 1975 the San Diego agency was one of the most active model users, active in the sense that there was a very extensive set of models in addition to the PLUM activity allocation model in-house, data collection was very detailed and well-organized, and many staff members were directly and indirectly involved in model development, preparation of inputs, evaluation of outputs, use of model analysis for other purposes. Four alternative comprehensive plan simulations had been carried out, each of which involved large numbers of policy variations. The model had also been used to estimate the impacts of a new airport location.

In 1980, despite major changes in key personnel, "PLUM is alive and well in San Diego". Although, here as elsewhere, no separate policy studies had been carried out, the regional plan had been redone in 1978, with many policy alternatives tested in the process and a new set of sub-regional forecasts produced. It is reported by CPO that these forecasts are widely accepted and used by local governments and in regional planning activities. This revised regional plan was a response to the 1975 effort which was criticized for inadequately representing local plans. As a prelude to the 1977 revision, a series of base line forecasts was made in 1977 in which the simulations reflected "best guesses" about future policy, given current policies and plans.

At present, data base and policy inputs are being updated once again for a new plan simulation. It was stressed that the model has achieved general acceptance. Whereas in 1975, and even in 1978, there was much discussion and often challenge to the model itself, now the discussion among the local jurisdictions and within the regional agency concentrated on policy assumptions to be used as input data for the new simulations.

MIAMI VALLEY REGIONAL PLANNING COMMISSION AND TRANSPORTATION CO-ORDINATING COMMITTEE

When visited in 1975, the first run of the PLUM model had only recently been completed. The decision to develop PLUM had been made in August of 1973. It was chosen after elaborate comparison with six other models. PLUM was viewed as an operational model, available from the federal government and capable of being adapted to local use by agency personnel. It was thought to be inexpensive to run and therefore highly attractive since, "A model that sat around on the shelf or was run infrequently because of high costs was undesirable". This suggests that repeated analyses, updating of plans, and evaluation of the impacts of local policy changes on the region were planned.

In contrast with these early views, when contacted in 1980, it was determined that PLUM had not been run since the initial simulation in 1975. Contrary to initial expectations, based upon review and comparison of many models and intensive data gathering on PLUM itself, the agency had found it too difficult to revise the policy inputs and too expensive to run individual simulations. Although the model was still considered to be "in the arsenal", it was not being used. Indeed, a totally different model (Newling) was being used for annual update of the sub-regional growth allocation forecasts. The model is considered much more transparent than PLUM, easier for users to see its assumptions, much more widely accepted. It is strictly a forecasting model, not capable of policy analysis or regional plan simulation. The model posits an inverse relationship between population growth and density and requires data only on past growth rates and actual density for estimating its parameters. Density changes can then be forecast as a function of time and forecast density multiplied by land area yields forecast population.[6]

METROPOLITAN COUNCIL OF THE TWIN CITIES

Although the EMPIRIC model was estimated for the region and a set of sub-regional forecasts of population and employment was produced with it, the model never really received any acceptance within or without the agency. Only two simulation runs were carried out: one unconstrained, the other subject to density constraints in individual jurisdictions. There was much local disagreement with the forecasts, as well as disagreement within the regional planning agency itself. The predicted land use pattern was described as donut-like, an emptying out of the centre and major filling in of the periphery. The model forecasts were subjected to extensive hand adjustment before a set of official forecasts were ultimately adopted. Not only were its forecasts considered generally unreasonable, the policy analysis capability of the model was disappointing. "Those responsible for policy formation felt that the model had been oversold by the consultants and staff members directly involved with the effort. The model was not helpful in policy simulation because many relevant policy variables were not built into it, and those that were were not significant..."

The model has not been used since its initial runs. An alternative activity allocation procedure has been developed by the transportation division. Cohort-survival forecasting techniques combined with trends analysis are used to derive municipal control totals. The land use and housing data base is updated annually to derive the growth trends for particular sectors and rings of the metropolitan area. The sub-municipal zone allocations are based upon 1979 development in each zone, land area, base population and households. The allocation program assigns growth to where it was before, to the extent that vacant land is available. When a zone is filled up, growth is allocated to the next zone with the highest previous allocation in which there is vacant land available.

Several reasons were given for a general lack of interest in comprehensive land use and transportation models: (i) transportation analysis is no longer very meaningful, at least not of highway expansion, since so little highway construction is possible, (ii) comprehensive planners are not as interested in long-range planning and forecasting as they once were, but are concerned primarily with monitoring trends and short-run forecasting, (iii) little money is available now for data collection.

WASHINGTON, D.C. COUNCIL OF GOVERNMENTS (WCOG)

By 1975 three sets of alternative Year 2000 comprehensive plans had been simulated by the Washington COG. These proceeded in an evolutionary way, with subsequent simulations being based upon what had been learned from earlier simulations. Local governments were highly critical of the forecasts and of the policy assumptions and data with which they had been generated. The officially adopted forecasts were substantial hand-adjustments of the model outputs, which left everyone dissatisfied. The Year 2000 simulations had been completed in 1973 and at the time of our visits in 1975 no additional use had been made of the model and none was planned. There was still, however, a need and demand for regionally consistent local activity and land use forecasts.

Later in 1975 a Cooperative Forecasting Program was established which has undergone considerable changes, but is in place in 1980, and has, between 1975 and 1980, been put through two forecasting rounds. The forecasting process has two principal parts: the regional projections and the jurisdictional forecasts. Given its importance, "federal employment in the region based upon statistical trends in the region's share of total federal employment," is estimated first. Non-federal employment, population, and households are projected on the basis of regression equations in which the independent variables are national projections of population and employment, and the previously derived projections of federal employment. Initial local forecasts are made by the jurisdictions themselves, based upon a common set of procedures and assumptions. These are summed and compared with the overall regional forecasts. Where differences exceeded 2-3 percent, local forecasts are subject to peer reviews of methodology, assumptions, and implied growth rates. Projections are modified as appropriate. These peer review deliberations take place in weekly meetings of a committee of the key forecasting persons from each of the jurisdictions.[7]

Thus here, as in the other agencies, regional forecasts and consistent sub-regional forecasts of population and employment are considered absolutely necessary. When the EMPIRIC model was judged to be inadequate to the task, no alternative comprehensive model was considered. This appears to be due to a loss of faith in such models and the view that to be politically acceptable there needed to be far more local involvement in the forecasting process.

The only mention of possible future use of the EMPIRIC model came in response to a question about whether or not any analysis of the recently opened METRO (subway) was being undertaken. Here it was acknowledged that a model, perhaps the short version of EMPIRIC estimated earlier, would be required to analyze land use and development impacts throughout the region.

SOME IMPRESSIONISTIC INFERENCES

With the exception of the Puget Sound Council of Governments, when land use models continue to be used, it is for forecasting purposes, generally conditional forecasts based upon regional development plans. The only policy analyses are embodied in the plan simulations and the revisions and updating of these plans. As noted earlier, this is not the policy analysis envisioned when the models were developed and adopted. Then the expectations were for relatively continuous use of a new planning tool; a tool which would have great flexibility for determining the impact of transportation proposals, land use controls, e.g., zoning, sewer moratoria, major industrial developments, and many more, on population and employment distributions throughout a region. Such uses are seldom observed, even where models are not abandoned, even where they are generally held in high esteem.

This pattern of very limited model application is entirely consistent with what was observed in 1975. At that time, however, many of the agencies visited had only recently calibrated and begun to apply their models. The first use was generally the one most representative of the agencies' missions, namely the development of comprehensive regional plans and the conditional activity forecasts associated with them. Upon these plans and forecasts regional and local activities were to be based. In 1975, for most of these agencies, it was still too early to conclude that there would be little or no further policy application. Indeed, several of the agencies had plans for further use, for analysis of more specific policies, and most had undertaken model development initially on the assumption that the models would become a major part of their analytic apparatus. As proposals were made, by their constituent jurisdictions, by state or federal agencies, within the regional planning agency itself, the model would be available for impact analysis -- the appropriate policy variable(s) would be adjusted and the model run to determine the implications for the spatial distribution of activities, of population and employment, throughout the region.

Why were these applications rarely realized? Why have the models really not been institutionalized, that is not become regularly applied instruments of policy analysis within regional planning agencies? In 1980, as in 1975, it appears that four factors were important influences on the application of the models.

First, political factors, in particular inter-jurisdictional policy disagreements, inhibit the use of the models. One of the most direct ways in which such disagreements were manifested was in the specification of the values of policy input variables to make up the regional development plan. If local jurisdictions were brought into the process early, they could argue about the reservation of open space, about the definition of vacant available land for development, and about the classification of land for residential or industrial purposes. If they were brought into the process later, they could object to the conditional forecasts as not reflecting the policies or plans of their jurisdictions. At this stage disputes often confused the forecasts and the conditions upon which they were based. Thus, where conflict was great, it was tempting to avoid it by limiting the number of policy simulations, by not experimenting with politically unacceptable alternatives, by keeping the analysis within the limited range on which agreement was easiest to obtain.

The regional agencies are required to obtain the local jurisdictions' approval of a regional plan. In the past these have been made up of relatively broad policy outlines with which a set of regionally consistent population and employment forecasts for subparts of the region can be associated. Both of these -- the plan and the forecasts -- are then to be used to evaluate the consistency of local plans and specific proposals (eg. the extension of sewer lines, or the construction of a new hospital) with regional plans, forecasts, and goals. The need to quantify, for model simulation, what were heretofore only broad policy outlines, in some cases exacerbated local political conflict.

Second, the models themselves, that is their relevance for policy analysis, often received greater scrutiny at the point at which policy analyses were actually being proposed. It was then that analysts had to consider in far more concrete terms how a particular proposal would be formulated with the variables included in the model and how adequate a representation that could be. Moreover, this generally occurred after the calibration of the model. In many cases the calibrated models had indicated that activity allocations were insensitive to many of the policies (ie. to variations in them) thought to be important for regional spatial development. As a result, it was either concluded that the models were improperly specified for purposes of policy impact analysis or that the policies themselves were not as influential as had been supposed. In either case, the conclusion was that further policy simulation was not worthwhile.

Third, the difficulty of using the models, that is the effort involved in adjusting policy inputs for re-simulation, the expensive running times for each simulation, and the time and cost involved in keeping the data base current also inhibited the application of the models. What had been envisioned was a relatively simple process for adjusting policies. In practice any but the most trivial adjustment of land use or transportation policies involved extensive consultation with local jurisdictions and recalculation of densities, areas, times, distances, accessibilities. The differences in the quality of locally available data and the frequency and the form in which they were collected, also made the model less useful over time without considerable expense and effort to keep the data base current.

Fourth, the regional and local agency staff, in particular their attitudes and interest in the models and the continuity of their tenure were also found to be important. This was especially so in influencing model adoption and development and, in 1975, we expected these to exert some influence as well on subsequent applications. The studies of others had shown that often the ability to use the model was embodied in only one or two persons and that the "failure" of early modelling efforts was associated with their departure from the agency. Although we too found in 1975 that in many of these agencies only a few persons were well enough acquainted with the models to carry on their use, we also noted this was not universally the case and there was, in several agencies, an explicit effort being made to broaden the base of persons steeped in the model's operation.

In our recent follow-up, all of these reasons have been cited for the lack of application of the models, although the last, the continuity of personnel, is cited least often. Indeed, in three agencies, the most "active" model users, model application has survived the loss of key personnel. In all of these, there was relatively greater involvement of agency personnel in the preparation of model inputs, the evaluation of outputs, and the general discussion of the model's use in the regional plan simulations. Key personnel have also left agencies which no longer use the models, but only one cited these losses to justify the absence of model application.

The first three factors, political conflict, the suitability of the model for policy analyses, and the difficulty of re-running the models, were cited repeatedly, although with varying emphasis, by users and non-users alike. Those who had not run, and did not plan to run, the models since 1975 generally cited all three reasons. The agencies which had used or planned to use the models were generally interested in keeping policy simulation within the context of the revisions of the regional plan. They did not stress the difficulties or expense of use -- the impression they give is that the continuous updating of the data base and rerunning of existing models with their well developed software is the most efficient way of revising their regional plans and activity forecasts.

While what we are reporting here is based neither on comprehensive surveys nor on detailed case studies, there appears to be some basis for pessimism about the extent of application of urban activity models. They seem to be little used after their initial calibration and the uses observed are far more restricted than had been anticipated.   The findings are consistent with, although not directly predictable, from our earlier more comprehensive and detailed study.

The decline in the intensity of interest in the use of urban models can be explained in the terms used earlier; doubts about the models' plausibility, the costs of setting up and carrying through policy impact studies, and so on.   These intrinsic problems, combined with and/or intensified by a general retrenchment in all areas of government over the severe recession followed by inflation of the 1975-80 period would seem to be sufficient to explain the decline, if not the demise, of active modelling efforts.   However, all of these features, while plausible, may be too narrow in perspective, ignoring important changes in the intellectual milieu occurring over the period.

In our earlier study we found that the major factor in the decision to adopt a model was the presence on the agency staff of an energetic model enthusiast, often recently graduated from a university.   Despite complex organizational explanations of the decision (or lack thereof) to adopt a model, the "advocate's" role seemed to be decisive.   Such advocates (as we will continue to refer to them) were clearly the product of a specific confluence of intellectual currents, all roughly reaching their apogee during the formative years of these advocates, in the US roughly from the election of John Kennedy in 1960 to the end of the decade.

In the early euphoric days of the Kennedy administration all social problems seemed manageable with the correct application of "systems analysis", an umbrella term for a variety of techniques ranging from program budgeting to cost-benefit analysis to a variety of techniques borrowed from operations research.   All problems (according to the increasingly ascendant view) could be quantified and reasonable people (read technocrats) could derive optimal solutions.   Though the intellectual underpinnings of this entire view almost immediately encountered substantial scepticism, at least among thoughtful academics (and occasionally literary intellectuals), this new Weltanschauung nevertheless gained ascendancy and, in many ways, the highest stamp of approval in the 1966 executive order by President Johnson ordering all federal agencies to utilize PPBS.

A reinforcement of the perception of the possibility of technical solutions to many social problems came from other directions as well: the spectacular success of the 1964 US federal tax reduction and, in the private sector, the ever greater employment of the newest operations research results ranging from optimal inventory models to complex production control models.   And behind all of this was the increasing capability of each new generation of computers.

Almost surely this was the Zeitgeist in which the "advocates" were educated and matured.   Not only was there "an end of ideology" but even of politics.   John Lindsay implied, as part of his 1965 New

York City mayoral campaign, that his proposed adoption of program budgeting systems (embodied in super-agencies incorporating <u>all</u> functions providing one kind of service) would enable New York City to be run, if not as well as Tapiola, Finland, at least as well as London. Program budgeting, combining agency files and personnel so that each delivered one service, whose level was measurable, was the solution. Civil service tenure rules, enormous divergences in neighbourhood interests, patronage requirements, could all be forgotten once the programs were debugged.

For newly minted planners entering regional agencies the prospects were even rosier. Often they were dealing with relatively undeveloped areas; their decisions would have limited impact on existing interest groups; federal agencies, often staffed by adepts of models were anxious to proselytize, often with the enticement of large funds. And finally, attractive models were being developed which, according to the developers, could perform such marvels as predicting the impact on industrial location and housing choice of a change in zoning, the extension of sewers, the introduction of an airport, or a new rapid transit system.

Just as PPBS could not win the Vietnam War or the War on Poverty, at the regional level the urban model could not generate adequate predictions or solutions. One major flaw was identified from the beginning, namely, the inability, despite much ingenuity, to model industrial location. Similarly, they could not adequately cope with residential location without better specification of the determinants of housing demand and supply, most critically the role of prices. But even if these were corrected, the reasons given in 1975 and 1980 for limited use of calibrated models would hold, namely, the expense of using them for policy analysis, their inappropriate geographic disaggregation, conflicts over assumptions, and since it is assumed the results will matter, over outcomes.

Many agencies have come full circle: the current technical methods are those which would immediately be recognized by those whose knowledge was gained from standard textbooks of the early 1960's and who have never heard of EMPIRIC or PLUM. Nevertheless, there is a residue among former advocates and their disciples who recognize that broader analyses of both direct and indirect effects of policies must somehow be undertaken. However, unless they are impervious to the times, the lower level of respect, indeed the absence of hoopla, concerning systems analysis must leave many demoralized, unwilling to be the entrepreneurs of a new model generation, and often the not very enthusiastic executors of the possibilities offered by the earlier models. This is unfortunate since many of the issues faced by regional agencies are more tractable than the objectives of the larger scale systems failures had been and there may be increasing public acceptance of some guidance in some of these areas: with $3.00 a gallon gasoline on the horizon, the analysis of the implications of more compact development or transit expansion must have an audience. The development and use of improved models, surely has some merit.

NOTES AND REFERENCES

1. Pack, J.R., 1978, Urban Models: Diffusion and Policy Application, Regional Science Research Institute, Philadelphia, Penn.

2. State of the art is defined as the analytic possibilities envisioned by the original model developers. Whether, in fact these objectives could be achieved with the models is a difficult issue to resolve. (See Pack, op. cit., Chapters 2 and 8.)

3. Memorandum of January 26, 1979 from King Subregional Staff of the Puget Sound Council of Governments, "1990 Forecasts of Population and Employment".

4. Formal documentation is not available on these analyses. They have been described briefly to me in a letter from Jan Pilskog, Associate Planner, Land Use Planning/Research, PSCOG, July 8, 1980.

5. Pack, J.R., op. cit., 253.

6. Miami Valley Regional Planning Commission, 1979, Population Projections for the Miami Valley Region: 1980-2000, A Technical Report, Miami, Florida.

7. This process is nicely documented, including the formal techniques, the estimated equations, the actual sub-regional forecasts, in Metropolitan Washington Council of Governments, 1979, Cooperative Forecasting: Round II Summary Report -- 1979, Washington, DC.

ACKNOWLEDGEMENTS

Conversations with Howard Pack were very helpful in formulating the concluding section.

APPLICATIONS OF CORPORATE PLANNING IN URBAN POLICY ANALYSIS AND

DECISION-MAKING

Juri Pill

Toronto Transit Commission
1900 Yonge Street
Toronto, Ontario, M4S 1Z2
Canada

INTRODUCTION

Although the term "systems analysis" is of fairly recent vin-
tage, the ideas it embodies are not new. Essentially, systems
analysis is a technique for focusing attention on the particular
element of the world one is interested in, and bringing to bear
certain logical constructs to model its behaviour. The objective of
this exercise is to understand cause and effect, and thus to be able
to predict how the system will interact with its environment, and
how its various component parts will behave in different situations. A model which enables prediction can also be used for manipulation, and thus can be a very powerful tool politically. This
paper is concerned with the latter issue: the relationship between
systems analysis and politics, in the urban context.

There is a trade-off in systems analysis between comprehensiveness and analytical rigour. The real world is infinitely complex,
and any model of it will necessarily be incomplete. With a given
amount of resources for carrying out the analysis, one can cast the
net wide with fairly general assumptions but lose on depth and
detail, or one can make very tight assumptions, define the problem
quite explicitly, and use specific mathematical techniques for
analysis, often arriving at a theoretically optimum solution.

Both approaches are valid, and the choice depends on the problem at hand and on the objectives of the analysis. Sometimes, however, an analytic model takes on a life of its own in the academic
and professional literature, and its devotees tend to forget that
the model is merely a set of hypotheses whose connection to the real
world may be quite tenuous.

Systems analysis must not be confused with the scientific method, although the temptation and the incentive to do so are often overwhelming, given the astounding success of the latter in allowing us - for better or worse - to control and shape the physical world. The principles of the scientific method are so disarmingly simple as to belie their revolutionary nature:

1. Observation
2. Hypothesis
3. Experimentation
4. Development of a verifiable theory

Valid scientific theories are those which, within the limits of their assumptions, allow perfect predictability. In a sense, the theory summarizes the results of an infinite number of experiments which need never be performed. This is the major utility, for instance, of Newton's Laws.

There have been numerous attempts to apply scientific method, either implicitly or explicitly, to man and society. There are many pitfalls in such an approach, especially if it is based on some idealized concepts of human nature rather than on mankind's track record. In some cases, theories about the negative characteristics of specific cultural groups have been presented in a scientific guise to camouflage a rationalization for the exploitation or oppression of these groups. In other cases, theories about economics, human nature, or God and man are put into practice at enormous social cost, as various coercive strategems are applied to force the people to fit the theory.

Science is a fairly recent invention, while systems analysis, under a variety of names, has a much longer history as it applies to human behaviour. Plato, Machiavelli, and Shakespeare were not scientists, but they probably told us more about human systems than any modern social scientist could hope to. Similarly, Eastern philosophers and thinkers have been observing human behaviour for thousands of years, and during that time, humans have changed very little, despite the invention of automobiles and microprocessors.

If nothing else, urban planners should at least be familiar with the legend of Icarus. There are numerous cases where the reach of urban systems analysis has exceeded its grasp because it pretended to have a theory based on sound scientific principles, but overlooked some fairly basic characteristic of the real world. It is time to be realistic: comprehensive urban systems analysis can currently deal only with the first two steps in the scientific method, ie. observation and hypothesis.

There is a great temptation at times to present a hypothesis as a validated theoretical law, with all of the requisite genuflections to the computer, and thereby to embark on some grand experiment -- with the pretense that the results are predictable. They are not, except in very constrained (almost trivial) cases, and systems analysts should acknowledge this fact.

The temptation to act the sage is very real, since there is always a thirst for certainty in a time of rapid change. Moreover, the analysts' recommendations are much more likely to be accepted if he predicts pleasant consequences; and since he derives as much pleasure from affecting world events as any other reasonable person, the trap is seductive indeed. It can be almost irresistible if he is willing to say what the political decision-maker wants to hear.

However, this paper is aimed at a different kind of planner -- one who sees things as they are, has a reasonable understanding of what is likely to occur if existing policies continue, and wants other options to be considered within the democratic framework. This is a much more tentative approach than used by the planner-as-scientist, and is less likely to lead to publications and professional esteem. However, it is much more likely to lead to positive results in any particular city.

For those who accept the need for more objective observation in the urban sphere, and reject any hope for a comprehensive theory in the foreseeable future, this paper addresses two hypotheses, and provides some evidence to support them. The hypotheses are as follows:

1. That there is a need for a dynamic, ie. time-sensitive, perspective in dealing with urban policy-making, as contrasted to a one-shot plan.

2. That for systems analysis to be effective in an urban setting, its practitioners must bridge a number of cultural gaps among the actors in the decision-making process, namely the academics, the bureaucrats, the politicians, and the mediacrats.

THE NEED FOR A DYNAMIC PERSPECTIVE

Despite its obvious superiority over any other system of government in most respects, democracy has one flaw: it has a sometimes costly tendency toward wishful thinking and over-optimism, and an avoidance of brute facts and unpleasant consequences. In a great many cases, the optimism is justified and rewarded, as the incentive structure and the diversity of perspectives lead to the eventual solution of most problems. Major breakthroughs occur without much thought about the consequences, and in this sense a democracy with a market economy can be characterized as playful and almost childish in its view of the future. This playfulness of course leads to the sorts of innovations that rarely occur in the much more serious atmosphere of a planned economy and a totalitarian regime, and it is important that this attitude be maintained.

However, the assumption that problems can be resolved when and if they occur, rather than anticipated before they become serious, will increasingly become a luxury in the developed world. The pace of change is becoming so rapid that by the time the problem becomes serious, there may be no time to do anything about it. The urban systems analyst's major role must be to stretch the time horizon, to somehow force decision-makers and the public to acknowledge the future consequences of present actions (or inactions). However, the planner must not be permitted to <u>make</u> the decisions, because he is not accountable for them.

In some ways, the planner's role is analogous to that of the artist, ie. to tell the truth as he sees it about the direction of society. Although the medium is very different, planners might learn from art like Christo's nylon fence in California, whose commentary was on process rather than substance. Control engineering, with its paradigm of flows and feedbacks, provides another process model for the planner.

The current decision-making process has a very short-term horizon for some perfectly rational reasons. The democratically elected representative serves for a fixed term, and his survival depends on keeping the electorate happy in the present. He wins few votes by advocating short-term pain for long-term gain. Moreover, a version of Gresham's Law applies: short-term panaceas drive out long-term realities.

Market forces tend to operate the same way as the political system. For instance, the US auto market demanded V-8 cars even after the 1974 oil embargo and despite the handwriting on the wall regarding future oil supplies. Since the market, like politics, is a competitive arena, the best short-term policy is always to give people what they want. Successful marketing, like successful politics, requires an understanding of people as they are, not as we wish them to be.

If the planner is to be a vanguard for change and to anticipate potential problems in this context, he must examine the tools at his disposal, and the paradigms employed by his colleagues. Most of the paradigms do not include a sense of process, or much respect for the flow of time. The classic planning tool is <u>the plan</u>, which by definition is an end. The means often receive only secondary consideration, if any at all. Ends and means must be considered together in planning, using a <u>dynamic</u> model which not only incorporates the physical world but also the decision-making process itself, however sketchily.

In the financial sphere, which is always the key to action, the basic professional perspectives are those of the accountant and the micro-economist, both of whom by and large view the world through a static paradigm. The accountant is concerned with the budget, and the economist with his equilibrium model. Both of these are basically snap-shots of the world, and provide few semantic tools for considering the connectedness of events over time.

This static perspective, allied with the short-time horizon of the elected representative and the market, performs very well in a time where change is slow and predictable. The necessary incremental adjustments are made through trial and error, and the system optimizes itself in almost a random walk. Systems analysis provides only marginal benefits in this sort of situation, as demonstrated by the experience with operations research after the Second World War. Operations research had been remarkably useful as an aid to decision-making in situations where past experience was unavailable -- the application of radar, for instance -- but it was less effective (though still helpful) after the war in industrial settings where many decades of cumulative experience and competition had led to practices not too far from the optimum.

The incremental approach, and extrapolation of experience, do not work well, however, in a time of rapid technological and social change. It's fair to say that cities and urban systems are currently faced with rapid change -- demographic shifts, the energy situation, and the impact of international economic instabilities. There is a need for more analysis of urban options in the context of how decisions are really made, and more importantly, for a dynamic perspective: an awareness of process, feedback loops, and the implications of existing trends. The systems analyst can provide this perspective. To be effective, he must also communicate it.

THE RELEVANT CULTURES

The urban policy analyst must be able to cross cultural boundaries, and to understand each relevant culture as it affects urban decisions. As C.P. Snow pointed out in his original "Two Cultures" lecture (Snow, 1959), the underlying assumptions and thought processes inherent in a professional or vocational culture are instrumental in determining its strengths and weaknesses, its views of the world, and the potentials for misunderstandings with other cultures. Communication among professional cultures is at least as fraught with difficulty as communication among nations.

First, one is usually unaware of one's own cultural axioms, as pointed out by A.N. Whitehead (Whitehead, 1927). The axioms lead to an inherently biased view of the world, and the native is usually unaware of this bias. Secondly, the outside observer cannot apprehend the perspective of any complex culture without becoming immersed in it, a situation which inevitably causes a further bias. So an Indeterminacy Principle must always apply in cross-cultural communication.

Snow focused on the schism between the sciences and the arts, but his ideas are highly relevant to urban systems analysis, which

involves many cultures of its own. The four principal cultures are
as follows:

1.   The academics
2.   The bureaucrats
3.   The politicians
4.   The mediacrats

Each of these groups has its own internal rules, regulations,
perceptions, language, and reward structures. (There are of course
overlaps among these groups, and the list is not exhaustive; this
framework is sufficient, however, to outline the major issues.)
Systems analysis can be effective, in the sense of having a positive
impact on action, only if there is communication and cooperation
among these four groups, despite the inherent conflicts and tensions
involved.

Each culture has a complementary role in the urban decision-
making process. Many of the better ideas are generated by the academics, but they are not effective if they remain isolated. The
bureaucrats may not be innovative, but they constitute the backbone
that keeps the system standing, and very little can be achieved if
they oppose it. The politicans have the nominal decision-making
powers, but their options are always highly restricted. Finally,
the media have the power to influence the outcome of any public
decision at the local level. The successful policy analyst must
understand and work with all four groups -- bridging their cultures,
but not being overwhelmed by the values of any one of them.

The academic culture in urban systems analysis has a number of
obvious strengths: it carries out the most rigorous analytical work,
represents outstanding intellectual ability, and has a detachment
(the Ivory Tower) which implies objectivity. Academic ideas can
have widespread influence, and totalitarian regimes are especially
fearful of academic freedom. However, much academic work is self-
propogating and is often carried out for its own sake. The reward
structure sets priorities which allow for demonstrations of intellectual flamboyance, and do not lead to much congruence between the
areas that are studied and the areas where the greatest real-world
payoff might obtain. The interesting always drives out the useful
in academia, and the systems analyst must dismiss entire fields of
study as useless because he knows from experience that their axioms
are absurd when related to the real world. The ideas that are
useful are often not very complex, on the other hand.

The academic culture's major failing is its lack of respect for
(or ignorance of) the importance of experience and straight common
sense. When an academic theory exists, there is of course great
satisfaction to be derived from seeing it applied, and academics are
no less susceptible to the blandishments of raw power than anyone
else. However, the track record of applied theory in education,
social services, and even in urban planning (witness "urban renewal"

in the US) has not been positive. And the private corporate world
is littered with the wrecks of companies that have been taken over
by MBA whiz-kids and run into the ground with theories. The ideas
generated by the academic culture must always be balanced by the
conservatism born of experience.

The bureaucratic culture -- and one can include consultants
here as an appendage -- has conservatism to spare. For better or
worse, bureaucrats are responsible for keeping the system running,
and therefore have an admirable pragmatism, a respect for experience
and "the way things are done", and a classic reluctance to rock the
boat. The reward structure is usually such that size rather than
quality of accomplishment determines the pecking order, and the
result is often under-utilization of resources. However, the para-
military hierarchy leads to great efficiency in division of labour,
and most bureaucracies are in fact quite effective in meeting well-
defined goals.

It is in setting the goals that problems often occur in bureaucracies, as they gallop off very efficiently in the wrong direction. The strategies are often developed by individuals who have
achieved power through infighting skills within the bureaucracy, and
continue to use these criteria in developing policy and in advising
their political masters. For instance, skill at defending one's
budget often counts for more in a bureaucracy than a deep-felt com-
passion to build a better city. This is not said by way of condem-
nation; that's simply the way it works, always has, and always
will. And good ideas without the infighting skills are useless.

Some private corporations, notably General Motors, acknowledge
that good line managers and good strategists are both necessary, and
that they are not the same people. The most successful large organizations are the ones that can blend them, drawing on both the academic side and on experience. The urban systems analyst must always
be aware of the need for this blend in a bureaucracy, and to listen
to both sides.

The politician is the ultimate pragmatist, since his reward
structure is so clear-cut: to remain effective, he must be re-
elected. He is quite willing to do the right thing, but not at the
expense of getting kicked out of office. There are exceptions --
the one-term politicians who break the accepted rules and significantly jolt the system, often for the better -- but most of the
time, the politician wants to remain in office and therefore will
not displease his constituents. At the urban level, keeping constituents happy often involves so much time that he rarely has a
chance to consider all important decisions in detail. Therefore,
intangibles, media coverage, and advice from bureaucrats have much
more influence than at the more senior levels of government. The
higher levels of government are of course also involved in urban
decisions, and their positions tend to be more disciplined and
cohesive.

Although it is easy to disparage the politician's crass pragmatism, the effective urban systems analyst must have some empathy and
respect for it. Being a politician involves a regular personal risk
-- the loss of one's livelihood and status -- to which the academic
or bureaucrat rarely subjects himself.

Finally, at the level where policy is made, the media play an
important role, since they reflect and influence public opinion, and
thus often define the options open to the politician. In many ways,
the media have parallels with the academic community: a reporter is
by and large judged by his peers, prides himself on his objectivity
and detachment, and is rewarded for publications (or broadcasts).
The media often have a herd instinct, reinforced by their increasing
concentration (at least in North America). The process of how news
is defined is a fascinating and extremely complex subject, and cannot be fully treated here, but suffice it to say that news priorities are usually set not by the objective importance of a subject in
terms of its actual or potential impact on people's lives, but by
the readers' or viewers' own interest in the subject. Like politicians, the media must keep their constituents happy. Therefore, it
is very rare that a major policy issue becomes news, except in a
sensational form: it's usually either too complex for glib exposition, or so obvious as not to be newsworthy. Television has probably had a major role in trivializing policy debate, but that issue
is much better left to Marshall McLuhan and other experts in the
field.

From the above discussion, it is clear that great potential
exists for frustration and misunderstanding in the urban decision-
making process, since the four relevant cultures are so different.
The policy analyst must see through the misunderstandings, and build
bridges. One of the more effective ways of building bridges is
through the function of corporate planning, usually within the
bureaucratic structure.

CORPORATE PLANNING

Corporate planning is the function in an organization whereby
the longer-term perspective is brought to bear on major decision-
making, not simply through a strategic plan or a set of specific
proposals, but as a process. It is categorically not based on
optimization methods, which in the real world are merely tactical
tools, and it is not "muddling through" either, ie. incrementalism
with no underlying strategy.

Corporate planning basically involves figuring out what business an organization should be in. Quite often, this role is
carried out implicitly by the chief executive and the board of
directors. When the role becomes a separate function, it remains a
top management activity and becomes quite complex, since its objective is to keep the organization on course without any line auth-

ority.   The role of the corporate planner is well expressed by
Taylor (1976):

> "To be effective the Corporate Planner today has
> to see himself not as installing and maintaining
> a planning system -- though this may be part of
> his objective -- but rather as a Policy Analyst
> working with the decision-makers whether they be
> a single chief executive, a management team, or a
> complex coalition of interacting groups.
>
> The test of his success is not whether he pro-
> duces impressive plans but whether he helps the
> decision-makers to identify key issues and to
> make these important decisions systematically and
> in good time.
>
> If planning is to develop as a profession, plan-
> ners must see planning as a process of control,
> innovation, organizational learning, political
> bargaining, and as a choice involving basic human
> values and the future shape of society."

The corporate planning process achieves its success by bridging
the culture gaps, forcing debate on the real issues, and developing
strategies that can be understood and accepted by the decision-
makers.   Some ideas are borrowed from the academic sector, but many
of these strategies are not complex -- they often involve fundamental choices among very basic and simple policy options -- and
therefore the analysis is rarely sophisticated enough to result in
academic publications.   A great deal of the effort in corporate
planning involves translation from one culture to another, and the
use of a wide variety of communications media to structure the
debate.   It must be a multi-prolonged approach, and the computer can
be very useful -- as much for communicating ideas as carrying out
substantive analysis.

It is possible for this process to work, and the point can best
be made by describing two qualified planning successes in Toronto
during the past decade.   One was the Metropolitan Toronto Transportation Plan Review, and the other has been the corporate planning
process at the Toronto Transit Commission.   (The author was deeply
involved in both, so the reader must weigh the richness of observation against the potential bias in reporting.)

THE METRO TORONTO TRANSPORTATION PLAN REVIEW

The Metro Toronto Transportation Plan Review (MTTPR) was a
planning exercise lasting three-and-a-half years, from 1971 to 1975,
and was unusual in that it more or less fulfilled its mandate: it
reviewed an existing transportation plan, found it wanting in a

number of respects, and effected a change in direction that basically altered the community's perspective as to what Metro Toronto
will probably look like in the year 2000. More importantly, a
number of major substantive decisions have resulted which are consistent with this new policy.

The process whereby this impact was achieved was unusual and
innovative in a number of respects (see Pill, 1979a), and was in
many ways closer to corporate planning than to urban planning as it
generally taught and practised. That is, there was an awareness
throughout of the importance of implementation, and less emphasis on
technical elegance and transportation modelling than in most planning exercises of this type. This is not to say that technical work
was lacking; it was in fact much more comprehensive than the classic
computer modelling approach, and covered environmental impact,
financing, land use, phasing, income distribution, social effects,
as well as travel demand analysis. Given the wide area of subjects
covered, however, the analysis was less detailed and rigorous than
it could perhaps have been if only a few of these areas had been
examined. The higher priority was given to examining a wide range
of options and issues, instead of digging deeply into a few.

For instance, the travel demand analysis was carried out in a
broad-brush fashion. The "sketch" planning model developed by the
MTTPR became known as STAP -- the Simplified Travel Analysis Procedure. Its character was summed up in one of the MTTPR reports:

> "STAP is by nature an approximate method, and does
> not claim great accuracy in predicting future
> travel flows. Indeed, the results of all travel
> simulation models are limited by the accuracy of
> the input land use and travel data and of the
> travel behaviour assumptions; some models imply
> an unwarranted degree of accuracy by the level of
> detail in their calculations. The advantage of
> STAP lies in its relative ease of operation,
> enabling the travel consequences of several land
> use and transportation assumptions to be studied
> in a comparative manner. However....the formula-
> tion of the input and interpretation of the out-
> put require a group of experienced planners, and
> cannot be performed by purely technical personnel
> according to preset rules."

MTTPR (1975)

Although this sketch procedure was criticized (with some validity) by local transportation planners, the range of options tested
was much wider than in most transportation studies. And the inclusion of a wide range of criteria made the results credible to a much
more diverse audience.

Another key aspect of the MTTPR analysis was the overall evaluation approach used. After a thorough review of experience elsewhere, it was concluded that the "definition of objectives" stage, which is usually seen as the critical first step in a planning process, is a waste of time. It was seen as meaningless because the objectives that everyone could agree on were too general or too vague to be translated into concrete actions that fulfilled all the objectives. All significant real-world decisions involve difficult choices and conflicts, and most of the conflicts resolve around trade-offs among the objectives, rather than around which objectives might seem worthwhile in a world of infinite resources.

The first step in the MTTPR approach was to generate alternative courses of action, and then to allow the implicit trade-offs to surface in the comparison of these feasible land use and transportation problems. The comparison of the options became a medium for structuring the debate, making it more difficult for decision-makers to avoid the choices. Another important aspect of the technical work was that the panaceas -- which are always proposed and receive publicity - were examined in detail, and shown to have flaws.

There are some sound arguments for taking a comprehensive sketch planning approach rather than a focused and narrow one. First, any twenty- or thirty-year plan must be conjectural, and honing it to very fine detail reduces its credibility: no one could possibly know the traffic flow at a specific intersection in the year 2000, because there are too many major uncertainties. A twenty-year plan is an abstraction which provides some direction and consistency to near-term decisions, but it is likely to be reviewed and altered as circumstances change. The essential value of the different long-term options is to function as a Rorschach test for the community and the decision-makers regarding their land use and transportation values, and to indicate the longer-term implications of current trends in decision-making.

The technical work carried out in the MTTPR had a profound effect on the conventional wisdom of the community with respect to transportation and land use form. The most important result was a shift away from a radially-oriented single-centre urban form to a multi-centred network form. At the time the MTTPR began, it was assumed that the next subway line would run radially downtown, and that most major office growth would occur in the central area. The Review proposed a "sub-centres" concept, with more two-way use of the transportation facilities, and an east-west transit facility serving all of Metro, rather than a line for downtown trips. This concept became part of the Metropolitan Official Plan, and the sub-centres have now achieved the "critical mass" stage.

There were a number of other concrete results from the MTTPR. The Scarborough Expressway, a radial link to downtown from the northeast, was stopped as a result of a special study by the Review involving public participation, and a light rail transit link to the

Scarborough Town Centre is now under construction, as recommended by the MTTPR. The Review also recommended that the Spadina Expressway, a northwest link to downtown which had been stopped in 1971, should be extended slightly toward the central area as an arterial road, and this was carried out.

The technical work was successful in affecting action for two reasons: the director of the MTTPR played a high-profile political role and used the media to good effect, and the credibility of the Review was significantly enhanced by its influence on a major project decision -- the Scarborough Expressway. As a result, the final report of the MTTPR received a great deal of media coverage, and had considerable political credence, which is unusual for a long-range planning document.

The key to the MTTPR was this process of establishing political credibility, which made it possible to outline the basic choices once the community was listening. As Daland and Parker (1962) expressed it in another context:

> "Acceptance tends to increase to the degree that planning activity is considered "useful" by the actors in policy making. The reputation of usefulness is gained less by the quality of long-range planning than by "practical" work on short--range projects. This suggests that there may be an innate necessity to build the planning program from short-range spot-planning work, through projects with larger implications and intermediate perspectives, and ultimately to the long--range comprehensive plans which the planner is trained and anxious to provide."

This approach requires a great deal of patience, and an understanding by the planner of the other cultures he must deal with. The objective of the process is not, however, simply to produce a plan, but to use the discussion of options as a medium for structuring the debate. In the MTTPR, the structuring of the issues involved the planners, the politicians, the academics and the media, and the result was a fundamental change in direction.

CORPORATE PLANNING AT THE TORONTO TRANSIT COMMISSION

In the MTTPR, corporate planning principles were applied within an urban planning framework. At the Toronto Transit Commission, corporate planning has been applied in a more explicit fashion, also with some success. As with the MTTPR, the principal characteristics of the TTC corporate planning approach have been the bridging of the technical, political, and media cultures, and to some extent the academic culture, and an explicit consciousness of the time parameter, ie. the use of a process model in analysis and strategy development to demonstrate longer-term effects of certain courses of action.

Like most other urban transit systems throughout the western world, the Toronto Transit Commission has undergone rapid changes over the past three decades. Its role has shifted drastically, from a basic necessity for many people, which could support itself out of the fare-box, to a social service that requires financial support from the community as a whole, and whose riders often have the option of using their automobiles instead. Transit now operates as a mixed economic venture, in that a great deal of its revenue is derived from its operation in the competitive market-place on a fee-for-service basis, while another part of its funding is drawn from general tax revenues, to be spent in meeting certain community goals which are usually not well-articulated. The transit industry is in a sense at the forefront of a more general trend toward a mixed public/private economy, given for example various industrial corporations which have been nationalized or propped up over the past decade, so the solutions developed by transit systems to blend business and political criteria may have application in other spheres.

In Toronto, as in many other cities, the transit system is an important element of the urban structure. The essential objective of the TTC corporate planning process over the past four years has been to determine what role the community expects transit to perform, and the criteria by which that performance should be measured. As Deep Throat expressed it in the Watergate capers, the most effective way to understand a complex situation is to "follow the money". So the corporate planning exercise started with an examination of TTC operating finances (Pill, 1977).

The TTC's financial autonomy, and its ability to determine its own policies, have been steadily declining since 1950. The five Commissioners, who meet every two weeks and effectively function as a board of directors, are appointed by the Metropolitan Toronto Council, and since all subsidies are routed through the Council, the Commissioners are subject to certain political pressures. Between 1940 and 1949, an operating surplus was generated each year, and these funds were used to build Toronto's first subway line. The Commission was of course quite independent at that time. After 1950, with increasing auto ownership and dispersal of the population, ridership started to decline, and although increasing operating costs were covered by fare increases, public subsidies had to be used to finance the subway expansions carried out in the 1960's.

In 1971, the TTC received its first regular operating subsidies, split 50/50 by the Metro Toronto and Ontario (provincial) governments, and amounting to $2.9 million. The fares were frozen from 1969 to 1974, a time of rapidly increasing inflation, and by 1974 the operating subsidy had increased more than ten-fold, to $34.2 million. Then came the crunch: with a slow economy and government spending curbs, subsidy increases were severely limited in 1975, and a 33 percent fare increase was implemented, with a second 20 percent increase eleven months later. So after almost six years of frozen fares, there was a cumulative 60 percent increase in less than a year.

In control engineering parlance, the financing system was characterized by dynamic instability. It tended to go from one extreme to the other, and to over-correct rather than anticipate. In part, these extreme positions resulted from the almost schizophrenic political attitude with respect to transit financing: transit was seen either as a social service and a mechanism for helping low-income people, or as a business venture which was expected to keep its "deficits" under control, but almost never as what it actually was – a blend of both, with constant trade-offs necessary between these two necessarily conflicting perspectives. The political process tended toward the extremes because that was where the political hay could be made. In urban politics, individual publicity is important, and the first one on a bandwagon always gets his name mentioned. The media prefer a flamboyant quote over statesmanship at deadline time, so there is little premium placed on compromise.

Needless to say, it was difficult for the TTC to carry out long-term planning under these circumstances. The TTC's financial partners, ie. the Metropolitan Toronto government and the provincial government, who split the operating subsidy, held a ritualistic poker game each spring to determine who would get the credit for staving off a fare increase, or take the blame if there had to be one. Due to the timing of this poker game, TTC management would usually be halfway into its fiscal year before a firm operating budget was approved. In 1976, the Commission staved off the debate by simply increasing its fare early in the year, and going on with the business of running the system.

In 1977, a combined technical and political process was initiated to stabilize operating finances, and to clarify the role of the TTC in the community. The process established three fairly basic principles with the Metropolitan Toronto Council and with the provincial government:

1.   That operating costs must be covered by fare revenues plus operating subsidies, and that as wages and other costs go up, total revenues must keep pace or services must be cut back. This may seem like an obvious point, but many of the financing proposals made at the political level failed to acknowledge the reality of even this simple arithmetic.

2.   That transit subsidies should be viewed as a payment by the community to enable the transit system to meet certain social objectives, such as saving energy, reducing air pollution, re-distributing income, and reducing the need for road construction. They should not be viewed as a business loss. It was pointed out, not as a threat but as a matter of fact, that

the TTC was in a position to eliminate its deficit by increasing its fares substantially and allowing a ridership drop.

3.   That the most acceptable way to avoid the financing instabili-
     ties of 1969-1976 was to have small incremental fare increases
     that kept pace with increases in operating costs, ie. to retain
     fare revenues at a fixed percentage of operating costs, aside
     from service expansions.

These are fairly basic points, but about six months was
required to make them at the political level. First, a five-member
senior political committee was appointed to review the issues, and
they worked very closely with technical staff. The main medium of
communication was a computer model which tested different financing
options over a five-year period, with a variety of assumptions. The
implications of freezing fares for that period, combined with 10
percent annual increases in operating cost, quickly became crystal-
clear to the politicians. The subsidy implications tended to reduce
the momentum of the fare-freeze bandwagon.

A wide variety of such scenarios were reviewed by the policy
committee, and at their recommendation, the Metro Council on
November 1, 1977 adopted the "user's fair share" formula, after
reviewing a technical presentation on the options. The formula
"involved an acceptable compromise between the need to keep a rein
on public expenditures and the desirability of maintaining ridership
on the TTC" (Pill and Warren, 1979). It involved small periodic
fare increases as required to maintain fare revenues at seventy
percent of expenditures, and a small fare increase (eight percent)
was implemented on January 1, 1978. It was projected that this
policy would lead to a slight ridership growth over the five years.

Unfortunately, ridership in 1978, which had been projected to
hit 359 million, reached only 338 million, due to reduced reliability as a major subway extension was debugged, a general economic
slowdown, and a four-day transit strike. There was a municipal
election campaign in the fall of 1978, and pressure grew for a fare
freeze to counter the declining ridership. TTC staff debated the
pros and cons of a fare increase in 1979 at some length, in parallel
with the political debate, but neither was able to reach consensus.

Early in 1979, a TTC staff report entitled Transit in Metro:
Some Tough Choices (Pill and Warren, 1979) was produced; it acknowledged the need to review the 1977 financial formula. The conclusions of the Tough Choices report were as follows:

1.   The reasonable and acceptable compromise of 1977 between finan-
     cial constraint and ridership no longer works, and some very
     tough choices must be made in 1979.

2.   Unless Metro and the area municipalities take some deliberate
     steps to promote transit, ridership will probably decline gra-
     dually over the next few years.

3. The question of whether or not to increase the fare in 1979 must not be decided in isolation or out of context. Ideally, the decision should not be made until Metro decides what the proper long-term role of transit should be.

4. The best first step would be to appoint a Metro/TTC policy committee to make recommendations to the Metropolitan Council on the appropriate long-term role for the TTC, and to review the options for 1979 in the proper context.

On January 30, 1979, the Metropolitan Council appointed a Metro/TTC Transit Policy Committee. It consisted of five politicians, including the Chairman of the Metropolitan Council and the Mayor of the City of Toronto, as well as the Chairman of the TTC. The Committee was a microcosm of the Metropolitan Council, with both the city's fare-freeze and the suburbs' fiscal responsibility perspectives represented. It embarked on a major corporate planning exercise which involved fifteen working sessions with technical staff drawn from eighteen public agencies involved in transportation planning and operations. These working sessions averaged two to three hours in length, and effectively forged a link, for the individuals in the exercise, between the political and technical cultures. The media were not involved in the sessions, despite some pressure, so posturing was minimized. The importance of these working sessions cannot be overemphasized.

There were no major revelations or technical surprises from the process, even after a wide variety of publicity-touted simplistic solutions were examined in detail, and in an objective manner. The outcome was a technical staff report entitled Transit in the 1980's: A New Direction (Pill, 1979b), which was discussed in detail at two public hearings after its release, and dissected by the media.

On January 15, 1980, the New Direction recommendations were adopted unanimously (albeit after much debate) by the Metropolitan Council. Their essence was that transit should receive higher priority in Metro Toronto, with a slight increase in subsidy (from 30 percent to 32 percent of operating costs) to be used to finance a monthly pass and experimental service improvements, plus steps to give transit some preferential treatment on Metro's road system.

Thus, in a period of four years, the TTC's operating environment has improved considerably. Its financing is relatively stable with the continuation of the "user's fair share" formula at 68 percent, and it has received explicit guidance from the political level regarding its role in the community. Most of the changes have resulted from initiatives taken by the TTC, initiatives which have combined technical fundamentals with an awareness of political realities. The surprise in the process has been the actual implementation of more or less coherent policies; it was possible to bridge the technical and political cultures, and to effect reasonable compromises.

The two keys were an awareness of process and of changing circumstances -- and thus an inherent flexibility and a certain timeliness -- and an awareness of the cultural gaps among the various players. As in the MTTPR, the principal staff person, the Chief General Manager of the TTC, played a political role as well as a technical one and was firmly identified with specific policy positions in the media. All of the reports discussed above received considerable media coverage, and there was very much of an interplay involving the politicians, the media, and the management staff. The technical analysis was important, but its main function was to structure the political decisions that needed to be made; it was therefore a necessary element in the process, but not a sufficient one for a satisfactory resolution. The results had to be translated and communicated.

The changes wrought through corporate planning at the TTC have been incremental rather than revolutionary, but they have all had a common thrust: to clarify the role of transit, and to make the policies that affect transit as consistent with this role as possible. As in the MTTPR, the discussion of role has not been explicit, but has occurred through the process of considering different concrete options. However, since the environment is constantly changing, it should be fairly obvious that no final solution has been achieved. Ridership in 1979 was above expectations by 9 million trips, despite the 15 percent fare increase that was finally implemented in March of that year after much debate, and in 1980 it is 5 percent to 7 percent above 1979. No fare increase will be necessary in 1980, partly due to the buoyant ridership. The TTC is currently carrying out market research to determine the reasons for these variations in ridership.

Most of the corporate planning at the TTC has focused on the medium-term, ie. up to about five years, and the longer-term questions remain. The main long-term issue is the rapid transit program, whose shape is highly uncertain after 1982. This will be the main area of corporate concern in the future, taking into account possible energy scenarios, changing lifestyles, and other such factors.

CONCLUSIONS

Two cases of more or less successful planning have been described, with success defined as actual impact on decisions in a generally "positive" (this is always open to debate) direction. Both involved an explicit awareness of the decision-making process on the part of planners, and attempted to bridge the gap between the political, technical, academic, and media cultures at the urban level. The two cases do not provide conclusive evidence, but they do provide some reasonable guidelines.

One might argue that the political process in Metropolitan Toronto is somehow more rational than in other cities, enabling more coherent policies to be formulated. While the Metropolitan form of government does provide some advantages, most objective observers of Metro Council would doubt that it is exceptionally different from any other democratic decision-making body; at times it is not very efficient, often it is chaotic or downright silly, but it gets the business done and by and large it represents the views of its citizens. The planning success in Metro Toronto has been achieved by accepting this reality, and working with it. Mutual respect between planners and politicians is possible and desirable.

REFERENCES

Daland, R. and Parker, J., 1962, Roles of the Planner in Urban Development, in F.S. Chapin, Jr. and S. Weiss, (Eds.), Urban Growth Dynamics, John Wiley and Sons, New York.

Metropolitan Toronto Transportation Plan Review, (MTTPR), 1975, Report 63-2: Travel Demand Models for Person and Truck Trips, Toronto, Ontario.

Pill, J., 1977, Transit Revenue Policy Study, Toronto Transit Commission, Toronto, Ontario.

Pill, J., 1979a, Planning and Politics: The Metro Toronto Transportation Plan Review, MIT Press, Cambridge, Mass.

Pill, J., 1979b, Transit in the 1980's: A New Direction, Toronto Transit Commission, Toronto, Ontario.

Pill, J. and Warren, R. M., 1979, Transit in Metro: Some Tough Choices, Toronto Transit Commission, Toronto, Ontario.

Snow, C.P., 1959, The Two Cultures and the Scientific Revolution, Cambridge University Press, Cambridge, UK.

Taylor, B., 1976, New Dimensions in Corporate Planning, Long Range Planning, 9, 80-106.

Whitehead, A.N., 1927, Science and the Modern World, Cambridge University Press, Cambridge, UK.

POLITICAL REALITIES AND THE IMPLEMENTATION OF URBAN TRANSPORT

POLICIES

John Bonsall

Ottawa-Carleton Regional Transit Commission
1500 Saint Laurent Boulevard
Ottawa, Ontario, K1G 0Z8
Canada

INTRODUCTION

This contribution to the NATO Advanced Research Institute on Systems Analysis in Urban Policy Making and Planning offers some observations on the use of systems analysis in transportation planning decision-making at the municipal level. Rather than comparing alternative approaches among different municipalities, it discusses the experience of a single medium sized urban area that has achieved a measure of success in the implementation of its transportation objectives.

BACKGROUND

Ottawa-Carleton, the municipality on which this paper's observations are based, is a regional municipality within the Province of Ontario in Canada. It is an upper tier municipal government, made up of eleven area municipalities of various sizes and degrees of urbanization and includes the City of Ottawa, the capital of Canada.

The present population of the Region is 541,000, of whom 90 percent live in urban areas which have densities of more than 400 people per square kilometre. However, it is only in the last twenty years, and more particularly, the last ten years, that the urban population of the Region has grown to represent such a high proportion of the total population. While the population increase between 1961 and 1971 was 18.3 percent for Canada, and 23.5 percent for Ontario, the population increase within the region over that same period was 31.7 percent, almost all of which occurred in the urban area. Future projections assume that this rate of growth will slow

327

significantly over the next two decades, but the area population is still expected to continue to increase to reach about 625,000 by 1991 and 750,000 early in the next century.

The region's employment is dominated by the presence of the Canadian Federal Government, which accounts for 31 percent of the total of about 270,000 jobs. The non-government jobs include manufacturing industries, of which printing and publishing and food and beverage industries employ the greatest numbers of workers. Efforts are currently being made to diversify the local economy and to reduce its dependence upon the presence of the Federal Government. In this regard, one of the employment sectors experiencing the most rapid rate of growth is the high technology sector of electronics and computer research and development.

Within the region, there is a considerable degree of job dispersal. Only a little over a quarter of all employment exists within the central business district, and the proportion of new employment opportunities which are being created in the suburban areas continues to increase.

The automobile is the predominant mode of transportation, accounting for more than 75 percent of all vehicular person trips, and with an average occupancy of 1.5 there are about 600,000 automobile trips made on the municipality's road system each day. Transit accounts for 23 percent of all trips within the municipality and 26 percent within its urban area on a daily basis. Transit usage is, however, more peaked than that of the automobile, with 17 percent of the daily total occurring in the peak hour and 67 percent in the peak six hours, compared to 10.6 percent and 48 percent respectively for the transportation system as a whole. The modal split to transit, therefore, ranges from a high of 41 percent for the PM peak hour to a low of 17 percent for the average off-peak hour for trips within the urban area. For core oriented peak hour work trips though, transit usage is now over 70 percent. This level of transit usage is high for a North American municipality, and has been achieved as a result of specific actions taken during the past seven years. These are now described in some detail by Bonsall et al (1979).

For travel to and from the region, the automobile is even more dominant, accounting for 95 percent of all persons with a daily total of about 101,000 vehicle trips. If travel to and from the rural area immediately surrounding the region is discounted, and the mode share of long distance inter urban travel is examined, the automobile passenger volume drops to a daily average of 10,100 trips, and represents 67.5 percent of the total. Of the remaining 32.5 percent of the travellers, 46 percent travel by air, 38 percent by bus and 16 percent by train.

Most goods are transported by truck, though rail services continue to meet some of this need. Typical screenline counts show that approximately 10 percent of the vehicles on the roads are trucks.

GOVERNMENT STRUCTURE

Administering this transportation system involves the activities of four levels of government to a greater or lesser extent.
Through the National Capital Commission and the Department of Public
Works, the federal government owns and controls a substantial portion (10 percent of the region as a whole and 29 percent of its
built up area) of land and facilities in the region, including a
greenbelt and numerous parks and parkways. Its influence on transportation planning occurs, not only because of its ownership of
existing parkways, which, while not specifically designed as commuter routes, are, nevertheless, integral parts of the urban transportation system, but also because its policies with respect to its
large land holdings can, and do, influence both the location and
effectiveness of new transportation facilities. The National
Capital Commission is directed by appointed Commissioners from all
ten provinces in Canada. Their responsibilities are usually
confined to questions of overall policy, and therefore, a considerable degree of decision-making authority with respect to individual
projects requiring the use of federal lands rests with the Commission's permanent officials rather than with the political process.

The Province of Ontario has direct authority over a few upper
tier roads in the region which function chiefly as inter-urban con-
nectors. While this is the theoretical justification for the
Provincial responsibility, it is also by force of circumstances
involved with the provision of commuter services because this has
become the dominant use on provincial highways within the urban¡
area. The Province of Ontario, through its local government funding
programmes, also exerts considerable influence over the provision of
transportation services at the municipal level. The Province subsi-
dizes both the construction and maintenance of the Regional road
system under a funding programme which is tailored to the Region's
ability to pay, and to its road needs. The rate of provincial subsidy varies with the optimum rate designed to encourage the Region
to maintain its road system in reasonable condition. On average,
the Province pays about 60 percent of the regional road costs, but
this subsidy is paid on a programme basis so that decisions on
expenditures between projects are made at the regional level.

A somewhat similar approach is taken towards public transit,
with the Province paying a fixed proportion of the operating costs
based on the achievement of a theoretical revenue/cost ratio. The
actual revenue/cost ratio is a matter of regional policy with the
difference between the total operating costs, revenues and provincial subsidy being paid by a levy on the local property tax.

The regional government of Ottawa-Carleton is an example of a
"two-tier" system, in which the upper tier (or Regional Council) is
given certain functions, while the lower tier (or area municipalities) performs other functions. As far as the urban transportation
system is concerned, the main responsibilities of Regional Council

are: the design and maintenance of regional roads, the operation of
the public transit system and the development of a regional plan.
It also has certain approval powers with respect to area municipal
zoning, sub-division plans and traffic by-laws.

The Council of the Regional government is composed of the
Chairman and thirty representatives from the local area municipal
councils. As a result, the primary allegiance of the regional councillors is usually to their area municipality, rather than to
Regional Council itself.

The fourth level of government involved in urban transportation
is made up of the area municipalities. They have a direct responsibility for the upkeep of their local or lower tier road system, but
all traffic regulations are subject to the approval of regional
council, as are the road systems for all new subdivisions. Because
of the method of representation on Regional Council, it is rare,
though, for it to overrule the wishes of an area municipality in
which the facility would be built.

Some of these details of the physical and political environment
of Ottawa-Carleton are undoubtedly peculiar to its particular circumstances. However, as far as a discussion of the place of systems
analysis in transportation planning at the municipal level is concerned, it exhibits many characteristics typical of municipalities
of its size. Of particular relevance to this discussion, and characteristic of other municipalities, is the fairly rapid suburbanization of both its population and employment, and a decision-making
process with varying levels of political and bureaucratic involvement.

THE PLANNING PROCESS

The formation of the Ottawa-Carleton Regional Government in
1969 can be considered as reflecting a systems approach to municipal
government. The City of Ottawa was, and still is, the largest urban
municipality within the region, but during the 1950's and 1960's,
significant urban development had begun to occur in the surrounding
rural townships. The traditional annexation approach to this problem was incapable of controlling this trend because the commuting
flexibility offered by the automobile meant that any regulation of
new development in one municipality would simply shift the development across the nearest municipal boundary. A comprehensive planning approach to an area large enough to encompass the total commuter area appeared to be the obvious solution. Thus, a second tier
government with specific jurisdiction in the provision of major
services and overall planning was formed by an Act of the Ontario
Government.

While the need for such an approach to local government may
have seemed self-evident, it only occurred in this case because a

senior government level (the Province of Ontario) imposed the solution on the area municipalities. Similarly, the requirement for an
Official Plan for the whole region was not a locally generated
process, but a legislatively imposed requirement by the Province.
Even ten years after its formation, one of the commonest political
stances at the local municipal level is to run against regional
government and to complain about an alleged reduction in local autonomy. This situation illustrates the essential incompatibility of
local democracy, and a systems approach to issues of other than a
purely technical nature at the municipal level. Political institutions may legislate such approaches on levels of government more
junior to themselves, but they will rarely willingly accept such
restrictions on their own freedom.

Future transportation service is considered in some detail in
the Region's Official Plan (Regional Municipality, 1974). Previous
transportation planning activities had occurred in the region, but
the Official Plan represented the first attempt at a comprehensive
land use and transportation planning process in which significant
variations in both were to be considered.

Since the Second World War, there had been three previous
transportation plans developed for this urban area, namely: Greber
(1950), Wilbur Smith (1955) and De Leuw Cather (1965). While these
plans show an increasing trend to a more sophisticated analytical
approach to both the interrelationship of transportation and land
use, as well as the need for specific transportation facilities,
most of the analysis was concerned with the extrapolation of existing trends. This is typified by the general acceptance of the
increasing use of the automobile, and therefore, as far as transportation matters are concerned, the plan recommendations were chiefly
concerned with the provision of bigger and better roadways. Another
common feature of these plans was an almost total absence of any
form of what has come to be called public participation.

The terms of reference for the Official Plan on the other hand
required a comprehensive examination of all land use and transportation alternatives, and included an extensive public participation
process. An examination of such background technical reports as
Progress Report No. 1 (Regional Municipality, 1972) and Travel Forecasts and Analysis (Regional Municipality, 1973) produced as part of
the Official Plan shows an honest attempt to fulfill this mandate.
The Technical work done for those various reports also employed
available rational analysis tools to discuss the transportation
implications of development options. Much of the technical analysis
was based on the results produced by a classic transportation model
of the 1960's style, which followed the four-stage, trip generation,
trip distribution, modal choice and trip assignment pattern. The
plan itself is also certainly more detailed than any previous plan
in its treatment of these questions, but whether in the end it
reflects anything more than a formalization of current trends and
attitudes is certainly debatable. While the plan indicates the need

for certain transportation facilities by the planning horizon, only in one case was the distribution and size of a proposed new development specifically limited because of the information produced by the technical analysis.

The plan development process itself involved the production of such technical reports as these, which attempted to explain the transportation implications of different land use distributions, and the likely results of different policies with respect to the use of the automobile and public transit. This technical information was then discussed and considered at a series of public meetings, which culminated in Regional Council debate and direction to staff to formulate a particular draft plan. The draft plan was then subject to a somewhat similar process, which resulted in further modifications of the plan and its eventual adoption, after considerable debate, as the Official Plan of the Region.

The plan is, as a result of this process, clearly a compromise document in which the urban planning philosophies and aspirations of the constituent municipalities are traded off against each other. As far as transportation is concerned, the plan adopted many of the road proposals of the earlier plans, but qualified this with a policy statement to the effect that priority would be given to improving the municipal transit system over all forms of road construction and road widening. At the time that the technical analysis for the plan was done, there was little evidence that transit-oriented policy of this sort could really achieve the modal split targets required to sustain the plan if road construction were to be limited. Consequently, the plan contained several references to the long-term need for even more roads than those specifically identified.

It is interesting to note that the political deliberations which accompanied the development of the Official Plan occurred at a time of increasing public disenchantment with the results of major new road construction. No rapid transit corridors had then been specifically identified in Ottawa-Carleton, and because they had no negative connotations, it was politically acceptable to downgrade the road elements of the plan and replace them with some, still to be determined, rapid transit improvements. Four years later, the selection of rapid transit routes, which is reported in the Rapid Transit Development Program (Regional Municipality, 1978-1980), created almost as much controversy as previous road proposals and resulted in the application of more stringent environmental conditions to the proposed transitways than had ever been applied to any regional road project. This experience suggests that the political motives for the adoption of the original transit policies were possibly only partially concerned with the urban development arguments usually used to justify transit-oriented policies.

The plan also contains only a minor discussion of energy issues or the benefits that might result from such policies as flexible

working hours. This is not surprising because, despite the greater
use of rational analysis, this plan still reflects the prevalent
technical and political opinion of its time in just the same way as
had the earlier plans. Despite their topicality now, such issues
were only minor items in the technical literature of 1970.

This natural tendency for transportation plans to become
rapidly obsolescent is further illustrated by the fact that even
before the Official Plan had gone through all its stages of approval, transportation developments were outdating much of the original
technical work. An aggressive policy of transit expansion increased
per capita usage by over 50 percent, and thereby greatly exceeded
the Plan's original projections. At the same time, the introduction
of flexible working hours among more than half of the downtown work
force reduced the peak hour trip volumes by more than 15 percent.

Re-analysis of the Plan's projections using the new status quo
has, as might be expected, led to a revision downward of many of the
original future transportation capacity needs. This change, it must
be stressed, occurred not because of any improvements in the techniques of rational analysis, but simply because the current performance of the transportation system had changed significantly over a
period of less than five years.

The performance of today's transportation system could not have
been forseen by the makers of the earlier plans, regardless of their
level of computational ingenuity, and it is more by good luck than
anything else that any of the transportation elements of these plans
are adaptable to today's transportation requirements. In fact, many
of the assumptions on which these earlier plans were based have
produced situations which are, today, counter-productive.   An
obvious example is the typical suburban street pattern and low
density, widely dispersed suburban job locations, designed for the
private automobile, which make today's transit-oriented plan difficult to implement. This is not, however, an argument for no planning, as the fact that earlier plans reserved certain corridors for
transportation purposes has meant that opportunities now exist that
would otherwise have been foregone.

It appears, therefore, from this examination of the major
transportation plans developed for this urban area, that the amount
of rational analysis employed had made little difference to the
success of the overall plan. On the other hand, there is evidence
that the recently increased use of systems analysis to examine individual elements of the land has been beneficial. As the dictum
says, "Those who do not study history are compelled to relive it",
and, therefore, the apparent lack of success of earlier plans is
worthy of greater study if the proper place for systems analysis is
to be determined.

A SYSTEMS APPROACH

It is, of course, tempting to believe that progress and the passage of time are the same, and, therefore, that today's transportation decision-making will stand the Region in better stead than those of a generation ago.   While the proposition is at best unproven, too much faith may be placed in the accuracy of the predictions of transportation planners, and not enough consideration given to the consequences that will ensue if the predictions are wrong.

Fortunately or unfortunately, depending upon one's point of view, the consequences of a poor transportation planning decision are not always particularly evident because the urban system as a whole has a great capacity to adapt over time.   Inherent inefficiencies that might have been avoided are rarely very obvious.   That such inefficiencies can have a significant impact on the future development of the Region is undisputable, and, therefore, in an era which continues to place great reliance on the products of systems analysis, it is worthwhile to explore the likely limitations of this technique in the urban environment.

As previously described, the process by which transportation planning is currently undertaken in Ottawa-Carleton includes some formal systems analysis complemented by technical professional judgement and political compromise.   The political component is an integral part of the decision-making process, and a reality that has to be recognised in determining the appropriate mix of the first two elements.   The political process responds first and foremost to the current perception of needs.   To expect a municipal government to accept short-term problems for the sake of a long-term gain is unrealistic, to say the least, unless a very convincing argument can be advanced.   Since the people who must be convinced are not only the politicians, but, more importantly, the public at large, a sophisticated mathematical justification for some long-term need is generally of little avail.   This is not only because of lack of understanding, but also because of the public perception of the poor track record of previous attempts at such predictions.   Long-term planning is also politically unfashionable because there is a general belief that we are living in a period of rapid change, in which any predictions based on past or present experience are unlikely to be fulfilled.   The first requirement therefore for systems analysis at the urban level is that it be understandable to the average layman.   The only exception to this are questions involving purely technical issues, which are usually of little concern to the public at large.

Effective systems analysis presupposes an accurate description of the problem to be solved, relevant data describing both the existing and likely future scenarios and a proper understanding of the appropriate cause and effect relationships.   It is the premise of this paper that, except for issues of a purely technical nature, all these conditions are poorly satisfied at best in municipal transportation planning, and that the inherent nature of the environment in which it occurs will prevent any substantial improvement in this state of affairs.

Considering first of all the formulation of the problem to be solved, any examination of transportation literature over, say, the past ten or fifteen years, shows that the nature of the problems discussed, while obviously all concerned with mobility, have varied in several important respects. It is apparent that, initially, the major concerns were such items as how to improve the overall level of service or operating speed of the transportation system at the least capital cost. As time went on, other considerations begin to complicate such simple initial objectives. The viability of downtown residential and employment areas, the mobility of the carless members of the population, environmental damage and energy consumption are all typical examples of the increasing complexity of transportation planning objectives. This trend is well illustrated by an examination of Ottawa-Carleton's various plans.

The transportation planning objectives set for a twenty-year planning horizon twenty years ago, therefore ignored factors that in many cases are now pre-eminent. If this situation were simply an example of better problem definition and understanding occurring as time progresses, there might still be some cause for confidence in a systems approach. Unfortunately, it appears that this is unlikely to be the case. Both in the planning process in Ottawa-Carleton and elsewhere, the inclusion of these new evaluation factors in the process came, not from the technical experts, but as a result of the participation of other interested groups. In other words, the inclusion of these new factors appears to reflect a change in society's values, not a better understanding of the problem by the technician. No amount of study in the 1950's could have predicted that the major issue that would decide the alignment of a transitway in Ottawa-Carleton in 1978 would be a trade-off between the use of open space and the environmental impact on a group of people in their home environment. In an earlier, more authoritarian age, such considerations were almost irrelevant.

Further reducing our ability to define properly the problem to be solved is the trend to involve more people and institutions in the decision-making process. As each of these new participants in the process may represent a different point of view, they will also perceive the objectives of the process in different ways. No single set of objectives is therefore likely to encompass easily all aspects of the problem in a manner acceptable to all the participants. Because the necessary trade-offs between these different objectives often involve value judgements, they can only be properly expressed through the political process. Thus, political issues will intrude right at the outset, and there is no longer a neat division between a rational technical process that produces alternatives and a political process that selects between them.

In the earlier studies, this situation was less complex, as the participants were usually limited to technical staff and the politicians who gave the final approval. Only in exceptional circumstances were the general public involved, and when they were their interests were very specific. As a result, the problems addressed by the study were almost completely transportation oriented, since this was the primary interest of the technical staff. Limited in this way, the process to be solved appeared to be tailor-made for a systems analysis approach. At the same time, the advent of the electronic computer, by providing a way of manipulating large quantities of data in an efficient manner, appeared to offer a method of quantifying future transportation usage. Because the issues were formulated in these terms only, the political considerations were similar.

The modification of the study process to include the general public has made it more responsive to current societal values and complicated the formulation of the study objectives. Experience has shown that, while some members of the public will become involved out of a general interest in transportation matters, the majority will be motivated by a desire to protect their own particular environment. Thus, their interpretation of the objective of the study will have more to do with discovering a rationale for either not constructing a particular facility or with identifying suitable remedial measures, than with solving a transportation problem. This public will therefore wish to broaden the basis of the analysis until some reason for not proceeding with the project can be found.

By its very nature, this broadening of the scope of the enquiry generally also entails further consideration of judgemental issues not capable of quantitative evaluation. Unfortunately, the importance of these issues will not be readily apparent until the public has become sufficiently involved and aware of the possible outcome of the study. Thus, the final definition of the problem, as far as the general public is concerned, may not be known until the study is nearing its completion. A systems analysis approach which has rigidly specified the ground rules by which the eventual decision will be made may well not be able to accommodate these last minute considerations. In fact, because of its inherent inflexibility, the approach may prejudice the possibility of any final decision other than one to defer the issue or do nothing.

This is particularly important because, despite the democratization of the process, the final recommendation by the study technical experts still carries the most weight, and will generally be supported by the political level unless concerted opposition to it occurs. Thus, the expert bears a heavy responsibility to be fully aware of his own biases and prejudices. Unfortunately, an inflexible systems approach which has difficulty in accommodating last minute changes can formalize these biases and make changing them even more difficult.

The involvement of different government agencies who may or may not be directly accountable politically can pose as many problems with respect to problem definition as does the public involvement. In some instances, the agencies may be neutral, but in others, where they perceive that one of their own objectives is threatened, they may not be prepared to make any concessions to trade-off their objective against the transportation objectives of another government.

Because trade-offs between different levels of government are by their nature political, they will likely involve issues quite unconnected with the transportation problem to be resolved. To talk of an optimum system solution in these circumstances is obviously totally inappropriate.

It is clear, therefore, that problem definition in urban transportation planning is not an issue independent of the time and place in which the study is done. It will reflect the society's values prevalent at the time. There is, thus, a considerable risk that studies of systems ten or twenty years in the future may fail to address either the right problem or the most significant decision-making factors of this future era.

Since it is probable that in formulating the objectives for our future urban transportation system, we may be completely ignoring the most important criteria, it is obvious that we may equally well be basing our decision on the wrong, or at the best, irrelevant data. Aside from this concern, however, there is evidence that even traditional factors such as mode share may be the victim of inherent data obsolescence. This is illustrated by a comparison of the mode share projections done as part of the Region's Official Plan less than ten years ago, and those done as part of the recent rapid transit route location studies which show a much greater use of public transit.

Some of this difference can be explained by different levels of transit service, but the majority is due to a different attitude towards the use of transit by the public. Attempts to model these attitudinal effects have been tried with some success, but actual experience in Ottawa-Carleton, where transit usage has increased significantly, suggests that such unquantifiable factors as the general public confidence in its transit system, are more important than the attitudinal variables that can be modelled. This level of confidence is a funcion of many factors to do with the relative reliability of the service and the results of various marketing strategies which cannot be accurately predicted.

The process by which the need for future transportation facilities is established involves the collection of a large quantity of data, its manipulation by a computer based modelling process and a final analysis step to interpret the computer results in a manner that allows for the planning and design of the appropriate facili-

ties.   Because it is essentially a numerical process, it is often
credited with being in some way more exact and correct than many of
the other inputs into the planning study.

In practice, every step is subject to the judgement of the
expert responsible.   The decision as to the data to be collected
depends upon a compromise between what the expert considers are the
explanatory variables for trip making, modal split, etc. and the
availability of current and future projections of the values of
these variables.   The model itself contains mathematical relationships between these variables and the desired outputs, which are in
turn based on a statistical analysis of the present relationship
between the variables and the outputs.   The whole process is usually
checked by attempting to replicate the existing situation.   Rarely
is even short time series information available with which to check
the consistency of the modelling assumptions.

The best that can be said for the process is that it may
explain the existing situation as far as the outputs under study are
concerned.   For the future, the outputs can only be judged in terms
of their reasonableness.   In other words, the result is considered
acceptable in terms of today's situation.   Any other test is
obviously inappropriate, since to postulate that the relationship
between the chosen variables and the required outputs will be different in a particular way from that that exists today, just com-
pounds the problem.

Further complicating the interpretation of the results is a
common institutional limitation.   Because the process usually
requires the collection of vast amounts of data, junior staff are
usually employed for the purpose.   The technical judgement on which
the quality of the model output rests may therefore be that of relatively inexperienced technical staff.

Experience, therefore, suggests that the fruitful application
of systems analysis to urban transportation planning is limited.
Not only are optimum solutions time and place dependent, but the
relative importance of the evaluation factors can often only be
assessed by a political, not technical, process.

While the foregoing discussion highlights the limitations of
systems analysis in the transportation environment, it is also
appropriate to review the achievements of an approach based more on
professional judgement than comprehensive systems analysis.

The Ottawa-Carleton transportation system has been judged a
successful system in terms of most urban transportation objectives.
Transportation policies set in its Official Plan are bearing fruit,
and thus, it appears to be demonstrating the benefits of the classical systems approach of goal setting, policy development and programme implementation.   A closer examination of the achievements of
the last seven years or so suggests, however, that on the contrary,

the results occurred quite independently of this process and were occasionally even frustrated by it.

The most significant results, such as the major growth in the use of public transit and the reduction in the peaking of commuter travel, occurred as a result of short-term innovation and not in response to any grand plan. It was because there was political support, money available and a local staff willing to innovate that these results occurred. The process was one of "cut and try" with the emphasis on short-term gain, rather than sophisticated study. Obviously, this approach is inappropriate where major capital expenditures are concerned, but in this respect, it is interesting to note that the results achieved have, in fact, obviated the need for a very large proportion of the long-term transportation investment previously thought to be necessary as a result of earlier systems analysis.

Because it is more in tune with the municipal political environment, and because in the final analysis it is what does or does not get implemented that is important, the philosophy underlying the approach to operational and minor capital investments has also begun to be applied to planning of capital intensive projects. Rather than attempting to develop a twenty-year plan in detail, and then embarking upon a staged construction programme, improvements based on the success of today's transportation system and justified in the short-term have been identified. Of course, some long-term planning has still taken place, but only to establish that the options being adopted today appear to be compatible with those future scenarios that a limited ability to predict suggest may occur.

When the recommendations of the various major transportation plans produced for the Region over the past thirty years are compared with what exists today, what stands out is the continuing presence of certain existing, and still to be built, facilities. It is, therefore, probable that over, at least, say the next twenty years, the major facilities that are likely to be built have, for better or worse, already been identified. The pertinent questions facing the transportation planners are not, therefore, ones of theoretical location, but more to do with the size and order of construction.

In the light of the limitations described above, this is perhaps fortunate, because systems analysis is probably capable of making a useful contribution in these circumstances, as long as the application is topical and limited in scope.

Whether, in the long run, this more flexible, and in some respects, less systematic, approach to the future will be any more successful, only the passage of time will tell. What is certain, though, is that today's transportation user will at least gain some benefit, rather than having to wait for a promised Utopia tomorrow.

No major revelations are obviously possible from this brief discussion of experience in one municipality. It is suggested, however, that there is clear evidence that the municipal planning process involves both so-called technical and social issues which cannot be traded-off against each other in any systematic way. To attempt yet further increases in the comprehensiveness of present systems analysis is unlikely to be worthwhile. What is worth doing is to improve the experts' ability to answer truly technical questions with respect to costs, level of service, environmental impact, etc. Questions involving human attitude such as trip generation, distribution and mode share can never be predicted with the same level of certainty, and care must always be taken not to give the impression of an unjustified level of certainty simply because the predictions are handled by a numeric process.

Finally, the transportation planner must recognize that his recommendations deal with political and social values, and his planning process must therefore allow the necessary trade-offs to be made by the political process and not be hidden by a pseudo-scientific systems analysis procedure. There is, thus, some truth to the claim that the process is more of an art than a science, and it would be a bold individual who would seriously suggest that a work of art can be created by systems analysis. On the other hand, such an approach might improve the choice of materials from which the work of art is created.

REFERENCES

Bonsall, J.A., Somerfeld, W.O. and Hue, R., 1979, Case Study on Ottawa-Carleton, Organisation for Economic Co-Operation and Development, OECD, Paris.

DeLeuw Cather and Company, 1965, Ottawa-Hull Area Transportation Study, Ottawa, Ontario.

Greber, J., 1950, Plan for the National Capital, National Capital Development Commission, Ottawa.

Regional Municipality of Ottawa-Carleton Planning Department, 1972, Progress Report No. 1, Ottawa, Ontario.

Regional Municipality of Ottawa-Carleton Planning Department, 1973, Travel Forecasts and Analysis, Ottawa, Ontario.

Regional Municipality of Ottawa-Carleton Planning Department, 1974, Official Plan Ottawa-Carleton Planning Area, Ottawa, Ontario.

Regional Municipality of Ottawa-Carleton Transportation Department, 1978-1980, Rapid Transit Development Program, Volumes 1-7, Ottawa, Ontario.

Wilbur Smith and Associates, 1955, Traffic and Transportation Plan for Ottawa, Canada, Ottawa, Ontario.

TYNE AND WEAR AND HONG KONG: CAN SYSTEMS ANALYSIS TACKLE THE

REALITIES OF DECISION-MAKING?

Tony M. Ridley

London Transport Executive
55 Broadway
London, SW1H OBD
United Kingdom

INTRODUCTION

There can be few urban areas of the world without severe transport problems. The problems vary, but there is general dissatisfaction with an inability to move easily and cheaply from one place to another. Public transport is too infrequent or too crowded, unreliable or expensive or, more usually, some combination of all of these.

The literature documents these problems at great length. Meanwhile, scientists and engineers are constantly on the lookout for some new technology, a breakthrough, which will overcome the problems at a stroke. But, as has been observed almost seriously, it is easier to take three men a quarter of a million miles to the moon once, than to take a quarter of a million three miles through the city every day. It seems likely that, while there may be interesting developments on the way, no new technology will dramatically change urban public transport during the rest of this century. The train, the bus, the tram will have to struggle on as the main carriers, albeit in improved form.

Where new developments do take place there are countless technical journals ready to provide a forum for articles outlining system designs and technical details. The financing of projects does not usually produce the same spate of professional papers, although the financial press is usually capable of producing authoritative coverage.

Comparatively little attention has however been paid to the process by which projects get underway in the first place. This is regrettable because, not only is the decision-making process of interest and worthy of study in its own right, the manner in which the decision to proceed is reached often determines the outcome of the project.

This paper compares the decision-making process for two projects, superficially similar but in fact quite different. They are the Tyne and Wear Metro (Metro) in the North-East of England and the Hong Kong Mass Transit Railway (MTR). It then goes on to comment on and question systems analysis in that light.

The two projects are superficially similar. Indeed, the first pair of cars for Hong Kong were tested in the North-East of England on the test track at Backworth developed and used by the Tyne and Wear Passenger Transport Executive (PTE) for the proving of their own prototype equipment. Tyne and Wear and Hong Kong both use British-made aluminium cars on German bogies, steel wheel on steel rail of standard gauge, and both use 1500V dc overhead supply.

On this evidence the technologist could well say that the two schemes are basically similar. They were, however, quite different in derivation transcending esoteric discussion of the mere difference between a 'supertram' on its own right-of-way, the Metro, and a heavy rapid transit system, the MTR.

Much transport debate has been hamstrung by devotees of one form of technology versus another, rubber tires on roads versus steel wheels on rail, for example. The devotees not only include scientists and engineers, but professional operators too. A busman is a busman and not a railwayman, and never the twain shall meet. When it comes to deciding whether the term 'light rail transit' means something closer to a railway or a steel wheel bus on reserved right-of-way, then the debate can be endless.

In fact, transport decisions are based, as they should be, on much wider issues. What was the quality of the existing transport service at the time of the decision? What view was taken of the future? What data existed on which to base a decision? Who were the decision-makers and what was the governmental structure? What was the role of the public, of the other operators? What professional organization was necessary, or already existed, to implement the decision? How could the project be financed?

These and other issues determine not only how and how quickly the decision is made, but often the manner in which the project is implemented, and even, to some extent, how successful a service it ultimately provides.

We turn now to each of the two cases in turn to examine these issues. From this it will be clear that the genesis of each project was quite different.

TYNE AND WEAR METRO

The first move which led to the decision to build the Tyne and Wear Metro was not made in that part of the world, but within Government in London. Doubtless, of course, no one had Tyne and Wear specifically in mind at the time. In the mid-sixties there was a new surge of interest in urban problems in the UK, particularly transport problems. There was a desire, moreover, to avoid the 'mistakes' of the United States.

The mood is best summed-up by the words of the 1967 White Paper "Public Transport and Traffic" (DOT, 1967):

> "Britain is basically an urban country; nearly half our people live in large towns and cities and about 80 percent of them in urban areas. The quality of urban life depends to a large extent on the excellence of the transport services available both to those who live in the cities, and to those who come there for work, shopping or pleasure. The freedom to move easily about the city — to go places and see people — is something of great value. Yet this freedom can, if ill-used, do great damage to the quality of urban life.
>
> The nature of urban transport systems must be based on our ideas of the kind of cities we want. But city planning must be based on a realistic appraisal of what can be provided in the way of transport investment. It is useless to devise a structure for a city on the basis that nearly all journeys will be made by car if this demands an investment in highways so huge that the resources of the country cannot provide it. So general planning and transport planning must be carried out hand in hand.
>
> All the studies carried out so far from the Buchanan Report onwards suggest that our major towns and cities can only be made to work effectively and to provide a decent environment for living by giving a new dynamic role to public transport as well as expanding facilities for private cars. Unless we recognize this we shall pull down the centres of our towns in an attempt to get rid of congestion; and at the end of the day we shall find congestion still with us, and the character of our towns destroyed. We have neither the physical space nor the economic resources to rebuild our cities in such a form that all journeys can be made by private car;

and in any case we must provide for the large
number of people -- particularly many of the
old, the young, the housewives, the poor -- who
will not have the use of cars.

To provide an attractive and efficient system of
public transport is therefore vital to deal with
the immediate and pressing problems in our
cities. We must also see how far public transport can offer new opportunities in the renewal
of our urban areas.

The pattern of the growth of London in the past
century was largely determined by the building
of the suburban railways and tubes; the structure of many provincial cities reflects the pattern of the electric tramways. New rapid transit systems for our major cities might provide
an attractive basis for new patterns of development."

The White Paper was followed by the 1968 Transport Act among
whose provisions was the creation (then) of four PTE's centred on
Birmingham, Liverpool, Manchester and Newcastle. These came into
being late in 1969 under the control of ad hoc bodies of politicians
nominated by individual local councils, the Passenger Transport
Authorities (PTA's) because at that time the reorganization of provincial local government had not yet taken place.

The White Paper had a clear idea of what was required of an
Executive, which, in the case of Tyne and Wear (then Tyneside),
consisted of a Director General and Directors of Operations, Planning and Finance:

"Its primary job will be to plan the public
transport system of the Area as a whole in the
context of the development and traffic plans of
the local authorities. The Executive must comprise men of vision and wide experience; and
they must employ staff skilled in the latest
techniques of transport planning and development, not only by road but by all means of
transport. It will be the job of the Executive
to work out with the local authorities a practicable balance between private and public transport, to integrate the bus and rail services in
the Area and to evaluate the costs and benefits
of major new investment in public transport,
whether in fixed track systems, reserved routes
for buses or by other means."

It would be fair to say that this enthusiastic view was not
universally held, particularly by those local councilors who thought

that the PTE's were 'stealing' their municipal bus companies, nor by
the established professionals in the newly formed National Bus Company (NBC) or in British Rail (BR), many of whom saw the PTE's as
rogue elephants sent to complicate their lives.

Prior to the creation of the PTE's, the Tyne and Wear area had
become one of the last metropolitan areas to follow the fashion of
the sixties by carrying out a comprehensive planning and transport
study. Commencing in January 1969 consultants produced the Tyne
Wear Plan in 1971, on behalf of and blessed by the various local
authorities in the area. The road proposals were essentially those
which had already been developed previously and were, by and large,
accepted and recommended by the consultants. The new concept,
indeed the heart of proposals for the improvement of public transport, was that a new underground tunnel should be built to carry a
rapid transit system so that local rail services could pass directly
through Newcastle and Gateshead, linking up with existing lines to
Tynemouth and South Shields on the north and south banks of the
Tyne. Even more significantly, perhaps, it was proposed to change
the balance of transport investment giving public transport a much
higher proportion than previously.

What no one realized at the time was that a combination of the
realization that the North-East had done rather well out of the road
programme in the non-urban areas earlier in the sixties, together
with the arrival of the 'motorway revolt' from San Francisco via
London, would kill the motorway programme some few years later.
This left the Metro with an even higher proportion of investment
than had been intended. This would not necessarily have disappointed the consultants for they had said:

> "We believe the Tyne Wear Plan Area is in a
> unique position to avoid some of the difficul-
> ties encountered by other urban areas in their
> transport planning policies. At the present
> time, car ownership is lower than in many other
> urban areas and the public transport system is
> significantly better than in many other areas.
> As we note in our report, there are clear signs
> of the drift away from public transport, but our
> analysis shows that the area cannot afford to
> let this continue as no reasonable level of
> highway expenditure could meet the resulting
> demand. The opportunity should be taken, there-
> fore, to maintain and improve the public trans-
> port system while ridership is still relatively
> high. While the bus system will always be an
> important element of the public transport
> services we believe that the only way to provide
> the level of service and accessibility required
> in the future, in the face of increasing road
> traffic, is by significant investment in a sys-

tem with its own exclusive right of way. This
system should be based on the current railway
facilities with appropriate renewal and upgrading of lines and stations and the improvement of
central area accessibility by linking of rail
lines north and south of the Tyne.

For the public transport programme the longest
lead time of necessity applies to the underground tunnel, but it must nevertheless be
completed at the earliest practicable time.
Consequently it is programmed for completion in
1979...Even if no further significant public
transport improvements are made during the
balance of the planning period, it will be
important to hold this objective."

Thus there came together a political PTA and professional PTE
in search of a plan and a plan which needed someone to drive it
along and implement it, in an atmosphere created by Central Government of trying to encourage and develop public transport backed up
with money available to assist in the purpose. Not quite the 'accident view of history', but it is certain that while men occasionally
create history they are immeasurably helped if the time is ripe.

The Plan was not accepted uncritically, but there was an additional reason for its attraction. Under the terms of the 1968
Transport Act, the local areas were to assume responsibility for
decisions about and payment of subsidy for BR's local passenger
services, albeit with a sliding scale of grant from Central Government. As the passengers carried represented only four percent of
the daily public transport passengers, and as the subsidy would have
to come from the rates (property taxes), this represented a 'no win'
situation for Tyneside.

To pay an increasing subsidy for a poor service on lines which
actually had been downgraded by BR (having been electrified they
were converted to diesel operation during the 1960's) would have
been politically controversial to say the least. Yet to have taken
the opposite course of telling Government and BR that the lines
should be closed would have caused an uproar. So the third alternative of development was pursued, not as a compromise, but as the
best available course of action.

First, however, three other major steps had to be taken. The
PTE had itself to be convinced of the wisdom and feasibility of the
Metro project. As a concept it looked attractive, but that was
clearly not enough. Secondly, a Government infrastructure grant had
to be obtained and thirdly, statutory powers to allow the scheme to
proceed had to be obtained by means of a Private Bill in Parliament. Thus the Excecutive sponsored the North Tyne Loop Study to
examine in detail one part of the proposed rapid transit system.

Working together with representatives of BR and the Government, the PTE made a comparison of upgraded rail, a busway replacing the railway, a conventional all-bus system and a light rapid transit system. Particularly relevant to the ultimate decision to endorse the latter, the light rapid transit development, was the fact that, under the 1968 Transport Act, Government had for the first time made financial support available at the same seventy-five percent grant rate as for highways.

Next, application for such a grant had to be made to Government. The approach was at three levels. First, there was close cooperation between transport planners and economists within the PTE and their opposite numbers in Government so as to ensure that the presentation matched the style required for assessment. Secondly, close contact was maintained between the PTE Directors and the senior civil servants who received the recommendations of their technical colleagues and then had the responsibility of deciding whether to recommend the project to the Minister.

Finally, a united political stance was taken by councillors of both the majority Labour and minority Conservative parties, as well as by MP's of both parties. Then, in December 1972, less than two years after it had been first conceived, the Metro received the blessing of the then Minister of Transport, Mr. John Peyton, in the form of approval of the grant application. Undoubtedly, two factors helped the application; first, the need to take decisive action either by closing or developing the existing rail system rather than simply let it carry on in its existing unsatisfactory fashion. Equally important was the fact that the PTE had gone a long way in discussion with British industry towards working out proposals for the development of prototype equipment. Indeed, at this time the PTE also sought and Government agreed a research and development grant which allowed the conversion and equipping of an existing unused railway line for use as a test track for the prototype rolling stock built by Metro-Cammel of Birmingham, the manufacturer of London's underground stock. Here the urge to seek ways of assisting British industry so that they could turn their sights to export orders undoubtedly played a role in Government thinking.

Now the Executive took the next step of obtaining Parliamentary powers for the construction of the Metro and the Tyneside Metropolitan Railway Bill was deposited in November 1972 shortly before Government grant approval was given. The Bill received the Royal Assent in July 1973. There were a number of petitions against the Bill, as is usual, from both private and public bodies, but all were satisfactorily settled.

Importantly, the resulting Act included provision for agreements with British Railways Board as follows:

> "The alteration or conversion of any part of any
> existing railways of the railways board to form

part of the rapid transit railway and the maintenance, use and operation of rapid transit
railway shall, as between the Executive and the
railways board, be carried out and regulated in
accordance with such terms and conditions as may
be agreed between the Executive and the railways
board, and any such agreement may relate to the
whole or any part or parts of the rapid transit
railway and may contain such incidental, consequential or supplementary provisions as may be
so agreed including provisions with respect to
the defraying of, or the making of contributions
towards, the cost of the matters aforesaid by
the Executive, or the railways board, or by the
Executive and the railways board jointly, and,
without prejudice to the generally of the foregoing, any such agreement may provide for the
exercise by the railways board jointly, of all
or any of the powers of the Executive or the
railways board (as the case may be) in respect
of any part of such railways or of the rapid
transit railway."

In retrospect, it can be seen that these words were not nearly
definite enough to allow easy solutions in a situation where a new
system proposed and intended to be operated by the PTE needed to
take over a substantial length of right-of-way in the ownership and
use of BR, further complicated by the fact that some, at least, of
the right-of-way was currently carrying freight traffic. Furthermore, the first apprehensions were being expressed by railway unions
as to the effect of the Metro on the jobs of their members. In the
event both they and BR supported the scheme, if half-heartedly, perhaps because it represented the first new boost to rail-based local
public transport in many years.

Thereafter, detailed design began and, with the first contract
being let in October 1974, tunnelling works commenced less than four
years after the first thoughts on the scheme had been developed by
the consultants who produced the Tyne Wear Plan. By any standards
this represented very rapid progress.

HONG KONG MASS TRANSIT RAILWAY

In the mid-1960's Hong Kong, although far from having overcome
the housing problem created by the massive influx of immigrants from
China over a fifteen-year period, began to examine its transport
problem. Following survey work by the (then) Road Research Laboratory from the UK, a Mass Transport Study (MTS) was commissioned with
the object of developing:

"the best solution to Hong Kong's long term mass
transport problems, consistent with planning

goals, development plans and a level of mobility that will allow the Colony to continue to prosper."

The Study reported in September 1967 and contained:

"recommendations for a long range mass transport programme, the principal item of which is a 40-mile rail rapid-transit system. Planning and bringing to reality such a system ab initio is a highly complex and expensive process -- much more so than for the necessary new or extended system of bus services which will form the other essential part of the programme as a whole. We have therefore devoted a major portion of the report to this underground and overhead rail system.

Our proposals for this system take the form of a six-stage development plan aimed at opening the first section of line in 1974 and the last in 1984. To achieve this, detailed planning and design must start in 1968 and construction in 1970. The scheme has been worked out so as to achieve the shortest completion period consistent with economy."

At that time:

"approximately 75 percent of all trips made in Hong Kong today are by public transport; about 50 percent are by bus, 14 percent by ferry and 11 percent by tram. The Kowloon-Canton Railway accounts for less than one percent and the remainder are divided among cars, taxis (legal and illegal), public cars, lorries, dual-purpose vehicles, private buses, etc. In 1954 approximately 50 percent of the public transport trips were made by bus. Now 67 percent are by this mode.

The people of Hong Kong are using their public transport system more each day. Ten years ago the average person made 244 public transport trips per year; now he makes 335. This reflects a rising standard of living among the lower income groups. Many people who had to walk in the past now ride. There are also indications that trips are getting longer as new residential, commercial and industrial developments spring up in formerly vacant areas.

One feature of public transport in Hong Kong which contributes to its high profitability and low fares, is that riding is spread much more uniformly over the hours of the day, days of the week and months of the year than in most other cities. The peak hour of travel on an averge day is less than 10 percent of the total, and daily travel varies from 13.7 percent of the weekly total on Thursdays to 15.4 percent on Saturdays. There is even less variation on a monthly basis, with 8 to 9 percent of the total annual trips being made in each month.

Although the number of motor vehicles in Hong Kong compared to the population is small it is growing rapidly. In 1956 there were 11 motor vehicles for every 1,000 people, and now there are 25. There are 10 times as many private cars in the Colony as there were in 1948. The number of vehicles per mile of road is extremely high by world standards. Only Monaco and Gibraltar have more.

Consideration of these and many other factors has led to the conclusion that the travel needs of Hong Kong cannot continue to be wholly served by surface transport. Increasing congestion is already evident at many places and it is likely to get much worse. As it does, public transport vehicles will have to operate on slower and slower schedules which will require many more vehicles just in the attempt to maintain a constant level of service. The additional vehicles will, of course, increase the cost of providing service and cause more road congestion. On the other hand, these very factors will ensure the success of a grade-separated rapid-transit system.

Public mass transport is the most efficient means of moving people. The limited road mileage in Hong Kong requires such efficiency, and since a large proportion of travel is still by public transport, every effort should be made to retain it by improving and expanding the system."

The estimated total cost of the system (at 1967 prices) was some HK$3,400m. While very substantial revenues were forecast

"some assistance from public funds in the early years is likely to be necessary. This can be

justified in light of practices in other cities,
but still more important it can be justified in
consideration of the many benefits to all the
people of Hong Kong. The benefits will not be
limited to those who use the system but will be
spread among all who travel. The routes, stations, trains and fare structure have all been
designed to attract the maximum number of people
away from the streets and thus to leave them as
uncongested as possible for the free movement of
the Colony's commerce."

It is noteworthy that the report also said:

"the surface and rapid-transit systems should be
complementary rather than competitive, each
serving the type of travel it is best suited to
accommodate, and producing together an effective
and efficient transport system."

The MTS was followed by the Long Term Road Study. Thus, public
transport had been put forward in advance of road proposals,
although, of course, a proper road system was essential for the
important bus element. But the MTS proposals were quickly followed
up by a detailed feasibility study which was reported in August 1970
in Hong Kong Mass Transit Further Studies (FS).

A full Preferred System of 52.6 kilometers was recommended
costing some HK$4,400m. (at 1970 prices) to be developed in nine
stages. This cost did not include fees, financing costs or Government administration costs. An Initial System of the first four
stages 20 kilometres long would cost some HK$1900m (at 1970 prices):

"If Government gives instructions to proceed with
the detailed design and preparation of contract
documents by mid-1971, construction could be
started at the beginning of 1973 and the first
stage opened early in 1976. The Initial System
could be completed by the end of 1978 and the
full Preferred System by 1986."

Several financing schemes were described:

"One assumption is that finance at 7 percent
repayable over 12 years is obtained for plant
and equipment as these are possible terms of an
external loan. On this basis two schemes are
suggested. The first provides for all the plant
and equipment for the Preferred System to be
supplied under such a loan, the remaining cost
being paid out of current Government revenue.
The second is similar, but the commercial loan

would be for the Initial System only. The whole
cost of construction and equipment of subsequent
stages would be financed by Government loan on
which the interest would be the balance of the
net revenue after servicing the Initial System
loan."

A period of two years now passed, during which Government considered the proposals. Then in May 1972 Government decided that
construction of the Mass Transit Railway should be proceeded with,
subject to satisfactory arrangements being made for financing and
for letting contracts. This followed the appointment of a new
Financial Secretary who took considerable personal interest in the
project. Indeed, one of the unique aspects of the MTR is that it
depended considerably on initiatives taken by the Hong Kong equivalent of the Chancellor of the Exchequer. Of course there is not a
strict parallel with the UK because there is not a political system
in the sense of elected representatives. It has been described by
Harris as an administrative no-party state:

"Depending on the observer's perspective Hong
Kong is technically a colony or a part of
China. To use an overworked word it is unique.
Perhaps it is the only state in the world whose
future is programmed (in terms of the end of the
New Territories lease in 1997) and unprogrammed
in that its terminal date could be any point
between now and infinity...The machinery of
government would prefer not to be involved in
the market. When it does act however, it acts
either to suit itself or in conformity with the
various interest groups and elites which cling
to the Hong Kong body politic. Processes of
popular consultation are still in an elementary
rudimentary stage."

However, it is true that the financial arm of Government was
responsible for pushing the project forward. Thus, the manner in
which it proceeded was largely determined by financial thinking. It
is worth noting in detail the words of the Financial Secretary to
the Legislative Council in June 1972:

"The Government has decided, after the most careful consideration that, if the demand for movement in Hong Kong in the late seventies and
thereafter is to be satisfied, the surface public transport system must be augmented, if at
all possible, by an underground mass transit
railway. The alternative would seem to be an
unacceptable degree of congestion of the road
network -- unacceptable in both economic and

social terms. But, as I stressed in both the Budget Speech and in my speech winding up the budget debate, the addition of a mass transit railway must be coupled with the expansion and improvement of the road network, with a programme designed to increase the carrying capacity, efficiency and comfort of surface public transport facilities and with policies designed to optimize the use of available road space. This is because a mass transit railway cannot, on its own, constitute a solution to our emerging transport problem; and, in particular, it will not mean that our road network has been fully developed in accordance with the Long-Term Road Study. Present plans provide for expenditure on road reconstruction and development of $1,800 million spread over the eight years ending 1979/80 and, undoubtedly, this amount will have to be increased as the programme is updated. The fact is that the mass transit railway when fully completed will cater for only one third of public passenger journeys that people are expected to make by the mid-1980's (about 2 ½ million journeys out of a total of 7 ½ million) although, admittedly, it will carry about half of all passengers in the more densely populated areas where the railway will run. In doing so, efficiently and swiftly, it will enable us to maintain that degree of mobility of people and goods vehicles without which our economic and social life would be endangered.

Secondly, the Government has decided that there is a good chance that three of the four questions I posed in my winding up speech in the budget debate, and which I said had to be answered before a decision could be taken whether or not to go ahead -- the Government has decided that three of these four questions can be answered affirmatively. That is to say, we believe, first, that a substantial sum of public money can be committed to assist in financing the project; secondly, that the railway can be constructed without undue disruption; and thirdly, that the system can be operated in such a way as to generate a sufficient cash flow to service the investment and cover operating costs. So the way is now clear to see whether an affirmative answer can be given to the fourth question, namely, whether sufficient outside capital finance can be obtained on appropriate terms as regards interest rates and repayment

arrangements. To some extent the possibility of
raising finance will depend on how much finance
has to be found. On the basis of mid-1970
prices, we reckon that the minimum net capital
requirement will be $5,000 million. But it is
bound to be higher and, indeed, according to our
calculations, the net capital requirement could
be considerably higher than this depending on
the assumptions made as regards cost escalation
and hence of average costs over the whole construction period (and we can do no more than
make certain assumptions until tenders are invited and contracts let). The net capital
requirement will be less than total capital
expenditure, including interest on the accumulated debt, because revenue from fares on the
initial stages, as they are opened, will start
to flow in from early 1977 onwards and these
will provide some offset from that year on.

The financing problem is closely linked with two
other problems, namely, (1) the form of tender-
ing and the letting of contracts and (2) the
most appropriate operational arrangements.
These issues will have to be looked at separately but, eventually, they will need to be
brought together in a coherent scheme for the
financing, construction and operation of the
system.

To work out the best methods of resolving these
three problems with a view to assembling a package on the basis of which a final decision can
be taken whether or not to go ahead -- that is
to say, whether it is possible to go ahead -- a
small Steering Group has been established under
my chairmanship. This Group will go fully into
all the possibilities of financing, constructing
and operating the Mass Transit Railway, including consultations with financial interests and
groups of construction companies and manufacturers of equipment which may come forward with
ideas. It has already held its first meeting
and is under direction to complete its work and
to present its findings to Executive Council
within six months. The Steering Group will be
greatly assisted by the ready willingness of our
bankers -- the Hong Kong Bank -- to assist us in
exploring the various ways and means of raising
non-Government finance and a senior official of
the Bank has been appointed to serve on the
Group.

Concurrently with this work, and subject to the
provision of funds by the Finance Committee, the
Consulting Engineers are to be re-engaged and
will be instructed to push ahead as rapidly as
possible with all the necessary detailed route
investigations, preparations of designs and the
drawing up of contract documents for the first
two stages of the railway, that is, a line commencing at Choi Hung, linking the Resettlement
Estates of Wong Tai Sin, Lok Fu and Shek Kip Mei
with the Nathan Road corridor and then running
under the harbour to the central business district of Hong Kong to terminate at Western
Market.

Additional surveys and extensive ground investigation will also be undertaken by firms of specialist consultants and contractors as a matter
of urgency, as this will be an essential prerequisite to the preparation of detailed designs
for the civil engineering works.

It is also intended to accelerate work on certain associated public works, such as the recla-
mation of the area at Kowloon Bay intended for
the Mass Transit maintenance depots and the
construction of relief roads, to ensure that
they are phased in with the Mass Transit construction programme.

Finally, work will be starting on drafting
legislation to provide for the compulsory acqui-
sition of property (where this is unavoidable),
the alteration of streets and the compensation
of those people whose property will be interfered with in the course of construction. The
preparation of this and other enabling legislation will be supervised by the Steering Group,
which will also keep a watching brief on the
consultants' activities and the progress of the
ground investigations and the various associated
works I have just mentioned.

In other words, Sir, certain essential pre-
construction work is being undertaken now so
that no time will have been lost should the
financial explorations prove successful and a
firm decision is taken later on this year to
proceed with construction.

As I have said, such an eventual decision on
whether to go ahead definitely with the construction of the railway will be taken in the

light of the findings of the Steering Group as
to whether firm possibilities exist of raising
the requisite finance on reasonable terms. This
decision will be taken before the end of this
year and, if it is favourable, every effort will
be made to start construction early in 1974. If
it is not favourable, either because the finance
is not available in sufficient quantities and on
appropriate terms or because the cost turns out
to be inordinately high or both, all work on the
project will cease and it will not be proceeded
with further.

Sir, while the Government has taken a firm decision, in principle, to proceed with the Mass
Transit Railway and has every intention of
exploring all possible sources of finance, we
are conscious of the size of the tasks -- in
physical as well as financial terms -- we have
set ourselves. The complete system will take
many years to build, though the first stage will
be operational within 3 years from the time
construction begins and the second stage 15
months later. We are confident that it is a
feasible project to construct, that being capital intensive it will not make excessive demands
on the labour force in the construction industry
and that it is the best system of all the alternatives examined. But at mid-1970 prices it is
estimated to cost over $6,000 million, including
accumulated interest after the construction
period -- or, as I said earlier, just over
$5,000 million after ploughing back the revenue
earned after the opening of the early stages.
Construction costs are, however, rising and can
be expected to rise further, so the actual cost
will probably be much greater than this.
Clearly, an enormous sum of money will have to
be raised. And then there is the problem of
ensuring that the system is financially viable
in the sense that the capital cost and accumulated interest can be repaid from the revenue
from fares and other sources over a given period
of years.

Clearly, it is incumbent on Government to be
very sure of all the financial aspects of the
project before committing the community to such
tremendous expenditure. In particular, we must
not put either our ability to finance our other
commitments or our liquidity or our fiscal policies at risk.

> Sir, if in the end it does not prove possible to
> build the Mass Transit Railway, I must make it
> clear that the measures needed to ensure that
> people can continue to get to and from work and
> for goods to continue moving will need to be
> more and more severe as time goes on. They will
> certainly not be popular and will be greatly
> criticized by some. Yet they will be necessary
> in the general interest if Hong Kong is to con-
> tinue to function as a viable community in the
> years to come. It is because the Government
> would like to avoid having to introduce such
> severe measures that it has taken a decision in
> principle to build the Mass Transit Railway and
> I can assure honourable Members that every ef-
> fort will be made to resolve the financial prob-
> lems involved."

Thus, the tone of the development was set. A commitment to
financial viability had been made. One earlier consideration was
noticeably missing, however. There was no reference to the surface
and rapid-transit systems being "complementary rather than competitive", not perhaps because this was considered unimportant, but the
concept of a total transport system had become lost in the consideration of the financial aspects of the project.

It is not intended in this paper to rehearse in detail the history of the next two years, during which international consortia put
forward competing proposals for the construction of the Initial
System under a single contract. This culminated in the signing of a
Letter of Intent with a Japanese consortium in February 1974. The
attraction of the proposal, which was not matched by the international competition, was that the Initial System would be built for a
fixed price. This was clearly a seductive proposition where the
decision-making process placed financial control ahead of almost all
other considerations.

The need for financial control had been illustrated during
those two years during which the estimated cost of the Initial
System increased twofold, and by September 1974 had reached three
times (at 1974 prices) the Further Studies estimate of HK$1,900m (at
1970 prices). Nor did this take any account of future escalation.
To keep pace with the cost increase the assumed fare range for the
Initial System had been pushed up from $0.20-1.00 (at 1970 prices)
to $0.80-2.40 (at 1974 prices). There is, however, no evidence that
specific account had been taken of fare elasticity -- the tendency
of numbers of passengers to decrease as fares in real terms are
increased.

At the end of 1974 the Japanese consortium, amid some recrimin-
ation, withdrew from the commitment in their Letter of Intent. The
impact of the inflation which followed the 1973 Middle East conflict

and the difficulties of organizing a consortium of more than fifty
firms had made life impossible for them. With commendable speed the
project was resurrected when apparently on the point of collapse.
It was being carried forward by the Mass Transit Railway Provisional
Authority (MTRPA) which had been formed in March 1974 and was
chaired by the Financial Secretary. It consisted largely of Government officials, but also financial advisers, and later, future Executive Directors of the Mass Transit Railway Corporation.

Government had decided that it would not itself build and operate the MTR. Harris says:

> "The Mass Transit Railway Corporation Bill was
> placed before the Legislative Council for debate
> in April and May 1975. The Mass Transit Railway
> Corporation was the 'first public statutory
> corporation of its type in Hong Kong', to quote
> the introductory speech made by the Acting
> Attorney General.
>
> The bill was seen as creating a new species of
> administrative organization in Hong Kong, and it
> was described by one lawyer as a statutory
> rather than a public corporation. In fact the
> bill setting up the Mass Transit Railway Corpor-
> ation described the administrative apparatus
> deemed appropriate and in very basic form. The
> British model of public enterprise derived from
> the Herbert Morrison model of a public corpora-
> tion (devised in the 1930's), was not applicable
> for several reasons.
>
> Firstly, a British public corporation was the
> normal vehicle for taking over an industry into
> public ownership and control. The Hong Kong
> Mass Transit Railway Corporation Bill apparently
> envisaged public enterprise as a preliminary to
> the restoration of the project to private enter-
> prise. The conventional justification for the
> British model is in the need to provide a public
> service, a public utility ('gas and water
> socialism') taking over an enterprise which was
> failing the nation or failing 'to capture the
> commanding heights of the economy'.
>
> None of these things was applicable in Hong
> Kong's case. The Mass Transit Railway Corpora-
> tion was given to it afresh ab initio. 'Having
> to create an efficient management team as
> opposed to keeping efficient an existing team is
> not wholly disadvantageous. There are no inher-
> ited inefficiencies to eradicate.'"

In January 1975 a revised proposal was put to the Executive Council for a smaller, 16 kilometre, Modified Initial System (MIS). It was to be carried forward on a multi-contract basis and was estimated to cost HK$5850m (at current prices) inclusive of design, supervision, land and compensation costs together with the cost of inflation (estimated at 7 percent p.a.) and finance charges until construction was complete. Cash flow calculations were made, based on a fare structure of $1.00 with $0.50 increments, which suggested that the MIS would have paid back its loans by 1992 and that operating cash flow would be positive by 1983. This was on the basis of a Government equity injection of KH$800m, the remainder of the money having to be borrowed by one means or another.

The Executive Council agreed to the project proceeding, but with two substantial qualifications in that the Council were to be consulted again when initial representative tender prices were known, as well as arrangements for financing of the system. Happily, the first tenders confirmed the estimates and the MTRPA were able to put together a package of finance -- export credits and stand-by loans -- which was enough to cover the cost of the MIS.

Thus, in September 1975 the project was approved, the Corporation was formed with two executive and six non-executive Board members, and first contracts were let. Because of recession in Hong Kong from mid-74 to mid-75, inflation was effectively zero. The new cost estimate, upon which new cash flow calculations were based, was HK$5,200m since the escalation assumption of 7 percent p.a. commenced a year later than previously. However, the prudent precaution was taken of assuming no more than that the previously stated cost of HK$5,850m of the MIS was confirmed. Public statements, in fact, used a figure of HK$5,800m.

SYSTEMS ANALYSIS

Very interesting, perhaps. But what has all of this to do with systems analysis? The aim simply is to generate discussion on the role of systems analysis in the two cases where the motivations for action and the constraints, together with the political atmosphere, were decidedly different. Moreover, the two cases are unencumbered by significant technological differences.

It is clear from the literature that even the critics of the past, or perhaps they were just the puzzled hiding their inferiority complexes, had a point. Systems analysis cannot comprehend all of the issues or model the roles of all of the actors. Intuitive thinking is more fashionable, compared with analytic thinking, than it was twenty years ago. We talk more about finding or defining the problem, as compared with analyzing or solving the problem, than we used to. We are more humble.

One significance of the Tyne and Wear and Hong Kong projects is that they happened. Not without problems, of course. The inter-union problems of the former have been well chronicled. Whether the latter will 'pay for itself' as intended remains to be seen. The Tyne and Wear Metro came into operation in August 1980, while the Hong Kong system was officially opened in February 1980. How successful they will be in the longer term remains to be seen -- in the wider urban context.

Now, of course, there is no guarantee that the successful implementation of projects must imply their innate goodness or that the decision-making process was 'correct' in some sense. Their mere existence does, however, contrast happily with the experience of many analyses, plans and project proposals which have produced nothing, at the expense of vast numbers of man-hours of effort.

Both projects were underlain by the kind of transport planning modelling which was very fashionable in the 1960's, but has faltered since then. Did this analytic work help the projects to go forward or did they proceed in spite of it? I believe that the work done, by others than myself, perhaps I should say, was as good as anyone was capable of at the time. To 'do a study' was an essential prerequisite for approval of projects. But the studies did, in both cases, have one great benefit of such studies. They brought together a disparate bunch of decision-makers who, even if they did not understand the nature of the models being used or the techniques of systems analysis, were nonetheless better able to focus on the complexities of the issues involved.

It is important to recognize that the analytic work carried out in each case in no way pretended to cover all of the issues involved, the political nuances and even some of the economic considerations. While people reflected on this there was no agonizing. Indeed, if miraculously, perfection could be achieved and all-embracing models could be designed, it is doubtful whether the final outcome for 'the public good' would be significantly, if at all, improved.

Why did the projects proceed. First of all, as in all such projects, there was a great deal of luck. The luck, of course, or the knack, is to get into the right place at the right time. The backgrounds of the Tyne and Wear Metro and the Hong Kong Mass Transit Railway were each very different, but they were each very timely. Yet either or both could have failed to proceed, for quite different reasons.

What does this experience have to say to the systems analyst, a question which is clearly of concern in the literature of the 1970's and, as I already observe, to some of the authors of papers for this Institute? Having been in policy making and project implementing for the last ten years, my needs and experience may be different from that of others. However, I am not alarmed if systems analysis does no more ultimately than significantly improve the discipline with which we address problems. That is not to say that I do not need or want models, merely that the discipline is more important to me than the 'answers' I derive.

Even there I have to be cautious for the psychologist will tell me, and my wife will if he does not, that even disciplined thinking — by the analytic mind — can mislead in a world which is increasingly understood to respond to a multitude of pressures which seem to have little to do with the analytic.

There seems to me to be no immediate prospect of trying to do more than clarify the issues, which may or may not be the same as Batty (1981) requiring a perspective.

The question I would put is whether we think that the systems analyses can or should try to include more of the 'realities' within their ambit. Or whether the 'realities' are simply not analyzable in any scientific sense.

REFERENCES

Batty, M., (1981), A Perspective on Urban Systems Analysis, in D. Banister and P. Hall, (Eds.), Transport Policy and Planning, Mansell, London.

Department of Transport (DOT), 1967, Public Transport and Traffic, HMSO, London.

EDITORS' NOTE

This paper is based on the author's invited lecture to the NATO Advanced Research Institute.

SYSTEMS CONCEPTS IN GOVERNMENT PLANNING: EXPERIENCE AND PROSPECTS

Douglas T. Wright

Ministry of Culture and Recreation of Ontario
Government of Ontario, Queen's Park
Toronto, Ontario, M7A 2R9
Canada

INTRODUCTION

At the end of World War II there was an apprehension in the West that the economic distress and depression of the 1930's would resume. Instead, an era of unprecedented economic growth and prosperity developed. This economic progress was widely acknowledged to be a function of the conjunction of science and technology with the discipline of modern theories of management. With increased prosperity, expectations arose for improvements in social and urban conditions. Governments were able to respond to these expectations because economic progress had swollen government revenues.

In the 1950's and through the 1960's, in every Western country, there arose a variety of programmes aimed at the improvement of the human condition. With increased productivity in the primary sector of the economy, employment growth was concentrated in the secondary and service sectors which in turn reinforced a trend to urbanization. Government commitments to the development of social and urban systems increased at a phenomenal rate. Public sector expenditures, for capital development, for programmes of service, and for redistributive transfers to individuals, increased to levels constituting a greater percentage of GNP than had been consumed by governments in war time. But unlike war time, these high levels of expenditure were not transitory. Their scale and the increasingly complex nature of the responsibilities assumed by government came quickly to beg review of a sort that had never been attempted before.

Starting about twenty years ago, notions arose in senior levels of government in most Western nations that public sector planning and resource allocation could generally be made more systematic. It

was thought and hoped that objectivity and rational analysis could
come to balance, if not even outweigh, the play of politics. It was
argued that rational analysis could both reduce uncertainty and
improve performance. This paper attempts to identify what has been
learned and achieved through these efforts as they affect central
government policy-making and resource allocation.

The intellectual challenge was set out by Daniel Bell, in 1973,
in the following optimistic passage (Bell, 1973):

> "the methodological promise of the second half of
> the twentieth century is the management of organ-
> ized complexity (the complexity of large organiz-
> ations and systems, the complexity of theory with
> a large number of variables), the identification
> and implementation of strategies for rational
> choice in games against nature and games between
> persons, and the development of a new intellec-
> tual technology which, by the end of the century,
> may be as salient in human affairs as machine
> technology has been for the past century and a
> half."

Against this stirring challenge, stands Lindblom's remarkable
paper, "The Science of Muddling Through" (Lindblom, 1959). Lindblom
seemed to anticipate and prejudge many future efforts when he termed
"rational comprehensive" approaches to decision-making "impossible"
and "absurd" because, amongst other weaknesses, they "assume intellectual capacities and sources of information, that men simply do
not possess". The alternative, reflecting actual practice in
government he termed "the approach of successive limited comparisons".

As has been so evident, Bell's challenge was taken up, and
Lindblom's caution was disregarded.

SYSTEMS GALORE: ACRONYM SOUP

An amazing variety of efforts of greater or lesser comprehensiveness, each usually characterized as a "system", have been put
forward and taken up by various governments, almost as in a quest
for the Holy Grail. Because a fairly extensive if often obscure
literature exists describing the individual methods, only a very
brief description of some salient features of the individual systems
and their impacts will be given here.

Although somewhat restricted in its scope, cost-benefit analysis must be acknowledged because of its long history and its fundamental importance, and because some advocates have at times directed
its use on a fairly comprehensive basis for general expenditure
budgeting. Although this technique has only come into prominence in

fairly recent years, the notion of assessing the utility of public
works in terms of costs and benefits was put forward in France as
long ago as 1844. By 1902 the American congress had required the
Corps of Engineers to report on the benefits and cost of proposed
navigation projects, (Prest and Turvey, 1965).

The notion of measuring cost-benefit is disarmingly simple. In
practice, it has become clear that the greater the care with which
analysis proceeds, the more insuperable become the problems of the
measurement of benefits through a haze of externalities and imperfections, and the hazards and uncertainties of assessing costs.
Experiences with such diverse undertakings as groundnuts in East
Africa, the Concorde and urban expressways, all make the point that
this apparently simple and technical methodology has only a limited
utility. The Nowlans (1970) present an interesting analysis of the
costs and benefits of a controversial urban expressway, and illustrate the use to which cost-benefit analysis can be put in political
debate. Yet, with whatever caveats may be imagined, what better
than cost-benefit analysis exists for assessing the "chunnel"?

It is probably appropriate to acknowledge at this point that
environmental assessment, technological assessment, and social
impact assessment are all extensions of cost-benefit analysis. Each
of these subsidiary methodologies has arisen as governments have
tried to cope with public concerns about spillover effects and
externalities. Experience has shown that quantitative measures in
these areas are particularly elusive and/or misleading; but even the
enumeration of "impacts" has none the less been shown to have considerable value in resolving political problems.

Perhaps the first of the "modern" systems methodologies were
PERT (Program Evaluation Review Technique) and CPM (Critical Path
Method), (Miller, 1962). The former was created for the US Navy to
manage the development of the POLARIS System in 1958, and the latter
was developed, roughly at the same time, by the Dupont Corporation.
Although differing somewhat in detail, the two systems both provide
the same kind of management technique for the planning of very large
and complex one-off physical development projects. As well as
providing important improvements in time-scheduling, PERT and CPM
have also facilitated communications amongst those responsible for
various portions of large undertakings. CPM and PERT have become
essential in the management of large projects, but efforts to
develop analogous systems for recurrent operations have not been
successful.

The notion of applying successful private sector management
techniques and discipline to the problems of the public sector
brought Robert McNamara to the Pentagon and spawned the first of the
truly comprehensive public sector management systems: PPBS (Planning
Programming Budgeting System), (Lyden and Miller, 1968). Although
invented for use with military systems, PPBS was widely touted and
quickly taken up by government as a kind of panacea, with President

Lyndon Johnson ordering all US Federal departments and agencies in 1965 to employ the PPBS methodology.   It is astounding, in hindsight, to acknowledge what was intended: that all government expenditures and undertakings would be classified in programme format, that explicit quantifiable objectives would be specified for each programme and that each year there would be an enumeration and analysis of all the options open to government from which the most efficient and effective policies would be chosen.

Enormous efforts were made, and thousands of man-years were devoted to trying to implement this methodology, first in the US Federal government, and then in numerous state governments as well as governments in Canada.   But by 1971, the Office of Management and Budget (OMB) in Washington announced the end of PPBS.   It seems clear, in retrospect, that PPBS as a comprehensive system never had a chance.   Perhaps its most fundamental flaw was its failure to accommodate political values and conditions.

In practice, the overwhelming amount of analysis required to do justice to the intentions of the system overwhelmed the capacity to do competent work, quality suffered, and analysis became discredited.   Although ordained by the President and OMB, PPBS was not genuinely integrated with budgeting except at the department of defense.   Yet another critical factor was the circumstance that, by the late 1960's, the costs of the "Great Society" and the burden of war in Vietnam were such that the issue was no longer whose ox should be fattened, but whose ox should be gored.   The most if not the only continuing benefit of PPBS has been the presentation of estimates of government expenditure in programme format, in place of the previous, traditional, classification of categories of expenditure which had been quite meaningless.

Out of the PPBS debacle came a conviction that neither the policy analysts nor the politicians could contemplate more than a few variables at a time.   It was concluded that analysis would have to be piecemeal, and change disjointed (Wildavsky, 1969).   This was characterized as a system by yet another acronym, PAR (Program Analysis and Review).   It was contemplated that only a few programmes at a time would be subjected to PAR, but that all programmes would be subjected to such a discipline on some periodic basis.   In the event, the PAR concept was taken up fairly generally, on both sides of the Atlantic (Fletcher, 1972; Diamond, 1975).   In Britain, in the early 1970's, a dozen or so programmes were subjected to PAR, one of them being higher education.   In North America, the acronym was fairly quickly forgotten, but the methodology continues in regular use.

Instead of functioning as a centrally directed, periodic, management review of programmes, the PAR notion has been used most frequently to cope with policy needs in political issues where fundamental review became essential.   (An example is the 1975 review and subsequent fundamental restructuring of services for children with behavioural problems in the province of Ontario.) It may be argued that a system that becomes so episodic is no longer a system, and that the entire process is little different from traditional policy reconsideration, as most frequently conducted by government commissions, royal and otherwise. But experience in the 1970's strongly indicates that the PAR technique has offered significant improvements over the ways in which policy reviews had previously been undertaken. Where, in the past, experts and analysts had determined "best courses of action" and offered them to government on a take-it-or-leave-it basis, the new methodology provided for a more insightful analysis of the available options for government, with a sequential decision-making pattern arising out of a continuing dialogue between analysts and politicians. The tradition of confidentiality of cabinet processes, designed to ensure that final decisions can, no matter how protracted and bitter the debate that preceded them, be supported publicly at the end by a unified cabinet, has unfortunately not yet permitted sufficient understanding of PAR-type methodologies in practice.

In Britain PAR was employed in tandem with PESC (Public Expenditure Survey Committee) (Diamond, 1975; Heclo and Wildavsky, 1974). The extravagant enthusiasm for PPBS as a comprehensive system for both policy review and budgeting which had seized imaginations in North America had had no counterpart in Britain. In a much more comfortable fashion, it was argued that if PAR could provide fundamental review from time to time of policies and programmes, the balance of the annual budgeting and resource allocation process within government should be made as tidy and inoffensive as possible. PESC was invented to avoid debates on policies, priorities and on the appropriateness of any particular provision of resources. Instead, it was supposed to produce agreement amongst officials on current and future costs of existing policies and commitments. It was assumed that these could be fixed without particular reference to Ministers. To assist in longer term planning, always regarded as desirable but rarely achieved by governments in explicit terms, projections of the costs of maintaining current policy and service levels were carried five years forward on a rolling basis. Because of uncertainties about inflation, wages and the like, planning came to be carried out in terms of volumetric indicators, eg. numbers of school teachers, nurses, overpasses, miles of expressway built, and so forth.

With hard commitments based on such measures of volume, expenditure levels became the final consequence of the planning process, rather than a starting point. Much more critically, inflation and the insensitivity to price of the volumetric planning indicators drove costs in a frenzied spiral. The share of the GNP consumed by government was pushed to unanticipated levels. The entire process seems to have become a mockery of planning, with PESC itself perhaps the most important single factor contributing to the astonishing inflation in Britain in the mid-1970's. It was only with the greatest difficulty that the cabinet was able to restore annual expenditure limits in traditional terms.

This litany would not be complete without a reference to ZBB (Zero-Based Budgeting). Although first developed for use in industry (Pyhrr, 1970), at Texas Instruments, this system came to public attention as an adjunct to the popularity of Georgia Governor Jimmy Carter. Before the (one is tempted to write pyrrhic, but the dictionary suggests the fit is not too good) results of the Georgia experience had been chronicled, Carter was in the White House and ZBB had been proclaimed by the Office of Management and Budget for all departments and agencies, (US Office of Management and Budget, 1977). In a fashion disturbingly reminiscent of the PPBS experience, innumerable short courses were scheduled and over-subscribed, and mountains of analysis prepared. It is easy to guess that the principal beneficiaries of the ZBB phenomenon were those who convened the short courses. After several years of budgetary containment, the opportunities for radical reductions in programmes and activities in ways that were politically admissible were few and far between. ZBB appears to have been quietly abandoned without even the courtesy of the death notice OMB accorded to PPBS.

The notion of formalizing management by objectives (MBO) is attributed to Peter Drucker. In the private sector, readily quantifiable and unambiguous objectives such as market share, volume of sales, and profit, could and did indeed provide a basis for planning and performance review. Notions of causality did not need to be fully developed before all the responsible players could understand and be judged by their performance in measurable terms (Brady, 1973). When taken up by government MBO was crippled by the absence of measurable outcomes – short of election results. The notion of defining comprehensive objectives was as impossible under MBO as it had been with PPBS (Rose, 1977). What was different, perhaps, was that the fanfare and commitment of resources to the implementation of MBO were much less than had been the case with PPBS.

What was much more valuable was that MBO mutated into MBR (Management By Results) which has turned out to be a good deal less ambitious, but far more useful. MBR does not involve comprehensive policy analysis and review nor does it demand the identification of an inclusive statement by objectives. But, by focusing on unit and total costs, on service volumes, turn-around times, and so forth it enables a frequent orderly scrutiny of efficiency and effectiveness of management, in a context in which individual managers are responsible and accountable for performance.

The last "system" to be acknowledged in this section is so new that it does not yet have an acronym, and at first sight may not seem to deserve to be included. It is comprehensive auditing. The audit function in government rose to importance in the early years of this century, primarily to inhibit misappropriation of public money. In recent times reviews of public accounts by legislative committees have, with the assistance of government auditors come frequently to identify embarrassing cases of waste. The audit function in government has been substantially refined and broadened in the last few years to embrace notions of "effectiveness and efficiency", and "value for money". This enlarged scope, along with the tendency to consider not only expenditures but procedures, has come to be defined as comprehensive auditing. Although remaining nominally retrospective it already appears that this new approach to the audit function may have a more powerful if gradual influence on the way decisions are made than the more ambitious so-called comprehensive planning systems. The reason is that individual accountability is imprecise with the grand scale of comprehensive plans, while the burden of individual decisions, examined on a statistical sampling, is unambiguous. There is some irony in the fact that a traditional function, such as audit, may have such a powerful influence.

NEW STRUCTURAL SYSTEMS

Also forming part of the effort of the 1970's to establish new and "rational" systems for the management of government were two kinds of structural developments that stand in distinction to the procedural methodologies already described. One of these was the development of new machinery in government intended to institutionalize policy review and debate, particularly on issues that cut across ordinary departmental boundaries. The other was the development of institutes, more or less independent of government, devoted to policy research. In both cases, the motivation arose from the awareness of the difficulties governments were facing in coping with unprecedented demands, and an awareness that many problems could not be dealt with adequately within traditional structures. Moreover, the traditional role and capacity of a President or Prime Minister to co-ordinate important policy issues and initiatives had become overwhelmed by the volume of major issues.

Herbert Morrison analyzed the political and administrative failure of Churchill's experiment with the cabinet of "superlords" in 1951 (Morrison, 1954). The two tier system aroused the ire of Parliament because it undermined the direct accountability of Minsters. Morrison's practical advice on how to manage cabinet process has had obvious applications. Several variations have arisen. In 1972 the Organization for Economic Cooperation and Development convened a conference on innovation in the procedures and structures of government, and the papers prepared provide a catalogue of innovations tried (OECD, 1972). The most valuable, by far, of the structural systems has been new forms of cabinet committees, with special staffing.

In Canada, first in Ontario (Fleck, 1973; Webster, 1976) and then later in other provinces, there have been established standing committees of cabinet corresponding to fields of provincial jurisdiction (administration of justice, social development, and resources development). In Ontario and Quebec these committees have been chaired by a member of cabinet who was freed of ordinary operating responsibilities. Small staffs were established to assist each chairman and committee. In these structures, no executive relation prevails between chairman and minister members. The accountability of individual ministers is unimpaired. The main disadvantage has been the generally low political profile of the ministers who have carried the responsibility for chairing such cabinet committees. Balancing this is the utility these structures have had in coping with the burdens of policy co-ordination and development during periods of financial constraint. Although not anticipated in the initial proposals, these structures have also come to fulfil an important role in the annual resource allocation process, operating within global expenditure targets. An important feature of these systems has been the breaching of the traditional monopoly of a minister's own departmental civil servants in providing him with policy advice. In 1979, such structures were introduced in Canada at the Federal level by the newly elected Conservative government, and in 1980 appear to be continuing with the return of the Liberal party to power.

In Britain, the CPRS (Central Policy Review Staff), quickly dubbed the "think-tank", was established in 1972 with a small staff and given direct access to the Prime Minister and Cabinet. It seems to operate vis-a-vis Cabinet committees rather as the policy field secretariats operate in Ontario and Quebec. Working partly in private (hidden by the tradition of cabinet secrecy), and partly in public, (with the publication of discussion papers on certain issues), the CPRS has been much criticized but has survived several changes in government.

In Washington, in 1973, President Nixon promised a structure of cabinet committees, with special support staff, but this was never pursued. (President Nixon didn't convene a single meeting of his full cabinet during his last fourteen months. To one accustomed to working with parliamentary cabinet forms of government this is nothing short of astonishing.)

The increasing interest in policy analysis in the 1970's led observers to acknowledge the obvious; the Brookings Institution in Washington had served effectively for many years as an adjunct to government and the legislative process in analysing options for public policy. In an effort to transfer this experience to other capital cities, policy research institutes were developed in the Netherlands and Canada (Ritchie, 1971), and proposed in Britain (Ritchie, 1977). Such institutions seem to operate in a nether world between government and the universities. Not surprisingly, the Brookings Institution continues in a league by itself. Although the notion of policy research has gained some respectability it remains an anomaly in most universities (Dror, 1974). Perhaps because of the difficulty of penetrating the shroud of Cabinet secrecy, most of the journals that have arisen in the field tend to be filled with ethereal abstractions.

RETROSPECT: THE TWENTY YEARS 1960-1980

The above review of the management and planning systems that have been introduced in efforts to "rationalize" public sector planning, budgeting and decision-making in the past twenty years is not complete. But it is complete enough to understand what happened, and for conclusions to be drawn and implications studied.

There has been considerable improvement in the capacity to plan and manage the development of primarily physical sub-systems and systems, even when these become very large. CPM and its variations are in wide use in private sector contracting, and probably constitute the only important example of the spin-off and transfer of a management system from public to private sector in the past twenty years.

Improvements made in the structure of government, primarily in the form of institutionalized sub-committees of cabinet, have made fundamental change possible in the harmonization of competing interests in government. Policy planning and the co-ordination of policy have also benefited significantly from these new structures. In the context and jargon of "systems", it is correct to define these new structures and processes as constituting new "comprehensive management systems".

Substantial improvements in government management processes have also been achieved in the past decade through improvements in control and accountability, operating primarily at less senior levels of decision-making. Inasmuch as these seem to be by-products of more ambitious "systems" reforms, they deserve to be acknowledged.

But grandiose notions of comprehensive systems through which government policy-making and budgeting would be rationalized have been failures. Lindblom's caution has apparently been confirmed, while Bell's optimism has nowhere been fulfilled.

Perhaps this discussion could be concluded here. But it would leave uneasiness and uncertainty. Why did political leaders, unashamed to be called practical, succumb to the theories (some would say ideology) of the systems experts? Are we left in a slough of despondency facing nothing better than simple empiricism for the future?

PROSPECTS AND POSSIBILITIES

The enthusiasm with which various forms of systems analysis were proposed, taken up somewhere in government, touted by the media, and then spread contagiously from government to government, has been truly amazing. There may be more of interest in this process itself as a political phenomenon than in the substance of any of the reforms proposed.

There was undoubtedly a perceived political need, as the scale of government increased rapidly in the 1960's, to demonstrate some new form of synoptic control and direction. There is little evidence that many of those involved in proposing or deciding to adopt the various systems that have been described had read or even heard of either Bell or Lindblom. It also seems true that it would have made little difference if they had. The argument can be too easily made by political opportunists that comprehensive systems failed, not because they were naive, technocratic or impracticable, but because they were defeated from within, by a trahison des clercs. There was and there remains a political need to take initiatives demonstrating control. And notwithstanding the evidence of so many monuments left standing in the desert, it is quite possible that some new system, naive as all the others, with some yet unimagined acronym, will be put forward and taken up with no less enthusiasm. Could any of those who had experienced PPBS imagine that ZBB could have been taken seriously? Perhaps the most promising sign is the move, by some governments at least, to impose tough expenditure limits.

Less pessimistically, there seems to be good reason to hope that something better than Lindblom's "successive limited comparisons" may be possible as a basis for central government decision-making. The key to this would seem to lie in improved understandings of the nature of the actual systems that we confront. Systems, that is, in the sense of the dictionary definition: "complex wholes". Such systems may in toto never be subject to mathematical modelling or any of the other textbook techniques of physical systems analysis, but they may be understood, and as systems. Between the naivety of comprehensive systems analysis, the political myopia of most physical planning, and the adhocracy of incrementalism, there is certainly scope for understanding. Understanding that acknowledges and embraces political and social factors as well as the more convenient parameters.

From experience and analysis it is possible to identify a comprehensive inner logic of large "systems" such as urban transportation and land use, and health and social services. It seems to be possible to use such understanding as a basis for prediction, and in turn it seems to be possible, at least to identify the circumstances under which certain outcomes would be possible or impossible, and perhaps in time to devise initiatives that will produce desired outcomes. If correct, such a prognosis suggests far more than empiricism or incrementalism.

Few analyses of the sort that seem to be possible have yet been completed and published. One is that by Allison, on the Cuban missile crisis (Allison, 1971). Allison sought to construct a logic of events and actions. he tried first a textbook notion of rationality (catalogue alternatives/define consequences/rank consequences/select best option), and found this to be very imperfect. He extended his analysis to include the effects of bureaucratic and administrative procedures, with many delegated decisions and imperfect communications. This succeeded in explaining much more but was still inadequate. The third and most satisfactory cut acknowledged as well the role of individuals and the play of personalities and the councils around the leaders. This proved to be remarkably close.

But the Cuban missile crisis took only thirteen days. It had a beginning and an end. It involved foreign policy, so that the number of players was limited and decisions were relatively central-ized. That would not be the case, say, in any major urban or social programme. Notwithstanding, it is possible to contemplate analyses of this same sort extended to such situations, producing better understandings than seem generally to prevail today.

In considering the "logic" of such systems, certain paradoxes arise. What may seem to be incontrovertible evidence may often be of no relevance to the determination of public policy. Also per-plexing is the lack of satisfaction with real gains. Closer inspec-tion reveals that such phenomena are not quite so surprising as they seem to be at first sight.

The "rules of evidence" in the political domain in democratic societies reflect public opinion. This is not the trivialization of government by the Gallup Poll. Rather it is simply and fundamen-tally that the burden of widely-held public opinion (on such diverse issues as the utility of rent control or of surgical interventions) is so great as to overwhelm any kind of contrary evidence as a basis of formulating public policy. Perception is not only more important than reality, it is reality.

The apparently widely-held view that government is or should be omniscient and omnipotent is of recent origin. Geoffrey Crowther, who did much to elevate The Economist to its present high status, noted in his farewell editorial some twenty-five years ago: "the most striking aspect of post-war politics in Britain has been that both parties seem to accept without demur a responsibility for main-taining the economic health of the community in the face of the evi-dence that neither party has had, at best, more than the most rudi-mentary control over the economic climate".

The compulsion with which democratically-elected governments are driven by political and media pressure to intervene, even where interventions are not likely to be beneficial, or may be counter-productive, has been clearly identified by Jacques Ellul (1967).

The other paradox that causes much concern is how little satis-faction many recent achievements seem to provide. Notwithstanding inflation and other economic troubles, real per capita income and purchasing power have risen in recent years in most western coun-tries to levels that were unimagined twenty years ago. Yet dissa-tisfaction with economic circumstances is widespread.

Hirsch (1976) suggests that the explanation lies in the fact that consumption has changed in character and now has a large social component. No matter how excellent the automobile or the express highway, the satisfaction derived depends upon the conditions under which they are used -- which depend primarily on how many other people are trying to use them at the same time. The democratization of access to educational opportunity, sought with such fervour in the 1960's has been so fully achieved that university graduation no longer ensures an elite status. What some can achieve, all cannot. Congestion is a form of scarcity that seems to be at least as important as traditional absolute measures.

CONCLUSIONS

From all of this certain conclusions may be drawn. Systems analysis, in the sense of the ability to analyze and design the operating characteristics of purely physical systems, offers neither utility nor legitimacy in dealing with systems that involve social and political factors.

But the analysis of systems that do embrace social and political, as well as technical factors, does seem now to be practicable. Analysis in the sense of understanding the "inner logic" of such systems. While such systems may never be susceptible to mathematical modelling, it may still be quite precise, in the style of Allison's analysis of the Cuban missile crisis. Much more complex circumstances must be dealt with, however, than in that case which was time-limited and relatively coherent. Civil systems are more complex, and more extensive, and almost invariably involve paradoxes in which objective evidence is at odds with public opinion and the compulsions of politics. Notwithstanding, such systems do indeed seem to have an "inner logic", and are therefore comprehensible and likely predictable.

Some analyses of the sort described have been completed or tried. But very little has yet been published. One example, though incomplete, is Wildavsky's treatment of the health field in one chapter of his most recent book (Wildavsky, 1979).

In retrospect, at least, it is obvious how overly ambitious and inadequate were the technocratic planning systems of the 1960's and 1970's. The well-deserved failure of those efforts should not inhibit efforts to develop new and more subtle analyses of civil systems, as now seems possible.

REFERENCES

Allison, G.T., 1971, Essence of Decision: Explaining the Cuban Missile Crisis, Little, Brown and Company, Boston, Mass.

Bell, D., 1973, The Coming of Post-Industrial Society, Basic Books, New York.

Brady, R.H., 1973, MBO Goes to Work in the Public Sector, Harvard Business Review, 51, 101-110.

Diamond, Lord, 1975, Public Expenditure in Practice, George Allen and Unwin, London.

Dror, Y., 1974, Policy Sciences: Some Global Perspectives, Policy Sciences, 5, 83-87.

Ellul, J., 1967, The Political Illusion, Knopf, New York.

Fleck, J.D., 1973, Restructuring the Ontario Government, Canadian Public Administration, 16, 55-68.

Fletcher, Stephen M., 1972, From PPBS to PAR in the Empire State, State Government, 45, 198-202.

Heclo, H. and Wildavsky, A., 1974, The Private Government of Public Money, Macmillan, London.

Hirsch, F., 1976, Social Limits to Growth, Harvard University Press, Cambridge, Mass.

Lindblom, C.E., 1959, The Science of Muddling Through, Public Administration Review, 19, 79-89.

Lyden, F.J. and Miller, E.G., 1968, Planning Programming Budgeting: A Systems Approach to Management, Markham, Chicago.

Miller, R.W., 1962, How to Plan and Control with PERT, Harvard Business Review, 40, 93-104.

Morrison, H., 1954, Composition of the Cabinet: Overlords; Co-ordinators, in Government and Parliament, Oxford University Press, London.

Nowlan, D. and N., 1970, The Bad Trip: The Untold Story of the Spadina Expressway, House of Anansi, Toronto.

Organization for Economic Cooperation and Development, 1972, Innovation in the Procedures and Structures of Government, OECD, Paris, France.

Prest, A.R. and Turvey, R., 1965, Cost-Benefit Analysis: A Survey, The Economic Journal, 75, 683-735.

Pyhrr, P.A., 1970, Zero-Base Budgeting, Harvard Business Review, 609, 63-71.

Ritchie, R.S., 1971, An Institute for Research on Public Policy, Government of Canada, Ottawa.

Ritchie, R.S., 1977, A Brookings for Britain, The Economist, June 11, 18.

Rose, R., 1977, Implementation and Evaporation: The Record of MBO, Public Administration Review, 37, 64-71.

Webster, N., 1976, Streamlined System Pays Off, Globe and Mail, Toronto, Nov. 30, 7.

US Office of Management and Budget, 1977, Zero-Base Budgeting, Federal Register, 42, 22342-22354.

Wildavsky, A., 1969, The Analysis and Evaluation of Public Expenditure: The PPB System, Joint Economic Committee US Congress, 835-864.

Wildavsky, A., 1979, Speaking Truth to Power: The Art and Craft of Policy Analysis, Little, Brown and Company, Boston, Mass.

PART 5

PHILOSOPHIES OF SYSTEMS ANALYSIS

SYSTEMS ANALYSIS AND URBAN POLICY

E.S. Savas

Center for Government Studies
Columbia University
New York City, NY, 10027
United States of America

This paper makes the following points about urban systems analysis, in the spirit of improving its usefulness and extending its range:

1. Systems analysis is encumbered with certain dysfunctional myths, which impede its effective use in policy-making.

2. It is applicable at different levels, which present different environments, and call for different methods of analysis.

3. There are other competing and complementary disciplines and perspectives that can profitably be embraced by systems analysis; hence, the education of systems analysts needs reform.

THE MYTHS

The Myth of Decision-Making

Perhaps the most mischievous myth that afflicts the field of urban systems analysis is the very notion of decision-making. As conventionally taught, alternative choices are analyzed, and the consequences are arrayed before the decision-maker, whereupon decision-making occurs. In this idealistic scenario, decision-making is a neat, crisp act which follows quite automatically from the analysis. In fact, however, in the urban policy arena, policies are rarely made; at best they emerge from a vague, prolonged, diffuse, pluralistic and evolutionary process. They are often the accumulation of sequential ad hoc choices in environments characterized by a multiplicity of conflicting sources of partial information

379

and disparate value judgements. They require persuasion and consen-
sus-building, not merely decisions.   The situation is described
eloquently by Kash (1968):

> "Decision-making...is a process. Phrases like the
> "decision-making process" and the "process of
> policy formulation" are not mere incantation:
> they refer to the continuous flow of decisions,
> large and small, that make up the seamless web of
> policy formation and administrative action in...
> government. The dynamic flux of the policy pro-
> cess makes the job of the advisor particularly
> difficult.   It means that there is no orderly
> procedure whereby the advisor can state his views
> or explain his research, and then retire from the
> scene, confident that his advice will receive
> systematic consideration.   There are numerous
> distractions and competing demands on the deci-
> sion-maker's time and span of attention.   Deci-
> sions once made can become unmade a week later.
> The advisor may face a difficult task to secure a
> full hearing for his views in the first place,
> and then must struggle to keep attention focused
> on his recommendations for a long enough period
> to assure action of some kind.   Continuity is
> thus an essential attribute of effective communi-
> cation of policy-oriented research.
>
> A corollary of this is that the advice cannot
> simply be given to the top levels, if favourable
> decision and effective implementation of advice
> is desired.   Consider the case of a high level
> decision-maker accepting the recommendation of an
> advisory group and making a "policy" decision
> designed to implement the advice.   Unless the
> subordinates carry out the decision effectively,
> the whole intent can be defeated. Comprehension
> of the basis for the decision reached at the
> higher level can be a vital factor in winning the
> consent and enthusiasm of those who must execute
> the decision and, in doing so, make a myriad of
> other decisions which can determine the success
> or failure of the original decision."

The Myth of Problem-Solving

A second major myth inherent in the conventional view of systems analysis is the idea of problem-solving.   The quantitative
roots of systems analysis nurture the belief that problems and solutions constitute a dyad; after all, every problem in mathematics
textbooks comes with a matching solution, usually in the back of the
book.   Use of the right technique will produce the right answer.

Given this view of the world, urban issues are treated as problems with answers, and the entire armamentarium of problem-solving weaponry is wheeled into position and brought to bear to "solve the problem". Again, this is a false and dysfunctional view of urban policy issues. More often than not, it is far more useful to approach these issues as conditions to cope with, or situations to be ameliorated, rather than as problems to be solved. (For example, poverty is not a "problem" that can be solved; some people will always be labelled poor, with all the stigma that implies, regardless of their absolute level of income.) Incremental improvement, and even "muddling through", are not to be sneered at when "solutions" are unattainable. As Wildavsky (1974) points out, problems involving government aren't solved, they're merely superseded.

The Myth of Optimality

A third myth that obstructs effective urban systems analysis is the myth of optimality. Of course, optimization is a useful construct, but it plays a surprisingly small role in the exasperating urban environment where, in linear programming parlance, there is usually no feasible solution. In this complex setting, optimization consists of trying to determine which of the multitudinous constraints can be forced back, at minimal political cost, far enough to create one feasible -- hence, optimal -- point.

The Myth of the Moon

The fourth myth holds that if we could successfully accomplish something as difficult as getting to the moon, we should be able to...revitalize cities, solve social ills, etc.

The journey to the moon was indeed a triumph of systems analysis, but it offers little encouragement to urban analysts. Getting to the moon was relatively easy, for two important reasons: (i) there are no contituents on the moon; (ii) there was no identifiable constituency on earth that could be organized to oppose the effort of going to the moon.

The total absence of a lunar constituency meant that the public did not care where the astronauts landed, whether in the Sea of Tranquility or on the Heights of Asininity, or what they did once they got there. Therefore, the decisions about lunar landing sites could be based entirely on scientific and technical considerations. However, with respect to more terrestrial decisions, the public cares a great deal indeed about where a waste disposal site is to be located, for example, or where an airport is to be built, and therefore, technical analysis plays a relatively minor role in such decisions. An algorithm which seems to prove that the "best" place for a drug-treatment centre is just down the block from a councilman's home, proves instead that the analyst is a poor one who should be ignored, or worse.

No group felt that its vital interests were threatened by a large-scale national effort to send a man to the moon and bring him back alive. On the other hand, many interest groups saw positive benefits to themselves from such a programme. Under such circumstances, a national consensus was easy to achieve, and a massive programme could be mounted without opposition.

But this fortunate conjunction of circumstances rarely arises on the urban scene. In particular, there is no consensus about constructing transportation networks, building subsidized housing, or improving the productivity of municipal workers.

The Myth of Attributable Achievement

When a course of action -- recommended after analysis of a complex issue -- is adopted and proves successful, many forces will have conspired and converged to bring about the observed outcome. Not all of them can be identified, nor can their impacts readily be assessed. The causal chain cannot be traced back, for it is more like a net than a chain. Therefore, each of the many participants can legitimately claim credit, for each can orient the net uniquely, and sight alone one particular thread. In short, success has many fathers (while failure is an orphan). The analyst who was trained to revere the creativity of the solitary scientist, and who seeks his professional rewards through attributable achievement, is likely to be frustrated in this setting.

The Myth of Model-Building

This myth, the sixth in our list, also merits comment. The builder of a mathematical model sees his work as culminating in a product, a theoretically sound and comprehensive tool of quite general applicability. But, if the activity that the model presumes to represent is a very complex one, such as an urban system, a product, no matter how complex -- and particularly if complex -- is inadequate for improving that activity ("solving the problem"): a process is needed.

Ideally, model-building is called for -- nay, demanded -- as part of the improvement process. Too often, unfortunately, a model is constructed under the builder's implicit assumption that its mere presence will initiate the necessary process. While the existence of such a tool is necessary -- but not sufficient -- for its use, it is neither necessary nor sufficient for initiating the improvement process.

Certainly a model can be useful in the process, under favourable circumstances. Sometimes a model can even be used to initiate this process. However, there are also other possible initiators of the change process. Therefore, the question can rightly be asked for any given issue: are there more cost-effective ways to initiate or accelerate the change process than by constructing a costly mathematical model? Rarely is this question asked.

The Myth of Planning

How could any rational person, let alone a systems analyst, oppose planning? The initial list of five subject areas circulated by the organizers of this Institute clearly put planning at the heart of our concerns by listing the areas as public services <u>planning</u>, physical network <u>planning</u>, built-form <u>planning</u>, strategic land use <u>planning</u>, and corporate <u>planning</u> (emphasis added). While I would be among the last to deny the virtues of planning -- I do attempt to practice it myself -- it does not merit quite such an exalted position. In the first place, emphasizing it in this way painfully accentuates the distinction between planning and accomplishing. Secondly, the emphasis on planning exaggerates our ability to foresee and forfend.

With respect to the well-known abyss between planning and doing, it is well to remember that problems not perceived by the public are problems not acted upon. A problem has to exist before it can be addressed effectively (ie. coped with). Detection of a problem by an intellectual elite, such as analysts, is not sufficient to induce action, for the public does not reward long-range political decisions which prevent problems from arising, nor does it punish its politicians for short-sightedness.

As an example, consider the issue of birth control. For years, demographers had been vainly calling attention to the problem of over-population that they foresaw. Yet, in 1960, the President of the United States enunciated the policy that family planning was a private matter, and government would have no role in it; public funds were not to be used, even for research on fertility and birth control. Just ten years later, however, abortion was available virtually on demand in New York hospitals, paid for by public funds. What had happened in the interim was a raising of the public consciousness to the "world population bomb", the spectre of a "population explosion", and other such popularizations of what had hitherto been exclusively a concern of an obscure scientific community. The change in public awareness made possible this dramatic reversal of public policy.

As to our ability to anticipate and correct, it is important to note with humility that our predictive models for urban phenomena are poor, and are not likely to become dramatically better. It is difficult to forecast the effect of a disturbance on a system, and it is difficult to calculate the kind and quantity of anticipatory corrective action that should be taken to counteract that effect. The combination of this factor, and the preceding one, conspires to limit the applicability of feed-forward control to urban problems.

In other words, planning is of limited potential in this arena and we should not expect too much of it.

As a modest example, consider the following, which is also drawn from the New York experience. It was predicted that legislation to provide free medical care to low-income families would have a certain budgetary impact. The prediction was wrong, and the cost to the taxpayer turned out to be much greater than had been expected. The reason: planners assumed a relatively low enrollment rate, due to ignorance and apathy, but welfare rights groups, neighbourhood groups, and legal clinics in poverty areas were effective in making contact with, and educating, eligible patients, and helping them enroll in the programme.

In addition, actions in urban government often have quite unintended consequences. For example, consider the convoluted workings of New York City's civil service system, which was designed to hire the most meritorious job applicants into the public service. In fact, the process results in an _inverse_ merit system: the _lower_ an applicant scores on an entrance _examination_, the _more_ likely he or she is to be hired, and the _higher_ the score, the _less_ the probability of being hired! This perverse and unintended consequence comes about because of the long, drawn out, bureaucratic procedures that are built into the hiring and selection process to assure that the merit principle is followed scrupulously. The end result was a median delay of seven months between the time someone applied for a job and someone was hired to fill it. What happened was that the better people, who scored well, found jobs elsewhere, while those lower down were still available when the inexorable, but ponderous, mechanism of the "merit system" finally produced a job offer (Savas and Ginsburg, 1973).

The Myth of Technical Elegance

This eighth myth results in excessive value being accorded to technical elegance by systems analysts. This is a well-known perversion that afflicts many service professions, wherein the norms and values of the insiders come to dominate those of the public at large. The irrelevance of technical elegance is abundantly obvious to political leaders and government executives, if not yet to analysts.

The Myth of Decomposition

The ninth myth on this list derives from the fact that the analytic disciplines are very good at abstracting a problem from its natural state, dissecting it, idealizing it, applying simplifying assumptions, and then forcing it into a well-defined category. Unfortunately, the distinguishing characteristic of complex social systems, such as urban systems, is that they are aggregates with strong bonds between components, and with values playing major roles. Such systems incorporate -- totally and indistinguishably -- elements which historically have been segregated artificially into the man-made intellectual compartments that we call the fields of knowledge. Urban policy issues do not lend themselves readily to such decomposition, for they are robust and resistant to abstraction and fractionation.

The Myth of Irrational Politics

A common lament by urban systems analysts is that decision-making is too political, and not sufficiently rational. All this reveals is the analyst's failure to understand the objectives of the decision influencers. The implicit objective functions of the latter include terms, often unarticulated, that the analyst is either unaware of or unable to handle, and therefore, prefers to ignore.

LEVELS OF APPLICATION

It is useful to differentiate between two very different levels of application, the operational level and the institutional or strategic level. (Other, finer distinctions and sub-levels could be identified, but it is not necessary for the purpose here to do so.)

The operational level can be illustrated by an analysis to determine the appropriate replacement policy for fire trucks. Given certain well-known data on purchase costs, repair costs, repair rates, etc., one can readily utilize the standard tools of systems analysis to arrive at a reasonable technical conclusion, even an optimal one. Nevertheless, even in this case, there is no simple, universally acceptable, technical "solution". Vendors have a major stake in the outcome, and will attempt to influence it. The fire-man's association can also be expected to press for early replacement. On the other hand, political leaders campaigning on a platform of fiscal austerity are likely to favour a longer replacement cycle. Other agencies may see themselves as competing with the fire department for scarce resources, and will, no doubt, suggest better uses of the municipal funds. Thus, even at this simplest and purest technical level, the factors cited above come into play and complicate the analysis, or, more precisely, affect the policy outcome which the technical analysis was intended to inform.

At the broadest level of institutional or strategic analysis, profound societal issues are involved. For example, it has been argued that the relatively high crime rate in American cities is an inevitable consequence of the high status we accord individual, as opposed to group, rights, and of the weakness of our informal institutions (family, church, neighbours) in controlling deviant behaviour (Bayley, 1980). Reducing urban crime through gradual societal

change at this level is very different from a programme to reduce
crime by better allocation of police resources; it requires a very
different sort of analysis; one fraught with the difficulties, and
subject to all the dysfunctional myths cited above.

The difference between the different levels of systems analysis
can be further illustrated by reference to residential refuse collection. At the operational level, one can conduct an analysis to
choose the most economical vehicle for the job; one whose capacity
is properly matched to the route and the trip to the disposal site.
(If the vehicle is too large, there will not be enough working time
available to fill the truck completely, and hence, a smaller, less
expensive truck would have been adequate. Conversely, if the truck
is too small, much of its time will be spent travelling to and from
the disposal site, and therefore, the working time of its crew will
be squandered; the larger the crew, the more costly this option.)

At the institutional level, one can conduct an analysis to
determine whether eliminating the city agency and turning over the
entire function to a private, profit-making contractor is best;
there is impressive evidence from three different countries which,
indicates quite conclusively that substantial economies can generally be obtained by this action (Savas, 1979a).

Both of these approaches have the potential of improving this
service.   Clearly, the operational analysis requires more data,
lends itself more readily to modelling, is more complicated technically, and yields a relatively unambiguous answer.   On the other
hand, the question of contracting for service is enveloped in ideological issues, abounds in intangibles and incommensurables, is
perceived as a major threat by established bureaucracies, poses
transitional problems, and appears to be risky in that it requires a
new, untried relationship.   Technical analysis plays a relatively
minor role in this decision, and aside from some simple economic
calculations -- but not too simple (Savas, 1979b) -- there is no
need for formal models, elegant constructs, or extensive data collection.

The foregoing examples have been drawn from the subject area of
public services planning, but other examples could just as easily
have been drawn from physical network planning.   For instance,
deciding whether or not to build a bridge is a strategic question;
designing it is an operational one.

EDUCATING FOR URBAN SYSTEMS ANALYSIS

In urban systems analysis applied at the strategic level, one
may confront such fundamental issues as intergovernmental relations,
the appropriate degree of centralization and decentralization in
municipal government, the role of neighbourhoods vis-a-vis regional
governments, and the choice of delivery systems for providing

services: government, private contractors, the marketplace, franchises, grants, vouchers or voluntary citizen associations. It is evident that, at this level, the tools and training required by the systems analyst will differ greatly from those he needs to work effectively at the more technical operational level.

At the strategic level, there are other competing and complementary disciplines and perspectives that can be and are employed. To name but a few, they include law, political science, political economy, economics, sociology, organizational behaviour and psychology. These provide alternative decision-making paradigms, and it behooves the urban analyst to appreciate and be conversant with, if not fluent in, these approaches. Systems analysis should embrace the aspects of these fields which are particularly appropriate for urban policy and planning purposes. Analysts should be made sensitive to conflicting value judgements. The conventional education of systems analysts is sadly lacking in this respect.

As part of his training, the urban analyst should be made aware of his exposed role; he is a participant and not a detached observer. Thus, the traditional analyst, while comfortable with the abstract concepts of resource allocation, is frequently disconcerted when he finds himself caught up in the unfamiliar political turmoil of which group gets what, when and where. After all, this is what the allocation of public resources is all about.

Another role in which the urban analyst must become more comfortable through better training is that of a change agent. The analyst rarely recommends maintaining the status quo; almost invariably he recommends change. But at the strategic level in the urban policy arena, change means altering institutional relationships, work patterns, and the relative power of people. Such changes are bitterly opposed and difficult to effect. Not surprisingly, the role of change agent in this setting is a hazardous one. Socrates, an early analyst who questioned the established ways of thinking, was prosecuted by a public official, and given poison by a civil servant.

SUMMARY

A number of dysfunctional myths have the effect of obscuring the proper role of systems analysis in urban policy-making, and inhibiting its effective use for that purpose. These are the myths of:

1. Decision-making

2. Problem-solving

3. Optimality

4. The moon

5. Attributable achievement

6. Model-building

7. Planning

8. Technical elegance

9. Decomposition

10. Irrational politics

One must recognize that systems analysis can be applied at two very different levels: the operational level and the institutional or strategic level. The environments at these two levels differ greatly and require different tools, approaches and emphases. The education of urban systems analyst should stress these distinctions, and should familiarize the analyst with other complementary disciplines and viewpoints, such as those of law, political science, economics, sociology and psychology.

NOTE

1. For example, the model of world dynamics (Meadows, et al, 1972), while of limited scientific merit, was a very effective tool which successfully captured public attention and focused it on the issue of the limits to growth in a world of finite resources. In this respect, it has probably hastened the process of change — viz., societal acceptance of this newly-perceived condition and gradual public adjustment to it.

REFERENCES

Bayley, D.H., 1980, Ironies of American Law Enforcement, The Public Interest, 59, 45-56.

Kash, D.E., 1968, Research and Development of the University, Science, 160, 1313-1318.

Meadows, D., et al, 1972, The Limits to Growth, Universe Books, New York.

Savas, E.S., 1979a, Public vs. Private Refuse Collection: A Critical Review of the Evidence, Urban Analysis, 6, 1-13.

Savas, E.S., 1979b, How Much Do Government Services Really Cost?, Urban Affairs Quarterly, 15, 23-38.

Savas, E.S. and Ginsburg, S., 1973, The Civil Service: A Meritless System?, The Public Interest, 32, 70-85.

Wildavsky, A., 1974, The Politics of the Budgetary Process, Little, Brown and Company, Boston, Mass.

Savas, E. S., The Structure, Aid, etc., of a City: Organization of New York City, Public Interest, 22, 1-83.

Winkler, etc., T., The Science of Managing Organized Technology, Brace and Company, Boston, 1966.

REFLECTIONS ON AND IMPLICATIONS OF SYSTEMS ANALYSIS AS A

SOCIOLOGICAL PHENOMENON

Ida R. Hoos

Space Sciences Laboratory
University of California at Berkeley
Berkeley, CA, 94720
United States of America

"Here is Winnie-the-Pooh, coming down the stairs
now, bump, bump, bump, on the back of his head,
behind Christopher Robin. It is, as far as he
knows, the only way of coming downstairs, but
sometimes he feels that there really is another
way, if only he could stop bumping for a moment
and think of it."[1]

How we became Pooh-Bears, how this has affected life in the
latter half of the twentieth century, how this could shape the
future -- these are matters of sociological interest, for systems
analysis, its components and derivatives, are, in essence, a sociological phenomenon reflecting our Zeitgeist. In T.S. Kuhn's termi-
nology, they constitute our "dominant paradigm".[2] In order to
ascertain how this came about, we must understand the attribution
and continual re-attribution as well as the social process through
which re-affirmation of the methodology has occurred. Important,
and demanding review in the social context, are the essential elements of the methodology, especially as they emerge in and impinge
on present and future planning. Finally, we must consider the
social consequences of the continued domination of this technical
paradigm.

ATTRIBUTION OF THE METHODOLOGY

Anachronistic as it may seem to some of us[3] who tried to offer
foresight in the early days of the transplant of systems techniques
from military to peaceful pursuits and from outer space to earth-
bound planning, genealogical retrospective is still the customary

way of establishing credibility. It is remarkable that even the clarity of hindsight cannot dispel the aura of magic that has preserved the image of "precision" and "rationality" despite two decades of experience that should, at the very least, have shaken and undermined some of this confidence. That it has persisted is a manifestation of the Zeitgeist.[4] As Sorokin told us long ago, ours was to be a Technological Civilization; we were going to value the technical, the scientific, the "rational"[5], and the quantifiable. It was entirely consistent with the spirit of the times that operations research techniques, demonstrated to have won the Battle of Britain and won the war on the Potomac, imbued McNamara and his Rand band of systems analysts with attributes of managerial know-how so ubiquitously useful in both war and peace.

Noteworthy in the process of accreditation was the linking of two quite separate entities: the managerial techniques and a "delivery system". The techniques, which were a bastard offspring of economics and engineering, were quickly adopted by some segments of both parent disciplines. The idea of the delivery system was the "think tank" and carried much appeal, stemming from an unarticulated respect for the arcane activities at Rand, as well as a gallimaufry of inchoate longings for interdisciplinary approaches. The association between systems analysis and think tanks is historical rather than functional, but it is nonetheless real. The centrality of the "biggest bang for the buck" as an element in systems analysis was the occasion for re-attribution of a technique long known but, like the prose of Moliere's bourgeois gentilhomme, little honoured until it was given a name (cost-benefit analysis).[6] This authority laden context strengthened the accreditation process. Later to offset the dangers of denouement posed by Senator Proxmire's disenchantment with Department of Defense managerial practices came NASA's space spectaculars. The mighty Apollo programme's successes were a tribute to management know-how in general and systems analysis in particular.

The systems approach was acclaimed as "the way to go". And in the subsequent era of retrenchment in both military and aerospace activity, "systems capability" became the vehicle for diversification and deployment of personnel. A slogan was born: "The nation that can put a man on the moon can ..." get Dad to work on time, clean up its air, clear its streets of crime, keep people healthy. No task was beyond the skills of the systems analysts; their chores were given names, among them social engineering, public technology, civil technology. Accompanying events gave an urgency to the process of accreditation in the social milieu. A war-weary country sought diversion of effort to peaceable pursuits. Cutbacks in major industries were causing severe unemployment among engineers. Harnessing this presumed reservoir of management talent and putting it to work "to improve governance" sounded convincing and was often exploited as a plank in the platform of presidential and gubernatorial campaigns. Rational planning, scientific management, and a systems approach won ready acceptance in semantic juxtaposition with

bureaucratic bumbling, horse-and-buggy methods, and piece-meal frag-
mentation, quite irrespective of the performance of the former in
actual practice.

Not to be overlooked as a source of reaffirmation of the technical paradigm is, strange to relate, the education establishment.
In some ways, this may be understandable in that the education
system is both cause and consequence of the society in which it
exists. However, imposition on education planning of the industrial
model and assiduous adherence to the management techniques have had
severely adverse effects on an institution that might once have been
expected to influence rather than be influenced by current fashions. Systems analysis and its components are taught in almost
every department of institutions of higher learning. Whatever curriculum one may scrutinize -- engineering, education, business
administration, public policy, environmental design -- one finds
that this is the method. In many disciplines, the PhD dissertation
is little more than a cost-benefit exercise in which numbers are
gathered and juggled to prove a point. The degree then becomes the
stepping-stone in the Gradus ad Parnassum to a career devoted to
doing more of the same. For persons schooled in a particular way,
ie. those who learned the fundamentals of their field from the same
concrete models, subsequent experience will seldom evoke overt
disagreement over principles.[7]

Most students must eventually compete in a job market, and
because shrinking opportunity has made them vocation-minded, the
"practical techniques" are popular. And everyone involved seems to
benefit. Those who teach the courses can claim enrollment numbers
in their fight for survival in academe; the graduates acquire a tool
that makes their vitae attractive to a prospective employer. And so
along with the affirmation process, there is self-perpetuation.

The Pentagon has proved to be no paragon and systems techniques, in their various manifestations, have actually been pinpointed as factors in gross mismanagement of the military enterprise. The kind of systems management performed in space exploration has been shown to have little relevance to social planning.
Millions of taxpayers' dollars have been spent on systems studies
that only add to the saga of unfulfilled promise.[8] And yet the
systems syndrome prevails. This should come as no surprise. The
social framework is one in which the dominant paradigm not only
condones but dictates the technical definition. All problems can be
"managed"; management "science" provides the know-how. The
"rationality" perceived in a society like ours rationalizes these
approaches. Winnie-the-Pooh, at the bottom of the stairs, thought
that there might be another way of coming down, but he was so busy
bumping that he could not think of it. Perhaps we are more like
this hapless bear than we care to acknowledge. We may have been
bumped into a condition resembling a hypnotic state. Psychologists
tell us that this is one in which heightened suggestibility prevails, where there is highly focused attention, where peripheral

distractions are blocked out, and where only selected events seem
real. We exclude the others from consciousness. A psychiatrist
tells us that the trap of "group think"[9] occurs whenever circumstances promote concurrence-seeking.

ESSENTIAL ELEMENTS OF THE METHODOLOGY

While manifestations of the methodology are numerous and
ubiquitous, and the name of the game changes with the times, the
core elements remain much the same, as do the game-players. Where
once we had a plethora of cost-benefit analyses, we now see environmental impact statements, technology assessments, and risk analyses. For all of them, the model is essentially the same. The analyst defines the system; this is the universe that he will address.
Anything else is, by definition, ruled out. Thus, a Battelle study
of the impacts of a hypothetical nuclear waste dump could properly
collect numbers relating to the "boom town" aspects and systematically and systemically ignore the difference between a nuclear waste
depository and a Disneyland. (In the context of this kind of analysis, the "boom town" is used as a whipping boy. In other contexts,
as when the Mt. St. Helen's volcano has brought hordes of tourists,
the boom aspects are counted as a blessing.)

The element of cost-benefit analysis, sometimes translated and
somewhat re-cast into cost-effectiveness terms, is inherent no matter what the game is named. This, of course, implies trade-off,
which, according to the rules of the game, requires calculation by
the same measure -- mainly monetary. Here we still find evidence of
the contribution to the state-of-the-art by Hitch, who, as Assistant
Secretary of Defense under McNamara, legitimized the already existing liaison between engineering and economics. Defining the problem
of national security as "one big economic problem", with strategy
and cost "as interdependent as the front and rear sights of a
rifle",[10] Hitch designated economics as the dominant desideratum in
systems analysis, cost-benefit analysis, programme budgeting,
environmental impact assessment, and all the mutations that
followed. It is apparent, as we review developments over the past
several decades, that economics became the imperialist of the social
sciences. As a consequence, early aspirations toward interdisci-
plinarity were submerged. All systems analyses were dipped in the
same vat and came out the same colour. All manner of problems were
reduced to numbers; the only "rational" view was that emanating from
an econometric model, however remote from reality.[11]

The information base remains an essential ingredient. In fact,
much activity is devoted to gathering data to fit the model, for
validation in the application of these techniques is a curiously
incestuous affair. The data drive the model and the model specifies
the data. In Kuhn's words, the model provides "a context of justification".[12] This should indicate that, contrary to claims about
the "rationality" and "objectivity" of these approaches, as con-

trasted, presumably, with irrationality and subjectivity, the methods prescribe that information be gathered for a purpose. The result is inevitably a drunkard's search.[13] Since one rarely finds matters relating to environmental impacts, future effects of technology, or risk analyses to be free of controversy, there is usually an approved position, not always made explicit. This has an architectonic thrust, for the definition of the system, its parameters, the costs, the benefits, the trade-offs — in a word, the model — are virtually <u>givens</u>. The arrow has been shot; all that is left is for someone to paint the bull's eye on the target. This is done by eclectic use of data. Those that fit the model are taken into account. The rest do not count. Advocacy parades in the trappings of analysis.

Information gathered for a purpose becomes a case of inviting foxes into the henhouse when a government agency contracts out a study, such as one having to do with regulation of pollution, to an industry that has a vested interest in the outcome. That this is far from uncommon practice suggests that here is a situation where the "technical study" produces not guidelines but a smoke screen for egregious violation of the tenets of public interest. Justification for contracting out can be attributed directly to the anxiety of government agencies to improve their own cost-effectiveness ratio by hiring outsiders to perform as many of their mandated duties as possible. The use of consultants as a way of circumventing personnel ceilings and procedures has been a matter of considerable concern and the object of extensive congressional inquiry.[14] The incidence of abuse has nowhere been made as explicit as in a series of articles in the <u>Washington Post</u>.[15] Citing widespread waste, conflict of interest, the revolving door between public policy and private industry, and procurement favours, the investigative reporters provide documentation in abundance.

For our purposes, it is important to note the mechanism that encourages public agencies to turn to affected industries for information that will be used in formulating policy, programmes, and regulations. Justification stems largely from the management syndrome that has pervaded every level of government, from township to the White House. The watchword is "cost-effective operation". Translated into action, this still means "the biggest bang for the buck". Thus, by some calculus, it has been decided that the United States saves money by hiring outsiders to do its work. To save even more money, facility of screening and of managing contractors are criteria in the selection process. Organizations known to deliver in quantity and on time are likely to receive preference. If quality considerations enter in, it is more by chance than by design. Part of the customary "boiler plate" is the roster of "experts" who will perform the work. And here convention favours private industry or the think tanks which regularly service it, for where better to seek a specialized and ready reservoir of brainpower?

With objectivity and rationality the most frequently-flaunted attributes of the technical paradigm and of the methodology it promulgates, it is important that we scrutinize the role of technical experts in a matter of superordinate concern to our society, more or less developed, urban and rural, present and future -- namely, energy. In the final analysis, here are the decisions that will determine the quality of life on this planet. The experts who provide the information and the methods which they use have enormous bearing on the kinds of conclusions reached. When the paradigm is technical, ascription of expertness goes almost unquestioningly to the scientific-technical community. Where there is controversy, it is the approved position which is most likely to be vindicated through the intrinsically incestuous process of modelling. The use of systems techniques in the analysis of energy options provides us with a telling example of the way the elements set forth in this section of the paper interact to determine the course of public decision-making.

"Facts" are the key; "experts" gather and interpret them in accordance with their model's requirements; the methodology allows, indeed calls for, enumeration, calculation, and a balancing of hazard against benefit always with economic considerations uppermost. Whose conception of economics, whether in the short or the long run, who pays the costs, who gets the benefits are questions never addressed. The goals have been presented; the arbitrary focus condones a tolerance for incongruity that other, more realistic, approaches would not allow to go unchallenged. The "nuclear option" is a case-in-point. When the course of atomic power was dictated by the Atomic Energy Commission, the game plan was to promote and preserve the image of the safety of the silent-atom-peacefully-at-work providing so much cheap energy safely that one day the metering of it would become an anachronism. In order that this worthy goal be achieved, only such information as fostered the notion of safety was allowed to reach the public. The reassurance emanating from the AEC constituted a strong determinant of the public's perception of the risks associated with nuclear energy. That this perception markedly influenced acceptance of nuclear energy was a circularity not unnoticed by its advocates. The objective in the dissemination of information, whether about reactors at Three Mile Island or the movement of spent fuel through New York City, has always been to make the public feel safe, whether they were or not. This accounts for the standard response to all incidents involving radioactive releases, large or small. "The levels of radiation do not exceed safety standards" is contrived more to assuage public fear than to protect it from harm. In 1958, Lauriston S. Taylor (then Chief, Atomic and Radiation Physics Division, National Bureau of Standards) made the statement, "aside from medical applications, no radiation is good for man".[16] But by 1977, the possible deleterious effects of medical x-ray were causing serious concern.[17] "Safe standards" have shifted over time; the more we learn, the less reason we have for sanguine complacency.

When cracks in the facade of certainty about nuclear safety were revealed by the opening of secret AEC files,[18] the game plan changed. The strategy of the Department of Energy and the Nuclear Regulatory Agency, lineal descendents of the AEC, was to admit the inevitability of uncertainty. The logic then went this way: life itself is uncertain. But, since uncertainty is a reality, we need only manage this uncertainty. And for this, there are techniques. Moreover, there are experts, specialists in the methods presumed to provide "rational" assessments of the probabilities of danger. While these methodological Merlins could not be expected to recapture the fool's paradise of ignorance that had preserved the image of the atom as a cheap, clean, and safe source of energy, they could supply seemingly foolproof models that would make nuclear power seem cheaper, cleaner, and safer than other sources. Their "rationality" was given credibility because it rested on "hard", ie. quantified, data — about comparative costs and benefits. But the fact is that the data were gathered for a pre-determined purpose; the authoritative sources of information were in a high-placed but not neutral position. And it was the pro-nuclear view that was inculcated in the calculations which led to conclusions designed to influence public opinion and policy.

The methodology used in the managing of uncertainty is a familiar one, grafted from military and space roots. It is, in essence, a transfer of systems analysis from the present to the future tense. Under the name of fault-tree analysis, the techniques were used to estimate the probability of a variety of accidents that might occur in the operation of nuclear reactors.[19]

A word of description might be in order at this point.[20] A tree-like schematic design is used to depict relationships within the system. The typical branch point would have a safety system either responding or failing to respond. The "safety system fails" branch might then lead to a branch point for "plant evacuation alarm" which would not sound. The probability of a "pathway to disaster", the chain of events in which everything goes wrong, is computed by considering the probability of each of its constituents' failures. The risk associated with a pathway is then determined by multiplying its probability of occurrence by the magnitude of its consequences. The sum of the risks associated with each pathway represents an estimate of the total for the systems.

Professor Norman Rasmussen of MIT and his team of technical experts employed decision-tree and fault-tree analysis, "based on judgement of risk at each step of a specific sequence of events leading to an accident". The conclusions were that "the actuarial risks (sums of the probabilities of consequences multiplied by the severity of consequences) are very small" and that "the chances of severe accidents that would cause large numbers of casualties are extremely small — so small as to be within the range of risks we hardly deign to consider".[21] Using this method of risk calculation, technical experts found that the health risks associated with the Three Mile Island reactor in Pennsylvania on March 28, 1979 were "negligible". In their view, "a single automobile accident during

the evacuation would have done more damage".[22]   Even though subsequent review[23] found the WASH-1400 probability estimate low, perhaps
by a factor of as much as 500, and the expected number of cancers
for a given accident several times higher, the methodology emerged
unscathed.   Professor Harold Lewis had this comment at the time
(January, 1979) that, due to his critical review conducted under
auspices of the American Physical Society, the Nuclear Regulatory
Commission revoked its endorsement of WASH-1400's findings:[24]

> ....they tried to find numbers with greater pre-
> cision than the data base available, the informa-
> tion available, and the statistical tools that
> they had available would permit. So they came
> down with results whose precision was over-
> stated. We commended them for a good try, and we
> also commended the methodology. (my emphasis)

Lewis's criticism is one most frequently levelled at the
Rasmussen Safety Report. But it has not undermined the authorita-
tiveness of that document, which remains a Bible in the hands of
proponents of nuclear energy. Notable as an example of and factor
in the perpetuation of the faith in fault-tree techniques and their
"proof" of the safety of nuclear reactors is the honourable mention
of them in high and purportedly neutral councils, by high and presumably neutral authorities. Thus, we find a publication by the
National Academy of Sciences of the Proceedings of the National
Academy of Engineering's 1979 Technical Session on "The Outlook for
Nuclear Power".[25]   The Academy "felt compelled" to publish and
distribute the presentations, "which are intended to contribute to a
better public understanding of the problems and promise of nuclear
power". Five of the six papers were distinctly pro-nuclear. One of
them, by Harvey Brooks, addressed "future nuclear systems technology",[26] and came to an all-systems-go conclusion, significant
because of the assurance supplied by the Safety Report, and in spite
of the official NRC revocation ten months earlier. Brooks sounded
an optimistic note about prospects for improvement on the safety,
health, and environmental effects of nuclear power. He foresaw
"improvements in design" and "reduction in the width of the uncertainty band" and attributed both to the techniques of calculation.
"The greatest value of the fault-tree methodology developed in the
reactor safety study lies in its capacity to identify priorities for
improvements in safety through pinpointing the most likely accident
sequences and concentrating design improvements on them."[27]

It is the certainty of conclusions about uncertain questions
that should concern us. More than a function of style and volume,
it is a methodological matter and one that must be addressed
because, just as in the case of the parent methodology of systems
analysis, there is an aura of emperor's new clothes about the whole
business. Of prime importance is the fact that the honest use of
the techniques is as dangerous as their abuse, what with their

potential for incorporating and obscuring bias and providing validation for predetermined conclusions. A nice bit of irony resides in
the fact that theirs are the conclusions that are "scientific" and
"rational", while others, which do not fit this model, are "emotional" and "irrational". The latter are, according to official policy,
impediments to achieving "rational energy policy", and must somehow
be overcome. It is by this calculus that hazards at the Three Mile
Island plant are dismissed as trivial, even though the full extent
of the danger to human beings is not yet known.

Basic to the theory underlying tree analysis is the Bayesian
theorem, an abstract statistical principle facetiously called the
"Bayesian haze",[28] otherwise known as "the equi-probability of the
unknown criterion". This states that where there is no firm information available about relative probabilities, we must assign equal
probabilities in the calculations, and then adopt a particular strategy. Such an essentially subjective creation of probability values
is an approach that is arbitrary and likely to be based on personal
preference or intuition. Noteworthy in this connection is the
revived interest by social psychologists[29] in the way personality
differences function as a subtle element in these projections.
Presence or lack of skepticism is a quality long known to influence
perception.[30] Cautiousness has similarly been pinpointed as an
"anchoring form of bias".[31]

Methodological superstructure obscures the essential subjectivity of the Bayesian premise and, in fact, lends an aura of logic
and precision. But the Bayesian foundation supports what may be
a seriously erroneous assumption, viz. that all possibilities are
equally probable, and thus, that any starting point for the analysis
is allowable. The natural inclination on the part of the analyst is
to follow his own drunkard's search. A perilous fallacy lies in
overlooking the point, perhaps more socially than statistically
significant, that for every variable there is a distribution. And
every distribution has its extreme, as well as its mean values.
Thus, a contingency that might be calculated as having low probability can and, in the case of Three Mile Island, did occur. As to its
consequences, they may be great or trivial, depending on which
expert one chooses to believe. Applied to significant policy
issues, the tree approach has a fatal weakness in that the aspects
of the world ignored are usually equally or more important to determining how the process evolves than those parts included in the
analysis.[32]

The cost-benefit element, so gemuetlich to econometric modelling, plays a strong role at all stages of energy decision-making.
It is clearly apparent in the way the "nuclear option" has been
rationalized as a desirable one. The starting point is often embedded in the "energy crisis" which has the United States paying up to
$200 billion for oil in 1995, and a global inflation rate predicted
to continue indefinitely.[33] At best, the US can expect continued
disruptions in oil supplies and price-manipulation from suppliers

abroad and at home. Supply-demand projections bolster the nuclear case; we need the energy, and the atom, while no longer universally regarded as cheap and safe, comes out in some cost-benefit analyses looking cheaper and safer than other energy sources. The benefit balance improves as the disamenities of other sources are subjected to concerted aggregation. Thus, strategy among nuclear proponents is shifting from making the case for nuclear power to making one against other kinds. The environment, cavalierly omitted in calculations at the various stages of the nuclear cycle, becomes an object of great concern in the coal scenario. Here, carbon dioxide in the atmosphere, boom towns in the hinterlands, and black lung in the miners contribute to a necromantic conclusion, while the constant and repeated insults to the environment, the mining towns built with uranium tailings, and the health of uranium miners do not appear in the calculations.

The cost-benefit notion is insinuated into the process of trade-off, in which the nuclear option is made to appear good just because other demonstrably dangerous activities are acceptable to the American public. The hazards of exposure to ionizing radiation are shown to be minimal when compared to other routine risks. Gleefully recounted are the foibles of a people willing to expose themselves to all manner of hazards. Statistics especially favoured have to do with death on the highways or the number of cigarettes smoked per year. The "benefits" of nuclear energy are generally couched in terms of energy availability. And so, in the international context, there is a new version of the old white-man's-burden argument. America will presumably win friends and influence the less developed nations if we can supply them with food and other means of sustenance. Their well-being is tied to American prosperity, in which energy is a vital ingredient. The power-for-progress notion is, of course, related to that of power-for-peace, since a depressed state of the US economy undermines its world position. This line of argument fortifies the benefit estimates and, coincidentally, conveys the sly implication that an anti-nuclear posture is both anti-humanitarian and anti-American.

Slid with subtlety into the cost-benefit equation is the debt owed the piper. Through invocation of the inexorable truth that there is no such thing as a free lunch, the public is reminded that it has enjoyed the advantages of nuclear power and, therefore, must face the costs. This is the rationale that has been used to counter resistance to having radioactive materials transported through cities or waste repositories placed in certain locales. New York City, which tried to ban shipments of spent fuels, and California, which has tried to prevent off-site-storage of wastes within the state, have been accused of selfishness for "refusing their share of a national responsibility". More often than not, this ploy to transform the benefit into a kind of guilt and imbue it with a moral obligation makes much of the medical wastes from laboratories and hospitals, in their argument of their service to mankind. The latter may be over-estimated; the amount is insignificant, compared with the quantities being generated by commercial reactors and in defense-related activities, about which the public knows nothing. There is a sad parallel observed in the case of the Italian village, Seveso. The inhabitants were faced with terrible illness and forced to evacuate their homes because of an explosion that had contaminated the countryside. They were burdened with the realization that they had been willing beneficiaries of the chemical plant's operation. They had, in effect, been party to the poisoning, because for years they had brought their sick and dead chickens to the back door of the factory and been paid off in thousands of lire, many times the worth of the poultry.[34] The message is clear: prosperity is dear. We all enjoy it and someone must be made to accept the costs.

SOCIAL CONSEQUENCES OF THE TECHNICAL PARADIGM

Why, a reader may ask, is it appropriate to concern ourselves with such applications as these in a symposium on systems analysis in urban policy and planning? The answer should serve as a two-alarm warning. In the first place, persistent and uncritical use has contributed to the entrenchment of the techniques as the "powerful tools for rational decision-making", ubiquitously applicable. They are already part of, and party to, an emerging post-industrial society ideology which should be of vital interest to us. Ominous as it may seem in some of its present dimensions, this mental set is even more portentous in its implications for the future. As long as a decade ago, Brzezinski[35] foresaw interesting political possibilities. It was to post-industrial politics that Huntington devoted his attention. He viewed it as "the darker side",[36] having hinted elsewhere that more authoritarian and less democratic modes of government might be desirable for meeting the challenge of Communist threat, inflation, unemployment, shortage, and frustrated aspiration.[37]

It might be well for us to remember McNamara's unequivocal assertion that these tools serve the strong master. Used in public decision-making, they do not _inform_ the process so much as they _confirm_ it. In fact, Gross's thesis that indicators are, in fact, vindicators and that manipulation of them is a manifestation of what he describes as "friendly facism" deserves a place in our warning system.[38]

Where, as is more and more often the case as we move into the greater interdependence, uncertainties and complexities of the post-industrial era, decisions involve science, technology and politics, technology and science become increasingly the tools of politics. Already to a degree, and ultimately overall, these will require the imprimatur of officialdom. We see this happening in the current allocation of research money and support of selected institutions. Those who toe the invisible line are bound to prosper.

The second alarm to be sounded comes from our observation of the way the methodology has been used and found useful, with respect to energy. The decisions being made will have a profound effect on the shape of our society and on every facet of our existence. We all live near Three Mile Islands. Whether we reside in city or country, radioactive materials by the ton are being carried past our doors. Whatever our location on this planet, outer space is being scanned through the lens of cost-benefit analysis as the home for transuranic wastes. In all of these matters, the pro-nuclear cause has been cloaked in "rational", "technical" studies; in all of them, the consequences are becoming clear. Fault-tree analysis of reactor safety yielded underestimates of danger; this encouraged improper operating procedures and inadequate regulation. Limited "safety models" emboldened federal agencies to ignore local protest and preempt states' rights in the shipping of spent reactor fuel through towns and cities from coast to coast. The relative benevolence of radioactivity having been established through accommodating research, disposition of wastes in land, sea or air has been adjudged "technically feasible". A major problem associated with nuclear power thus "solved", the nuclear "option" becomes an overpowering reality. In all of these considerations, the element of coercion associated with the decision-making apparatus is one that we overlook with dire consequence.

There is a distinctly techno-global flavour to the kind of thinking associated with the systems approach. It creates a kind of surrealism that divorces past from present from future, cause from consequence, design from implementation, and act from responsibility. Besides being inherently anti-democratic, the methodology is inherently anti-intellectual. It does not open inquiry; it forecloses it. It provides clean and simplistic solutions to problems so messy and complex that they have engaged the mind of man since Biblical days. But now they are being laid to rest in a Procrustean bed. The portions that are discarded because they do not fit lie festering in the dark of both history and the future. As the seed bed for upheaval and revolution, they will make their presence known quite irrespective of some technically prescribed and sanctioned model.

NOTES AND REFERENCES

1. Milne, A.A., 1948, Winnie-the-Pooh, E.P. Dutton and Company, New York.

2. Kuhn, T.S., 1962, The Structure of Scientific Revolutions, University of Chicago Press, Chicago, Illinois, defines a paradigm as "an accepted model or pattern", page 23. The dominant paradigm is the constellation of beliefs, values and techniques shared by the members of a given community.

3. Boguslaw, R., 1965, The New Utopians, Prentice-Hall, Englewood Cliffs, New Jersey.

    Hoos, I.R., 1972, Systems Analysis in Public Policy: A Critique, University of California Press, Berkeley, California.

4. Sorokin, P.A., 1957, Social and Cultural Dynamics, Porter Sargent, Boston, Mass.

5. The notion of the "rationality" of technique persists. "Techniques of Rational Planning" is the title of a review by Milch, J., June 27, 1980, in Science, Page 1449, of a book on the role of forecasting in public policy: Whiston, T. (Ed.), 1979, The Uses and Abuses of Forecasting, Holmes and Meier, New York.

6. The Flood Act of 1936 had stated the requirement that the costs of Federal projects were not to exceed benefits.

7. Kuhn, T.S., op. cit.

8. Lilienfeld, R., 1978, The Rise of Systems Theory, John Wiley and Sons, New York.

9. Janis, I.L., 1972, Victims of Group Think, Houghton Mifflin Company, Boston, Mass.

10. Hitch, C.J. and McKean, R.N., 1961, The Economics of Defense in the Nuclear Age, Harvard University Press, Cambridge, Mass.

11. See, for example, discussion of a proposed London airport in Adams, J.G.U., 1972, You're Never Alone with Schizophrenia, Industrial Marketing Management, 4, 441-447.

12. Kuhn, T.S., op. cit., pages 8-9.

13. Kaplan, A., 1964, The Conduct of Inquiry, Chandler, San Francisco, page 51, tells the story of a drunkard, hunting under a lamppost for keys dropped some distance away. When asked why he does not look where he lost them, he replies, "Here is where I can see".

14. Contracting Out, Hearings Before the Subcommittee on Human Resources of the Committee on Post Office and Civil Service, House of Representatives, Ninety-Sixth Congress, September 7, 11, 20 and October 4, 1979.

15. Neumann, J. and Grip, T., An Epidemic of Waste in U.S. Consulting, Research, The Washington Post, June 22, 1980; The Regulators: A Study in Conflict, The Washington Post, June 23, 1980.

16. Taylor, L.S., 1958, Radiation Exposure as a Reasonable Calcul-able Risk, Health Physics, 1, 63.

17. Bross, I.D.J., Ball, M. and Falen, S., 1979, A Dosage Response Curve for the One Rad Range: Adult Risks from Diagnostic Radia-tion, American Journal of Public Health, 69, 130-136.

18. Pollard, R.D., January, 1979, (Ed.), The Nugget File, Union of Concerned Scientists, Cambridge, Mass.

19. US Nuclear Regulatory Commission, 1975, Reactor Safety Study, WASH-1400 or NUREG-75-014, Washington, DC.

20. I am indebted to Baruch Fischhoff for this capsule explanation taken from his article, Cost-Benefit Analysis and the Art of Motorcycle Maintenance, Policy Sciences, 8, 177-202, (1977).

21. National Academy of Sciences, 1979, Energy in Transition 1985-2010, Final Report of the Committee on Nuclear and Alternative Energy Systems, National Research Council, W.H. Freeman and Company, San Francisco, California.

22. Ibid.

23. Panofsky, W.K.H., (Director, Stanford Linear Accelerator Center), Testimony before Subcommittee on Energy and the Envi-ronment, Committee on Interior and Insular Affairs, Washington, D.C., June 11, 1976.

24. Lewis, H., January 23, 1979, Nuclear Reactors: How Safe?, The MacNeil/Lehrer Report.

25. Perkins, C.D., (President, National Academy of Engineering), Foreword, The Outlook for Nuclear Power, Presentation at the Technical Session of the Annual Meeting, November 1, 1979, The National Academy of Sciences, Washington, DC.

26. Brooks, H., (Professor of Technology and Public Policy, Harvard University), November 1, 1979, The National Academy of Sciences, Washington, DC.

27. Ibid., pages 72-73.

28. Feinstein, A.R., 1977, The Haze of Bayes, The Aerial Palaces of Decision Analysis, and the Computerized Ouija Board, Clinical Pharmacology and Therapeutics, 21, 482-496.

29. Loye, D., 1980, Personality and Prediction, Technological Fore-casting and Social Change, 16, 93-104.

30. MacGregor, S., 1938, The Major Determinants of the Prediction of Social Events, Journal of Abnormal and Social Psychology, 33, 179-204.

31. Armstrong, J.S., 1978, Long-Range Forecasting: From Crystal Ball to Computer, Wiley-Interscience, New York.

32. Taylor, V., 1979, Subjectivity and Science: A Correspondence About Belief, Technology Review, 81, 49-57.

33. The Petro Crash of the 80's, Business Week, November 19, 1979.

34. Whiteside, T., 1978, Contaminated, The New Yorker, September 4, 34-81.

35. Brzezinski, Z., 1970, Between Two Ages: America's Role in the Technetronic Era, Viking Press, New York.

36. Huntington, S., 1974, Post-Industrial Politics: How Benign Will It Be?, Comparative Politics, 6, 163-191.

37. Crozier, M., Huntingdon, S. and Watanuki, J., 1975, The Crisis of Democracy, New York University Press, New York.

38. Gross, B., 1980, Friendly Fascism, M. Evans and Company, New York.

MODELS, METAPHORS AND THE STATE OF KNOWLEDGE

Sam Cole

Science Policy Research Institute
University of Sussex
Brighton, BN1 9RF
United Kingdom

INTRODUCTION

Policy analysts are often troubled by the lack of attention paid to their work and to the harsh critique which it sometimes encounters. This is especially true of systems modellers and those making use of the more sophisticated analytic techniques of the policy sciences. Several of the papers from the present seminar on Systems Analysis in Urban Policy and Planning indicate the doldrums, and even disrepute that modellers feel. In the present paper I shall explore this issue and ask to what extent the concern is legitimate, or whether it is not in some ways misplaced. I will argue, for example, that criticism is inevitable, given the social role of policy-oriented systems research. I will be concerned most directly with the use of large-scale models, drawing on my own experience in long-range forecasting and urban planning research.

Despite the concern with the use and misuse of analytic techniques, I do not want merely to provide a chronicle of horror stories. Perhaps because of this, I will not offer a strong "pro" or "anti" sentiment as my conclusion. As a substitute, at the end of the paper I offer, as a metaphor for the future of modelling, the history of photography as a descriptive art.

USES AND ABUSES OF MODELS

Recurring debate among modellers, centres upon the "usefulness" of their efforts. What contributes a "use" or "misuse" is not easy to define. Clearly one person's use of models to further knowledge or fulfil his own self-interest, may not necessarily contribute to social good. We might accept the furtherance of human knowledge for its own sake as something close to an unquestionable good, although many people would disagree and most people would accept counter-examples. But more particularly, as many people have pointed out a model may convey a spurious sense of authenticity which exaggerates its true value. This may be unintentional on the part of its author or a deliberate attempt to mislead. Whichever, the result may be the same although the latter clearly comes closer to any moral definitions of misuse than the first. The point here, however, is not to lead into a discourse in moral philosophy but to indicate my difficulty with the topic. Clearly in any practical situation the criss-crossing of self-interests and seeking for clarification is more complicated than simplistic moralistic arguments might allow.

The moral issue is not, of course, confined to social modellers although the impossibility of separating the various functions which scientific activities have in society becomes especially acute there. To provide an extreme example of the conflict of objectives we need only to look at Einstein's famous equation between mass and energy, $E = mc^2$. Arguably this has been the most significant break-through in human understanding of the century. The model has indeed been used and its possible applications viewed with euphoria, such as the promise of unlimited supplies of cheap energy, or the dismay at the threat of nuclear holocaust. Nuclear physicists themselves have reflected all these concerns even though their initial task was innocently directed towards seeking a better understanding of nature.[1] However, historians in the future ultimately see this gain in human knowledge, the example demonstrated clearly the need to distinguish use, in the sense of application, from usefulness, in the sense of being of lasting social value. Certainly we cannot merely be content to consider that simply because some government department chooses to "use" the results of a policy analysis that this deems it "useful".

I shall escape the dilemma of defining what constitutes "use-fulness" in a model by accepting some measure of reality and adding to this equation a measure of moral overtone. "Reality" about the context of modelling in this seminar, ie. the planning process, is well described by the following quotation from an anonymous French Planner,[2] "The drawing up of the Plan is an extraordinarily complex affair compounded of good intentions, compromise, playing with words, guile, wariness, arbitrariness, imagination, good nature, misunderstood agreements, faith and base interest". This captures well the essence of the planning process and indeed many other social activities, and emphasizes the need to escape oversimplified definitions of "use" and "misuse".

In this context of planning, "large-scale" models especially provide a useful focus of discussion for the present paper. They display, on the one hand, the attempt to incorporate sophisticated analytic tools; a systematic approach, mathematical and statistical techniques and large computing facilities. On the other hand, it is exactly the apparent over-sophistication and complexity of these models which draws criticism. For this paper, I want to give a particular angle to the term "large-scale". In the context of systems theory we are often concerned with mathematical models. It is common, therefore, to define "large-scale" models as being those which require a computer for their solution, as opposed to, say, models which have simple algebraic solutions. Thus we could differentiate simulation models from analytic models.

With simulation models one is generally trying to get as close as possible to reality, even if this means that the statistical significance of representation will be reduced by the additional parameters. With analytic models one is usually trying to obtain general, if unquantified, descriptions with as few parameters as possible. But for the present purposes this is not the meaning I wish to attribute to the term "large-scale" — rather I would like it to stand for the idea of "overwhelming". Thus the definition is directly related to the complexity, sophistication, the use and pos-sibly the authenticity of the model. We should be careful here, because a truly mind-blowing model may be extremely simple, for example, Einstein's equation has had fantastic repercussions in more than one sense, and, as far as our present knowledge extends, it provides authentic pictures of the real world. To use the idea of an overwhelming model links well, however, to the other part of my story and the idea of use and misuse of models. I want to consider models as part of a collective experience which is part of the on-going negotiations between actors in societies.

With this overture I now wish to recount a few personal exper-iences with modelling and forecasting. Most are based on my own, and my colleagues, critiques of other authors' works. The purpose here, however, is not to revive old arguments or add fresh critique, but to review the process of modelling as part of the individual's and society's attempt to gain greater understanding and control over its destiny and its environment.

MODELS OF DOOM

A useful example to begin with is the debate between our own forecasting group as Sussex and the MIT Systems Dynamics Group and their sponsors, the Club of Rome. This is an especially useful example, since when the catastrophe prone "The Limits to Growth" model was published,[3] the debate about "growth" and "no growth" was relatively simple. The issues appeared clear-cut and, more impor-tantly, the debate was not as institutionalized, as for example, is certainly the case with policy-oriented national econometric model-ling exercises or many urban planning models. The "Limits to Growth" model was clearly "mind-blowing". Few people can have failed to be impressed with the powerful authenticity of its spag-hetti flow-diagram which, together with the well-publicized computer output spelled the end of civilization. What the authors argued on

the basis of the model was, that if world development continues
along its present path, the world population and economy will suffer
a rapid decline sometime in the next century either through lack of
resources, foodstuffs or over-pollution. Analytically the model is
one of exponential growth coming up against fixed or only geometric-
ally expanding constraints -- exactly as the similarly controversial
model of Thomas Malthus, put forward long before the advent of
computers.

There are two points about this. First the model was not
really very complex -- most people could understand the underlying
assumptions even if they did not choose to believe them. For many,
however, the model confirmed a conviction that "environmentalists"
such as Paul and Anne Erlich, were right. The model merely appeared
complicated, and hence, authentic. Perhaps it was intended to
appear that way. I have had private discussions with the author of
the model -- Dennis Meadows, and he tells me that at one point he
even considered omitting the computer exercise from their book. Had
they done so, however, it is doubtful whether the book would have
its undoubted impact or sold over two million copies.

If we look at the origins of the exercise, we see it is very
unlikely that the model would not have been used. The reason for
this is to be found in the writings of the founder of the sponsoring
Club of Rome -- Aurelio Peccei. Peccei and his colleagues, as a
genuinely concerned group, had put much effort into attempting to
persuade high ranking people that the world was on some kind of col-
lision course.[4] Meeting with little success the "club felt that
nothing short of shock treatment could do the job".[5] So they deli-
berately set out to "search for a device capable of opening a breach
in the hearts and minds of people, of arousing their awareness of
the complexity and seriousness of the world problematique. After
long consideration, a commando operation was decided upon, in the
hope that its tactical success might have strategic consequences."[5]
Was this a use or misuse? It clearly was an attempt at persuasion,
possibly the model was unnecessary for the advancement of human
knowledge achieved. It was not really "counter intuitive" in the
sense often advanced for systems type models. The result was an
expensive revamping of Thomas Malthus. It attracted a lot of attention to the Club of Rome and the important issues with which they
were concerned, and generated the whole new discipline of global
modelling.

Our Sussex critique of the "Limits" model was equally simple
and was emphasized by using the original model.[6] Relax the constraints in the "Limits" model and the catastrophe disappears. The
contraints in the model arise because it is assumed that there will
be little or no increase in productivity arising from technical
change. If technical change is included at two percent per annum in
the crucial resource, agriculture and pollution abatement sectors
all appears to be well, after all. Firm estimates for the historical rate of technical change are, however, unreliable. We, there-

fore, tried to find the best evidence available for our critique. The most satisfactory appeared to be Solow's estimate for the United States over a fifty-year period, which we used. But, and this is the most important point, even with this, it now seems that this data may be wrong. Solow's estimate was made essentially by fitting a production function to historical increases in output. Everything which could not be explained, for example, by new investment, he put down to "technical change". A more recent study by Jorgenson and Gilchrist, using the same approach, find approximately zero increases in output arising from technical change. Clearly there has been significant technical change this century but, if we are to believe Galbraith in "The New Industrial State", it may have been as much a contribution to organization and control as to efficiency. The point is, however, that our knowledge of the world is very poor, especially for many of the major parameters used in world models and, indeed, many models used for forecasting. Before turning to these models, I want to critique further our critique of "Limits".

The most widely quoted aspect of our critique from Sussex was not the obvious experiment above -- but the notorious "backcasting" test. Our argument was that if a model is to project the future a century ahead, then it should also retrodict the past for the same period. Unfortunately, the Limits to Growth model runs back only twenty years. To have the model retrace its post-1900 behaviour we must assume there to have been a near infinite world population in the year 1880. To start the model from, say, 1800 with a reasonable population would merely have brought forward the predicted catastrophe to the present day. The point about this backcasting is not that it is invalid, but that the arguments as to why it is a valid test are elaborate and not easily explained in "lay" terms. Much heated and misguided debate took place. Attention was drawn away from the more important aspects of the critique; such as the discussion about technical change and environment and other more significant experiments. In one experiment we divided the model into "developed" and "developing" regions and showed that, on the basis of the Limits authors' own assumptions, the catastrophe was entirely the products of the developed world. Hence, although backcasting provided a gimmicky, and attractive critique, it sold the book but not the message. In that sense our attempt to emulate the Club of Rome's sensationalism backfired.

At this point, since this is an after-dinner paper, I may throw in a "dirty" story which indicates that computers may take a very serious interest in the programmes that are run through them. Ray Curnow and I first reprogrammed the Limits model only with considerable difficulty, but after a few days we were ready to run. On loading the programme, the computer immediately stopped. After a further two days the cause was apparent; pollution in the cooling system of the computer, the largest non-military machine in the United Kingdom at that time, had blocked the filters and so ended prematurely our attempt to forecast the end of the world for the same reason.

THE STATE OF KNOWLEDGE

In addition to illustrating the usefulness of models as instruments of debate, the above discussion throws up the major question of "What is the State of Knowledge?" How much do we really know about the world? Not much, if the above insight into variations in assumptions about technical change is any guide. Clearly to some extent the Limits exercise was a trick — it gave a misleading impression of the "State of Knowledge". Such exercises can mislead people about the amount of knowledge decision-makers and scientists, upon whom they rely, have at their command. Obviously, this is a game which politicians play all the time, but, if the earlier quotation from our anonymous French planner is correct the, the habit goes much further down the line. This is an important issue which I will take up again later, but before doing so, we may examine the "State of Knowledge" more closely.

Supposing, for example, we take our ability to predict the future in different areas as a guide to the level of human understanding in those areas. In doing so we should also learn something about the value of large-scale computer models.[8] Immediately, we see tremendous variation. We can forecast the position of the planets thousands of years ahead, although even here we must take caveats, such as "barring unforeseen astronomical catastrophe", eg. from a wayward black hole. In some other physical sciences such as climatology, there is much controversy; while in others such as meteorology our prediction is more reliable and improving. In all these areas, very elaborate computer models are acknowledged to be an essential component. In the social sciences, short-run econometric models and location models are widely used. The former are believed to have a useful track record. In the United Kingdom, for example, rarely are major policy decisions taken without at least checking out the implications in the Treasury model and often, cross-checking soon afterwards with the more independent university or research institute based models. I shall say more about urban location models. In other areas, such as population forecasts or commodity price projections, the catalogue is long, varied and often dismal. Global modellers argue that one reason for this is that most of the above areas are interrelated and one cannot be forecast without knowledge of the other. To discuss this and other issues in more detail, I now take population forecasts as a typical example.

MODELS OF REALITY OR MODELS OF PREDISPOSITION?

Estimates of total world population in the twenty-first century, from a wide range of future studies, vary between four and forty billion.[9] They are made using a range of techniques — extrapolation, simulation and inspiration. Two questions to ask are (i) why is there such a wide range and (ii) do models improve the prediction?

To answer (i), it is useful to return to the question of how the forecast is to be used. To be specific we take the highest estimate of Heilbroner.[10] In "The Human Prospect" he suggests a figure of forty billion for the long-run total world population. The only way societies could survive such an overcrowded environment, he suggests, is through either social "anarchy or authoritarianism". Of the two he prefers the latter. Indeed provided governments respond strongly enough, there may even be some hope of ensuring that world population does not exceed nine billion. Looking more closely we see that Heilbroner is using a Malthusian argument, the world cannot support a population of forty billion, as a lauching pad for his own abhorrence of social disorder, especially in developing countries. Looking back at Heilbroner's earlier writings, at a time when he believed that by the end of the century people would typically be working only a twenty-hour week to satisfy all their expanding needs, he still recommends strong regimes in developing countries to cope with the disorder that rapid economic change will bring. Heilbroner is not alone in this technique of posing a straw man to introduce an argument. Spengler, another neo-Malthusian, has argued that in the post war period, birth rates in developing countries should be cut to avoid over-population. Pre-war, he argued that birth rates in the developed countries should be increased. His apparent concern therefore is to maintain a racial balance. Arguably the Limits study discussed earlier is a fine example of this rhetoric, again a common tool of persuasion by politicians. To have predicted catastrophe very far into the distant future or next year would have lacked the desired impact. Nobody cares too much about catastrophe many generations ahead. We may all be on other planets by then. Conversely, if catastrophe is forecast to be too imminent there is nothing that can be done anyway, and so we won't have time to implement the recommendations of the modeller.

The art of successful doom-watching is to pose a problem which is not too difficult to solve in the manner you intend to suggest. It would be unfair to single out only Malthusian authors in this respect.[11] Even my favourite anti-Malthusian study by the Fundacion Bariloche in Argentina employed a similar tactic.[12] This group wanted to show that, when developing countries are "delinked" economically from the industrial countries, they would more easily be able to satisfy the "basic needs" of their populations. To do this the group was obliged to cut by half the historical rates of technical change estimated with their model for purposes of projecting into the future. If technological change was not adjusted in this way then growth in the developing countries would have proceeded rapidly under almost any assumed international links without need for the policies they advocate. In order not to appear too critical, I should say that despite their model I have a great deal of sympathy for the Bariloche theory and policy, and consider theirs to be the most sophisticated exercise even though the model has certain defects.[13] This leads into the second question (ii) of whether models improve the accuracy population prediction. Several of the studies in my review employ mathematical extrapolative

models, cohort models and integrated socio-economic models. The
Limits to Growth and the Bariloche models were both of the last
kind. The Limits projection apart, the model studies examined seem
to be less extreme. But before jumping to conclusions, it may be
said there is a social pressure among the peer group of global
modellers to make their forecasts "consistent". For example, at
international meetings of global modellers, the question of consistency between forecasts is given precedence. Thus, whether any eventual reduction in uncertainty will reflect coherence or coercion is
open to question. Understanding the context of forecasting may help
us to make better personal and collective judgements about their
reliability. Certainly there will still be a lot of uncertainty,
and arguably, a knowledge of this uncertainty is at least as important as agreement about the future, especially when that agreement is spurious.

This leads us to consider more directly the nature of scientific endeavour. Most scientists would accept the earlier description of French planning as a fair approximation to the average university department or funding committee meeting. Thus, it may be
said that scientific endeavour comes closer to a political scientist's view of politics that it does to their view of science. I
have argued elsewhere[14] that this is one reason why social scientists appear often to be so disconcerted by attempts to use mathematical and "scientific" methods in the policy sciences. The human
element is as vital to an understanding of what is happening in the
physical and related sciences as it is in the social sciences. An
extreme example would be the "Lysenko" affair; a more common example
is the above-mentioned research funding committee. Anyone who suggests that scientific research is simply the advancement of human
knowledge must be laughed at. Whether it should be so is another
matter, but clearly, strategic, industrial, social and personal factors are all important in determining what science is done, and
also, what is published.

A rather nice example of the human element in the social
sciences was related by the Nobel Prizewinners, Yang and Lee, famous
for their discovery of the elementary particle, the pi-meson. It
was found that successive measurements of the mass of the pi-meson
showed discontinuities. For a few years the measurements would all
be more or less the same, but then there would be a jump to a new
level (up or down) for a few more years and so on. The explanation
could have been profound for physics, but appears to have been more
human. The leading measurement on each new step was made by senior
and well-known researchers who were not as prone as their less confident colleagues to discard results which did not conform with
prior expectation!

As pointed out earlier, mathematical models have been essential
to the development of the physical sciences, and very accurate
models of quite complicated systems (eg. the solar system) have been
made. But these systems are very simple compared even with most

other physical systems.   For example, if we wish to know the properties of a new alloy or semi-conductor, we experiment rather than calculate.   Models give us insights into the properties and predict and explain in an approximate fashion; some materials even provide us with limiting cases of the theories.   But in the main, experiment rather than theory, provide the data for building up practical knowledge.   The same is true in engineering; we usually build a pilot plant or a model of a new reactor or aircraft, although there are increasingly sophisticated computer simulation techniques.

In economics too, formal theory has had a tremendous impact on thinking, for example, much policy on trade and development comes from the Hecksher-Ohlin "gains from trade" and Stolper-Samuelson "factor equalization" theories.   International agreements like GATT are premised as such theories.   The theories are represented (for example) as very simple two country -- two good algebraic models which are, practically speaking, untested and untestable in the real world.   Nevertheless, by the accepted standards of society the models have been deemed useful and the Nobel prizes duly awarded. Many people, especially from developing countries, believe the theories to be wrong, or at least only partially true, and merely an instrument of the dominant actors in the world economy.   Hence, they see them, like the Limits to Growth models, as an instrument of persuasion or at best, of rationalization.   The rituals borrowed from the natural sciences by, for example, the high priests and wizards of economic policy are probably not as spurious as we believe those of the witch-doctor calling up rain to be.   Nevertheless, it is apparent that the social roles played by "policy analysts" and witch-doctors are not completely dissimilar.   Clearly there is one essential difference between models of physical systems and models of social systems.   As far as we know, no model we make of a physical system, whether "right" or "wrong", will affect the laws of behaviour of the elementary particles.   Our knowledge of those systems may, of course, affect us.   For example, Einstein's E $= mc^2$ has presented us with the possibility of the imminent doomsday omitted from the Limits of Growth model.   But social theories can change the rules by which social systems operate at some level.   As I have argued above, they are often actually designed to do so. Some models are designed to indicate policy; others like the Limits model were designed to change "hearts and minds".

THEORIES FOR ALL SEASONS

I have mentioned earlier that the "Limits" debate was fairly uncluttered in terms of actors and argument.   With urban planning, formal models are arguably much more institutionalized although, as many papers at the present seminar have made apparent, there is a widespread air of disillusion as to their value in the urban planning process.   World models seem to have weathered the storm in the wake of the Limits to Growth exercise and several new national and international exercises are being conducted. Hence they too are now institutionalized. Whether these models have so far had any impact on "policy" is quite another matter.

Arguably, however, urban models do affect policy. For example, most planning of new shopping proposals does employ some form of shopping model. Often a "developer" will engage a planning consultant to demonstrate to the satisfaction of the inspector ruling the enquiry (i) that there is a need for the new proposal and (ii) that no existing interests will be seriously affected. Providing these two conditions are fulfilled, clearly the proposal may be accepted. In general the two objectives are mutually contradictory, except in "green field" planning. Often the consultant engaged will adopt a particular mathematical model making use of whatever data there is to hand. Although, in essence all shopping models are rather simple, the documentation of the consultant's analysis may run to several hundred pages, giving it the same air of authenticity that the spaghetti flow-diagram gave to the Limits to Growth exercise.

Often, however, the results of these models similarly rest on one or two key assumptions. For example, in a public enquiry with which I was personally concerned[15], the consultant had demonstrated the "need" for a proposed new large supermarket on a local yachting marina by adding into the locally generated demand for groceries, the casual "ice cream" purchases of many thousands of holiday-makers. This provided the argument for a large sales area. To avoid the inconvenience that the predicted impact of such a large centre would have when the usual assumption of centre attractiveness being proportional to floor space was included in the "gravity" model employed, all centres, whatever their size, were assumed equally attractive. Thus the case for the new centre was "proven".

In criticizing this analysis it was easy to demonstrate that single alternative assumptions completely changed the predictions using the model. However, the public enquiry system is a rather unsatisfactory forum for engaging in such "academic" debate. When the debate is between a well-documented consultant's report and a little old lady with a shopping basket, "responsible" and "documented" decisions are easily taken. When the true uncertainty and arbitrariness of even apparently weighty evidence is made clear, then even more arbitrary decisions are made.

As an academic exercise, a research student, Roy Turner, and I, have explored for a range of shopping models, how variation in the model used, the data base and the method of estimating a model affect the predicted impact of new shopping centres.[16] By comparing five shopping models, data for four towns and a range of estimation techniques, we show that within reason practically any desired result may be obtained in a forecasting exercise. Even when the models are used "correctly", each of the three variables, "theory", data and estimation technique lead to approximately equal sources of variation and leave plenty of scope for malpractice. However, we should not necessarily conclude from this type of exercise that models are useless, only that they are very easy to misuse. If we examine more closely the underlying theory of the model, or the way that a given estimation technique weights evidence, then we can, in principle, ascertain how a given model and experiment will bias the forecasts made with it. If we want to protect certain interests, we must be watchful of both the model and the way in which it is used. Clearly, to achieve this sort of awareness at a public level may be slow process, but as I suggest below it is not improbable.

A METAPHOR FOR THE FUTURE

    Unfortunately, as indicated in my introduction, I have few conclusions to draw about the use of models beyond the obvious, that models have both uses and abuses, and that even abuses are likely to be to somebody's advantage. I could repeat the by now almost plati-tudinous modeller's defence that it is not models which are undesir-able, but the way in which they are used. Even if this is true, there is certainly some need for "social control" in modelling. It is obviously impossible to ensure that critique is totally "disin-terested". Technical critique, however, by people who wish to be seen as technically competent, and who are not directly involved with the political issues at stake, exist to some degree, but more is needed. Nevertheless, in the same way that academic journals, however serious, reflect ideological and cultural prejudices, so will any similar form of peer group social control. I fall into a trap of my own making when I recommend at this point a dynamic, varied and liberal approach to the search for knowledge and policy analysis. However, I do not see a better alternative. To stop modelling would be like asking people to stop thinking. Modelling is, for many people, a way of approaching "problems" which is only one aspect of other thought processes. It is more reasonable to expect modellers and others to understand the limitations and possi-bilties in each other's approach.

    We may, nevertheless, construct images of the way in which models may be used in the future.

    If we accept my earlier definition of a state of knowledge measured by our ability to forecast the future, then it is surely doubtful whether that ability has improved since Malthus' day. Our knowledge of the world may have advanced, but in part, because of our new knowledge, the world itself has become more complex. Thus, we need new knowledge simply to maintain our present limited under-standing and control. At least one school of thought advances the position that modern policy problems are so "wicked" (ie. novel, complex and intangible) as to defy rational analysis.[17] I have argued elsewhere that the logic of this argument is that societies are ultimately directionless.[18] To accept that social decisions cannot be implemented successfully would surely be to deny policy analysis any role.

The ultimate "technocratic" nightmare is surely that of a society governed on the basis of some massive computer model. The unspoken fear of some critics of systems modellers may be that this is, indeed, the modeller's dream. In this particular nightmare there is a clear link between the organization of society and the state of knowledge. Such a model and such a society would only be possible if the "laws of motion" of societies were well-understood[19] and effective mechanisms of centralized social manipulation operated. In anticipating this society, it is assumed that as human knowledge increases, so will the possibilities for the exercise of state or corporate control. Much science fiction literature employs this model of the future, even though, in many respects, it seems closer to the social reality of the past.

I cannot prove that the nightmare is impossible, but I believe it to be unworkable. By comparison with the data needed for such an ultimate model, our understanding of social processes is extraordinarily weak, and likely to remain so. Thus analytic methods which attempt to internalize processes which legitimately belong in the wider political debate, and for which we have neither the measurement or theory, are not only bad models, but pose risks when applied. To say this is merely to point to a direction for policy analysis which I believe will lead to better preparedness to enter the future. To achieve this requires that a much wider body of experience penetrates the theories which society conducts about itself.

I would prefer to argue that as human knowledge expands so there should be a greater degree of social participation. The extent to which an artificial consensus is acceptable depends upon the nature of the society involved. However, in some societies, taking part in the debate, for many people, may be at least as important as its outcome. Whatever the findings of a theory, its application is bound to stumble if this debate has been omitted. In a nutshell, I am arguing that putting models into society is fundamentally more important than putting society into models.

The examples I have given show that modellers are right to be concerned about the value of their endeavours. We should be dismayed if there was not both social and self-criticism. Nevertheless, examples such as that of nuclear physics research or indeed other areas which I have not mentione_, such as bio-physics or everyday medical practice, indicate that the problem faced by modellers is one of societies at large. To accept that the world is, in general, far from ideal does not remove, however, the obligation on modellers to improve their contribution to it, but it does indicate the constraints within which they operate.

The metaphor of nuclear physics provides a useful illustration for this part of my argument. As a conclusion, I offer another metaphor which I think serves well to indicate a possible future for modelling.

Shortly after the seminar, I read a perceptive review[20] of an equally perceptive book, <u>Photography and Society,</u> by Gisele Freund. To paraphrase the book: one hundred years ago the first photographic reproduction appeared in press in the New York Daily Herald. (Significantly for the present seminar it was captioned "Shanty Town"). Tinsdall writes, "It seemed to presage truth, immediacy, a shrinking of the globe and a growth in understanding....Now when thousands of newspapers and magazines reproduce millions of photographs daily, we know that the camera can be the most skillful liar of all". Freund's book exposes the "illusory objectivity" of the photograph and calls for socially aware practice. Apparently, at the time of the Paris Commune, the French government attempted to make the new technology state controlled, and hence, to serve only the interests of the state. Today, practically everyone owns a camera, but nevertheless, as Freund points out, "photography, more than any other medium, expresses the value of a dominant social class". The parallels with modelling are obvious and striking. Possibly, therefore, we may read from the history of photography, a future of modelling.

No one, except perhaps a few "primitive" tribes, believe that to be photographed is to lose one's soul, although some might argue that standardization and conformity through the social models propagated via photography have in effect achieved this. This is precisely the fear of the anti-technocrats with respect to models.

Despite its ubiquity, few people, including the most prominent practitioners of the art, can be aware of all technical and scientific aspects of photography. A socialization of this knowledge is necessary to protect against misuse. We rely, in general, on a plurality of views and abilities in society to ensure that "unacceptable" deception will be revealed. "Standards" of advertising and news reporting are intended to control the application of the technology. In some societies these social controls work better than others and protect different interests. Most people recognize and allow for some degree of misrepresentation. In interpreting advertisements, particularly, we have become socialized to compensate for the "illusory objectivity" of photography. Nevertheless, photography remains a valuable tool of scientific endeavour.

We might argue likewise that, while today access to the technology of modelling is limited, in the future new technologies and raised levels of numeracy will make models a part of everyday life. The average child may literally have the world at her fingertips. Although this wil not necessarily enhance her control over society at large, it may improve the quality of mutual understanding. If this happens, then greater familiarity with models will also bring greater, rather than less, scepticism, but a certain usefulness will be recognized. Inevitably, models, like photography, will be used to "product market" the future, but also to increase social and individual understanding. The hope for modelling is that healthy scepticism does not become downright cynicism.

NOTES

1. Robert Jungkt's (1958) book, <u>Brighter Than A Thousand Suns</u>,
    Harcourt Brace, New York, remains the most human description of
    the growing awareness of nuclear physicists.

2. Anon, 1973, Letters from across the Channel, <u>Public Administra-
    tion, 51</u>, 185.

3. Meadows, D. et al, 1972, <u>The Limits to Growth</u>, Universe Books,
    New York; also Forrester, J., 1971, <u>World Dynamics</u>, The MIT
    Press, Cambridge, Mass.

4. See, for example, Peccei's (1977) own account in <u>The Human
    Quality</u>, Pergamon Press, New York.

5. Peccei, A., 1973, The Moment of Truth is Approaching, <u>Success</u>,
    December.

6. Cole, H.S.D., et al, 1973, <u>Thinking About the Future – A
    Critique of 'The Limits to Growth'</u>, Sussex University Press,
    Brighton, UK, and <u>Futures, 5</u>, 2-32, (1973).

7. See Meadows, D., 1973, Response to Sussex, in <u>Futures, 5</u>,
    135-152, and our own reply (1973): Backcasting with the World
    Models, <u>Nature</u>, 242, 147-148 (1973).

8. See Clark, J., et al, 1975, <u>Global Simulation Models – A
    Comparative Study</u>, John Wiley, Chichester, UK.

9. See, in particular, my review (1978) in <u>World Futures – The
    Great Debate</u>, Martin Robertson, London, and Universe Books, New
    York.

10. Heilbroner, R., 1975, <u>The Human Prospect</u>, Norton, New York.

11. See, for example, my review (1977) of Herman Kahn et al's
    (1976), book <u>The Next 2000 Years: A Scenario for America and
    the World</u>, in <u>Futures, 9</u>, 65-71.

12. Herrera, A., et al, 1975, <u>Catastrophe or a New Society</u>, IDRC,
    Ottawa, Ontario.

13. See my review (1977) in <u>Global Models and the New International
    Economic Order</u>, Pergamon Press, New York.

14. Cole, H.S.D., 1976, Long Term Forecasting – Emphasis and
    Institutions, <u>Futures, 8</u>, 305-317.

15. The Brighton Marina Enquiry is described in T. Whiston, (Ed.),
    (1979), <u>The Uses and Abuses of Forecasting</u>, Cambridge Univer-
    sity Press, Cambridge, UK.

16. This is reported in Turner, R. and Cole, H.S.D., 1980, An Investigation into the Estimation and Reliability of Urban Shopping Models, Urban Studies, 17, 139-157.

17. Rittel, H. and Webber, M., 1973, Dilemmas in a General Theory of Planning, Policy Sciencies, 4, 155-169.

18. In Cole, 1976, Futures, op. cit.

19. A similarly powerful model is conceived in I. Asimov's, (1951), Foundation Trilogy, Doubleday, Garden City, New York.

20. Caroline Tinsdall, on the book, Photography and Society, by Gisele Freund, in The Guardian, October 2, 1980.

EDITORS' NOTE

This paper is based on the invited lecture which the author presented to the NATO Advanced Research Institute.

ON SYSTEMS THEORY AND ANALYSIS IN URBAN PLANNING: AN ASSESSMENT

Michael Batty

Department of Town Planning
University of Wales Institute of Science and Technology
Cardiff, CF1 3NU
United Kingdom

"There are no simple answers here and I am not
going to give any. There are moments of history
when we simply must act, fully knowing our ignorance of possible consequences, but to retain our
full rationality we must sustain the burden of
action without certitude, and we must always keep
open the possibility of recognizing past errors
and changing course."

Kenneth J. Arrow (1974:29)
"The Limits of Organization"
New York: W.W. Norton and Company

INTRODUCTION

All periods of history have been characterized by critics
attempting to survey, evaluate and perhaps alter contemporary
approaches to human action, and it is somewhat naive to expect that
our times are likely to be very different. Yet, in one sense, there
is a clear difference. Never before have so many experts been
organized to pronounce upon and prescribe for the human condition,
and the development of such specialist roles has consciously influenced the course of social affairs. As Herbert Simon (1978) has so
cogently observed of modern management, it is no longer possible to
study management phenomena in the way it could be a generation ago,
for management science techniques themselves have had an enormous
impact, for better or worse, on the present style of such manage-

ment. Whether or not this change is one of kind or degree is of little matter here, for the point is that such communities of experts are now of some significance, and it is unlikely that in our self-conscious world, the previous situation will ever recur.

Thus, it is not really surprising that as soon as such expertise becomes significant, another community of experts begins to assess the contemporary wisdom which is being established in the light of goals pertaining to both of the specialist communities. Although I have not tried to justify this assertion in any comprehensive way, it is certainly borne out by a study of the experiences in a variety of fields pertaining to social action: in urban planning, business management, in economic policy, and in the various technical domains which support them -- spatial modelling, management science, econometrics and so on. Indeed, it is the purpose of this conference to articulate these types of response, in particular in relation to the use of formal analysis in urban planning, and it is as much to raise awareness in the original community of experts as to establish any new expertise to replace the old; for that in any case can only evolve through a longer term process of introspection and criticism within the community as well as without.

Rational techniques of analysis in urban planning have, as in many other fields, been developing during the last three decades. These developments have been single-minded in the extreme, and only of late has the field responded to external criticism, through the slow process of dawning awareness that the experience has not always been what it was anticipated to be, that the system of interest seemed more complex than was originally perceived. We are, as Hoos (this volume) has so vividly pointed out, like Winnie-the-Pooh bears, bouncing downstairs on our behinds, assuming that this is the only way to travel, but wondering whether there is any other way, yet never having any time to really reflect on this. It is this dawn of realization which quite unapologetically forms the concern of this meeting, and the substance of this paper.

There are many different ways to begin to approach these experiences, and it is doubtful whether there is any means to engender a total synthesis. Thus, this paper will attempt to view both diagnosis and prognosis for the field, through the lens of problem-solving, that is through the ideas which characterize the preparation of plans, policies, proposals, designs however they may be termed, using those systematic techniques of analysis which form the conventional wisdom of social action in the present day. Certainly the ideas presented here are bound by the paradigm of analysis itself, and it is contestable as to whether this approach will throw any light on the same approach it is designed to judge (Vickers, this volume). But as alluded to already, progress in learning about any approach to social action can only come from relating internal to external criticism, and this indeed is one of the central assumptions of this paper.

In exploring the idea of problem-solving, from within or without the field, it is possible to proceed using conventional approaches to gaining an understanding of this phenomena; that is, to either postulate a theory of the phenomena, and to attempt to see whether or not the experience is reflected in this, or to describe the experience with a view to eliciting some appropriate theory. Or as is more likely the case, to use a mix of both approaches. It would be attractive to first prescribe some theory of social systems which could then be used to judge the efficacy of formal problem-solving in the context of the example of this paper — urban planning; but this is manifestly impossible. Indeed, it can be argued that such a style was the one which dominated the original development of the field, and thus at this stage, it is the experience which is all important and should form the starting point; although as something as a style for recounting the experience, an attempt will be made to cull the key ideas from the experience, and present these first as pointers to a rudimentary theory of social systems.

In this introduction, it is worth stressing two themes which have been largely absent from the perspective of systems analysts in planning and in other forms of social action. But these notions are increasingly important, and clearly so here. First, in the development of problem-solving methods in these fields, it has been widely assumed that such analyses can be conducted in the manner of those in the physical sciences. That is, the analyst is detached, independent of the social context. But as experience has been gained, this assumption has been demonstrably shown to be untenable, and it is an indicator of the force of this point, that the community of experts themselves have gained such awareness.

Related to this, and to the original "a priori" reasoning about suitable methods of problem-solving in social systems, is the realization that such suitable methods can only be evolved through interacting with and learning about the nature of such systems, and this takes time. Thus a sense of history is being acquired by reflecting upon these experiences, but more important, such experience, such history is being used as a guide to the nature of social systems and henceforth, hopefully, to more appropriate modes of problem-solving. Moreover, to anticipate a conclusion of this paper, to analyse and interpret these experiences using anything but the systems analytic approaches which constitute the experiences themselves, is tantamount to an admission of the difficulties with these approaches. And it is the importance of history and of the social context, so essential in developing these themes, that emerge as essential constructs in any theory of the social system.

As a prelude to examining various modes of analysis, some notes on the structure of social systems will be first presented, emphasizing two main problems — the problem of complexity or variety, and the problem of system closure. These ideas form the basis of many emergent theories purporting to explain problems manifest in social systems, in particular theories of the growth of bureaucracy, and stability and turbulence of organizational environments. From these ideas, the increasing complexity of society and the concomitant difficulties facing problem-solving will be drawn out. These ideas set the backcloth for an examination of the styles of problem-solving and styles of analysis which underpin systems thinking. In particular, the differences between analytic and synthetic approaches will be stressed, and used as a pointer to more general difficulties facing the field. At this stage, sufficient material of substance will have been assembled to present a brief critique of systems analysis by turning this type of analysis in on itself to reveal inherent paradoxes and dilemmas. This will then be extended using the development of systems analysis in urban planning as exemplar. Finally, in the light of these ideas, a plea will be made for intellectual pluralism, liberalism if you like, as a way of enriching our understanding of and ability to confront social problems. Prospects facing the field and suggestions for the future will finally be presented.

By way of qualification before the main arguments are launched, it is worth noting alternative approaches, which in my view and quite consistent with my conclusions, mutually complement any critical approach grounded in the object of its criticism. Looking at systems analysis as ideology or philosophy (Lilienfeld, 1978), as a consistent mode of abstraction (Berlinski, 1976), as a sociological movement (Hoos, 1972), as an organizational phenomena (Brewer, 1973) or as a community of experts (Boguslaw, 1965), all present different facets of the experience. The intellectual pluralism advocated here is enriched by these alternative critiques, and a greater understanding of the phenomenon can only come through more studies of different facets of the experience.

NOTES ON ORGANIZATIONAL COMPLEXITY

It is often asserted that formal knowledge, indeed perhaps any identifiable idea, represents a compromise between the infinite complexity of the perceived reality, and the necessary simplicity required for understanding. As Paul Valery so eloquently stated (Linstone, Lendaris, Rogers, Wakeland and Williams, 1979:291): "Everything which is simple is false, everything which is complex is unusable". Thus, it is of some interest to explore why complexity "per se" is of such central importance to systems analysis. One quite convincing explanation is that systems theory evolved, in terms of its theory and certainly through its subsequent applications, in relation to phenomena which were perceived to be immensely complex, chaotic perhaps; and that systems theory was prescribed at a level of generality which succeeded in at least containing and charting such complexity. In short, in subject areas which had not evolved in the more traditional scientific mould, in areas which Kuhn (1962) amongst others, has referred to perhaps prejoratively as pre-scientific in form, systems theory appeared especially appropriate in emphasizing explicit order through countable states, interdependence between the parts and hierarchical classification.

Indeed, one of the central tenets of systems theory - variety - relates to a count of possible states or configurations in which a system might exist. An example is clearly in order to illustrate the point. Imagine a social system in which there are n actors. If each actor has the choice of remaining active or passive, then the number of possible states of the system (a count of the number of distinct configurations of active and passive actors) is 2n. However, if the system is made a little more realistic and each actor can choose to use his activity by interacting with another actor in the system, or not as the case may be, then the number of interactions (including self-interactions) would be n^2 and the total number of distinct states of the system 2^{n^2}. For ten actors, this system could thus exist in any one of over one million states. Even so, this is perhaps too simplistic a form of interaction, for in social systems actors form coalitions and the coalitions themselves interact. If the order of actors is neglected in each coalition, and each actor individually is regarded as a coalition, the number of distinct coalitions is (2n-1) which gives rise to $(2n-1)^2$ interactions between coalitions. The number of states such a system could now take on is $2^{(2n-1)^2}$ which for even only three actors is a truly astronomical number. In fact, the usual measure of complexity or variety, termed information is $\log_2$ of this number, and in these cases is the number of possible interactions. And these counts do not take account of differing strengths of interaction, the simultaneous relation of three or more coalitions and other reasonable elaborations.

The point is an obvious one. Where there are few guides to simplification through theory, complexity assumes awe-inspiring proportions. Moreover, such combinational explosions fly in the face of certain basic ground rules which have been well-established for human systems. Miller (1956) in a classic paper, argued that the limit on information processing capacity was something in the order of the 'magical number' 7 ± 2 distinct elements which could be processed simultaneously. More recently, Simon (1974) has argued that this number is closer to 5 than to 9. Clearly these natural limits on individual coping extend to organizational coping. It is possible to decompose a system hierarchically into small groups of elements of no more than, say, 7 or so, and process these sequentially according to the hierarchical order, but this still puts our ability at processing complexity at a much, much lower order than that associated with the combinational demonstration sketched above.

An extension of these arguments to systems theory has involved explorations, first into the way systems naturally handle complexity and second, into the way systems might be artificially designed to do so. Most demonstrations have been made using examples from the physical sciences, of quite well-adapted and stable systems, and this makes their conclusions of doubtful consequence when generalized to the social domain. First there is a natural law which must apply to all systems and this relates to the obvious truism that any system can only survive if it manages its complexity by meeting its natural limits. In other words, anyone attempting to process simultaneously an order of magnitude of distinct information greater than the magical number will fail: the system will not survive, or the number of items will be reduced to the natural limit, or some form of sequential processing will evolve. Thus the system will obey the law of requisite variety, or Ashby's (1956) law, as Beer (1974, 1979) has called it.

This brings us to the second point. Hierarchical schemes for processing information work well in natural systems, and there have been many suggestions that such organization is the appropriate model for handling variety in human systems (Simon, 1962). For example, Stafford Beer (1979) bases his arguments for efficient management structures on the principle of hierarchy. Christopher Alexander (1964) bases his scheme for efficient problem-solving and design in high variety systems on the same. A third point is more contentious and in some respects, represents the major stumbling block of systems theory in a social domain. In natural systems, it appears that the pattern of significant interactions between system elements can be largely restricted to the first-order. In other words, the variety in such systems (using the $\log_2$ definition) is of order n^2 rather than $(2n-1)^2$ or greater. Clearly this makes for enormous simplification in information processing, and in the potential stability of such systems. But in social systems, it would appear to be too gross a simplification, given the previous examples where higher-order effects -- coalitions -- were of the essence.

The last concept which ties these notions together is perhaps the most important, and this relates to the definition of the system. Once again, in physical systems, the number of critical elements (variables) n, seems small in most cases, and thus, it is relatively easy to distinguish the system from its environment or context. Moreover, the physical world seems well-organized hierarchically, and thus, it is quite possible to find breakpoints for which the density of connection within the system is much richer than between the system and its environment. Indeed, minimizing the energy interchange or number of inter-relationships between the system and its environment (all other systems), is a basic prerequisite for generating a successful understanding of systems phenomena.

The evolution of systems theory has by no means been a cutting edge for scientific understanding. Indeed, for many of its most dramatic examples, systems theory has evolved "post hoc": that is, such theory seems to fit existing theory quite well, after the event so to speak, and thus, has not necessarily been instrumental in generating new knowledge in those fields. Thus, there is some doubt as to whether the conditions which have been sketched here can all be met in systems other than in the physical domain, and this raises one of the great ironies of systems theory. If the tenets of the theory do not pertain to social systems, then why has there ever been the massive acclaim for such theory as a basis for social understanding and action? Before this question is broached, however, it is worth noting how these ideas do present a possible logic for explaining certain aspects of social systems organization.

SYSTEMS THEORIES OF SOCIAL ORGANIZATION

There is a problem which has recently caught the imagination of sociologists, and which provides an excellent illustration of the type of complexity posed by social systems. The problem, called by its originator Stanley Milgram (1967) 'The Small World Problem' for reasons which will become obvious in a moment, involves tracing the correspondence between geographical proximity and social proximity in large populations. It is assumed that the population of interest is mapped onto a social network which is strongly connected in that every member can reach every other member, directly or indirectly, were they to transmit a message through the network. Now the problem can be posed as follows: you are required to estimate the average number of links in the network between any two persons chosen at random from the population. Many examples of this method are now available, and for geographically extensive systems, the results are surprising. For example, using the United States population, the average number of links seems to be about five. Of course, the frequency distribution of links has a large variation due to the fact that there are substantial differences in interaction potential between different social groups, but the degree of social connectivity in contrast to geographical extent is still surprising.

It is perhaps more dramatic to pose the problem in a counter-form: to take a particular individual and to attempt to construct the 'worst case' (longest path through the network) one can envisage. In fact, the sampling procedure used to initiate these chains involves passing a message to an acquaintance whom the originator of the message at each step believes will transmit the message most quickly (in network link terms) to the destination. The problem can be attempted in a casual way, as a party game say, but the impact is still dramatic for it shows how densely connected we are in a global sense in terms of influence networks. The implications of this type of connectivity or first-order complexity as it is, are quite serious. Although it may be fairly easy to figure out the rationale for such influence patterns, the fact that the spatial and social worlds are so interconnected is a dramatic indication of the complexity of modern society. And this is by restricting the influence to first-order interactions. Indeed, it is no surprise that the study of networks in sociology is one of the major growth areas of that subject, due to the need to study social patterns which transgress traditional social, spatial and institutional bounds.

The small world problem only illustrates one facet of societal complexity, that which exists at present. In fact, it is quite clear that this complexity is increasing, probably exponentially, but just possibly on a global basis according to some limit. Increasing wealth has brought increased opportunity for interaction in a variety of ways. More time is available for interaction, technology in the form of machines to speed spatial and temporal transfer, have generated more variety and individuals have responded by setting up their own organizational structures to cope with such variety. However, at the same time, our collective ability to make sense out of this proliferating variety has not grown. The pace of change is now so great that it seems impossible to respond fast enough with appropriate strategies. In the past, whenever a problem was perceived, individuals acting singly or collectively, had time enough to respond with an action, which if then proved inadequate, could then be gradually modified until it became 'right' within certain acceptable limits. Alexander (1964) argues that this is the way good architecture evolved prior to the industrial age. In systems terms, the regulator which steered the system to a new equilibrium had enough time to respond before the next perturbation occurred, knocking the system out of equilibrium once again. The variety of such systems remained manageable in the sense that the system could evolve its organization to cope with changing variety before the next change in variety occurred.

The present proliferation of variety in social systems and our increasing inability to cope with this in individual and institutional terms, can be easily explained using this model. When a problem is perceived which in this case the social organization cannot handle, then the organization must reflect on the problem and adapt to meet it. Failure can occur in several ways. The time available for response might be too short, or the problem may be misperceived through information overload or reduced channel capacity, and so on. In fact, what normally happens is that when a problem is perceived which an organization cannot cope with, the organization sets up another organization specifically adapted to cope. However, this represents a proliferation of variety which brings its own problems. In terms of the example of the previous section, if organization A perceived a problem relating to organization B, it might set up C. But C then has to relate to A and B at the first-order, as well as at higher-orders and thus variety proliferates, as does the possibility of information overload. The simple diagram illustrated in Figure 1 makes the point more cogent. Using a conventional style indicating positive feedback (+), it is clear from this cycle of action -- reaction that variety proliferates and problems increase. The solution to this vicious circle of growth in the bureaucracy is obvious enough. It is to invent more ingenious organizational structures which constrain variety to its natural limits. Of course it is easier said than done, but the principle is clear. As Beer (1974, 1979) has argued, when there is little time to reflect on and adapt organizations to handle problems, organizations are seldom changed or modified, rather new ones are usually created, or new technology is introduced into existing organizations which amplify and exacerbate the problems,

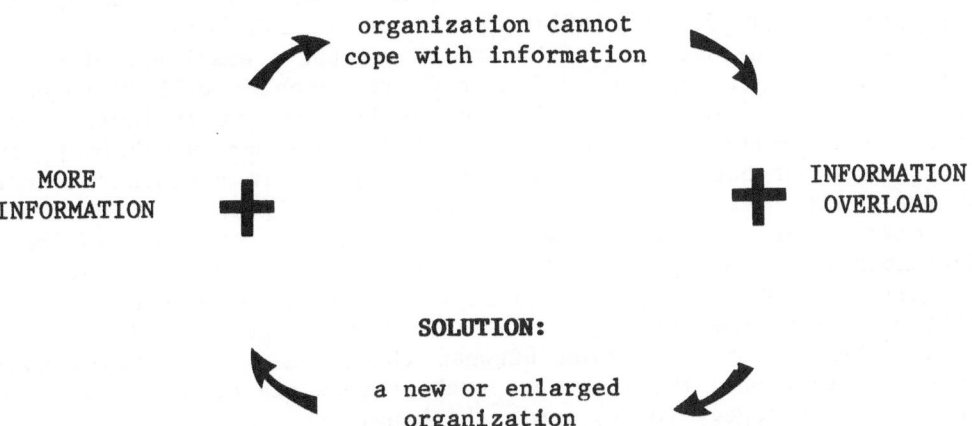

Figure 1 - Expanding Complexities of Problems and Solutions

rather than reduce them.   In short, new organizations appear, or existing organizations grow.   Variety is amplified where it should be attenuated, and in the critical places where there is information overload and messages lost, variety is attenuated rather than the organization amplified to cope with the overload.

What has just been sketched is a fairly well-known systems theory of organizational change.   Together with the increasing inter-relatedness of traditionally separate systems due to increased opportunities for interaction, these theories are consistent with the arguments of the learning theorists such as Toffler, Schon, Dunn, Michael, Friedmann, Webber amongst others who argue in diverse ways that we must learn about complexity and respond by learning to live with it, or consciously learning how to reduce it.   There are examples of our failure to do so everywhere.   The present British Government (at August 1980), intent on pruning the public sector organizations which have grown like topsy during the last two decades, are creating new organizations to effect the cuts, and where cuts are being made, they are in organizational areas where variety should be amplified rather than attenuated.   For example, in social and educational services, it is easiest to cut the worthwhile and tangible products of these systems, while the organizational bureaucracies which 'support' these remain.

In all modern governments, the growth of administrative instruments has been fantastic in recent years as decision-makers and politicians alike have realised that existing strategies have been sadly lacking in efficacy.   Policies which are designed to address one set of problems lead to others without in any sense solving, resolving or even dissolving the original set. For example, the all-pervasive problem of inflation in the British economy is being dealt with on the assumption that there are but a small set of variables to control, and once influenced, the problem will disappear. But it is quite clear that inflation is endemic, it reflects every facet of economic and social life, it has its own psychology, in fact, it is probably of psychological origin. It certainly cannot be dealt with using the closed system model. Traditional economic policies treat the problem of inflation as if the economy were analogous to an orange -- that inflation might be squeezed out, but a better analogy is that of a jelly: if inflation is squeezed, it will pop up in some other, or indeed the same, guise elsewhere due to the density of connection between the parts. In certain contexts, as this one, the idea that such systems are counter-intuitive is a useful pedagogy for systemic thinking.

STYLES OF PROBLEM-SOLVING

Virtually all explicit social action designed to alleviate widely recognised problems is based on the tacit norm that such problems are soluble, and thus susceptible to analysis. Closely related to this universally held assumption is the notion that such problems are soluble by experts, that explicit specialist knowledge is more likely to lead to solutions than common knowledge or even personal knowledge. In fact, the notion of a solution is quite deeply ingrained in modern society, but as Vickers (this volume) in his introductory address to this meeting has so cogently explained, such assumption is only a characteristic of industrial, perhaps even post-industrial, societies.

The sources of this view are not hard to trace. The apparent success of science in applications to closed systems and the resulting technology, has fostered and indeed continues to sustain this view in our present culture. Moreover, the resulting technical expertise is quite consistent with the way modern society is organized in its division of labour. In some senses, the apparent success of science has been only partly instrumental in bringing about this state of affairs, for the seeds of professionalism were sown in pre-history; indeed, they are a natural consequence of social organization. But, in the present day, the notion that an expertise can formally engender social solutions through technical analysis is somewhat of a departure. The difficulties which pervade such professions, even in times when their expertise is most valued, are not something which society-at-large has found very palatable.

The second related characteristic of technical problem-solving embodies the assumption that the problem-solver is in some way outside the system of interest. This too reflects the heritage of the physical sciences, in which to all extents and purposes, the goal of objectivity has been attained. In short, the problem-solver treats the system of interest as passive, amenable to considered diagnosis and prognosis; or rather, as passive during the period when problem-solving occurs, for there are few who would suggest that the system is passive "per se": this would be too great a contradiction. Nevertheless, the community of problem-solvers, especially systems analysts, consider themselves immune from the obvious activity of the system itself.   If the system fights back when solutions are implemented, the problem-solvers invariably assume a neutral role.

A quite remarkable conclusion emerges when this manner of problem-solving is approached using principles drawn from the theory of systems.   Clearly, closed systems manifest problems of under-standing which can be successfully solved through the production of appropriate knowledge, even though such knowledge is contingent upon the prevailing scientific paradigm.   In such systems, it is less than useful to use the term problem-solving for the problems which are solved pertain to knowledge of the physical world, not the human condition.   However, in social systems, given the points made earlier about the openness and the difficulty of their closure, it is immediately clear that solutions in the traditional sense do not exist.   Solutions require closed domains which do not exist in systems composed of elements with some unpredictable control over their future states.

Furthermore, the definition of such a closed domain is relative even in the physical world.   Stafford Beer (1979:36) has put this very clearly in his recent book when he says:

> "The observer is always part of the system; he
> determines its nature, its purpose, its variety."

Yet, in the development of systems theory in a practical context, that is of systems analytic techniques which purport to explore the nature of social systems, the assumption is quite the reverse.   Like so many ideas, their application in practice leads to subtle, but important, transformations made largely out of convenience, intel-lectual or otherwise.   As will be emphasized throughout this essay, systems theory is an eminently useful vehicle on which to evaluate systems analysis because of this difference in style.

These observations on social action are hardly new; indeed, they are as old as the hills, but with respect to the community of systems analysts, and the present condition of this field, they take on a new urgency.   For as noted in the introduction, these tech-niques are having a substantial impact on social affairs.   The impact is difficult to judge, which is a characteristic of the social domain in any case, but where relatively unambiguous evidence is available, such analysis does appear for a variety of reasons to lead to more problems than are resolved.   A host of critics of the systems movement have seized upon these points, and have character-ised such problems as wicked, squishy, ill-defined, ill-structured, and so on, implying that they are of a different nature from those found in the relatively closed domains of the physical sciences (Batty, 1981).

In fact, the idea that systems analysis leads to solutions to problems of any kind in a social context is now regarded with some disdain by social scientists, with some amusement, even.   Gall (1979:72) in his little book "Systemantics" says:

> "A system usually represents someone's solution
> to a problem.   The system does not solve the
> problem." (my emphasis)

In more serious vein, Dantzig (1979:4) reinforces the point when he says:

> "Modern technological societies are confronted by
> a vast array of problems. They are interlocked,
> one with another, forming a vast web. The solu-
> tion to any one problem will not necessarily ease
> the functioning of the whole -- indeed, it can
> often make things worse."

Yet, opinion is divided in systems theory. After the critiques of the early 1970's, anti-critiques have emerged which suggest that systems analysis has been an abuse of systems ideas which naturally lead to such insights into the connectivity and complexity of problems.   In fact, it is a consequence of being non-systemic which has led to the present dilemma. Churchman (1979), in particular, argues this point with some conviction, but before such a critique of analysis through theory is initiated, it is important to dwell a little longer on the particular styles of analysis which characterize the experience.

STYLES OF ANALYSIS

The dominant style of problem-solving -- that based on the idea that the system of interest is closed, hence its problems definable and soluble -- directly translates itself into a style of analysis on which its methods have developed. If systems are well-defined, that is, assumed to be bounded and knowable in the given sense, then it is possible to develop logically tight theories and models, and to embody these ideas in mathematical terms. Thus, the power of conventional analysis can be brought to bear on the system of interest, but the consequence is that the style of analysis converges towards the precise, the consistent, the quantitative -- in short, towards hard science.

There is a certain irony in all of this for it could be argued that such a style was only made possible by the development of the computer. Social systems, with the kind of perceived complexity demonstrated earlier, resisted such analysis on a widespread scale until the appropriate technology (computers) for such extensive analysis appeared. Only then was this style made possible, and it was obvious why those working in a similar style in the physical domain were attracted to this area, thus reinforcing and establishing a domination. There is considerable evidence for this, for example as shown by Hoos (1972) in her study of corporate and military systems analysis, and by Brewer and Shubik (1979) in their study of war games.

Such analysis has provided what can only be seen as a caricature of social problem-solving. The vehicle for theory-building in the system of interest has been the mathematical model, while the "modus operandi" for problem-solving has been optimization theory, as applied to such models. However, this tight duality of model system and its control/solution through optimisation does not suggest the origin of problem definition and the question of value, which are central to such activities. In physical systems where this style has been successful, there is little controversy over what should be optimized. Normally, as for example in engineering systems, some goal of efficiency, perhaps only workability or operationality, is met using this mode of analysis. However, in social systems, the simplest way in which this form of analysis can be implemented is to embed it within a wider social context from which spring the values and goals which the technical analysis is to optimize. This distinction is clearly uncomfortable, for it results in a separation of technical expertise from its context, a separation which is reinforced by the fact that decision-makers in the wider context inevitably have a different view of what can be accomplished than the analysts themselves. The dilemma might be presented as one in which the analysts have only the technical skill to provide results of a kind, while the decision-makers are more aware of the nature of the system of interest. Often the fusion or complementarity of both sets of knowledge is lacking, and conflict results.

The question is often posed as to whether or not such technically inspired modes of analysis are useful as forms of analogy; although in the development of systems theory, it has been tacitly assumed that this theory can be more than mere analogy, that social systems should be actually organized as theory suggests. A consistent line of argument then runs through the modes of analysis useful for suggesting the type of organization -- the 'solution' -- to the system problem. But this argument has certainly weakened as systems theory has matured. In an urban planning context, Harris (this volume) in his paper to this meeting presents the dualism of positive and normative systems analysis in a form which implicitly (my emphasis) suggests that the value of such thinking is for its analogy, for its power of suggestion, and for the focus it provides on where to search for knowledge and action in social systems. This question is by no means settled, for on the other hand, Beer (1979) implies a much more fundamental role for the use of systems theory in practical affairs, consistent with the earlier philosophies of this theory (Lilienfeld, 1978).

There is also a sense in which the predominant style of analysis has become victim to its own rationale. Every definition of theory or model implicitly or explicitly accepts that simplification is of the essence. Yet, the dominant style of systems analysis has been to attempt to elaborate existing simplifications, rather than to search for more appropriate ones. Urban modelling presents a classic case. Simple structures first suggested two decades or so ago and found wanting in various ways, have been elaborated by disaggregation of spatial and topical detail, by the quest to move statics to dynamics, and by the quest to embed control or optimization mechanisms into such models. Early recognition that models and methods were lacking in some substantial way, did not lead to their rejection but to their elaboration, but as Box (1976) says: "Since all models are wrong, the scientist cannot obtain a correct one by excessive elaboration". In the physical sciences a strong case could be made that most scientific research is simply filling in the details yet to be worked out, of some grander paradigm which emerged in a quite different way (Kuhn, 1962). Thus, in systems analysis, the search for better models in the conventional way is doomed, if the paradigm is inappropriate in the first place.

These are strong words against the dominant style of analysis in planning social systems, and a rather more constructive view of an appropriate style is therefore in order. There is, in fact, an obvious alternative to the precise closed, perhaps rigorous, largely quantitative, elaborative style of systems analysis. It has rarely been thought about in this field, but is quite consistent with systems theory. Systems analysis emphasizes analysis, but it is well-known that problems cannot be perceived, solutions found, without some related phase of synthesis. Indeed, analysis and synthesis are different sides of the same coin, an essential dualism which finds its expression in all human thinking, indeed in the very way the brain itself works (Blakemore, 1977). There are many ways in which this dualism has been described, but what is also widely evident is that the styles of thought characterizing analysis and synthesis are quite different. Analysis tends to be penetrating and incisive, synthesis wide-ranging and speculative. Analysis is concerned with detail and precision, synthesis with structure and pattern. Vickers (1978, this volume) refers to the distinction, this essential dualism, as cause and context.

That different modes of thinking are required for each and its corollary -- that both modes are essential for understanding and action, has been neatly described by Boulding (1958) in his discussion of approaches to economics. He suggests that economic theory is informed by both analysis and synthesis. Analysis consists of exploring problems in great detail, with considerable rigour and almost invariably by some form of partial theory in which most of the known influences on the theory are assumed constant. He refers to this as essentially "slicing" the problem into manageable, analyzable parts. In contrast, economic synthesis, macro-economic theory, consists of aggregating the problem, extracting its essential elements, and exploring its pattern and order in a high-level approximation. This is a method of "squashing" the problem to a form in which its structure is manageable and understandable. As Boulding (1958) says: "This is the economics of the bludgeon rather than the scalpel".

Systems analysis has been peculiarly deficient in deriving and applying techniques for squashing problems rather than slicing them for the predominant style of slicing has given only the most partial picture. Indeed, the emphasis on quantitative rather than qualitative is a consequence of this style. As Vickers (this volume) says:

> "I am a doubter about quantitative modelling
> because I believe that the effort to quantify...
> will always tend to obscure the importance of
> factors which are not quantifiable...."

In one sense, these dualisms involve extremes which never quite capture the essence of what is felt. The issue, in my view, is not about quantitative or qualitative, objective or subjective, partial or general, or whatever, but about more suitable styles of analysis which accept the complementary, yet non-comparable modes of analytic and synthetic thinking. This point will be developed in a later section, but it is worth concluding this argument concerning style by noting the ideas presented by Kane and Verlinsky (1975) in their discussion of appropriate methods of analysis. They argue that systems analysis has so far been based on a style akin to "arithmetic", whereas there is an urgent need for a mode of thinking based on the "geometry" of the situation, on its pattern and structure. Arithmetic and geometry emphasize "a common mathematical bond", "affirm one another and each extends the other's truths". (Kane and Verlinsky, 1975:115). A change of direction towards the development of 'geometrical' methods for problem-solving is clearly necessary, or this would inevitably open systems analysis to its wider context alluded to here.

A CRITIQUE THROUGH SYSTEMS THEORY

So far, in this essay the idea has been developed that systems theory and systems analysis have diverged, that the original tenets of systems theory have been abused in the practice of systems analysis. Thus, it is possible to begin a critique of the experience of systems analysis using systems theory. There is considerable dispute about this strategy, for much of the literature on systems theory itself has been so single-minded that it is difficult to use the literature concerning theory as "the" lens through which to view analysis. In some respects, this difficulty accounts for the space devoted earlier in this essay to a sketch of the rudiments of systems theory itself, its principles, so to speak. Yet, as might be expected from a view which emerges from the paradigm itself, to save the paradigm it is necessary to employ such a method of critique.

It is useful to proceed by presenting a typical critique from outside the field, and contrasting this against the critique to be developed here. An example is provided by the ideological analysis produced by Lilienfeld (1978:191) who says: "Systems thinkers exhibit a fascination for definitions, conceptualisations, and programmatic statements of a vaguely benevolent, vaguely moralizing nature, without concrete or specific references to historical, social or even scientific substance."

And later he supplements these 'besetting vices' with (p. 227):

> "...a fondness for abstract schematic formulae and diagrams having little practical reference; a fundamental begging of questions that takes the form of an unstated and presumably invisible shift from concrete world 'systems' in their fullness and complexity to closed formal models based on convenient 'simplifying assumptions', a shift we are not supposed to notice....."

Several of these points are hard to refute for the experience of systems analysis, as already observed, bears them out. Moreover, it is hard not to escape the fact that the vast majority of people involved in systems analysis have been merely content to elaborate in the narrowest sense, rather than to really examine the limits to their knowledge. A great deal of systems thinking, so-called, is guilty of what Box (1976) refers to as "mathematistry", mathematics for mathematics sake. Indeed, this criticism can be levelled against a great deal of formal social science, as well as physical science, and it involves a type of intellectual entrepreneurship of the worst kind. Nevertheless, in other fields it seems more tolerable, largely, I suspect, because in those fields there are more philosophers and advocates working within the paradigm who are prepared to defend its basic tenets, and to argue that its practice represents an abuse of its theory. Only is this situation beginning to emerge in systems theory.

Some of Lilienfeld's critique quoted above might be excusable, or at least explicable in the early days of any movement, due to the exigencies of simply 'getting started'. However, where the abuse of the systems ideal is clearest is in his point that the "fullness and complexity" of the system is invariably reduced to a "closed formal model based on convenient 'simplifying assumptions'". Systems theory itself teaches the importance of proper system definition as a prerequisite to appropriate simplification, and of a clear purpose for systems thinking as a prerequisite to such definition. This has been so rarely the case in systems analysis. For example, in the development of operations research in a public policy context, many of the associated problems have been tackled using off-the-shelf techniques, often totally inappropriate to the system of interest. In short, these types of problems have been treated according to some assumed classification derived from other systems where the techniques might have appeared more successful. It is therefore somewhat ironic that these principles which depend so much on suggesting the need to clarify and define prior to system modelling, should be abandoned in the practice of systems analysis.

Several theorists have recognised this dilemma, although they have been hard put to leap to a defence of their own subject matter. Churchman (1979) argues that the systems approach is not something that has simply arisen in recent decades, but is to be found in the writings of most philosophers through recorded history. He suggests that the essential logic of the systems approach relates to its search for comprehensiveness, and this implies that the systems theorist is never satisfied, is never convinced that he has gained an adequate understanding of the system. For if the system is effectively of infinite dimension, then every different view will enrich the perspective. Indeed, the existence of these differences is the system. Thus, the systems theorist sees good in all things. On planning, Churchman (1979:36) has this to say:

> "So in systems planning, in any problem, are to be found all other problems.... This means that an appropriate style for the planner is to recognise the reality of specific problems in order to obtain insight into all problems or the 'whole problem'."

One final aspect of the critique is worth developing before the experience in urban planning is sketched. It can be argued that the systems approach has a transparency which others do not, for in the quest for comprehensiveness, theories and methods must be employed which attempt to capture, singly or in conjunction, all recognizable facets of the system. As more facets are recognized, the opportunity for an irrelevance of any single model increases, as does the opportunity for criticism. Often the most trivial of techniques escape perusal and are judged acceptable, in contrast to more elaborate models which attempt the same. Indeed, this is consistent with the Popperian view of science which asserts that as theory matures, it provides more opportunities for falsification; if the theory withstands such attempts, scientific progress occurs. Examples of this exist everywhere. In urban planning in the United Kingdom, naive forecasting methods based on trend extrapolation are used extensively, whereas more elaborate techniques which would enrich such analysis and provide more reliable results are strongly resisted: population forecasting is the prime example. The power of the systems approach is such that, combined with the foibles of intellectual entrepreneurship, it provides a coherent explanation of such events. To illustrate this further, a brief sketch of such approaches in urban planning will be presented, before the arguments of this essay are drawn together in conclusion.

URBAN PLANNING AS EXEMPLAR

The social sciences have provided extremely fertile soil for the development of the systems approach, primarily because the infinite complexity of social systems has led to several distinct disciplinary approaches which inevitably require synthesis. In particular, social action which always depends upon inadequate knowledge of the system, requires some integrating framework to fall back upon, and systems theory has provided this base. Urban planning presents a classic example of such action, and its history is dominated by an increasing expansion of its subject matter and scope, as planners and their clients have gradually learnt about the infinite complexity and connectivity of their system of interest. Indeed, it could be argued that the history of the social sciences is one of such gradual realizations.

In urban planning, however, more so than in most other distinct social behaviours and collective actions, the change in scope and substance has been quite enormous during the last two decades. Certainly in Britain, and probably elsewhere, in 1960 urban planning was, to all intents and purposes, a professional skill which was more akin to architecture than to any social science. Its concern was almost exclusively with the physical environment, in particular, the aesthetics and functional form of this environment, and it dealt with social processes only in so far as these processes were influenced by physical form. This was the critical point -- that environment affected social process, rather than the reverse -- and such an extreme view was highly questionable, as the changes in the 1960's demonstrated.

A variety of circumstances thrust urban planning from its image and concern as architecture-writ-large to its present day focus as a particular form of social and economic action. The environmental determinism of the previous era simply did not hold up, but more important, the visible influence of changes in the city, such as traffic, could clearly be treated as systems phenomena. Finally, the 'apparent' success of systems analytic techniques in other areas showed that systems theory could at least provide an operational mode consistent with the idea of a community of experts, namely the planning profession. Of course, the development of systems theory in planning was not without its critics from the start; many argued, as Vickers (this volume) has argued at this meeting, that the idea was sound but its operation might narrow, rather than broaden, the area of concern. This was the reaction against quantitative modelling, although in the early years of the systems approach, the critique was rarely put so eloquently as it is today. The story of systems theory in planning is well-known and will not be repeated here (see Batty, 1981), but certain aspects of the experience which relate to the present critique will be emphasized.

The great strength of systems theory -- the quest for comprehensive understanding -- has been its great weakness; that is, the notion of system provides a model which is comfortable in that to engage in systems analysis, the system must be closed. Moreover, the model of the system and its optimization must be consistent, and once consistency has been attained in such a closed domain, then the elegance of the closure often drives away any desire for improvement. In short, there is considerable reluctance to reject such a system in favour of something much less tidy as any new approach would inevitably be. This has the effect of internalizing the critique so that the closed system itself is perfectly engineered, yet its relevance can continue to remain in doubt. The elaboration of urban models referred to previously, rather than the search for more appropriate model structures, represents a perfect example. It is easier, and more obvious progress in the detailed analytic sense can be made, if this strategy is adopted. Consequently, whole schools now exist in the social sciences devoted to the formal elaboration of models which are regarded as irrelevant by others. This too can be regarded as part of the proliferation of societal complexity.

At the same time, the tension between what is possible in systems analysis and is demanded by a systems approach, has actually sustained the development of systems theory in planning. An examination of urban and transport modelling, for example, shows that there are waves of enthusiasm and disillusion with such approaches, quite clearly discernible over as short a span as thirty years. It appears that although the knowledge contained in such models is always inadequate, it can never be abandoned because it contains important facets of the system which cannot be captured by other approaches. One can speculate endlessly about the existence of such cycles or otherwise (Batty, 1979; Downs, 1972), and it is hard to generate perspectives on contemporary history. Nevertheless, these ideas are reinforced by work such as that presented by Pack (1978). Indeed, Pack's (this volume) paper to this meeting represents another surprising twist to the history, which implies that the development of models which seem to have been broached so enthusiastically in the United States in the 1970's, may well be over, once again.

Much more significant than changes in style have been changes in the types of urban problem which planners and their public have perceived of importance during the last two decades. In one sense, it is the types of problem which have led to the prevailing style of analysis, rather than the style of analysis or its theoretical paradigm being instrumental in problem definition. Urban issues in the 1960's were dominated by questions which appeared potentially suited to systemic analysis -- problems of inner city decline and suburban growth, of spatial economic efficiency and accessibility, of movement, and so on; in short, problems dominated by the goal of efficiency which seemed soluble in terms of local spatial systems. In the 1970's, and probably in the decade we are beginning, the predominant problems have been concerned much more with questions of equity, and consequently, these have rarely appeared soluble in a local spatial framework. Problems of energy, for example, are of global significance, as indeed are problems of income distribution between cities: even within cities, distributional questions only appear to be understandable in a geographical framework wider than the city itself. The list of such issues is long: pollution, the fiscal crisis in cities, the problem of local unemployment, housing markets infected by endemic inflation, are all issues which can only be handled outside their immediate spatial frame, in stark contrast to questions of spatial efficiency which systems analysis has most to say about.

In fact, systems theory would appear to have a great deal to say about such issues, but these issues remain remote from any of the forms of systems analysis so far developed. Because the predominant questions in planning are now so much concerned with broad issues pertaining to the human condition, these issues do not fall within the domain of conventional modelling and analysis, hence the disreputable air which now surrounds such work. Indeed, in some quarters, certainly in planning practice, the rejection of systems analysis can be interpreted as a type of anti-intellectualism, which also appears rife throughout government and wider public attitudes to many forms of collective action at present. Perhaps the real insight given by systems theory is that such attitudes towards systems analysis are explicable in terms of the closed system thinking which has dominated planning expertise during the last thirty years.

Yet, the experience in urban planning with these forms of analysis has been partly salutary. The importance of systems theory is heightened by critical analysis of the experience, for the experience suggests that what has been learnt is the difficulty of operating and accepting the limits to human systems, perhaps even the difficulty of operating in the self-conscious expert orientated style which characterizes all contemporary approaches to human action. With this realization comes the momentum for change in style, and a degree of humility which is an essential recognition of the level of complexity to be dealt with (Webber, 1978). This conclusion is consistent with the quote by Arrow (1974) which served to introduce and focus this paper. And it provides the direction for a more appropriate style of planning.

THE LOGIC OF INTELLECTUAL PLURALISM

Systems theory in the form which Beer (1979), amongst others, refers to as managerial cybernetics, poses an extremely simple solution to the problem of proliferating societal variety. This is based on establishing a requisite variety at the level of the basic element of the system, at the level of the individual. Natural self-regulation, argues Beer, would lead to an organization giving maximum personal freedom. In short, such a system would give back power to the individual, to let the individual decide, for with systems of infinite complexity, this is the only feasible way of meeting the law of requisite variety. The theory is not as yet well-worked out, for in a sense, it still represents a straw in the wind. Nevertheless, others concerned with such matters have come to the same conclusion (see Friedmann, 1973), and there are a host of similar indications from other commentators concerned with the dilemmas of contemporary social action.

In systems of infinite complexity, any consensus with regard to the most appropriate strategy and theory for understanding and changing the system, is not possible. That is why there are as many theories of society as there are individuals concerned with articulating such theories. This is notwithstanding the fact that a few distinct schools might emerge, approaches which appear promising but only in certain tractable ways. At the most elemental level, each individual holds a different theory, relating to his own life space. This is very clearly manifest in the variety of approaches which characterize social science, and in this context, it seems inevitable that each approach will enrich in some distinct way our collective understanding. Indeed, it would appear that the problem of theory in the social domain is one of gaining an ability to use the variety of approaches available to build a total understanding, not in aiming for one general theory to the exclusion of all others. The same may even be true of the physical world for knowledge is designed, man-made, and as Popper (1957) has so frequently emphasized, future knowledge is impossible to predict. In this context then, the logic of intellectual pluralism -- the acceptance that there is something to be gained from every style of analysis -- is inescapable.

It is interesting that those working in the practice of systems analysis -- those closer to the decision-making and implementation which systems analysts seek to inform -- should feel the need for this liberal view quite keenly. In this meeting, Pill (this volume) argues that the quintessential model for extracting knowledge and engendering action through urban governance should be based on a bridging of the various cultures involved: academics, bureaucrats, politicians and mediacrats. The different intellectual styles which each of these groups brings each contributes to our overall understanding of social action, and in a slightly different context, Allison (1971) comes to a similar conclusion. Indeed, Harris (1975) argues that such bridges between theory and practice should provide the rational and focus of relevant systems modelling, thus supporting this pluralism from a more technical stance.

All of these notes are pointers to the need to approach social action with a much more open mind, with the recognition that the system of interest, hence the problem or perceived difficulty, is uncertain and infinitely complex. With the rate at which variety and novelty are being proliferated in the present day, it is essential to approach planning with a humility and ability to learn as one proceeds. At no time in the past has it been more important to realize the uncertainty surrounding every type of behaviour, and of the need to explore the process of action. The logic of a 'learning society' as various theorists term it, is a natural extension to this argument. And in any critique of the experience of systems analysis in a social context, its future style must reflect these notions. As Hoos (this volume) has said: "It is the certainty about uncertain questions that should concern us."

CONCLUSIONS

One conclusion from this argument is that in suggesting a broadening of systems analytic thinking according to the original tenets of systems theory, it is essential to consider philosophy. The concern with method which has characterized the experience should be complemented with a concern for methodology. Thus, meetings such as this one represent an inevitable development in the quest to provide a more relevant style of social action. The cybernetic theory of society forces us to consider these broader questions, for although the quest to explain more and more is doomed, it emphasizes the need to bear in mind the limits to social action which are a perennial feature of such activities: as reflected yet again in Arrow's (1974) introductory quote.

The major theme of this paper has been that the experience of systems analysis is an abuse of systems theory, that the tenets of systems theory have not been considered in the application to social problems. Systems theory suggests that questions of definition, classification, extent and so on are essential prerequisites to understanding; that is, that such theory can only be developed if these issues are resolved first. Thus, systems theory tends to widen debate about what constitutes the system and problem of interest. In contrast, systems analysis and modelling has largely disregarded these questions. The problem of system definition has simply been ignored in most cases, as problems have been crudely sliced to fit available techniques. Consequently, the style characterizing such analysis has been focused on internal issues, on problems of better and better modelling of more and more restricted systems. In one sense, it can be argued that this is explicable, given the need to proceed rather than just reflect, that analysis can never afford the luxury of extensive speculation that theoreticians are able to engage in. But, in this context, the luxury becomes a necessity, as the experience suggests.

These arguments are also reflected in style, and a consequence of the woeful narrowness which has emerged has been an unremitting application of a limited set of techniques without any consideration of other possibilities. A much greater sense of learning when not to use techniques must evolve. As Churchman (1979) has said: "In most cases, a student of decision-making should learn formal mathematics, but should also learn when not to use mathematical language". Following Richardson (1956), because mathematics is so definite, situations should be clear where it is definitely inappropriate, and this represents the power of formal reasoning. A change in style is therefore definitely required. Much more thought should be given to context, to ways of effecting a synthesis of the factors affecting any problem, to ways of understanding structure and pattern. And this must be reflected in techniques which squash, rather than slice, which recognize the limits imposed by the intrinsic ambiguity and uncertainty of social systems. Using Kane and Verlinsky's (1975) imagery, what is required is much more 'geometry' to complement the precise modelling abilities which have been acquired during the last two decades. Given the present temperament and concern of the field, this will be no easy achievement in future years, but it is of the essence if systems theory is to have the impact which its logic provides.

REFERENCES

Alexander, C., 1964, Notes on the Synthesis of Form, Harvard University Press, Cambridge, Mass.

Allison, G.T., 1971, Essence of Decision: Explaining the Cuban Missile Crisis, Little, Brown and Company, Boston, Mass.

Arrow, K.J., 1974, The Limits of Organization, W.W. Norton and Company, New York.

Ashby, W.R., 1956, An Introduction to Cybernetics, Chapman and Hall, London.

Batty, M., 1979, Paradoxes of Science in Public Policy: The Baffling Case of Land Use Models, Sistemi Urbani, 1, 89-122.

Batty, M., 1981, A Perspective on Urban Systems Analysis, in D. Banister and P. Hall, (Eds.), Transport Policy and Planning, Mansell, London.

Beer, S., 1974, Designing Freedom, John Wiley, London.

Beer, S., 1979, The Heart of Enterprise, John Wiley, Chichester, UK.

Berlinski, D., 1976, On Systems Analysis: An Essay Concerning the Limitations of Some Mathematical Methods in the Social, Political and Biological Sciences, The MIT Press, Cambridge, Mass.

Blakemore, C., 1977, Mechanics of the Mind, Cambridge University Press, Cambridge, UK.

Boguslaw, R., 1965, The New Utopians: A Study of System Design and Social Change, Prentice-Hall, Englewood Cliffs, New Jersey.

Boulding, K.E., 1958, The Skills of the Economist, Hamish Hamilton, London.

Box, G.E.P., 1976, Science and Statistics, Journal of the American Statistical Association, 71, 791-799.

Brewer, G.D., 1973, Politicians, Bureaucrats and the Consultant: A Critique of Urban Problem Solving, Basic Books, New York.

Brewer, G.D. and Shubik, M., 1979, The War Game: A Critique of Military Problem Solving, Harvard University Press, Cambridge, Mass.

Churchman, C.W., 1979, The Systems Approach and its Enemies, Basic Books, New York.

Dantzig, G.B., 1979, The Role of Models in Determining Policy for Transition to a More Resilient Technological Society, IIASA Distinguished Lecture Series, 1, International Institute for Applied Systems Analysis, Laxenburg, Austria.

Downs, A., 1972, Up and Down with Ecology: The Issue-Attention Cycle, The Public Interest, 28, 38-50.

Friedmann, J.R.P., 1973, Retracking America: A Theory of Transactive Planning, Doubleday, Garden City, New York.

Gall, J., 1979, Systemantics: How Systems Work and Especially How They Fail, Fontana-Collins, London.

Harris, B., 1975, Bridging the Gap between Theory and Practice in Planning, in J.R. Pack, (Ed.), Proceedings of the Conference on the Use of Models by Planning Agencies, Fels Center of Government, University of Pennsylvania, Philadelphia, Penn.

Hoos, I.R., 1972, Systems Analysis in Public Policy: A Critique, University of California Press, Berkeley, California.

Kane, J. and Verlinsky, I.B., 1975, The Arithmetic and Geometry of the Future, Technological Forecasting and Social Change, 8, 115-130.

Kuhn, T.S., 1962, The Structure of Scientific Revolutions, The University of Chicago Press, Chicago, Illinois.

Lilienfeld, R., 1978, The Rise of Systems Theory: An Ideological Analysis, John Wiley and Sons, New York.

Linstone, H.A., Lendaris, G.G., Rogers, S.D., Wakeland, W. and Williams, M., 1979, The Use of Structural Modeling and Technology Assessment, Technological Forecasting and Social Change, 14, 291-327.

Milgram, S., 1967, The Small World Problem, Psychology Today, 1, 61-67.

Miller, G.A., 1956, The Magical Number Seven Plus or Minus Two: Some Limits on our Capacity for Processing Information, Psychological Review, 63, 81-97.

Pack, J.R., 1978, Urban Models: Diffusion and Policy Applications, Regional Science Research Institute, Philadelphia, Penn.

Popper, K.R., 1957, The Poverty of Historicism, Routledge and Kegan Paul, London.

Richardson, L.F., 1956, Mathematics of War and Foreign Politics, in J.R. Newman, (Ed.), The World of Mathematics, Volume 2, Simon and Schuster, New York.

Simon, H.A., 1962, The Architecture of Complexity, Proceedings of the American Philosophical Society, 106, 467-482.

Simon, H.A., 1974, How Big is a Chunk?, Science, 183, 482-488.

Simon, H.A., 1978, On How to Decide What to Do, The Bell Journal of Economics, 9, 494-507.

Vickers, G., 1978, Rationality and Intuition, in J. Wechsler, (Ed.), On Aesthetics in Science, The MIT Press, Cambridge, Mass.

Webber, M.M., 1978, A Difference Paradigm for Planning, in R.W. Burchell and G. Steinleib, (Eds.), Planning Theory in the 1980's, Center for Urban Policy Research, Rutgers University, New Brunswick, New Jersey.

WHAT REASONS FOR RATIONALITY?   IN SEARCH OF A FUTURE FOR RATIONAL

METHODS IN URBAN PLANNING

Helen Couclelis

Secretariat for Physical Planning and the Environment
Ministry of Coordination
Platia Syntagmatos, Athens
Greece

HOW RATIONAL IS "RATIONAL"?

Back in the early seventies, Echenique ended a review of urban modelling and its problems, very appropriately, with a quotation from Don Quixote:

> "Let the dogs bark, Sancho, it shows that we are
> moving forward."

Echenique (1975)

Few dogs are barking today. Either the critics now feel that further attacks on systems methods are no longer worth their ink, or there is no longer anything much moving to bark at. Surely enough, somewhere between general indifference and premature death, we still find the odd efforts to keep the tradition going; but these are getting few and far between, and they sound increasingly apologetic (Batty, 1979).

Depending on which side of the fence one stands, the situation as regards the present relationship of systems analysis with urban planning is cause for deep concern or for self-righteous glee. On the one hand, those who in the past twenty or so years have invested so much enthusiasm, skill, intellectual energy and material resources in the development of rational methods for use in planning are anxious to rescue what they can and define a viable direction for their future efforts. On the other hand, those who never would or could have any understanding for systems analysis are content to utter the most disagreeable four-word phrase in the world, "I told you so". And, most disturbing of all, there is also the case of the

449

former believers who have lost their faith: their negation has
devastating effects — and their ranks are still growing.

Indeed, the scene has never looked bleaker. Widening realization of past failure, vanishing present interest and increasing
pessimism for the future are the three phrases that seem to summarize the state of the art in systems planning. In most of the
papers written for this seminar which include a review of past
experience in the use of rational methods in planning, the pattern
of ascent, decline and stagnation comes out very clearly in spite of
brave efforts to end on an optimistic note. The few reassuring
voices no longer persuade.

Was then systems planning a dead end comparable to Lamarkism in
biology — a hypothesis that was attractive and plausible in principle, but happened to be defeated in the confrontation with practice?

For it is a fact that the idea of systems planning was both
attractive and plausible. On the one hand, systems analysis was
developed specifically in order to aid rational decision. On the
other hand, there is hardly a definition of planning that does not
involve the notion of rational decision explicitly or implicitly
(see Simmie 1974:161 for a collection of definitions). Planning can
be good or bad, efficient or inefficient, wise or misguided, but it
could hardly be irrational: like the smileless Cheshire Cat, irrational planning would be a definitional conundrum.

Things being so, the introduction of systems analysis in planning seemed a very obvious and natural move. Thus, if there is a
historical question about the relation of the two fields, this is
not about how they came to be associated, but rather about why that
association did not work any better than it did.

As a matter of fact, the affinity between systems analysis and
planning goes much further than the common concern with rational
decision. It is usual to distinguish two components in planning,
the "substantive" component which comprises the knowledge about the
facts planning is dealing with, and the "procedural" component
involving the consideration of the process of planning as a series
of interrelated steps (Hightower, 1969; Faludi, 1973; Camhis,
1979). Systems methods cover both these aspects: the former through
a wide range of urban systems models, the latter through methods
such as Cost-Benefit Analysis, PERT, Critical Path Analysis, PPBS,
Zero Based Budgeting and the like, as well as models of formal decision.

The problems arising through the application of systems methods
in urban planning and policy-making have been amply documented in a
vast body of critical literature accumulated over the years, and
several reasons for this perplexing mismatch have been identified.
These reasons have to do with issues that are technical, theore-

tical, pragmatic, psychological, political, ideological and even ethical, and there is little doubt that the causes of the present unsatisfactory situation with respect to systems planning are to be sought in a synthesis of such factors.

In this paper, I wish to focus on one particular aspect of the total problem, the question of the nature and role of rationality in urban planning and policy-making and its relation to the same concept as construed from the perspective of systems analysis. The idea of rational decision may be central to both disciplines, but are the kinds of rationality involved in each case necessarily compatible with each other? The answer could well be "no".

The point I wish to make is that the direct use of current systems methods in planning and policy-making is only rational within very particular and rather limited contexts. That is, I would argue that in many kinds of planning situations it would not be reasonable to use "rational" methods, even if available in principle for the particular task. Further, I will suggest that much of the present disillusionment with systems methods is due to the fact that over the years, implicitly or explicitly, these have been much too closely associated with the "rational model" of planning, a normative representation of the planning process as a case of "objective" rational behaviour, which for several reasons is no longer tenable.

Outside classical economics, it is fairly well accepted today that there is no absolute standard of rationality. The "objective" rationality of Economic Man has gradually given way to various conceptions of "subjective" rationality whereby the acting agent's behaviour is assessed in the light of the pragmatic constraints imposed by the environment in combination with the limitations deriving from the agent's own necessarily incomplete comprehension of the global situation. Herbert Simon's concept of "bounded rationality" and the notion of "satisficing" as opposed to "optimizing" behaviour are by now well established. As the success of ideas such as Kuhn's and more recently, Feyerabend's show, such watered-down conceptions of rationality have even permeated contemporary views of scientific methodology itself, which according to Pareto is the form of rational behaviour 'par excellence' (Winch, 1958:101).

It is quite clear that these more recent conceptions of rational behaviour are much closer to experimental fact than earlier ones, although they do render theorizing much more complex. The earlier "objective" view of rationality can eventually still be maintained as a convenient "null hypothesis" in some cases of abstract theory building (Buckley, 1967), but in disciplines such as planning which on pain of death must never lose concrete practice from sight, insistence on "objective" formulations of rationality would be counter-productive, to say the least.

At any rate, in much of today's social science, including (behavioural) geography, rational behaviour tends to be seen as relative, subjective and ad hoc. This is a far cry from the standards set by early systems planners who strived to develop a "science" of planning true to the classical ideal of scientific rationality and objectivity.

It would be unfair to systems analysts, and also quite inaccurate to state that they are not aware of these developments. A number of very thoughtful papers written for this seminar, and some recent work on the analysis of rational decision in the face of uncertainty (Holling, 1978) testify to the contrary. I believe that most urban analysts would subscribe to the view that while planning behaviour on the whole may be only relatively and subjectively rational, it should be possible by using systems methods, to build into it some patches of 'harder' rationality; and that the more extended and numerous these patches are, the better the overall result will be. I shall argue in the following that this "patchwork" view of planning is only tenable under very particular circumstances, and that on the whole the direct applicability of systems methods to even limited aspects of the planning process largely depends on the overall nature of that process, itself a function of the wider socio-cultural context of planning. In other words, the hypothesis is that there are several different styles of planning determined by factors that are on the whole much beyond the control of planners and decision-makers, and that systems methods are compatible with some of these, but not with others. It will also be argued that there seems to be a general trend towards the styles of planning that are less compatible with systems methods as currently known. I will suggest that a frank recognition of this situation, rather than condemning systems methods in the context of planning to death by irrelevance, could eventually open up a new perspective for the definition of some future more promising type of association between the two fields.

RATIONAL METHODS AND "RATIONAL" PLANNING

A point that is easily neglected when speaking of rational methods in planning is that methods in general cannot be rational per se, except in a very restricted and rather uninteresting sense. "Rational", just like "useful" or "meaningful", is a predicate that refers to a relational quality of the subject even more than to an intrinsic one. Just as a statement may be judged meaningful either in a narrow grammatical sense or when seen in the more general context of discourse, a method may be called rational either because it has been developed in accordance with accepted logical (eg. scientific) procedures, or because it helps rationalize the context of its use. Systems methods have serious claims to being rational in the first sense, but this is no indication at all as to how they may perform under the second criterion, that is, when seen in the context of planning. This may seem obvious enough in retrospect, but, in fact, systems analysts and decision-makers alike might have been spared much disenchantment over the years if only the relation of systems tools to the task of actual planning practice had been made a priori more explicit.

It is quite clear that systems planning methods did not arise out of a conceptual vacuum with respect to the potential context of their use. They were developed in the same intellectual climate that produced the "rational model" of planning, itself an outcome of early systems thinking in the field. The "rational model" sees planning as a procedure involving a sequence of four fundamental steps: problem definition (and goal setting), generation of alternatives, evaluation and choice, implementation and monitoring. In spite of any iterative loops that may be added to the basic flow diagram, these steps clearly form a logical -- if not always chronological -- sequence (Chadwick, 1971; Lee, 1973:3).

There is a marked correspondence between the above representation of the planning process on the one hand, and accounts of the "scientific method" as construed from a positivist, deductive-nomological, verificationist perspective on the other (Camhis, 1979: chapter 1). The admission that this is a normative, not a descriptive representation of planning, helped hold scepticism at bay, in particular, the obvious objection that "things don't quite work that way in practice". Thus, the case was clear: planning could and should be a rational (ie. scientific) procedure; systems analysis is intrinsically rational (ie. scientific); therefore, systems methods are, in principle, perfectly compatible with planning, if not a necessary part of it, and should be used unrestrictedly.

From this perspective no further questions concerning the role of systems methods in planning need be asked, apart from those leading to the solution of the numerous "puzzles" arising within the framework of rational planning itself. There is a question of improving the methods, of expanding the methods, of developing new methods, of making methods more operational: there is hardly ever question of questioning the methods at the level of the fundamental presuppositions that link them to the wider context of their use. Given the rational model of planning, systems methods can be used in all innocence: there is no room for self-reflection within established paradigms (Kuhn, 1962).

The problem with rational planning as a normative model is that it is based on a large number of presuppositions and assumptions, none of which is self-evident or necessarily desirable as an ideal, and yet it is easily seen that the applicability of current "rational" methods very largely depends on the realization of these assumptions in practice. All of them have been identified and attacked by critics over the years, but it may be useful to review some of these premises of rational urban planning in the present context.

   Planning is a scientific activity. As discussed above, this is
a fundamental presupposition of rational planning and it is the one
that makes the applicability of systems methods possible in principle.   What may be added here is that the notion of scientific
activity implied is that of empirical natural science as seen from a
positivist perspective. This leaves a large part of contemporary
science, and a large number of alternative epistemological perspectives outside the scope of this particular model of planning.

   Planning is problem-solving. This is really a postulate of the
preceding assumption to the extent that science is understood to be
the problem-solving activity "par excellence". The parallel with
scientific problem-solving is clear: planning problems are defined;
alternative solutions (hypotheses) are generated; these are evaluated according to certain rules and criteria (testing of hypotheses); the solution implemented is monitored, this leading to the
next problem formulation and the next set of possible solutions
(further observation and improved hypothesis formulation). Systems
analysis, on the other hand, is more often than not explicitly
defined in terms of rational problem-solving. In the words of
Stanford Optner:

         "Systems analysis should be viewed as the most
         recent and perhaps most comprehensive vehicle
         for complex problem solving."

                                        Optner (1973:10)

   Clearly, the problem-solving paradigm is another factor that
directly relates systems methods to rational planning. However, as
soon as notions such as regulation, control, conflict resolution,
societal steering and so on are substituted for problem-solving in
more recent conceptions of planning, both the idea of "rational
planning" and the comfortable place of current systems methods
within it suddenly become problematic. This is in spite of the fact
that the notions of regulation, control, adaptation and the like are
themselves integral part of another discipline in the systems tradition, cybernetics.   Apart from the commitment to the model of
rational planning, there is no obvious reason why systems methods
for planning should not have developed in that direction much more
than they actually have. The point is worth pondering and will be
taken up later in this paper.

   Planning is design. Design is a particular kind of problem-
solving where the answer to the problem is the creation of some
physical or conceptual form. Traditionally, urban planning has been
associated with the creation of urban form. More recently, the idea
of design in planning has expanded to include the definition of
structures and systems that are not primarily physical, such as
public services networks. At the same time a science of design has
developed where there was nothing but intuition, experience, hunch
and art, and in the development of this new discipline, the role of

systems analysis has been prominent. Even more than problem-solving, design is an "objective" activity, in the sense that there is a definite end product, an "object" of some sort to crown its efforts; it is also a more explicitly goal-oriented activity since the goal is a concrete end state to be reached, rather than the fulfillment of a number of general requirements. In fact, design appears to be the most rigid and static form of problem-solving.

Planning is cohesive. According to the standard systems adage, everything in a system is related to everything else. The immense success enjoyed by the idea of "comprehensive" or "integrated" planning until recently is a consequence of the belief that planning must deal with everything at once and be informed about everything at once in order to be scientific. Within the more restricted fields which they cover, systems methods must also deal with everything at once, hence the permanent problem with data bases which never seem to be good enough for the needs of modelling.

In the same vein one could point out a series of further presuppositions underlying the rational model of planning which in no way are a priori obvious: planning is bold, rather than incremental; planning is a formal, rather than an informal activity; planning is something to be done by experts; and so on. With all these presuppositions, systems methods are in full accord.

To conclude the preceding discussion, one may say that the "rational model" is acceptable as a normative representation of planning to the extent that (a) as to the substantive aspect, the issue at hand is (seen to be) mainly technical and (b) as to the procedural aspect, the planning process is strongly centralized. It is significant to note in this context that the success stories of systems methods applied to planning problems are practically co-extensive with the realization of the above two conditions. Thus, it would be irrational to deny the actual and potential contribution of systems analysis to the solution of traffic engineering problems, to the planning of physical networks, to the allocation and management of urban service systems and other such largely technical issues. In these cases, the second requirement of a relatively closed and centralized planning process is fulfilled almost automatically since the amount of expertise needed to approach such problems usually limits non-specialist involvement to a bare minimum, at least up to the stage of final proposals. These are the "hard" issues of planning, and to the extent that such issues remain topical, systems methods in their present or amended forms have a definite role to play.

This is a gratifying statement to be able to make, but the antilogue is all too obvious: what is the relative importance of "hard" issues in today's planning and policy-making? How often does one need to plan a metropolitan railway system, or a new main sewer network, or a comprehensive fire protection service in a given planning framework? And how well do systems methods fare when confron-

ted with the kind of issues currently addressed by planners and
decision-makers in the normal course of their efforts? The short
answers to such questions are well known, and they are cause for
concern in the field of systems planning.

To reiterate what is probably the most popular commonplace in
the past decade's urban literature, there is a general trend in
planning towards less technical, more social and political definitions of planning issues. By this is meant, on the one hand, that
more social, political and cultural issues than ever before somehow
come under the scope of urban planning, and on the other, that even
in the case of problems that are primarily technical, there is an
ever-increasing concern for the non-technical dimensions of these
issues. Unemployment, social malaise and the integration of minorities, for example, are concrete planning problems these days; and
increasingly often of late the cost-benefit nirvana of traffic engineering projects is troubled with questions such as those about the
ethics of road building in the face of human deprivation elsewhere
in town. The "snail-darter syndrome", the fighting of technical
priorities from the standpoint of some fully incommensurate system
of values, is a fact to be reckoned with by modern planners and
decision-makers alike.

To the extent that planning is understood to address societal,
rather than technical issues, many of the traditional conceptions of
planning systems and decision structures are no longer tenable.
With the drastic widening of the field of concern in urban planning
came a corresponding expansion of the field of formal and informal
participants to the process. This is reflected in both the planning
process itself which becomes diffuse and decentralized, and in the
structure of decision-making, which becomes equally fragmented. As
the sources of relevant information proliferate out of control, it
is no longer practically possible for anyone to have a comprehensive
understanding of either the planning system or the system planned.

As the reassuring basis of the rational model of planning
disintegrates, so does the clear sequence of logical steps that
makes the application of systems methods such a plausible venture.
Thus, as will be argued later in greater detail, increasingly often
we encounter situations where:

1. Goals cannot be set at the beginning.

2. There are no problems to be solved, only issues to be dealt
    with.

3. Alternative possibilities of action can only be thought of
    sequentially, as the situation evolves.

4. The criteria and rules for evaluation are not given a priori,
    but develop out of an ongoing dialectic.

5.   Implementation does not follow the goal (or plan), but may in
      some sense precede it.

      Confronted with situations like this, current systems methods
are powerless: they would not know why to start, where to continue
and how to end, or what they are meant to do there.

      Surely, not all urban planning today is as chaotic, but there
is little doubt that since the days when the "rational model" was
advanced as a likely ideal, a fundamental redefinition of what planning is about has taken place. The reasons for the marked shift of
emphasis have been sought in phenomena lying far beyond the bounds
of the urban realm, such as the presumed end of the era of growth,
and the beginning of that of redistribution. Some see in the current economic recession, the energy crisis and the other political
and military headaches of the modern world, the signs of a return to
the time when efficiency, rather than equity or welfare, was the
prime mover of planning policy, and when systems methods could
deliver the promise, if not the actual benefit, of really clever
solutions to clearly articulated problems. But this hope is futile. Efficiency may well have become once again a crucial consideration, but this is on top of, not instead of, all the other concerns that have permeated societal thinking in the past generation
or so. History may move in spirals, it does not move in circles:
what has been learned is not forgotten; it leads to the next step
up. In planning, as in many other fields, the age of innocence is
lost forever, and candid, unselfconscious methods have little chance
to survive. This is a fact that must be faced if systems analysts
wish to continue playing a role in urban affairs.

THE VARIABLES OF RATIONAL BEHAVIOUR

      On account of the developments discussed in the preceding sec-
tion, it is sometimes argued that planning is becoming "irrational". A more likely interpretation of things would be that it is
very difficult to apply current criteria of rationality to the outcome of decision processes that are inherently collective and diffuse. Most existing models of rational behaviour, whether "objective" or "subjective", are based on individual action and thus imply
all the qualities of coherence, cohesion, continuity, consequence,
non-contradiction and so on that characterise the psychological and
mental makeup of sane human individuals. It is a fallacy to expect
such models to still be meaningful when transferred to situations of
genuinely collective behaviour, as if the rational minds of all the
individual decision-makers could ever add up to define some collective rational mind with identical properties. It is true that Arrow
(1963), in his seminal work on social choice, stated the conditions
under which rational collective choice could be possible; but there
are many more aspects to rational behaviour than just the static act
of choice from among a closed set of alternatives, and to the extent
that planning is a continuing process of means-ends redefinition and

reconciliation, even Arrow's insight is not very helpful. Today's planning looks irrational to us because we still lack the models that will help us see it in any other way, so, with Buckley (1967:123) we say that:

> "Irrationality may mean a disorderly or inconsistent value or utility scale, or a faulty calculation of alternatives or of outcomes, or an inability to receive messages or to communicate efficiently, or the presence of random influences in decision-making or of 'noise' in the transmission of information. Or it may simply reflect the collective nature of the decision-making process among various individuals and subgroups, their organizations, values and communication systems." (my emphasis)

The value overtones of this kind of judgement are evident.

In contrast to this, the "rational model" of planning is very clearly based on a model of individual rational behaviour: it works fairly well as the representation of a design process involving one individual; it still is useful as a norm in the case of a small, closed, homogeneous design group, as successes in engineering applications show; it is inadequate in the case of larger, not so homogeneous, not so closed groups; it breaks down in the case of wider institutional involvement, as in corporate planning; and it is shattered to pieces when confronted with a really open, democratic, collective planning process. The verdict: open planning is irrational. But what is irrational in this case is rather the "rational model" of planning itself.

And what about rational methods? It would have been futile to attack the straw man of rational planning, arguably long dead and buried as an ideal, were it not for the fact that its demise has serious consequences for systems methods as currently known. In earlier days the context of systems methods, the rationale of their development and the justification for their use in planning were implicitly given by the rational model and its numerous presuppositions. Now, with the rational model a thing of the past, systems methods sometimes look like tools left over by an alien civilization, tools meant for tasks very imperfectly understood, and in any case, of no direct concern to us. To those interested in securing a place for systems methods in tomorrow's planning a new kind of problem presents itself, the problem of redefining the logical context within which such methods make sense. Whether "substantive" or "procedural", there is little hope for systems methods to assume a non-trivial role in modern planning unless the formal context of their validity, which no longer can be the "rational model", is made explicit.

Let us turn to the substantive methods first. These were developed to aid understanding of the systems which planners wish to

plan.   Such understanding presupposes factual knowledge on the one
hand, and on the other, some theoretical conception about how facts
may be causally related to each other.

Factual knowledge. The factual knowledge necessary for planning
is of two kinds: knowledge about the urban world one is dealing
with, and knowledge about the particular planning system one is
working in. The two are closely interrelated.

Knowledge about the external world is both "hard" and "soft".
That is, it is based on statistics, counts and measurements, but
also on first- and second-hand direct experience, on newspaper
reports, televised news items, photographs, communications, opinions
and the like. It concerns not only the present state of affairs,
but also much of the recent past; not just the area that is actually
being planned, but a much wider portion of the world (up to national
boundaries and beyond); and it covers issues of societal values and
political attitudes, as well as the more conventional ones of demography, income, employment, housing, infrastructure, transport,
energy and environment.

Knowledge about the planning system itself concerns the institutional and legal framework within a particular planning system is
"de jure" embedded, but also the more informal aspects of its setup
which determine its "de facto" functioning: the efficiency, the
inertia, the responsiveness, the sensitivity, the blind spots, the
weak and strong points, the robustness, the stability, the pace
characterising its behaviour. This kind of knowledge is hardly ever
conveyed by formal planning courses, though every young planner has
to grapple with it from the very first day of his or her practical
experience in planning.

A large part of both of these kinds of knowledge is tacit, to
use Polanyi's phrase. That is, it helps guide thought and action,
while remaining unexpressed and often unexpressable. Much of the
aura surrounding an "experienced" planner is due to such an accumulation of tacit knowledge, while in another sense a younger, better
trained planner may "know" more.

What particular blend of explicit and tacit, hard and soft,
internal and external knowledge is used in actual planning and
policy-making is a question that can only be answered in the individual case, if at all. What seems a plausible hypothesis, however,
is that all these forms of knowledge play roughly equivalent roles
in the determination of the planners' and decision-makers' "mental
maps", the subjective representations of the wider decision environment which are at the basis of concrete planning behaviour.

Only a very small part of the total information making up such
mental maps is amenable to treatment by systems methods, and there
is no indication at all that this is the most important part, or the
one that holds these mental maps together.

Theoretical understanding. Factual knowledge is not really knowledge until it can be structured into some abstract representation, however primitive, of the situation concerned. This involves the recognition of relations among facts, and the construction of conceptual _models_ of the situation.

Two main currents of thought offer ready models for the planners to use. The first is the _empiricist_ tradition which in the field of urban theory culminated in the development of mathematical urban models. This is probably the area where systems analysis came closest to being integrated into the mainstream of planning-related disciplines. Much of this seminar has been concerned with the critical appraisal of the relationship that has developed between such models and actual planning practice, and nothing more need be said at this point.

The second current of thought which has influenced theoretical understanding of urban affairs springs from the Marxist tradition in the social sciences. In relation to positivist theory, Marxist theory is characterised by a different universe of discourse, a different method, a different theoretical language, and, of course, different interpretations of events. The deep philosophical and ideological differences between the two types of theories make them incommensurate rather than contradictory, which means that although their factual concerns may be complementary in many respects, they cannot, at least formally, be combined.

Most planners of the younger generation have been exposed to one or the other or both of the above theoretical traditions at some stage of their intellectual development, mainly in the course of their university education. How much of that influence may still be operative after some time spent in planning practice is anybody's guess. It is a fact that irrespective of any questions of objective truth or subjective appeal, both these types of theories have the disadvantage of not being sufficiently practice-oriented. This is in spite of the formers' emphasis on "hard fact" and the latters' insistence on the unity of knowledge and action.

Of the many difficulties marring the efforts to make such theories practically relevant, I shall point out only two which are really opposite sides of the same paradox. Positivist theories set out to discover the basic facts and mechanisms of the urban system so that decision-makers may act accordingly, while these decision-makers know that they are themselves to a large extent the causes of these facts, and the creators of these mechanisms. Marxist theories on the other hand exhort to action on the basis of postulated abstract structures, the reality of which is taken to be on a par with that of concrete action. In the first case, the difficulty comes from trying to separate the knowledge about what is being regulated (the urban system) from the knowledge about the regulator (the planning system). In the second case, the attempt to analyze the regulator (superstructure) next to the object regulated (the "base" structure) leads to a confusion of consequences as regards action and theory which is detrimental to both. In each case the paradox involved seems to be that the two aspects of concrete action and abstract theory both must and must not be considered together. This kind of paradox cannot be resolved within the theories themselves.

As a result, the theoretical understanding guiding actual planning practice seems still to be based on subjective, empirical, intuitive theories not much different in kind from what has been traditionally available in the discipline. What has changed, of course, is the degree of sophistication of such "theories" which now, in most cases, seem to be made up of a blend of positivist, Marxist and other theoretical wisdom in varying proportions (this is presumably what is meant by the "intuition-structuring role" of urban theories). What is so interesting about these "private" models of planning in action is that they often manage to effectuate a working synthesis of fundamentally incompatible theoretical viewpoints, a feat most unlikely to be achieved by some rigorously rational treatment.

Furthermore, as the paradigm of pluralistic planning spreads and ever more agents become involved in the making of urban decisions, the hope for a "scientific" model of the urban world that will provide to one and all the understanding necessary for action recedes into the realm of mythology. From his own particular perspective, equipped with his particular mental map of relevant facts, each planner, each administrator, each politician, each informal decision-maker representing what is fondly known as "the collective wisdom", will bring his private mental model along with him, and given a chance, will act accordingly. Under such circumstances it is even quite irrelevant to ask whose model is more "true", the real question being, whose views are more influential.

In short, choices in planning are often made on the basis of not only individual mental models that may be internally inconsistent in rigorous terms, but also of a very large number of different mental models that more often than not are mutually incompatible. Some see in this development an instance of scientific reason being crushed by the vulgarity of democratic procedures. However, as some other application of scientific reason has shown, it is perfectly possible for a single, very complex reality to yield to sufficiently different observers partial models that are not only different, but even mutually incompatible (Ashby, 1956:106). Whether these models are expressed in formal terms or not is a secondary consideration.

This is why the laudable efforts to increase the scope of urban models so as to embrace further and further aspects of the urban world (often at the expense of vital qualities such as clarity, simplicity, intuitive appeal, and so on) do not seem directly relevant to the need for better understanding. If there are fundamental problems with the modelling approach as currently known, these are not to be solved by doing more of the same at a grander scale. The hard questions nowadays are of a different nature: how can we account as model builders for the enormous variety of different perspectives relevant to the process of collective decision-making in urban affairs; how can systems thinking make sense of the coexistence of an indefinite number of partly incompatible models of the same urban reality, a coexistence that is as important for the development of democratic planning procedures, as it is necessary in a strictly logical sense; and further, how can the internal riddles of modelling, like the persistent paradox of social gravity versus deliberate action, be resolved for the benefit of greater applicability, as well as better understanding.

Thus, the recognition of the necessity to account for a large number of very different representations simultaneously raises issues that were unthinkable a few years ago, when "rational" planning, which could only deal with one model at a time, was the ruling wisdom in the discipline. We now seem to need a new kind of model that will serve as a framework for the integration of the plurality of perspectives inherent in an unconstrained acknowledgement of the complexity of the urban world. Such a model should be able to relate the partial models to each other and to whatever we call the "urban system", the primary abstraction on the basis of which all other abstractions are derived; it should help plot the bounds of each model's validity and delineate the areas which are beyond the reach of current methods; it should facilitate the formulation of new models covering perspectives thus far unexplored; in short, it should define the logical domain of urban models, the context within which these disparate fragments of experience combine into a picture of urban knowledge that suddenly makes sense (Couclelis, 1977).

Despite the numerous riddles existing in the area of urban knowledge, the accusations of irrationality obviously refer to the procedural aspects of modern planning much more directly than to the substantive ones. The process of planning is normally understood to be a form of rational behaviour and very often, as in the case of the "rational model", conceptions of individual rationality are implicitly used for the representation of the planning process. Such models of rationality are, as we saw, particularly inadequate in the case of the less centralized, more open forms of planning in which there is by definition a very wide range of formal and informal participants. This is because a collection of modes of behaviour that may be fully rational at the individual level does not necessarily add up to collective behaviour that is rational by the same criteria. This is an important point for understanding how these less formal modes of planning may work. We may, therefore, devote a few thoughts to the question of what rational behaviour could mean in such cases as this will allow us to consider in turn the question of rational methods as they relate to these contexts.

Central to most conceptions of rational behaviour, whether "objective" or "subjective", is the idea of an adequate relation of means to ends. Planning and systems analysis share this concern. In classical economic theory, this relation is construed as a maximisation of some sort, in particular, the maximisation of some value function (Arrow, 1963:3ff.). More "subjective" versions of rational action, however, imply wider definitions involving a number of other considerations beside the objectively optimal strategy for reaching a given end. Thus, according to Max Weber, a form of behaviour is functionally rational ('zweckrational')

> "when it is end-attaining in its intentions and in full accord with factual knowledge and theoretical understanding within its means; where the choice of an end among other ends and the choice of means satisfies these criteria..."

MacRae (1974:68)

The interest of this particular formulation is that it is well compatible with all but the most extreme versions of economic rationality on the one hand, and with conventional wisdom on the other.

Accounts of the planning process in situations of pluralistic decision-making usually assume that difficulties arise mainly with respect to the choice of ends, and the choice of means for reaching these ends. It is then argued that compromises can be struck at the level of ends, and that some generally acceptable agreements can be reached at the level of means, a situation that will eventually lead to a sub-optimal solution of the initial problem. This representation leaves the notion of rational behaviour and the possibility for the application of rational methods intact, but fails to correspond to the reality of decision-making in highly decentralized settings.

In actual fact things appear to be a little more complex. A closer look at Weber's conception of rational behaviour will show that it entails at least the following six notions:

1. an end or goal to be attained
2. an intention to reach the goal
3. a universe of factual knowledge
4. theoretical understanding of some sort
5. a deliberate choice of means and ends
6. also, a temporal element, since end-attaining behaviour is by definition future-oriented

All the above elements are meaningful in the context of urban planning and policy-making. This is how they may be realized in actual practice:

Ends or goals to be attained. The possibility of having a large number of partly conflicting, partly incompatible goals is inherent in highly fragmented decision structures. An explicit compromise

over goals may be reached through bargaining and persuasion, but
this will often represent "the greatest common denominator", a
choice not really desired by any of the parties concerned. Besides,
such reasonable settlement of issues presupposes the existence of
well-developed channels of communication among all concerned, a
condition that is seldom fulfilled in practice. Under such circumstances it may be rational policy for the key decision-makers not to
formulate any goals explicitly, and to proceed directly to the discussion of concrete policy measures. Or it may be effective to disguise real goals under formulations so general as to be unlikely to
cause controversy. In both cases, a basic condition for a process
of rational problem-solving, a clear formulation of the problem to
be solved, will be missing.

An intention to reach the goal. It is normally taken for granted
that once a goal is set, an actor (or a system) automatically will
start moving towards its realization. This is often not the case in
planning when responsibility for a given goal is shared by many
actors. Several conditions are necessary for setting the goal-
seeking procedures in motion, conditions that have very little to do
with the material feasibility or the adequacy of the goal itself,
and much more with the will to reach it. These involve questions of
personal attitude among those responsible for its promotion (progressive goals in the hands of a conservative administration do not
get much chance on the whole), questions of side benefits in terms
of professional, political or material gains (and of who will reap
them), questions of priorities in the face of other commitments and
sheer routine work, questions of perseverance, credibility, or even
of external compulsion. Temporal urgency is usually a prime incentive for action: decision-makers are often blamed for not acting
until it is "too late", but this is often the only time when the
intention to reach the goal can gather enough momentum to break
through the threshold of inaction. For much of the other time, the
tendency to postpone effort and commitment, and the hope that the
problem situation will take care of itself (which it often does) can
freeze the pursuit of a goal which is outside the scope of day-to-
day decision. Thus, where the appropriate climate is absent, even
the most clearly formulated of goals can easily remain at the stage
of policy declarations or electoral promises. In many cases, the
formulation of goals without any intention to realize them may even
constitute deliberate policy, aimed at mobilizing the mechanisms of
self-fulfilling prophecy, or for the sake of reassuring an impatient
public. Unless, and until the intention to act is genuine by all
key actors sharing the responsibility for a given goal, the planning
system will not behave as decided, but will move towards some other,
unintended state.

Factual knowledge and theoretical understanding. As already
mentioned, the mental maps and models of decision-makers can differ
radically without any of them being necessarily wrong or irrational. On the whole, the wider the sociopolitical and cultural spectrum of participants to the decision process, the greater the diver-

sity of rational representations of the situation will be. In cases where the contribution of these various participants is limited to the act of choice, as in a voting situation, any resulting problems will be confined to the definition of ends and means. But to the extent that these "conceptual dissenters" also have the power to <u>act</u> in accordance with their mental maps, as in free-market societies, they constitute a "de facto" control mechanism whose behaviour is radically different from that of the "de jure" planning system. As a result, the urban system's evolution becomes <u>unpredictable</u> both causally and statistically because on the one hand it is largely determined by events that cannot be accounted for by the internal logic of the official regulator, and on the other, the most signifi- cant of these external interventions (important locational deci- sions, large investments, etc.) are almost by definition isolated and non-repetitive, and thus not subject to the laws of large numbers.

Choice of means and ends. In cases where the goals of planning are not clear, the will to reach them not to be taken for granted, and the urban system's response and evolution to a large extent incomprehensible, the means or planning measures to be taken can become an end in themselves: the process is the goal, the style is the design (Simon, 1969). Rational planning behaviour in this case may simply mean the promotion of measures giving the greatest imme- diate satisfaction at the least risk of foreclosing important future options. This is a very special kind of problem, not amenable to current problem-solving methods to the extent that the nature of possible desirable future options is not as yet known.

The temporal element. The average time elapsing between the inception of a goal and its realization is a crucial element in assessing the practical applicability of systems methods in the corresponding planning context. Systems analysis takes time, weeks or months or years in some cases, and in fluid, fast-moving modes of planning where decisions are made and amended on a daily basis, there is just no time for formal analysis. In such contexts, even major issues, which do not arise every day and are examined with greater care and for a greater length of time, are usually not amen- able to treatment by systems methods because the minor elements that define their environment are likely to change more rapidly than formal analysis can account for.

These reflexions on planning behaviour in the more "open" modes of planning seem to indicate that in such contexts current models of rational behaviour are inadequate, problem-solving methods are useless for lack of problem situations as defined by the systems paradigm, and formal prediction is impossible because the urban system's evolution becomes inherently unpredictable in both the causal and the statistical sense. In fact, the concept of system itself becomes less relevant for the definition of planning issues just when it would be most needed. For the concept of "system" to be applicable, the identification of a whole of means, ends, factors, relations, interactions and so on is required that is relatively stable over time. Even allowing for extensions of the concept such as "open system", "dynamic system", "adaptive system" and the like, the notion (and utility) of "system" gets increasingly blurred as the identity of an initial set of planning-related factors becomes less recognizable with every passing day. In short, for "open" planning situations of the type outlined above (and there are many), there is little hope for solutions to concrete issues to be given by systems methods as presently known.

On the other hand, what is needed now more than ever before is the development of the kind of general understanding that will allow decision-makers to make consistently reasonable decisions from within processes that may be irrational when judged by conventional standards. This will involve more research into the qualitative properties of urban systems qua systems, in the abstract, and more reflexion on how to behave in the face of uncertainty once the hope of being able to predict non-trivial elements of the future has been given up. We need to know more about the conditions under which stability, instability, inertia, resilience, sensitivity, responsiveness, non-linearity and other such aspects of abstract system behaviour manifest themselves in urban systems defined in various ways, even if such knowledge cannot be directly applied for the resolution of concrete planning issues. In this sense, the recent experimentations in urban modelling with new models of thinking, such as catastrophe or bifurcation theory or the theory of dissipative structures (see for example Allen and Sanglier, 1979; Wilson, 1979), is to be seen as a step in the right direction; so is the concern with the "robustness" of decisions which seeks to determine the conditions under which today's choices will not become tomorrow's handcuffs (Holling, 1978; Rosenhead, 1979).

There is, of course, also a need for models of planning that can capture some of the variety and richness of real-world planning processes. There are suggestions, for instance, that between the two extremes of the utopian "rational model" on the one hand, and chaotic "muddling through" on the other, the range of possible forms of planning is discrete rather than continuous, and that there exists a small number of distinct planning modes sufficiently different from each other as to correspond to entirely different models of the planning process. Berry (1973) gives a political dimension to this hypothesis by linking four distinct models of planning to four different kinds of socio-political context, ordered along a spectrum of increasing individual choice and decreasing public powers. From a more explicit systems perspective, Boguslaw (1965:9-21) distinguishes four different "approaches to system design" (actually, planning) which he equally presents in order of decreasing central control. There is such strong affinity between Berry's four planning modes and Boguslaw's typology that the relevance of the latter to a discussion of the planning process is very evident. The first planning mode or approach corresponds to what is also known as "blueprint", "formal" or "unitarian" planning, and is the one closest to the traditional conception of planning as design, (Foley, 1964; Faludi, 1973); the last one is so far from it as to be, in the eyes of many, unworthy of the name of "planning": some call it "muddling through" (Lindblom, 1973; Wildavsky, 1973).

The way these models of planning are presented, systems methods seem to be very well compatible with the first mode, conditionally so with the second, only in special cases with the third and practically not at all with the fourth one. This has as much to do with the characteristics of the planning system proper, which as a regulating mechanism becomes increasingly diffuse and ill-defined, as with those of the wider socio-political context which becomes increasingly complex, pluralistic and "unplannable" as we move from the first to the last mode.

Irrespective of the concrete empirical merits of the particular typologies mentioned above, their consideration shows how, in the light of the perspective provided by hypotheses of this kind, planners and systems analysts alike could reconsider the scope of their efforts and decide whether these are in harmony with the type of planning process entailed by the particular socio-political context they happen to be working in. The study of the formal properties of such context-dependent models of planning could yield more insights concerning the applicability of systems methods in corresponding contexts than the evaluation of any number of empirical applications could do.

What seems in any case no longer possible is the neglect of the issue of how the world of theory may relate to the many worlds of practice, and what it is that makes the hard rationality of our various methods such an unreliable and contingent thing. We cannot go on claiming that such methods are rational as long as we are unable to determine operationally the presuppositions of their validity as rationalizations of actual fact. For, indeed, what has discredited systems methods in practice is not the necessary truth that they do have limits, but the lack of awareness about where these limits may lie with respect to the intended contexts of their use.

This is why the reaction that has set in against systems analysis is quite disproportionate to any functional inadequacies such methods may exhibit in their intended tasks. It is no longer a question of systems analysis not being useful or good enough: it is considered wrong-headed, if not outright evil. This is unfair to the extent that systems methods can accomplish feats in the areas of data handling and analysis which no other approach can possibly parallel. Analysts deplore the irrational turn of things and hope that Reason will finally triumph. Yet, this is not really a question of Society gone mad. It is just that the context of rationality has changed, and naive, straight-forward, "technical" rationality no longer satisfies the needs of a discipline that has evolved from a branch of architecture and civil engineering into one of the most profoundly committed intellectual traditions of our time.

There is a distinction between "technical" and "substantial" rationality (Simmie, 1974) based on the kind of relationship existing between a form of rational behaviour and its environment. "Technical" rationality sets its own terms, while "substantial" rationality can only be seen in the context of the total situation within which it acquires a meaning. Systems analysis is an exemplar of technically rational thinking, while the times call for approaches that are at least substantially rational.

There could still be scope, I feel, for systems methods in planning, if only these could be put in a framework that will make the connexion between their crisp algorithms on the one hand, and the chaotic dynamics of particular socio-political and cultural contexts on the other. What kind of framework this is going to be is not a question I am ready to answer. But the general direction of search is, I think, already distinguishable. The few concluding remarks will be devoted to this emerging perspective.

RATIONALITY AND BEYOND

The discussion, so far, has centred on what Max Weber has called the functionally rational ("zweckrational") aspects of planning. This had to be so in order to address systems analysis in its own terms.

There is, however, a second kind of rationality defined by Weber which must be mentioned at this point: an action is value-rational ("wertrational")

> "where the ends are given by values, or where
> such values affect the choice of means"

>   MacRae (1974:68)

No model of planning that is not inherently value-rational is worth its salt. This is a very pragmatic requirement that has nothing to do with any lofty humanistic concerns. It is simply that values are facts in planning, since they causally determine other facts, and a "rationality" that leaves no room for taking into account their existence and role is hopelessly restricted. There was a time, not too long ago, where the progressing rationalization of society was expressed by a widening of the sphere of zweckrational actions, actions in relation to functional goals (Aron, 1967:188). Today we may be witnessing a reversal of the trend; or is it that the distinction between function and value is becoming increasingly blurred?

Whatever the case may be, for dealing with value-rationality, systems methods in their current form appear sadly inadequate.

And yet there is hope. It may just be a question of switching to the next higher level of discourse, and using the systems para-

digm not for making plans, but for learning how to think about planning. It is perfectly conceivable that systems methods might some
day be used to point out the kinds of situations where it would not
be reasonable to use "rational" methods; or to clarify the still
mysterious relationship between the urban systems of modellers and
the cities of planners; or to explain away the paradox of the
decision-maker who applies his free will to the creation of an urban
system necessarily run by predetermined laws; or to put into perspective the question of human values in their relation with the
hard facts of pounds, vehicles and heads as these move about in
geometric spaces (Couclelis, 1977).

What is really advocated here is a shift of emphasis from systems analysis to systems thinking. The former provides us with a
box of tools; in the latter we can find the instructions about how
to use them. If there are serious problems with the systems paradigm these days, this may well be because there has been too much
analysis and not enough thinking.

To search for the logical context of urban knowledge and process does not mean that we must lose empirical reality from sight.
On the contrary, a better understanding of what we can rationally
know about urban systems could guide the development of novel
approaches that will shed light on concrete phenomena actually
thought beyond the reach of "science". It is a perverse science
that tells us that prices are facts but values are not, or that
planning is irrational if it is not "rational" planning.

The look from the next level up will also make it clear that
there is much more to the systems concepts of relation and structure
than assumed by current modes of systems analysis. Freed from the
parametric obsession which is the legacy of a now obsolete frame of
thinking, systems methods will be able to expand to include more
"qualitative" formalisms which could take over at the point where
statistical methods stop. For the past ten years, for example, a
mathematical language has been available which has given exact
expression to the intuitive notion of "structure". "Polyhedral
dynamics" is equally at ease with traffic flows as it is with aesthetic values, with counting heads as with representing human relations (Atkin, 1974): what could be more useful in a discipline like
planning, where an old tree can be set against a new road as a problem in cost-benefit evaluation?

More importantly, perhaps, such a language of structure could
make it possible for us to undertake the structuring of our subjective understanding of urban phenomena rather than the description of
some "objective" reality which may not be there at all. This would
allow us to bring together numbers and qualities, facts and values,
things and ideas, science and philosophy - for are they not all
aspects of understanding?

This, of course, would mean a departure from some of the basic
tenets of the philosophy that gave rise to systems analysis in the

first place. It is well known that the study of abstract structures
has never been among empiricism's favourite themes. Should we see
the need for a <u>frame of understanding</u> to integrate our hopelessly
fragmented present experience, should we go on to study meta-
planning, meta-procedures, meta-theory, some other philosophy will
be required; a rationalist philosophy, perhaps: something quite
different from rationalist <u>methods</u>. For there is no escape from
philosophy, in spite of what empiricists in splendid self-
-contradiction may claim. To structure, interpret and operationa-
lize our knowledge of the urban world, all three of systems thinking, philosophy and mathematics are required (Gould, 1979). The
task is so difficult that nothing short of the full use of our
rational faculties would suffice.

This would also mean the end of the "scientific method" as an
ideal in our discipline. Not much will be lost, because the classical model of science has always been inadequate for our purposes.
Classical science is the study of something "a priori" given, while
our task is to develop methods for ordering and surveying human
experience. This is, incidentally, what Niels Bohr thought of the
role of contemporary physics (Bohr, 1963:10). It should be good
enough a mission for a <u>human</u> science.

And, in any case, the public image of science no longer is what
it used to be some twenty or thirty years ago. We might find ourselves much better off adopting a paradigm much more akin to the
sense of the term "Wissenschaft" in German, a sense that embraces a
much wider universe of discourse than that of some strict epistemological ideal. It is <u>rational knowledge</u> we need, who cares if it is
not "scientific".

Does all this sound all too abstract and remote? Maybe it
does. But our urban concerns are well worth the trouble of a longer
journey, and the systems approach is much too valuable for us to
lose in an all-too-probable dead end. The planners may have to live
without our science for a while, but this should not worry us too
much: for they have learned the hard way that, as Popper once put
it:

> "all science, and all philosophy, are enlightened
> common sense."

Popper (1972)

May such enlightened common sense guide our own quest for
understanding.

REFERENCES

Allen, P.M. and Sanglier, M., 1979, A Dynamic Model of Growth in a
    Central Place System. <u>Geographical Analysis, 11</u>, 256-272.

Aron, R., 1967, Main Currents in Sociological Thought – 2, Penguin Books, Harmondsworth, UK.

Arrow, K.J., 1963, Social Choice and Individual Values, Yale University Press, New Haven, Conn.

Ashby, W.R., 1956, An Introduction to Cybernetics, Methuen, London.

Atkin, R.H., 1974, Mathematical Structure in Human Affairs, Crane, Russak and Company, New York.

Batty, M., 1979, Progress, Success and Failure in Urban Modelling, Environment and Planning A, 11, 863–878.

Berry, B.J.L., 1973, The Human Consequences of Urbanisation, Macmillan, London.

Boguslaw, R., 1965, The New Utopians: A Study of System Design and Social Change, Prentice-Hall, Englewood Cliffs, New Jersey.

Bohr, N., 1963, Essays 1958-1962 on Atomic Physics and Human Knowledge, John Wiley, New York.

Buckley, W., 1967, Sociology and Modern Systems Theory, Prentice Hall, Englewood Cliffs.

Camhis, M., 1979, Planning Theory and Philosophy, Tavistock, London.

Chadwick, G., 1971, A Systems View of Planning, Pergamon Press, Oxford, UK.

Couclelis, H.M., 1977, Urban Development Models: Towards a General Theory. Unpublished PhD dissertation, University of Cambridge, Cambridge, UK.

Echenique, M., 1975, Urban Development Models: Fifteen Years of Experience, in R. Baxter, et al. (Eds.), Urban Development Models, The Construction Press, Lancaster, UK.

Faludi, A., (Ed.), 1973, A Reader in Planning Theory. Pergamon Press, Oxford.

Foley, D.L., 1964, An Approach to Metropolitan Spatial Structure, in M.M. Webber, et al, (Eds.) Explorations into Urban Structure, Pennsylvania University Press, Philadelphia, Penn.

Gould, P., 1979, Signals in the Noise, in S. Gale and G. Olsson, (Eds.), Philosophy in Geography, Reidel Publishing Company, Dordrecht, The Netherlands.

Hightower, H.C., 1969, Planning Theory in Contemporary Professional Education, Journal of the American Institute of Planners, 35, 326–329.

Holling, C.S. (Ed.), 1978, Adaptive Environmental Assessment and Management, John Wiley and Sons, Chichester, UK.

Kuhn, T.S., 1962, The Structure of Scientific Revolutions, The University of Chicago Press, Chicago, Illinois.

Lee, C., 1973, Models in Planning. Pergamon Press, Oxford, UK.

Lindblom, C.E., 1973, The Science of "Muddling Through", in A. Faludi, (Ed.), A Reader in Planning Theory, Pergamon Press, Oxford, UK.

MacRae, D.G., 1974, Weber, Fontana/Collins, London.

Optner, S.L., 1973, Systems Analysis, Penguin Books, Harmondsworth, UK.

Popper, K., 1972, Objective Knowledge: An Evolutionary Approach, The Clarendon Press, Oxford, UK.

Rosenhead, J., 1979, Aspects of Robustness in Planning, Department of Statistics, London School of Economics, London, UK.

Simmie, J.M., 1974, Citizens in Conflict, Hutchinson, London.

Simon, H.A., 1969, The Sciences of the Artificial, The MIT Press, Cambridge, Mass.

Wildavsky, A., 1973, If Planning is Everything, Maybe It's Nothing, Policy Sciences, 4, 127-153.

Wilson, A.G., 1979, Some New Methods for Dynamic Geographical Modelling: Catastrophe Theory and Bifurcation, Working Paper 254, School of Geography, University of Leeds, Leeds, UK.

Winch, P., 1958, The Idea of a Social Science and its Relation to Philosophy, Routledge and Kegan Paul, London.

PART 6

EMERGING THEMES

POSITIVE AND NORMATIVE ASPECTS OF MODELLING LARGE-SCALE SOCIAL

SYSTEMS

Britton Harris

School of Urban and Public Policy
University of Pennsylvania
Philadelphia, PA, 19104
United States of America

INTRODUCTION

My fundamental interest in planning and policy-making is well-known. This interest stems, in the first instance, from a career in city planning, transportation planning and regional science, but I have a strong and longer-standing concern with the economic, social, and political development of disadvantaged populations, especially in the Third World. In this paper, I shall argue that these interests in policy-making and planning extend by their very nature to an interest in design, and I shall try to disentangle some of the issues which surround questions of modelling, optimization and decision-making as a complex of activities which unite theory and practice in the field of social policy.

Planning and policy-making are future-oriented. Even in the simplest cases, where the decisions which are taken are executed immediately, such as chess-playing and warfare, the context of these actions is that of their future consequences. Many such decisions are irrevocable, and this too has strong implications for the future. Thus, it follows that no matter how adaptive and responsive a planning system may be to changing conditions and unexpected circumstances, it nevertheless has elements of long-term prediction and planning. The common assumption that long-term and future-oriented planning assumes perfect knowledge of the future, or is necessarily based on a comprehensive and immutable plan turns out in this context to be a straw man whose demolition does not in fact affect the realities of future-oriented planning.

I hold strongly to the view that policy-making and planning are optimum-seeking processes -- that is, there is a constant effort to improve plans or policies, and when, for a variety of reasons, further improvement appears impossible, then the improvement effort ceases. This is an exact analogue of optimization by hill-climbing, which is extremely effective for a large class of problems. This contention with respect to the nature of planning is subject to some qualification and some criticism.

First, hill-climbing as a method of optimization breaks down where there are multiple local optima, and this is frequently the case wherever there are externalities and indivisibilities. This is the essential reason why optimization methods "per se" fail as an instrument of planning and why, therefore, planning and policy-making have some of the characteristics of design, at least with respect to its emphasis on invention and discovery.

In the second place, the turbulence and uncertainty of modern social and economic development certainly increases the difficulties of deriving suitable policy sets for a given constellation of social and political objectives, but it cannot be said that this turbulence makes it impossible to speculate about the future, to see probabilities, and to make choices on the basis of a comparison of desired outcomes and their costs.

Third, however, there is a view of optimization in policy-making which is widely cultivated and beautifully expressed in Steve Savas' article for this conference (Savas, this volume). Savas discusses the situation under which "in the exasperating urban environment...there is usually no feasible solution". He goes on to suggest that "optimization consists of trying to determine which of the multitudinous constraints can be forced back at minimal political costs far enough to create one feasible...point". While I appreciate the argument, this interpretation is nonsensical from the point of view of mathematical programming. A problem which has no feasible solution suddenly turns out to have one feasible solution. If it is indeed possible to relax the constraints, then the first problem turns out to have been improperly defined, but if the problem without a feasible solution is indeed improper, then why did Savas define it this way, and why would he expect any analyst to do this? Indeed, the moment Savas begins to talk about "minimum political cost," one smells the existence of an entirely different programming problem which minimizes some combination of political cost and, say, capital cost. Then the political constraints cease to become constraints, and become penalty functions which are a measure of cost. Then we can begin to think of minimizing cost by changing a number of constraints, or in fact, allowing the constraints to become variables, and thus, we are suddenly in the midst of a much more realistic, and probably much more interesting problem, but one which is still a problem in optimization and mathematical programming. Of course, there is a severe difficulty in defining the political costs at all, and even more in such a way that they will be commensurate with money costs and imputed benefits, but perhaps the principal difficulty with the new problem is that it is likely not to be any longer a linear programming problem, and hence, not likely to be suggested by an operations research analyst.

Viewed as a professional activity, an essential characteristic of planning and policy-making is that they take effect and have consequences. The intention of the professional is to be effective. At the same time, the future orientation imposes on the professional, the need for an effectiveness whose consequences are not fully known. The professionals must base their prescriptions on prior experience, and generalizations of experience are intrinsically the domain of theory. On the other hand, because of his responsibilities for effectuating the consequences, the professional becomes involved in practice, and in the social process whereby the feedback regarding results of practice can be used not only to improve practice itself, but also to improve basic theory.

It is well-known, of course, that the best-laid plans of mice and men frequently go off course and produce undesired results. Certainly if plans are oriented toward very immediate or restricted objectives, their chances of success may be limited by the context in which they are developed. This leads us to consider when and how a plan may be called "best laid". I take the view, which is widely shared at least in a naive way, that one of the principal practical problems of the planning process in the process of striving to achieve social objectives is the necessity to avoid unintentional consequences.

If there were simple connections between problems and solutions or between difficulties and their cures, there would be no need to debate the issue of unintended consequences in any detail. We are, however, faced with a situation in which, first of all, the causes of pathologies are deeply buried and remote, so that symptomatic cures are frequently not effective; and second, where palliative cures have side effects which influence the performance of many other parts of the total system. Unintended consequences, therefore, can arise; first, from using inappropriate remedies, and, second, from the reverberations of cause and effect through different functional parts of the total system. These reverberations extend over space and time, so that an assessment of the difficulties generated by a policy measure may frequently be available only after the damage has been done, because the ultimate effects have been too far delayed or displaced.

Examples of unintended consequences are so numerous as scarcely to require much comment. They are advanced with particular glee by those opponents of planning who wish to establish the virtues of unregulated free enterprise by the use of counter-examples. But, illustrations of unintended consequences on both sides of this argument are easily adduced. The impact of DDT on many wild species, and the impact of aerosols on stratospheric ozone are examples of the externalities of private decision-making. On the other hand, the ecological consequences of the Aswan High Dam can be laid to public planning. In all instances, serious repercussions outside of the original objectives have been suffered.

There are both normative and positive aspects of the effort to avoid unintended consequences. In the positive sense, the consequences of various aspects and of combinations of actions must in some sense be predictable. As we have already suggested, such prediction requires an understanding of the direct effects of decisions, which is not always available, plus an ability to visualize and predict the long-term impacts as they are propagated through a larger and more complex system than is the immediate target of the policy initiative in its most elementary sense. For example, the introduction of DDT had a succession of impacts of wider and wider range. Most directly, it was not foreseen that the use of this insecticide would lead to the rapid evolution of resistant strains of target insects, so that the long-term effectiveness of DDT would greatly diminish. In a slightly broader sense, it was not recognized that the massive destruction of insect pests would also destroy the populations of a number of beneficent insects. Finally, and perhaps most seriously, the build-up of concentrations of DDT through the food chain was grossly underestimated, and the effects on carniverous animals, such as the raptors, were quite unanticipated. I have purposely developed this example from the natural ecology because it is simple and well-understood, and relatively free of value implications. Examples from human society, and dealing with more complex issues, such as energy policy, have the same nature and could have been used to illustrate the same point.

The normative aspects of the exploration of unintended consequences are obviously even more difficult than the positive aspects. Without a thorough, positive understanding of the phenomena in question, normative attitudes become contentious and strident. For example, in the absence of technological understanding of proposed measures, anti-technological, anti-interventionist attitudes contend in an irreconcilable way. Even if the technologist issues are well-understood, major normative issues remain. Once again, in the case of DDT, there is first of all a serious intergenerational issue: the heavy use of this insecticide was based on its immediate advantages and was continued even after some of its impacts were known. The users of DDT had high discount rates, while the conservationist view calling for the preservation of aspects of the natural ecology emphasized the rights of future generations, and is implicitly associated with a low discount rate. Issues of equity also arise, and perhaps largely outside of the United States, where DDT had its heaviest and most damaging use. The anticipated reduction of the incidence of malaria in Third World countries carried a clearly defined benefit for a seriously disadvantaged group; the disbenefits were more diffuse and might be imputed to some more prosperous groups. This sketch of some of the normative issues which would have to lie behind a cost-benefit or other analysis of technological innovation gives a small taste of the complexity of these issues in a larger policy-making framework -- a complexity which persists even if the positive issues can be resolved.

I now turn to a more detailed examination of the positive aspects of policy-making, and of the avoidance of unforeseen consequences.

There has been over the past two decades an ongoing discussion of the appropriateness of various types of models for policy analysis and planning. This discussion has arisen because of the facility with which the scope and scale of models can be, and has been, expanded through the use of existing computer technology. By creating ever larger data bases from administrative, census and survey information, computers have also contributed indirectly to the growth of modelling efforts. Modelling now extends from a variety of specific and well-defined phenomena in physics, chemistry and biology, and their applications through large and diffuse natural systems such as the global atmospheric system, to large man-machine systems such as metropolitan location and transportation and energy production and utilization. National economic systems and international trade systems are also widely modelled. Modelling development has been primarily furthered by engineers and economists, but is becoming much more widely diffused in a whole variety of disciplines and professions.

On account, especially, of the pragmatic orientation of engineers, it is widely believed that computer-based system models are substantially lacking in theory, and represent a largely descriptive "ad hoc" approach toward system simulation. While there is some merit in this view, there exist many examples of cases in which pragmatic exercises have turned out to have important theoretical bases. In what follows, therefore, I will take the view that properly constructed models represent a formulation of theories in manipulable form, and that there is no practical way to approach the analysis of systems and of policies designed to affect them without the use of models. Pure theory at the verbal or logical level is of extremely limited usefulness because of its lack of a quantitative basis. Highly simplified mathematical models are inadequate because they neglect the complexity and variety of large-scale systems. This does not mean, however, that systems analysts have yet found a completely adequate style in which to construct models for large-scale, diverse, and highly interconnected systems.

If a principal problem of planning and policy-making is the avoidance of the type of unintended consequences which I have discussed, then the principal means of discovering them calls for a specific modelling style. This style must provide means for accounting for the interconnections between a large number of significant sub-systems. It must be of sufficient functional or areal extent to permit the interactions to spread through a large system and possibly return to their point of origin. The interactions must also extend toward some ultimate condition which permits us to judge their long-term impacts; this ultimate condition might be found by dynamic simulation over time, or by an effort to drive a large system to one or more points of equilibrium.

There are many large-scale models of specific sub-systems of complex social systems which exhibit features often regarded as inimical to intelligent policy explorations. Typical examples would be large-scale models for simulating traffic systems in metropolitan transportation studies and similar complex models of the housing market, such as the NBER model. Putman has shown that if the interactions of two relatively simple models of this type are taken into account so that the build-up of congestion in the highway network influences the locational patterns of the resident population, then the outcomes are typically different from those of either model taken separately. Indeed, within each of these models there is a feedback phenomenon which provides a type of interaction not visible in most of the predecessors. A modern transportation model takes into account the build-up of congestion on strategically situated links, and the feedback which tends to redistribute trips away from those links. Similarly, the NBER model takes into account the build-up of residential congestion which results from the relative popularity of selected housing in selected zones, and which therefore affects prices and redirects demand. Both of these models are somewhat limited in that there is no very strong supply response such as the construction of new highway facilities or the imposition of tolls, and such as conversion, redevelopment, and new housing construction. The importance of including a reactive supply structure is illustrated in my model of retail trade, where the adjustment of geographically distributed supply to meet geographically distributed demand leads in a surprising way to the concentration of retail trade in centres, rather than to a more or less uniform distribution in relation to demand.

These examples of interaction have been rather small-scale. One could offer other examples with much more widespread consequences. The provision of Federal mortgage insurance, whose intention was simply to strengthen the housing market and to provide greater access to decent housing for a large part of the United States population, has had enormous effects on American life styles through the suburbanization of large cities. The effects of the interstate highway system, which are largely interregional rather than intraregional, are still working themselves out. This system appears to have undermined the national rail system, made the country much more dependent on automotive transport, and generally increased population mobility. The combined effect of this type of mobility, of the cotton-picking machine, of national and state welfare policies, and of the civil liberties movement has led to enormous shifts of population, both within the South and between the South and the North, and has shifted the emphasis on a wide variety of cultural, social and economic problems. These examples, in spite of their sketchy description, tend to point to the dense interconnection of a large number of aspects of American social life.

In passing, we should note that most of the examples which have just been adduced have strong spatial aspects. Other examples having to do with such matters as family structure and sexual mores, communications media, and labour force participation are perhaps less strongly spatially structured. Space becomes important in policy analysis for a variety of reasons. Certain resources are spatially distributed. Space itself is a resource in limited supply. National and internal jurisdictional boundaries partially define space, and jurisdictional areas are almost never non-contiguous. Perhaps most important, public policy is by nature greatly concerned with externalities and economies of scale, and these phenomena have strong spatial aspects.

We are now approaching a central dilemma with regard to the design and construction of models.

The real behaviour of a complex social system requires the manipulation of substantial internal detail. Let us emphasize that this requirement is not in response to the oceans of available data. It is not in response to the requirements of sub-system designers such as transportation engineers for high levels of detail regarding the performance of specific facilities. It is not in response to a statistical view of analysis which requires the proliferation of at least certain kinds of detail. In fact, a great deal of this obsession with detail is completely misplaced in the policy--making context. But there remains a real need for detail which arises out of the necessity to reproduce the inherent relevant qualities of the system in a theoretically sound way; in this case, theoretical soundness means that system behaviour is generated by a realistic representation of the behaviour of actors and sub-systems, and that system behaviour reproduces a number of salient characteristics of reasonably universal occurrence.

Realistically reproducing system behaviour seems to depend upon at least four different types of disaggregation. The first is a broad sectoral disaggregation: households, governments and businesses all behave differently, and within these broad divisions local government differs from Federal government, and services differ from manufacturing and agriculture. A second disaggregation is functional. Sub-systems or markets providing transportation, housing, recreation, and so on, and meeting a number of organizational or function needs cut across sectors and behave as partially independent sub-systems. Within broad functional and sectoral divisions, finer characterizations of actors and sub-systems are frequently required. Some distinctions are relatively continuous -- for example, the classification of households by income and businesses by number of employees; others are essentially discontinuous, such as categorization of households by presence of a male head or by race, and the differences within manufacturing industry between, say, job printing and steel production. Finally, there is a necessary disaggregation by area which is required to deal with many issues of agglomeration economies and externalities, as well as with the performance of transportation systems, and with those measures of the quality of life which may be based on accessibilities.

Disaggregated models are obviously somewhat complex to conceive, troublesome to specify, and onerous to run.

Independently of questions of accuracy and scientific validity
of models, there is a perfectly valid planning point of view which
argues for disaggregated models. This point of view suggests that
value systems which are involved in planning are complicated by the
presence of many disparate contending interest groups, and by the
impact of the total planned environment on a very wide variety of
human needs and activities. Consequently, simplified measures of
system performance such as average density, vehicle miles of travel,
modal split for various types of trips, and overall employment location are unsatisfactory for politically-oriented plan evaluation.
Not only do we need these categories broken down, but we need a
cross-classification in considerable depth which shows the impact of
the plan on the access to jobs of various types by populations of
various types, and so on. To the extent that this highly articulated type of evaluation is becoming a political necessity, the disaggregation of models springs not from the professional bias of
model builders, but from the necessities of the planning process.

In the face of important countervailing considerations which
require simplified models, serious questions arise as to which forms
of aggregation should be sacrificed, and to what extent.

Most countervailing considerations arise out of the fact that
the number of policy actions is very numerous, and the attempt to
configure policy sets out of combinations of these actions leads to
a combinatorial explosion. The number of policy sets which ought to
be examined in a serious planning effort is, in principle, very
large, but, in practice, the number of alternatives which are subjected to analysis is, in fact, very limited. The common planning
practice of limiting the alternatives examined to as few as one (!)
and at most to about half a dozen is a reflection of the difficulty
in exploring alternatives by any means whatever in the current state
of the art. Even an exploration on an intuitive basis using planning standards and practical wisdom is an exhausting endeavour,
while a full-scale exploration using the best available modelling
techniques is both arduous and costly.

In what follows, I shall explore more specifically what is
basically involved in this combinatorial explosion, but at this
point, I should simply like to emphasize three of the reasons why
extensive explorations may be necessary. First, it is not a trivial
matter to find a "good" plan. Plans of very different characteristics may have highly-rated performance measures, but we have no
idea how many such plans there may be available, and very little
idea of how to go about looking for them. An extensive and semi-
automatic search procedure could have the advantage of guaranteeing
that the search will not be dominated by a narrow set of planning
biases. Second, a plan once evaluated and deemed suitable for further consideration ordinarily needs a great deal of fine-tuning to

bring it close to the best possible performance for a given configuration.   The Chicago Area Transportation Study, for example,
explored over forty successive improvements on one basic plan at a
cost of well over $10,000 per cycle.   Finally, it is probably impor-
tant to produce a number of plans which are as basically different
as possible because there is no guarantee that the value system
which will determine political choices is identically reproduced in
the value system which the planners or policy analysts have used to
generate, evaluate and improve alternatives.   This lack of congruence of value systems is the most common reason why proposed plans
and policies are rejected by executives and legislators.   A large
number of characteristically different plans which rank very close
together on the analyst's value criteria may be much more widely
separated in the decision-maker's mind, and, characteristically, he
may choose a plan which the analyst believes worthy of consideration, but which he would not have ranked among his first few
choices.   The opportunity to explore a large number of alternatives,
therefore, greatly relaxes the political tension between analysts
and decision-makers, and enlarges the scope of their cooperation.

In line with my desire to de-emphasize very large-scale models,
some discussion of the necessity for large-scale data bases and
model validation is necessary.   I have long believed that the very
large-scale data bases which are used in transportation studies can
be greatly reduced and coordinated with census data, so that an
appropriate base of the total population and its distribution exists
for initiating many analytical studies.   At the same time, various
behavioural patterns of a number of different types obviously
require analysis in depth on the basis of small samples.   A special
small sample (the 1/100 sample) drawn from the US Census of Population and Housing provides an unusually important base for analyzing the joint distribution of large numbers of characteristics of
people and dwelling units.   Using this and other sample information,
joint distributions may reasonably be reconstructed where they are
not available, but where marginal distributions are — eg. for
census tracts.   Other sample studies permit us to update universe
data and to explore specific behaviours in considerable depth.

Just as there has been an overemphasis on data, so, I believe,
there has been an overemphasis on model calibration and parameter
estimation.   A model may be designed and its parameters may be somewhat arbitrarily selected partly on the basis of prior knowledge —
with the end in view, not of achieving a high level of statistical
sophistication and accuracy, but rather of being able to replicate
the principal features of the phenomenon of interest.   One should
also be able to satisfy certain fairly elementary considerations of
internal consistency and logic.   In the latter class of conditions,
one might suggest, first, that any model should have a performance
which is independent of the exact way in which its areal disaggregation (and perhaps some other types of disaggregation) is carried
out.   Examples of phenomena which need to be reproduced at least in
some modelling contexts are the clustering of retail trade activi-

ties, the slow and differential suburbanization of different types
of manufacturing, the segregation or self-segregation of household
types by neighbourhoods (but not to the exclusion of some mixture),
and a large variety of phenomena such as urban blight which display
the effects of contagion. Needless to say, this list could be
greatly extended, partly depending on the tastes of the model-
builder or the policy-maker. Very seldom, however, does one see a
number of these phenomena placed up front in model design. Con-
trariwise, there are a number of models in common usage which can be
fitted with high R^2's, but which can be certified never to produce
certain very simple realistic outcomes.

One special issue regarding logical consistency is a matter of
substantial concern to me, but perhaps less so to many others. I
believe that a dynamic model ought to be examined for its long-term
equilibrium states on the assumption that there are neither internal
growth nor external shocks. I believe that models which never
achieve equilibrium (except those which cycle for clearly understandable reasons) or which converge to logically unreasonable
equilibrium patterns are probably unsound models. Their dynamic
character, and the short periods of time over which they are
applied, may mask these failings, and the analyst therefore has a
special responsibility to search out the model's long-term equilibrium characteristics.

The serious analyst of public policy and private responses may
be observed to have a natural tendency to disaggregate the analytical problem itself at least by function, or by sector, or by both.
Disaggregation within sectors and across areas are somewhat more
difficult to use as a basis for defining a particular study. There
are three very brief points, then, which can be made with respect to
the exploitation of this natural tendency. First, convenience in
data collection and research administration, together with disciplinary predilections makes it much easier to manage detailed studies of small parts of the larger and more complex social system.
Every theoretical conclusion reached in such a study is potentially
a model of the phenomenon studied. Since we need more good models
of these phenomena, such studies should be integrated into the
framework of a larger and more ambitious goal. Second, the connections of such studies and their theories expressed as models with
other parts of the social system are quite simply inputs and outputs. The principal input is the change of the environment,
including a whole host of social, economic and spatial variables
which influence the behavour of the actors under study. Larger data
bases like the census, and a more open approach toward the relationship between a given model and other models, would be conducive to
this style of analysis. Similarly, the outputs anticipated or the
behaviours to be modelled should include the actor's decisions which
affect not only his own welfare and other direct objects of study,
but also the general environment in which other actors take their
decisions, and through which indirect effects ultimately feed back
to their origin.

Policy studies and scientific research have a wide area of commonality in that both intend to study the variations of outputs in response to variations of inputs. In policy studies, the predictive theories or models are generally assumed fixed, while one method of scientific inquiry is to vary the relevant theories or models so as to secure as close a congruence as possible between the projected and actual outputs which flow from a given set of inputs. These two modes of analysis are basically so similar and overlap so broadly that we can consider jointly some experimental methods which apply to them. In principle, experimenters of all types vary one or another of a set of assumptions. These assumptions fall into three broad classes. The first set deals with inputs, and these consist of real-life conditions, either actual or imagined, and policy variables, which if acted upon would become a part of real life. A powerful method of testing the generality and validity of models is to expose them to very different real-world inputs. A second subject of variation is the models themselves, and possibly, their parameters. Variation in this area corresponds with some forms of comparison of models and parameter fitting. It has an important policy implication, however, in that different policy analysts, and especially different decision-makers, have different views on how the world actually works. In the absence of complete scientific knowledge, these alternative views can be regarded as assumptions. Finally, different assumptions can be made as to the interconnections between different parts of the real world, as expressed in interconnections between models. These interconnections, if expressed as environmentally conditioned reactions, acquire a certain air of automaticity. The important assumptions which are involved with respect to interactions are the inclusion or exclusion or variables as endogenous to a set of models, and the inclusion or exclusion of specific models in the set.

There are obvious difficulties in operating in either of these experimental modes. Basically, neither model design nor computer systems are well-attuned to the experimental substitution of one model for another, and to the expansion or contraction of such a modular system. Most of the complex models which have been designed with a view to accommodating some or all of the considerations outlined above are large and monolithic. This is especially true of large land-use models, and somewhat less true of large transportation models. A detailed development of experimental standards in this field, and the initiation of experimental work which is potentially general-purpose, rather than special-purpose, is a crying need.

I turn now to a brief consideration of normative aspects of these large system models, and of social system models in general.

The foregoing discussion has pointed in several different ways to the intrinsically normative context in which the design and application of models proceeds. On the premise that planning and policy-making are optimum-seeking processes, economists and operations researchers are anxious to propose means by which the optimum may be sought. All such procedures inherently require the formulation of an objective function which can measure the goodness of performance of plans, and this formulation suffers from a host of difficulties arising out of the pluralistic nature of many societies, the division of labour in the analysis and decision-making process, and the difficulty in specifying in an operational way many cherished, but ill-defined, social goals. All of these difficulties are relatively small by comparison with two other aspects of trying to find a match between policies and their objectives.

In the first place, it is difficult to maintain a clear conception of the connection between the inputs and outputs of large systems such as can be used systematically to improve the performance of the system as measured by the objective function. Classical methods of hill-climbing are, in principle, only available for well--defined continuous systems, in which the partial derivatives of the objective function with respect to the decision variables can be calculated. Other methods of optimizing large systems, such as linear programming, have other special formal requirements. The attempt of the New Urban Economics to define systems which can be analyzed in this context leads, it is generally agreed, to the wholesale oversimplification of problems. The use of partial derivatives and direct optimization can be replaced for some systems by a form of sensitivity testing. However, where the decisions are lumpy or discontinuous and quite numerous, sensitivity testing is very cumbersome, and "ad hoc" methods are ordinarily developed to deal with the systems of interest. Application of these methods frequently involves large-scale computer runs. As a simple example, the network planning of the Chicago Area Transportation Study mentioned above was initiated on the basis of a continuous model of the optimal spacing of the expressways, but was then continued through a sequence of very large-scale simulations. Specific and sophisticated methods were developed to select improvements to be tested at each stage of the development.

The typical large-scale planning problem has indeed two specific aspects which make its analysis for optimum-seeking purposes difficult and intransigent. First, many planning decisions embody indivisibilities. Facilities of many types, both public and private, have minimum sizes or thresholds below which they are infeasible or inefficient. More generally, there are very substantial economies of scale in location, in programme development, and in bureaucratic management. Typically, such indivisibilities lead to formulations of optimizing procedures which have the character of integer programming. Although many integer programming problems turn out in practice to be relatively tractable, in principle such problems are much more prone to exponential combinatorial explosions of computational time than are various forms of linear programming and related methods. The discrete nature of planning decisions and their combinatorial complexity are frequently regarded as the principal cause for difficulty in optimization procedures.

In fact, however, planning decisions exhibit a much more serious difficulty. Most policy actions depend for their impact on
other policy actions. These interactions introduce extremely powerful non-linearities into the objective function, and lead to the
conclusion that there are many local optima -- that is to say,
situations which cannot be improved by small changes. (In the
continuous case, a local optimum is defined by the conditions for
the maximum or minimum of a function; in the discontinuous case, an
explicit definition has to be given to the idea of "small changes",
but this does not present any serious difficulty.) The enumeration
of multiple local optima and the selection of the "optimum opti-
morum" arises as a problem, both in problems with indivisibilities
and in problems with multiple optima resulting from strong interactions. In principle, these problems can be attacked either by
dynamic programming or by branch and bound programming. In practice, however, many problems remain insoluble for any practical
expenditure of resources, if they achieve any significant size.

Physical planning, in particular, is full of well-explored and
precisely formulated problems of these types. While a number of
them seem to have similar mathematical properties, the actual
performance of the best-known solution methods varies widely. Very
large problems regarding the shortest path through a network can be
solved extremely efficiently by a dynamic programming algorithm.
Large facility location problems can sometimes be solved by linear
programming, but their integer properties are often violated by
absurd fractional results; the basic LP structure of this problem
lays a basis for an effective subgradient search procedure. On the
other hand, two problems involved in New Town planning seem relatively intractable. The network design problem is combinatorially
explosive, and responds only fairly well to heuristic methods. The
assignment of activities to neighbourhoods (which may be formulated
as the Koopmans-Beckmann problem of quadratic programming) is
utterly intractable for more than about fifteen locators, and heuristic methods applied repeatedly to the same problem constantly un-
cover new local optima.

It is demonstrable on grounds of both theory and experience
that direct optimizing methods, and especially intuitive ones, are
not suitable for exploring issues of policy and planning. This
leads us to search for approximate methods, and especially for
approximate methods which are compatible with human capabilities as
applied to the solution of policy problems. These methods may or
may not coincide with established professional protocols; they
should borrow, not only from them, but also from the fundamental
understanding obtained through a study of optimization.

I feel that there are two fundamental procedures which are
shared by intuitive approaches to policy-making and planning.

Most obvious is the idea of successive improvements to a given
plan. This is the method by which a great deal of government's

bureaucratic and legislative policy-making proceeds, with marginal adjustments to a very large and complex system. Since many organizational, programmatic and legislative decisions are discrete, the definition of "marginal" is necessarily arbitrary. Some relatively large programmatic decisions may still be marginal in relation to the system as a whole. What is perhaps most essential in this context is that a very large proportion of such decisions are taken one at a time. This is done with regard to the interaction with other parts of the system, but largely without regard to the possibility that other decisions may make other and simultaneous changes in the system.

The essential problem with respect to improvement methods is to bear in mind that change does not necessarily lead to improvement, and that policy-makers engaged in an improvement process must find some way of determining whether an improvement will indeed take place. Since systems are large, complex and ill-defined, this determination is not always completely rational and scientific. It involves perhaps only a partially articulated conception on the part of the policy-maker and planner as to how the system works. As an extension of this idea, it also appears likely that the selection of possible improvements which may then be tested either by intuitive or by well-articulated means is in itself a process requiring insight and skill. The number of possible changes, even taken one at a time, is very large, and when more than a handful of changes are considered jointly, the number becomes astronomical. Because of interactions, multiple joint decisions are apt to have greatly heightened effectiveness, but it is a practical impossibility to explore in detail any of the very large number of such combinations.

Any improvement process carried far enough leads to a local optimum, but some local optima may be very much better than others. The value of the end point of an improvement process, therefore, depends on the selection of the starting point. Finding new starting points for plans and policies is another skill of a somewhat different type which plays a major role in the preparation for decision-making. Such starting points involve a fully elaborated policy or plan at the level at which the analysis is being conducted. Because all of the elements must be fixed, the combinatorial complexity of surveying all possible starting points would be enormous. It is, however, clear that the number of significantly different local optima is much smaller than the total number of possible plans. Consequently, if a good improvement process is available, it is only necessary to find starting points in each of a number of regions. The business of finding good starting points may be characterized as finding good configurations or good basic ideas; this is a function which is carried out by utopians, generals, chess champions and people in a variety of walks of life where originality is significant. This process, to be successful, rests on two principal qualities. The first is an ability to escape from the bounds of tradition, and to find policy configurations which are generally different from existing policy and from each other in the sense that they will lead upon, improvement, to different final plans. The second quality is an intuitive perception of the goodness of various starting points. This feeling permits the very early rejection of ideas which are either too far from their best specification, or which, after improvement, will still be relatively inferior.

The fundamental ideas of selecting a configuration and improving it to the point where it can be rejected or retained in a cluster of alternatives is basically similar to sketch planning in architecture and planning. The improvement process corresponds with hill-climbing in the theory of optimization. The selection of starting points is relatively neglected in optimization theory. A number of suggestions have been made for random sampling, and some experiments have been conducted, but it appears that the policy space is ordinarily much too large for a sample to be effective in the fashion which I have just described. More attention might be given, it seems, to the creation of constructive methods which lead to a variety of different solutions in a relatively controlled way; the sampling, for example, might be over the parameters of a constructive method. Such constructive methods in operations research are usually intended to produce a single "good" solution rapidly. They are not placed in the context of multiple starts for an improvement process.

The formalization of optimum-seeking methods as a part of policy-making and planning lies in the relatively distant future. Various efforts have been made to structure this process in terms of man-machine interaction — that is, cooperation between analysts and a computer. In its simplest form, such cooperation could accept plans and provide the analysis with an estimate of their outcomes using the kind of a predictive modelling structure discussed earlier. This capability could be improved by permitting the analyst to give a broadly described plan and to have it elaborated automatically according to a specified set of rules from which the analyst might select. Another level of cooperation would permit the systematic improvement of plans leading toward a local optimum. This could be done by computer, with perhaps some guidance from the analyst. The prospect of a computer search over likely starting points seems highly improbable. It is, however, possible to imagine starting points for constructing and improving plans based on these starting assumptions.

The alternative to methods of these kinds in the realm of supposedly scientific policy-making appears to be the massive analysis of a very few large system alternatives whose provenance is not well-understood, and whose fundamental characteristics may miss the mark entirely. Such a policy, common in cost-benefit analysis and in a number of other cases, may be regarded as the normative version of the misdirected preoccupation with detail, which is also observed in positive analysis, but it similarly reflects a neglect of the development of positive models which can give a rapid analysis of the performance of a system while preserving relevant levels of detail in inputs and outputs.

There are many other aspects of both positive and normative planning and policy-making, some of which are also addressed at this conference, which can be related to the contents of this paper. For example, we find frequent references throughout the literature to the necessity for providing models which can easily be understood by policy-makers. I have observed numerous cases in which the inexperienced policy-maker will say, "That is too complicated for me to understand", while the more experienced policy-maker will frequently suggest, "This is much too simple and leaves out essential phenomena". The most important feature of a good model is not that it be transparent, but that it be correct. Good science, however, suggests that simplicity is a desirable feature of theories. Furthermore, simplicity may be obscured by a proliferation of unnecessary detail. Clearly different levels of policy-making require different levels of model complexity, and broad-scale policy explorations require rapidly operating models. Finally, I have suggested that most of the models of behaviour can and should be very simple, and that the complexity of models, if it exists, derives from the interactions of behaviours. That simple behaviour patterns can result in surprising results which are not intuitively obvious is demonstrated in Forrester's Urban Dynamics, and much more rigorously in Thomas Schelling's analysis of the consequences of microbehaviours.

An important source of difficulty in integrating models into the decision-making process, which I re-emphasize from the preceding discussion, is the differences in value systems which may exist between the decision-maker and the analyst. When decision-making extends into the public arena, and especially into legislatures, this difference is widened to a difference between the analyst and the population at large. Clearly normative methods of policy analysis should attempt to take the true values of the political system into account, but since many of these values are unknown or poorly articulated, the production of single policies and their recommendation as the best available has at best only a spurious pretension to scientific truth.

It is very obvious that the objectives pursued in the design of the policy-making process in its positive and normative aspects are contradictory and conflicting. Not all the objectives can immediately be satisfied. Our short-run target should be to make an appropriate selection of emphases in the improvement of the total process, and to apply the best of the state of the art to these conclusions. More long-term goals are to discover new methods which resolve some of these conflicts, and to incorporate them into the process. The pursuit of both the short- and the long-term objectives requires an analytic appreciation of the difficulties which is based on a purposeful structuring of our overall understanding of the process. It should not depend on conventional wisdom or popular platitudes about the nature of these relationships. It is my hope to have made some contribution to this necessary insight, or to provoke an appropriate discussion which will do so.

PLANNING AND DECISION-MAKING IN HUMAN SYSTEMS: MODELLING

SELF-ORGANIZATION

Peter Allen

Service de Chimie Physique II
Universite Libre de Bruxelles
1050 Bruxelles
Belgium

INTRODUCTION

There is a famous phrase in Mrs. Beeton's cookery book which occurs at the beginning of the recipe for "jugged hare". It is - "First, catch your hare", a task which in her time generally meant more than simply placing a special order with her local butcher. The importance of the remark is, of course, that all the technical discussion of how exactly to proceed, is quite useless if you don't have a hare, and indeed, if you don't know what one looks like. This inevitably springs to my mind on hearing discussions on decision-making, policy and, indeed, politics in situations as complex as families, businesses, cities and nations, for without an understanding of the functioning of such systems, of the true inter-play between the various factors, we are in a situation worse than that of Mrs. Beeton's "hareless" cook, because we are discussing what sort of recipe we should invent for ingredients we do not know, and whose behaviour together we cannot predict.

In reality then, "policy" can only be formulated and discussed in the light of some model which relates the variables which are considered relevant, and clearly, even those who abhor the penetra-tion of quantitative analysis in human systems must admit ultimately that they would have to base their decisions on some kind of intui-tive model predicting different possible outcomes. So, says the "classic" type of systems analyst, it is surely better to specify clearly the model on paper, to quantify the variables, set up the equations and turn the computer handle, remembering perhaps to do a sensitivity analysis of the resulting evolution. And that is where the discussion has remained for a long time: is it better to do what you can do, whilst sometimes honestly admitting that the variables you are discussing may not be the ones that really matter, and that by basing interventions on them you are influencing the evolution in perhaps an undesirable way, or does one simply criticize such "reductionist" manipulation of the system and talk about the "quality of life" etc., without having any concrete reasonable alternative to put forward in response to the question, "okay, what should we try to do and how?", for the necessary pre-condition to a discussion of policy is to have a model.

In fact, the basic truth about complex systems such as towns and cities, as indeed is true of cookery, is that obtaining harmonious results remains something of both an art and a science. And yet, how can this be so? How is it that the science capable of the Apollo project cannot "crack" such problems as the prediction of urban change? The thesis that I will try to put forward here is that despite its impressive appearance, the great achievements of science have pertained to certain classes of system subject to two highly restrictive conditions. Either, they deal with a "simple" system involving only a few interacting elements, where the dynamical trajectory of each can be strictly followed, and is generally subject to some conservation law, or, if dealing with a complex system of many interacting elements, then such successful laws (as, for example, the hydrodynamical equations of motion) have only been derived for systems at or near to thermodynamic equilibrium. Here, the system microscopically is in the "most probable" state, and all interactions can be linearized.

In fact, all living systems fall outside these two categories, and this is probably the fundamental reason why the application of "scientific" principles to human systems has proved to be much more difficult and controversial than had been optimistically, and perhaps naively, expected. Thus, the reduced description of a fluid, for example, corresponding to the Navier-Stokes equation (a very much used equation of hydrodynamics) is only possible, and in a sense, complete, for a fluid very close to thermodynamic equilibrium, and arises because of the existence of the collisional invariants of energy, momentum and mass. When we attempt to construct a "model" of a human system, we are also looking for a "reduced" description, because the whole point of the model is that it should be simpler to investigate than the real system itself. However, to what extent may we hope for a reduced description, or model, which is closed, that is which predicts successfully the evolution of the variables over long times without requiring the reformulation of the model and the redefinition of variables? In order to explore this question, let us first examine the manner in which reduced decriptions exist or breakdown in physics and chemistry, particularly for physical systems which, as is the case for living systems, are subject to flows of matter and energy which maintains them out of thermodynamic equilibrium.

THE EVOLUTIONARY PARADIGM OF DISSIPATIVE STRUCTURES

In physics there are three basic levels of description. Firstly, there are the classical or quantum laws governing the motion of each particle, presumably constituting the "basic" or "complete" description.   Secondly, we have the probabilistic approach which supposes some "unknown" underlying motion of the particles, which is supposed to give rise, for example, to a "birth" and "death" equation governing the evolution of the system.   The third level of description is the macroscopic one of thermodynamics, hydrodynamics, chemical kinetics, etc. where the equations governing the evolution of the system are written in terms of variables which are themselves "aggregate" or "average" quantities.   This is the reduced description.   When is the passage between these levels entirely clear and well-understood?   This is only satisfactory for systems which are at, or very near to, thermodynamic equilibrium. Only then can the macroscopic, reduced, description be deduced from the "complete" one.   However, far from thermodynamic equilibrium description breaks down, together with its associated determinism, and the idea that we can understand and predict the evolution of the system simply in terms of the average values of certain quantities proves to be inadequate.

This breakdown is associated with the occurrence of bifurcations in the solutions of the macroscopic equations.   Bifurcations introduce an unsuspected wealth of new phenomena into the otherwise rather trivial evolution of the system, resolving the apparent contradiction in the traditional meanings attached to the word "evolution" in the physical and the human sciences.   In the former, it has referred to the movement towards thermodynamic equilibrium, the elimination of non-uniformities and the increase of internal disorder, while in sociology and biology, it has been associated with increasing complexity, specialization and organization.

In order to understand some of the basic points which arise from these new concepts, let us look briefly at a simple, but remarkable chemical experiment which demonstrates these striking properties.   It consists of the relatively simple chemical scheme:

$$A \rightarrow X$$

$$B + X \rightarrow Y + D$$

$$2X + Y \rightarrow 3X$$

$$X \rightarrow E$$

where, starting from A and B, X is produced, which produces Y, which in turn helps to produce X, and the final products of the reaction are D and E.   If we were "systems" minded, then in modelling this problem, we would draw the very simple interaction scheme shown in Figure 1.

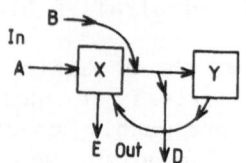

Figure 1 - The Schematic Diagram of Interaction of the Brusselator.

This particular reaction scheme has been studied in detail by the Brussels' school (it is even known as the Brusselator) and has been found to display various types of self-organizing behaviour. Thus, starting from the uniform, well-mixed, state of the system at equilibrium when the flows in and out of the system of A, B, D and E are zero, we can drive the reactions away from equilibrium by gradually increasing the rate at which A and B are pumped in, and D and E are extracted. At a certain critical distance from equilibrium, an instability occurs. This threshold marks the point at which the least fluctuation can cause the system to leave its uniform, well-mixed (maximum disorder) state, and move to some qualitatively different new state of organization. This can be perhaps a stationary pattern of the concentration of X and Y, or moving waves of concentration, even as complex and beautiful as expanding spiral waves (see Figure 2). Such patterns represent the coherent behaviour of billions and billions of molecules, organized over distances which are absolutely vast compared to that of the molecules. Where does the "information" for such organization reside? What are the necessary conditions that must be met in order to observe such phenomena? The intensive study of these structures, called "dissipative structures" has revealed the answers to these questions over recent years, and gives us a fascinating new perspective on structure and regularities observed in the world which I believe is of fundamental importance.

Firstly, then, what occurs in the system when a dissipative structure appears? At such a moment, what we witness is the instability of the previous macroscopic pattern of organization. Thus, for example, if we are moving away from equilibrium, we have initially a uniform distribution of the densities of X and Y. Each point in the system is doing the same thing, and because of this there are no internal flows of X and Y between different regions of the system, since there are no strong differences of concentration. However, as reaction rates are increased, the kinetic equations, the model, the reduced description, become ambiguous. That is, they permit potentially more than one solution. Thus, in the patterns shown in Figure 2, what we see is the realization of other macro-scopic organizations of the system, where high and low levels of production of X and Y in different regions of the system are balanced by flows of X and Y between them. A new structure such as this occurs then, when the old state becomes unstable. That is when fluctuations around the average values of the variables (remember that the kinetic equations are merely written in terms of averages, a reduced description) which are, of course, always present, will carry the system off to one of the new states of organization. Which one it is depends on the precise nature of the fluctuation, which is, of course, not controlled by the kinetic equations, which only discuss, deterministically, the evolution due to the average values of the variables (Glansdorff and Prigogine, 1971; Nicolis and Prigogine, 1977; Prigogine et al, 1977).

Dissipative structures, therefore, invoke both chance and necessity. The conditions required in order to observe them in a system are that there should be simultaneously more than one solution to the kinetic equations possible, and this in turn requires non-linearity in the interaction terms (linearity means only one solution, quadratic dependence two, cubic three, etc.). Thus, if our reduced description involves equations of change of the variables which are non-linear, then we may expect such a system to have bifurcations in its solutions at certain times, and for the fluctuations to play an important role in the evolution of the system, in choosing which of the possible branches of solution the system will take, in fact. Complex systems involving feedback will, in general, give rise to a whole series of bifurcations, as illustrated in Figure 3, and therefore, the understanding of its evolution necessarily requires the study of its passage, not just along the particular branch where it happens to be at the initial time, but the possibility of structural reorganizations corresponding to its passage through bifurcation points, and on to new branches, which are perhaps qualitatively different in character.

Such a point of view introduces several important points which I believe are particularly significant for the social sciences. Firstly, it introduces "history" into the explanation of structure, for if we look at Figure 3, we see that, for example, the fact that the system is organized in the manner corresponding to the branch C, necessarily implies that in a system growing from initial simplicity, it happened to take the fork B, and before that, fork A. No "explanation" can ever deduce from the reduced description of the problem the unique necessity of finding the system in the state C, for parameter value p, because of the ambiguity of the macroscopic description at this point. Secondly, even a system as simple as that shown schematically in Figure 1 has developed a certain autonomy. It is the system, not the experimenter, who chooses which branch of bifurcation it will take, since this depends explicitly on the particularities of the fluctuation that happens to occur at the moment of instability, and these fluctuations are "outside" the model. Thus, in addition to the fact that, of course, the external environment of a system may change according to a scenario in order

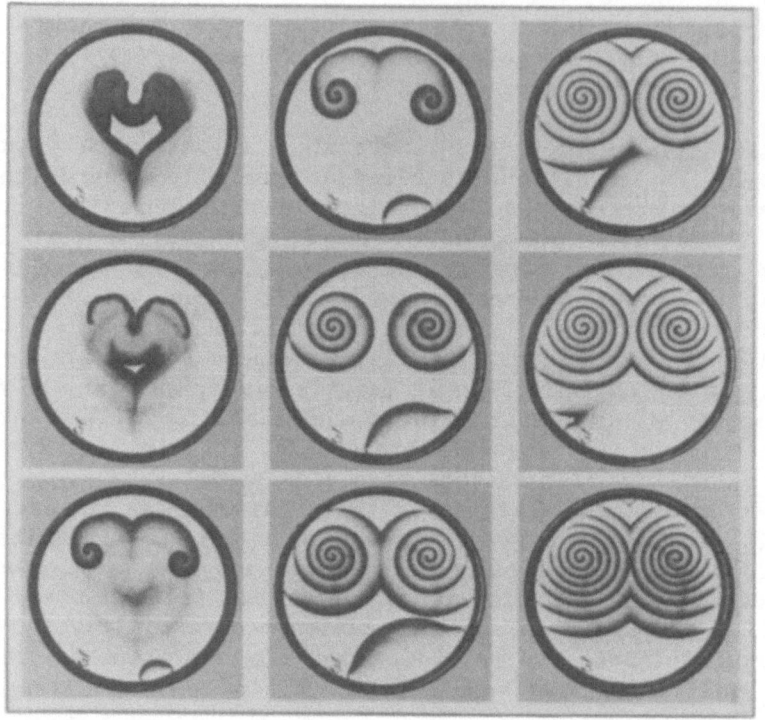

In the Belousov-Zhabotinski reaction, malonic acid is oxidized by bromate in the presence of cerium. When the reaction is performed in a shallow dish, spiral waves develop.

Figure 2 - An Example of Dissipative Structure

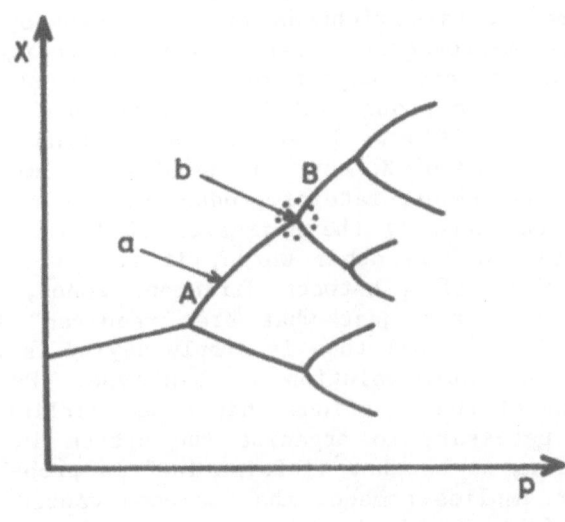

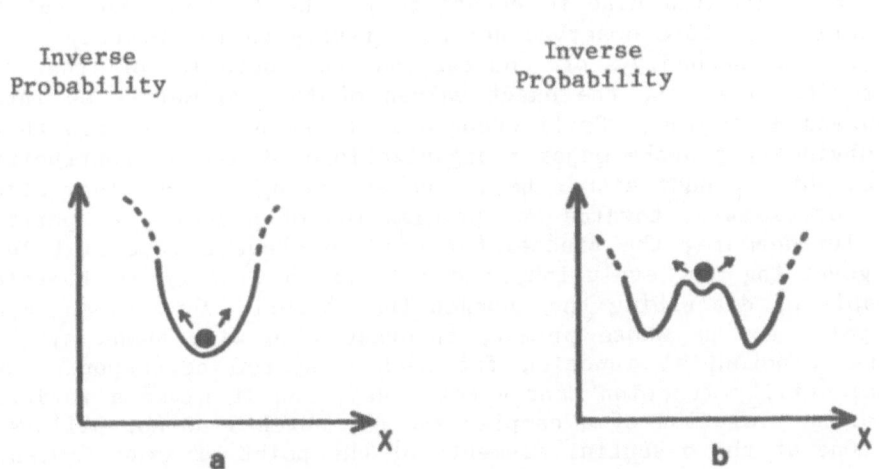

Far from bifurcation points (a), we have stability, in that small fluctuations are damped. Near a bifurcation point, however, (b) it is the fluctuations which decide which branch will be taken. The explanation of the structure corresponding to branch C necessarily involves the historical choices of A and B.

Figure 3 - A General Bifurcation Diagram for a Complex System to allow for the non-closure of the model to the outside world, also, we are obliged to admit that a non-linear dynamical system is not closed either with respect to its interior. This is another way of saying that the world is always richer than the model, and that at times, some of this richness of the microscopic level breaks through to the macroscopic level causing a reorganization. The third highly significant aspect of dissipative structures is that they already pose the problem of the "chicken and the egg", related to structure and function. If we ask why we find a particular band of high concentration of X, say, in one of the examples of Figure 2, it is that its non-linear rate of production (as a function of X and Y) is exactly balanced by the diffusion of X into the surrounding area with little X. In other words, if we now ask why there are strong micro flows of X between different zones, then the answer will be that they are just what are "required" to maintain the production of X. What all this is simply saying is that the kinetic equations have a stable solution of this type. We can now return, perhaps, to one of the questions that arose earlier. Where is the "information" necessary to organize the system in this way? The answer would seem to be that "information" is probably a misleading word, since it implies somehow that someone wanted to organize the system. The fact is simply that stable structured solutions are compatible with the kinetic equations of the system, and that the particular structure observed depends firstly on the history of the system, and secondly, of course, on the details of the non-linearities, that is, the exact values of the parameters of inter-action and diffusion. Small changes in these parameters can there-fore obviously provoke major reorganizations of the macrostructure, changes which cannot either be viewed as moving in any clear direc-tion, for example, towards an optimization of anything in particu-lar. Furthermore, the Brusselator does not have a potential func-tion governing its evolution, and catastrophe theory is therefore incapable of describing the changes that occur. This is an impor-tant point for us whose primary interest lies with human systems, because a potential function for such a system corresponds to a "global utility function" for a city, say, and it gives a rational-ity to the evolution of a complex system, which I do not believe it has. One of the essential elements of the point of view I wish to develop is that a complex system undergoes an "open" evolution, with new properties and new values emerging along an expanding tree of possiblities. This is quite a different view from that in which the system is governed by a potential, albeit one of an interesting shape, where, in a sense, everything that can happen is already contained in its specification.

The basic idea I wish to develop, therefore, is that just as the Brusselator can give rise to highly complex structures, arising nevertheless from a very simple, but non-linear reaction scheme (Figure 1), so perhaps the apparent complexity of human systems may be partly understandable in terms of a few non-linear interaction mechanisms. Thus, by supposing some simple form for the inter-actions between the actors of a system, we may produce, spontaneously, during the dynamic evolution of the system, a macroscopic self-organization into perhaps a highly complex structure, where structure and function will be enmeshed in the system, recording the particular course of history. Such a non-linear dynamic system is a collective memory.

A SIMPLE CITY SYSTEM

Having tried the reader's patience thus far with perhaps too much physics and philosophy, let me attempt to illustrate in concrete terms what the speculations of the previous section may be able to do for us when faced with the problem of modelling, for example, a city. The aim here is not to describe a completely realistic model, but rather to set out the basic framework, a matchstick drawing, as it were, of the "workings" of a city, in the hope of being able to explore the long-term evolution, involving structural changes. We hope from this to be able to build a model which, at least, can predict the sort of structure that may evolve under a certain scenario, with the accent on the qualitative features of that structure, rather than on quantitative accuracy.

The first step in the operation is then to choose the significant actors of the system, whose decisions, and the interplay of these, will cause the urban system to evolve. In agreement with much previous work, particularly, for example, the philosophy of a Lowry-type model, we first consider the basic sector of employment for the city, and, in particular, two radically different components of this; the industrial base and the business and financial employment. Next, we consider the service employment generated by the population of the city and by the basic sectors, supposing two levels, a short-range set of functions and a long-range set. The residents of the city, depending on their type of employment, etc. will exhibit a range of socio-economic behaviour, and for this we have supposed two populations corresponding essentially to "blue" and "white" collar workers.

The next phase of the modelling is in attempting to construct the interaction mechanisms of these variables, which in essence requires a knowledge of the values and preferences of the different types of actors represented by the variables, and, of course, how these values conflict and reinforce each other as the system evolves. Let us first discuss the mathematical representation of values and preferences in order to set up our system of equations (our kinetic equations which are our reduced description of reality).

The basic step, about which almost all multi-criteria analysts agree, is to suppose that each actor can, when faced with a choice, at least list his main criteria of decision: price, facility, prestige, time involved, etc. Let us assume that these factors define the "rationality" of that particular actor, without asking whether or not there is an objective definition of rationality. The second

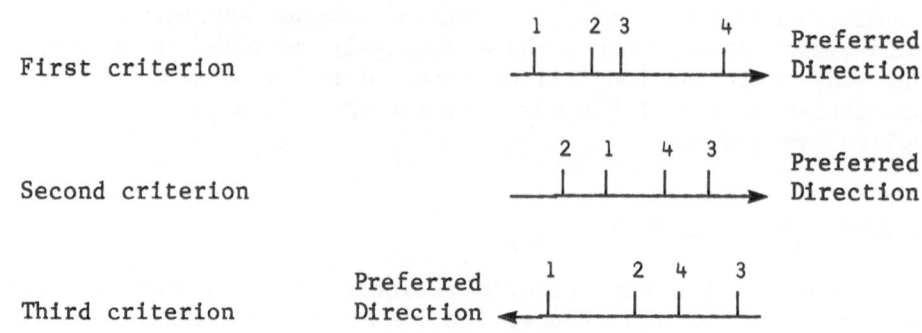

Figure 4 – The Ordering of Four Possible Actions on Axes Corresponding to Three Criteria.

step in attempting to characterize the preference for one or another
choice is to assign some appropriate "weighting" to the various
criteria already retained so that in some way their relative importance can be taken into account.   Let us stress here that we are
discussing the preferences of a single actor.   Figure 4 shows four
possible choices with three dimensions of preferences.

Clearly, if the three criteria were strictly quantitative, were
numbers, then it would be possible to simply "add" their weighted
values for each choice and identify the "best" choice.   Obviously,
this is simplistic in the extreme since it ignores uncertainties,
thresholds and non-linear reactions to equal steps, as well as the
possibility of purely qualitative values associated with each
choice.   Without going into the details, let us briefly sketch a
method which attempts only to extract information about preferences
which is certain on the basis of the information above.   In the
simplest version of this, the "pay-offs" believed to be associated
with each choice are compared by pairs for each criterion, and a
preference matrix is constructed.   In the most basic version, this
can be done using a simple Boolean response to each question, is
choice i better than choice j for this criterion?   This gives us the
three matrices below for our particular example.

	1	2	3	4
1		0	0	0
2	1		0	0
3	1	1		0
4	1	1	1	

first criterion

	1	2	3	4
1		1	0	0
2	0		0	0
3	1	1		1
4	1	1	0	

second criterion

	1	2	3	4
1		1	1	1
2	0		1	1
3	0	0		0
4	0	0	1	

third criterion

These now can be combined by multiplying each by its weighting factor.

	1	2	3	4
1		$\alpha_2 + \alpha_3$	α_3	α_3
2	α_1		α_3	α_3
3	$\alpha_1 + \alpha_2$	$\alpha_1 + \alpha_2$		α_2
4	$\alpha_1 + \alpha_2$	$\alpha_1 + \alpha_2$	$\alpha_1 + \alpha_3$	

Weight accorded to the three criteria is α_1, α_2, α_3, respectively.

From this final preference matrix a graph may be constructed indicating the relation between the four choices. For example, an individual who considers that the three criteria are of equal importance, $\alpha_1 = \alpha_2 = \alpha_3$ gives us:

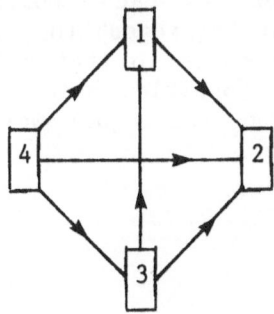

An arrow indicates net preference between the pair.

Clearly, choice 4 is the best. For somebody else, however, if for him the third criterion was very important, more important than the other two combined ($\alpha_3 > \alpha_1 + \alpha_2$), then we have the graph:

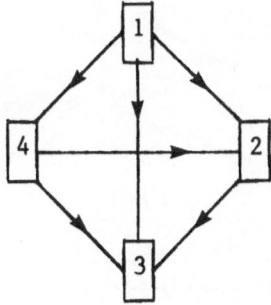

For this the choice 1 appears as the best.

Of course, this method may be considerably refined to consider weak preferences and a range of indifference, and it can be used to

analyze choices where considerable uncertainties exist as to the
possible "pay-offs" of each choice along the various axes, as well
as for purely qualitative criteria.   It has been given the name of
ELECTRE (Roy, 1968; Roy et al, 1975), and what is particularly
interesting for us is that it has been devised in order to attempt
to "model" the manner in which individuals really do make decisions,
and to capture the uncertainty, subjectivity, ambiguity, and the
role of qualitative criteria in people's behaviour.

A particularly clear way of visualizing the problem of choice
is to imagine that each actor is at the "origin" of a set of axes,
each of which represents a criterion involved in the evaluation of
the decision.   The origin represents the "ideal" choice for the
actor with the maximum imaginable pay-off in all directions, and, of
course, has nothing necessarily to do with what that actor actually
does, since the real choices presented to him will be somewhere out
in the space defined by the dimensions of his value system, a
"mental map" of imperfect offer he draws with the information he has
received.

In such a space, the "distances" of each choice along any
particular axis will be "stretched" or "squeezed" to a degree which
depends on the weighting the actor accords to that criterion.   In
such a representation then, the four possible choices of our previous example viewed by an actor who puts equal weight on each
criterion will look as shown in Figure 5.

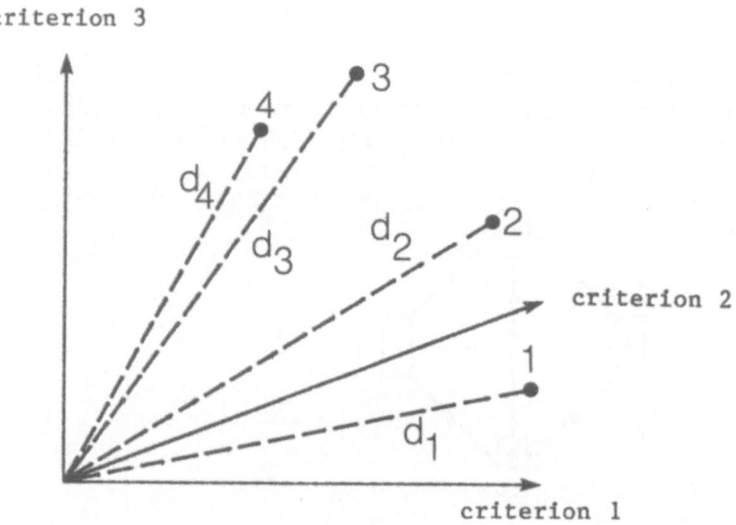

Choice 4 is the most probable because d_4 is shortest.

Figure 5 - Value Space of an Actor who Assigns Equal Weights to the
          Three Axes.

The same four choices, viewed by an actor who puts a much greater weight on the third factor, would appear quite different, as we see in Figure 6.

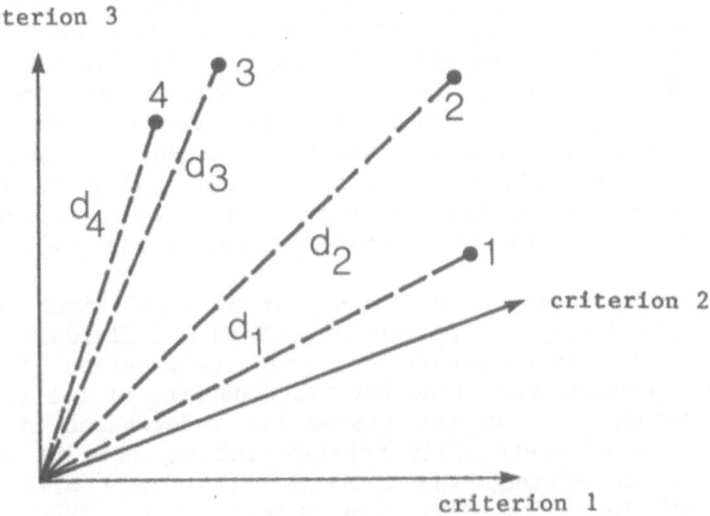

Now d_1 is the shortest and hence, choice 1 the most probable.

Figure 6 - Value Space of an Actor who Assigns Greater Weight to Criterion 3 than to the other Two Combined.

Thus, we may view the problem of choice under multiple criteria as the "distance" from the origin, in an n-dimensional space, of the various possible choices. Of course, the position of each point is uncertain to a degree, depending on the uncertainties involved in the estimation of the "pay-offs" associated with each choice, and also depending on the degree of precision one can give to the weightings assigned to each axis. Each possible choice is therefore associated with a "cloud" rather than a point, and the problem of decision reduces to that of estimating which choice gives rise to a "cloud" which is nearest to the origin. This corresponds to supposing that the "attractivity" of a given choice decreases as its distance from the origin increases.

Thus, for a single actor at a particular moment, we may suppose that the choice i will be selected with probability

$$\frac{(1/d_i)^I}{\sum\limits_{i'} (1/d_{i'})^I} = \frac{A_i}{\sum\limits_{i'} A_{i'}} \tag{1}$$

where I gives a measure of the informational precision of the distances. Thus, when $I \to \infty$ then the probability of choice is simply 1 for that nearest to the origin, and 0 for the others. In the opposite extreme, of extreme uncertainty, $I \to 0$ and we simply have equal

probabilities for all choices. Clearly, most decisions fall somewhere in between these two extremes.

An important point which we have not yet considered is that of
"time". In an evolving system, the "pay-offs" that characterize
each choice will change in time, and this evolution will be predicted by the decision-maker according to the "system model" he is
using. It is somewhat disquieting to realize that the model we are
going to build will contain the behaviour of actors, which will
depend in turn on the models available to them. However, that's the
way it is. He will estimate the distance of each choice at different future times, and consider which of these choices, according
to the scale of value he assigns to time, is his preferred choice.

If we look now at the behaviour of populations then, assuming
that we may define the probability of each individual making a particular choice in an interval is given by equation (1), then we can
construct kinetic equations for the behaviour of the system. If all
the decisions made in the system are independent of one another,
then we have an essentially trivial problem, but, if as is the case
in any human system, the decisions that have already been made
modify the "pay-offs" perceived by the actors, then we have a far
more interesting situation. What occurs is that the choices that
have been made, changes in the "real" system, are reflected in the
"mental maps" or "internal psychological value spaces" of the
actors, causing them to modify their behaviour. If we consider, for
example, a very simple problem of a homogeneous population, which is
growing in size, locating around a centre of employment, then
initially location occurs close to the centre, but later as the
density increases, the choice shifts to more distant locations. In
the value space of the population, the choice of the central location while remaining attractive in the dimension of spatial convenience, receded from the origin in dimensions associated with
crowding, and led to the adoption of the other more distant locations which seemed more attractive in comparison. Of course, with a
single population, and such a simple problem, the evolution is
trivial, but if we think of the interplay of decisions that are made
by the many different types of actor present in the city, say, we
see that we have a highly non-trivial relation between the decisions
that have been made and those that are going to be made, because the
dynamic interplay of the system evolution with the value space of
the actors is very dependent on the precise timing of events.

Returning now to our urban model, we show in Figure 7 the basic
interaction scheme for the six types of population which we have
supposed are most important in the evolution of the city. From
this, using the assumption that the probability per unit time of an
actor making a particular choice is proportional to

$$\frac{(1/d_i)^I}{\sum\limits_{i'} (1/d_{i'})^I} = \frac{A_i}{\sum\limits_{i'} A_{i'}}$$

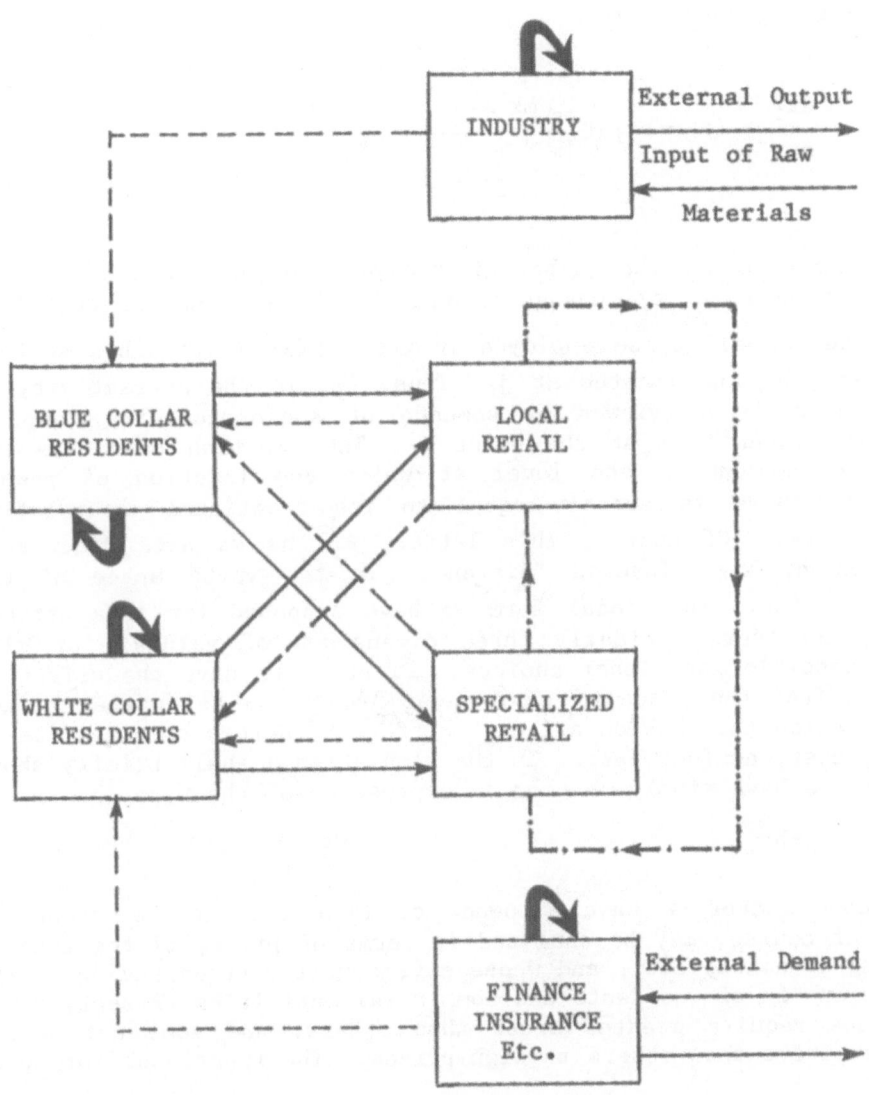

——————— Demand for Goods and Services

— — — — — Demand for Labour

— . — . — Cooperative Effects (Economies of scale, common infra-
                     structure, learning, etc.)

Figure 7 - The Interaction Scheme of Our Simple City System

then we can construct our kinetic equations expressing the evolution
of each variable in each locality. These are given in Appendix 1,
and have the general form

$$\frac{dx_i^k}{dt} = (a + bx_i)\left(\sum_{jm}\sum J_j^{mk} \frac{A_{ij}^k}{\sum_{i'} A_{i'j}^k} - x_i\right)$$

which expresses how the number of residents of socio-economic group
k, at the point i, x_i^k, change in time by the residential decisions
of the sum of all those employed in the different possible sectors
m, whose jobs are located at j. Thus, A_{ij}^k is the attractivity of
residence at i as viewed by someone of socio-economic group k,
employed in sector m at the point j. The equation describes the
growth of x_i up to the level at which the fraction of people
attracted there is exactly equal to its fractional or relative
attractivity. Of course, this latter, A_{ij}^k as we have discussed,
is given by some inverse "distance" in the value space of the
J_j^{mk}. The simple functional form we have supposed for this attractivity, considers basically three dimensions of values, for different possible locational choices. Firstly, we have the effect of
distance from the place of employment, which can be further broken
down to allow the consideration of different factors such as time of
travel, cost, comfort, etc. In the simulation I shall briefly show,
however, we have simply assumed an expression of the type

$$e^{-bd_{ij}}$$

The second factor we have allowed for is the effect of crowding,
which, of course, may be analyzed in terms of price, of the type of
building of noise, etc., and whose effect will vary depending on the
particular mix of residents and commercial activities present, since
some uses require greater areas than others, and some actors are
less sensitive than others to high prices. The functional form used
here is

$$\frac{\nu^k}{\nu^k + \sum_m \gamma^m J_i^m + \sum_k \gamma^k x^k}$$

Finally, we have also allowed for the attractivity of a particular
point to depend on its natural beauty, and also on the character of
the residential population already present. This would allow for
the possibility that, for example, people from the upper socio-
economic group prefer to live in an area where their own group is
already present. (It is the mathematical expression of "a nice
area".)

Putting all these factors together, we find the expression

$$R_{ij} = \frac{\nu^k(1 + \sigma^k x_i^k)e^{-b^k \delta_{ij}}}{\nu^k + \sum_m \gamma^m J_i^m + \sum_k \gamma^k x_i^k}$$

which we have supposed to express the value structure of the different types of resident, k, and also how this system of values "reacts" to changing possibilities.

We have written down similar equations for the other actors, which in brief express, for example, the need for industrial employment to be located at a point with good access to the outside, and for a very large area per job, as well as some 85 percent of their workforce being taken to be in the lower socio-economic group. We have also added the fact that the interdependence of many industrial activities leads to a preference for locations adjacent to established industrial locations. This term also covers many subtle effects of the infrastructure that grows around existing situations. The main effects are all noted on the interaction scheme of Figure 7, and the full equations are given in the appendix, and so here we shall simply proceed to discuss the evolution of our simple city system.

URBAN EVOLUTION

In this section we shall briefly describe some of the simulations that we have made using our simple model. In the first case, we have looked at the evolution of a centre which initially is only a small town, but throughout the simulation, due to population growth and expanding external demand from the industrial and financial sectors the town grows, spreading and sprawling in space as it does, and also developing an internal structure.

The initial condition of the simulation is shown in Figure 8, and the particular values of the parameters which we have used are given in Appendix 2. After ten units of time, the situation has evolved to that shown in Figure 9, where already, an internal structure has appeared. Industry, commercial and financial employment are all still located at the centre, but now we observe residential decentralization, particularly on the part of the upper socio-economic group. The centre is very densely occupied and is strongly "blue collar".

As the simulation proceeds, however, at around fifteen units of time, this urban structure becomes unstable. It is not a question of simply growing or shrinking: what is at issue is the qualitative nature of the structure. For, at this point in time, the very dense occupation of the centre is beginning to make industrial managers think about some new behaviour. For some of them the cost of con-

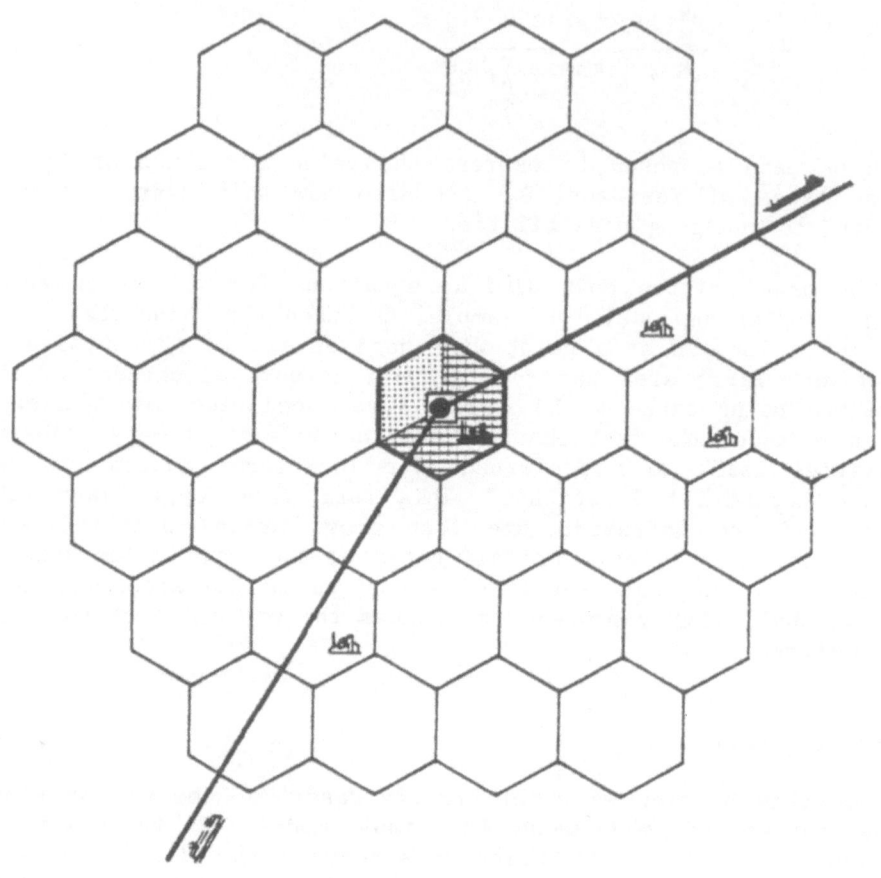

▦ The tightness of the mesh of the square lattice gives the
density of "blue collar" residents.

⠿ The density of points gives the density of "white collar"
residents.

● Size of point gives number employed in local retail.

☐ Size gives number employed in specialized retail.

🏭 Size of symbol gives number employed in industry.

Heavy hexagon defines CBD

Figure 8 – The Initial Urban Structure Condition. A small town,
unstructured as yet, and lying on a line of communication, begins to grow. The key to the symbols of this
figure and those which follow are:

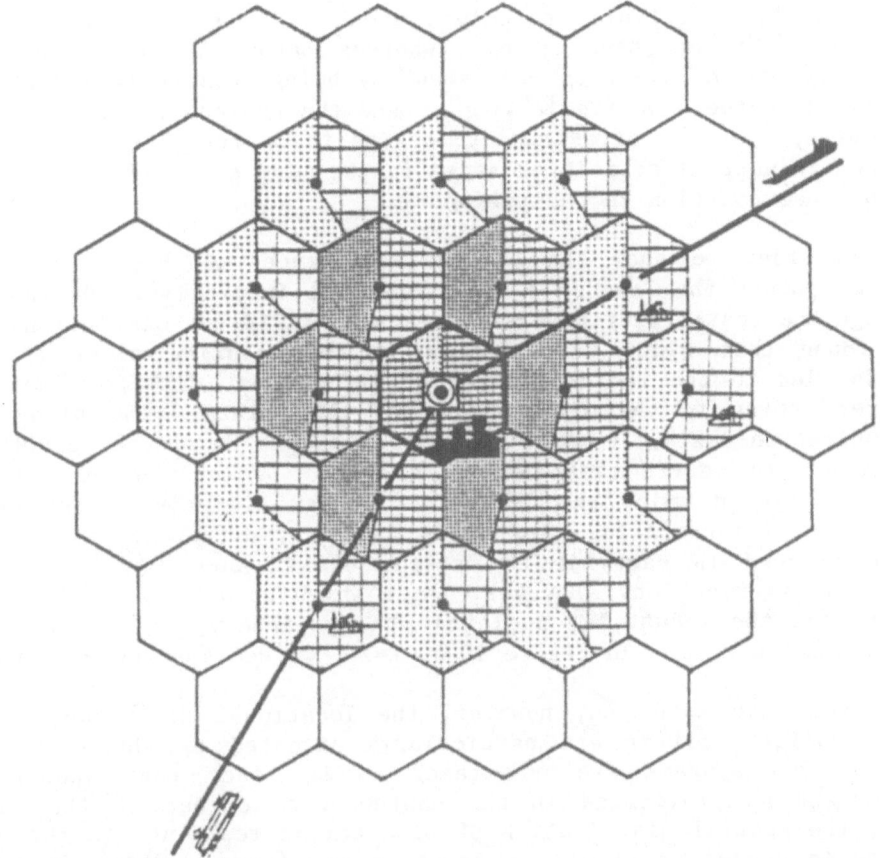

Figure 9 - The Urban Structure after Ten Units of Time. Residential
decentralization is already well-developed, particularly
by the white collar workers, who have an exponentially
decreasing density distribution, with a crater at the
centre. The blue collar workers have a shorter-range
exponentially decreasing distribution, reflecting their
lower mobility. The structure is still centralized with
all employment, except some very local retail, located in
the centre.

tinuing to operate in the centre, is making them contemplate the abandonment of the infrastructure and mutual dependencies that have grown up with time. At this point, as for a dissipative structure, it is the fluctuations which are going to be vital in deciding how the structure will evolve. At some point there is an initiative, when some brave individual decides to take his chance and try to relocate at some point in the periphery. Where, exactly, will depend on his particular perceived needs and opportunities. However, what is important is that whereas before this time such an initiative would have been "punished" by being less competitive, now around t=15, the opposite is true. Once the nucleus is started, and of course, its own infrastructure begins to be installed, so almost all the industrial activities decentralize, and establish themselves in this new position in the periphery.

In order to show the effects of chance, we had present on several points the "seeds" of an industrial initiative. By passing through the instability several times with different simulations, it was found that minute differences in the relative sizes of the "seeds" led to the reorganization of industry at different points. However, owing to the attractivity of the points lying along the communications axis, it was much more difficult to provoke growth of an industrial centre away from this axis. Summarizing the effect then, at around t=15, the hitherto circular symmetry of the urban system becomes unstable. At this point, many different initiatives could succeed in carrying the system off to some particular new state of organization. However, those which succeed with the least effort are the industrial nuclei in the periphery, lying along the communication axis. In Figure 10 at t=20, we see the new structure.

From this point on, however, the locational decisions of the "blue collar" workers are particularly affected by the fact that their value systems are now based on the fact that industrial employment has relocated in the southwestern corner of the city. Thus, the spatial distribution of blue collar residents in the city starts to change, having in a sense, a new focus. This, in turn, acts on the locational choices of the white collar workers, who find space easily in the regions of the city less favoured by the blue collars, and whose spatial distribution adjusts accordingly. Changes in the distribution of local service employment also then occur, and the whole structure evolves to the pattern shown in Figure 11 by time t=40. Here, we see that we have actually displaced the centre of gravity of the urban centre, and have an urban structure which resembles two overlapping urban centres of different character. In the southwest we have a predominantly working class, industrial satellite, while, the original city centre is a CBD and important shopping and commercial district, with predominantly white collar suburbs stretching away from it on three sides. In this part of the city, it is the second ring that has attracted the local shopping centres, while in the industrial satellite, it is the heavily populated, industrial district itself that has become an important shopping centre.

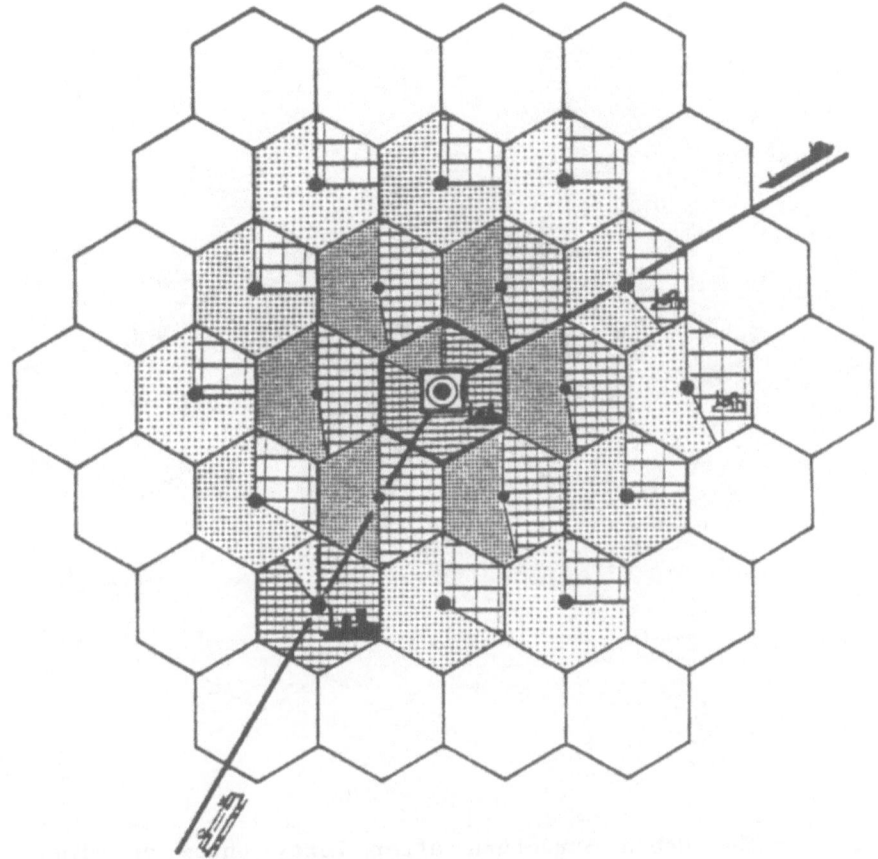

Figure 10 - The Urban Structure after Twenty Units of Time. Indus-
          trial employment is leaving the centre and relocating on
          the communications axis in the southwest periphery.
          Already, the distribution of "blue collar" workers has
          changed, affecting the evolution of all the other vari-
          ables.

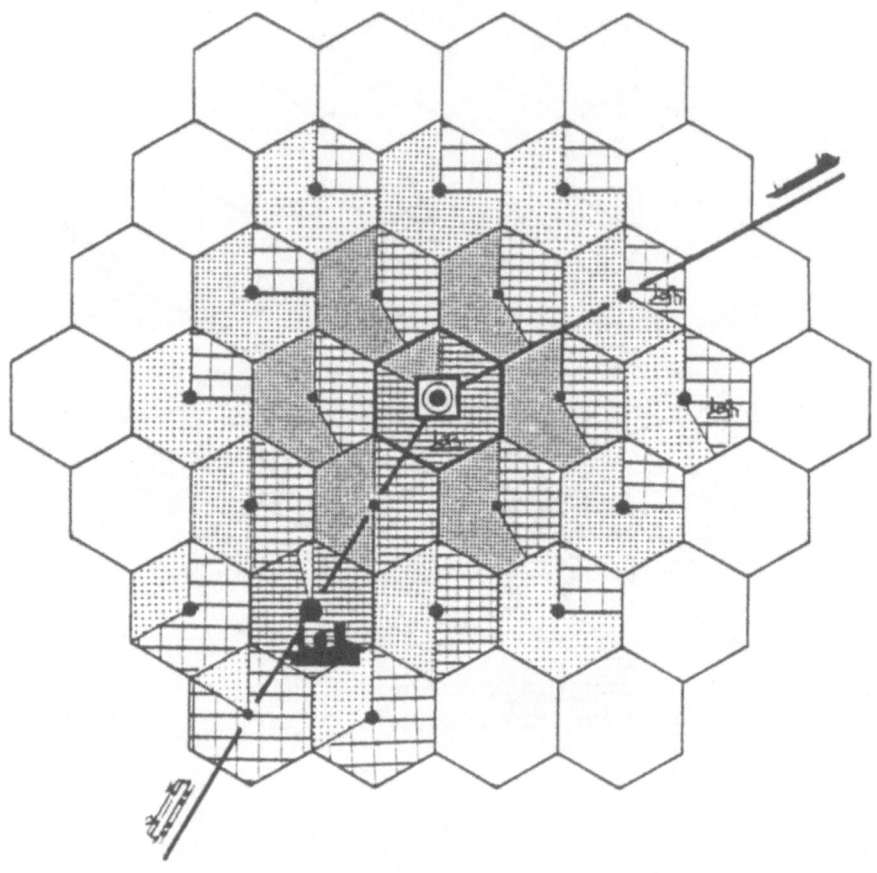

Figure 11 – The Urban Structure after Forty Units of Time.   The
            urban structure has changed qualitatively from that of
            Figure 9.   It has developed a second focus, and has
            structured functionally.   That is, one centre is essen-
            tially an industrial satellite, while the traditional
            centre has become largely a CBD and the important shop-
            ping centre.   We may also note that in the traditional
            centre, the retail employment has moved outwards to the
            second ring, (suburban shopping centres), while in the
            industrial centre the retail employment is still
            centralized.

During the simulation we can calculate a great deal of interesting information concerning the urban structure, and its "running
costs".   For example, we can calculate the total number of jobs
available in each sector and the total travel generated by commuting workers.   This can be calculated separately for "blue" and
"white" collar workers, and if necessary, can be calculated for each
residential location.   Similarly, from the location and size of
shopping centres, together with the distribution of residences, we
can calculate the total travel involved in consumer shopping trips.
Clearly, the energy consumption of the urban centre is related to
the sum of the total travel of commuters and shoppers, and this can
therefore be calculated.   What is particularly interesting here, is
that the usual procedure is to simply divide the total distance
travelled in the city by the population and discuss the average
distance travelled per inhabitant.   When we look, for example, at
the average distance commuting to work of "blue collar"workers, we
find that the trend changes when the city restructures at around
t=15, as it does also for "white collar" workers.

This shows us the dangers involved in global modelling, for on
that scale what we see is an apparently inexplicable change in
behaviour, in which the distance travelled per person, and the average energy consumption per person stops rising, and even decreases.
Only a model which can describe the internal restructuring of the
city could have predicted such a change, and linear systems theory,
and input-output flow models would have to be re-calibrated at this
point.   This also highlights another aspect of modelling method
which is sometimes used incorrectly.   The important point about say,

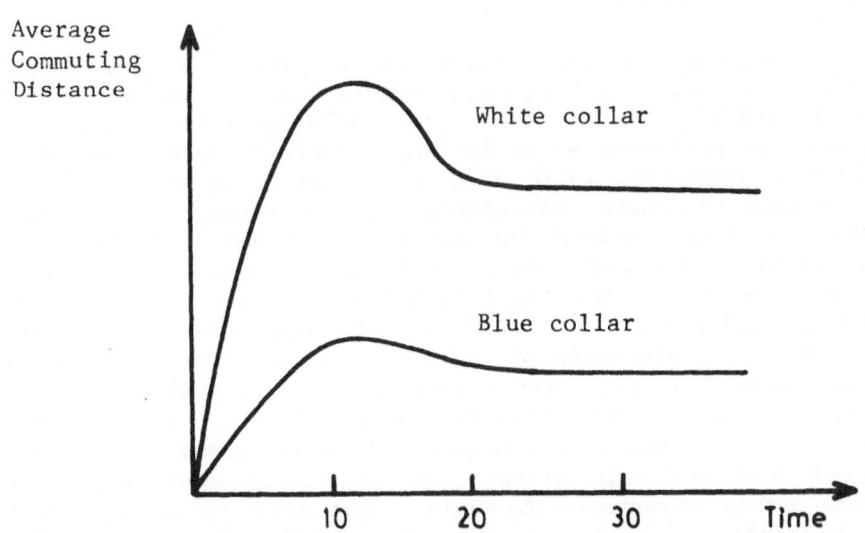

Figure 12 – Changing Trend of Average Commuting Distances with
            Structural Reorganization

the energy consumption of urban travel is that it results from all the travel that is taking place in the city, and hence, is an "observable" which has the value it does, because the city has the distribution of residences, jobs and shops that it does. It would be quite incorrect to use this total urban travel in order to model the system, or as part of a global model, because as we have seen, changes in say, total urban population can lead, through the type of internal changes that we have discussed, to modified values of the average travel requirements. In other words, relationships between global variables of complex systems are nearly always non-linear and a systems analysis which assumes linearity will only be reasonable in the short term, or in a neighbourhood of the calibration.

As a final example here, let us briefly describe a series of simulations which were performed in order to investigate the impact of rising travel costs in a city. In this case we started from the same initial condition as for the previous simulation, but with a slight change in the value of parameter $\ell p'$ (a systematic examination of the effects of the various parameters is given in Allen et al (1980). At t=20, the situation is that shown in Figure 13; characterized by circular symmetry, with a CBD, industry and main commercial and shopping concentration in the centre, and with a "white collar" residential suburbs surrounding it. At this time, we have performed two simulations starting from this particular state. In one case, we allowed transport costs per unit distance (costs being in time or money) to fall, and the other to rise. The first case corresponds to a policy of heavy investment in order to continue decreasing these costs in a still growing city, while in the other case, there was perhaps little investment and travel costs were allowed to rise. Possibly also, the first case corresponds to a heavy subsidy on rising fuel costs, and the second to simply passing this on to the consumer.

After running the simulations for a further twenty units of time, the structures which evolved were examined. They are shown in Figures 14 and 15. We see that they differ qualitatively, in that the simulation performed under falling transport costs still retains its circular symmetry, while that performed under the scenario of rising transport costs has become asymmetrical, as industrial activity has decentralized and nucleated in the periphery. The transportation and energy requirements of the two urban structures are quite different. The total travel generated in the "low-cost" city is approximately twice that of the "high-cost" city, and the average commuting distances of "blue collar" workers is three times as large, while that of "white collar" workers is doubled. In fact, the average "cost" of commuting for "blue collar" workers is greater in the first case than in the second. In other words, the reduction in travel cost per unit distance of the scenario causes a quite different urban structure to evolve, and this is such that blue collar workers on average must travel much further to work than in the other case. This means that although the cost per unit distance decreases, it is more than compensated by the increase in travel distance that the urban structure requires.

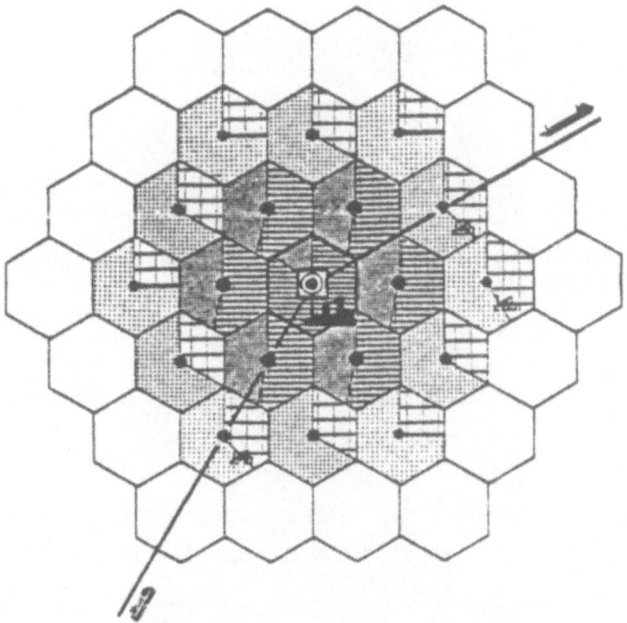

Figure 13 - Structure after Twenty Units of Time

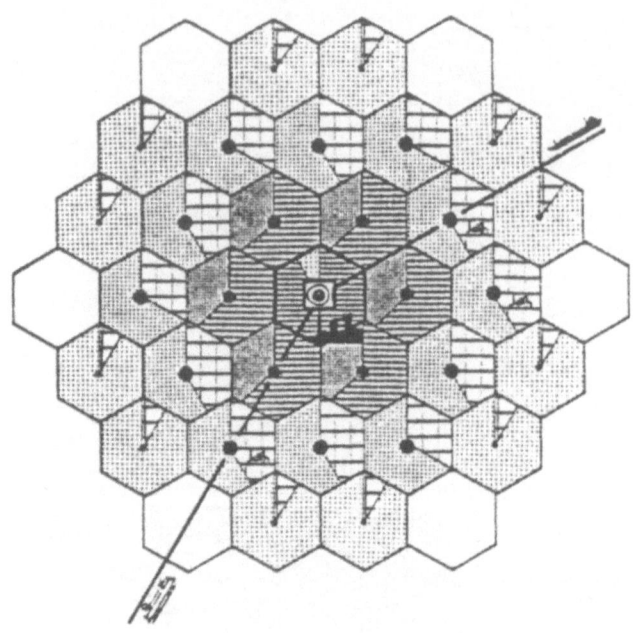

Figure 14 - Evolution of Structure with Low Transport Costs after
Forty Units of Time

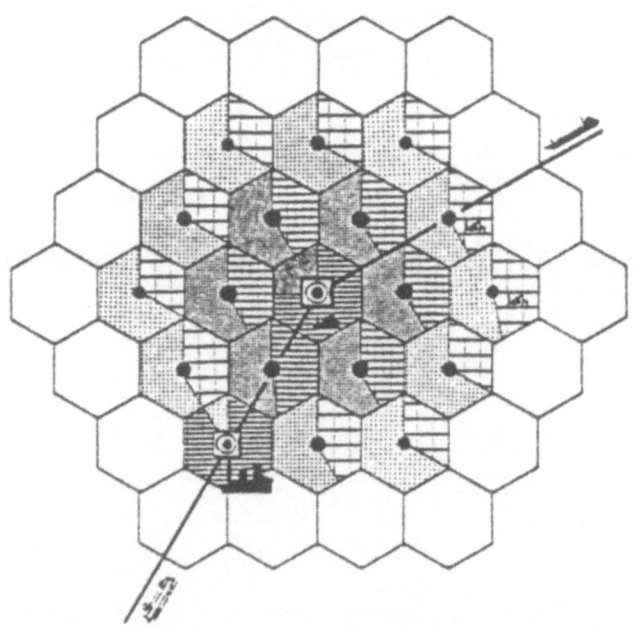

Figure 15 - Evolution of Structure with High Transport Costs after
Forty Units of Time

It is interesting to note that the "GNP" of the city requiring or generating greater total travel would probably be higher than that of the second city, although the actual consumption of goods and services is smaller. Again this points out the dangers of using such global indicators for complex systems, where structure and function are inextricably mixed, and where evolution and changing conditions can lead to internal reorganizations.

Various other problems can be examined, such as, for example, the effects of regulation of industrial and commercial location, or of changing patterns of external access for goods and raw materials, changing productivity in industrial or office employment. Similarly, a study can be made of the effect on the urban structure of the introduction of a metro line, including the chain of events that it sets-up in the long-term involving modified land prices, and changing commercial and residential attractivities. Our final simulation (Figure 16) shows one of our preliminary simulations of this problem, where we see that apart from distorting the urban space by causing greater residential densities along its path, we can discern the beginning of two new commercial centres which are forming at each end, and by becoming employment centres themselves, they lead to further modifications of the residential location pattern. Thus, the simple decision concerning the building of a metro line sets off a whole series of events, leading to the formation of sub-centres, and a change to a polynuclear structure, although intuitively, the effect of a metro line running to the centre of town would be to reinforce and preserve the status of the latter.

CONCLUSIONS

The simulations which we have briefly described illustrate the self-organization of our simple urban system. They show how policy decisions concerning transportation, housing regulations, industrial location, etc., can modify the evolution of the system qualitatively, leading to new spatial patterns and to changes in the trends of macro-variables related to the whole city.

Of course, a city model can only be used to explore future possible evolutions by placing it in its regional and national context, and a similar model, based on the analogy with "dissipative structures", has been developed to describe the evolution of a region and of an urban hierarchy as a result of such, historically dependent, self-organization processes (Allen et al, 1979; Allen and Sanglier, 1979a,1979b; Badinter, 1980).

An important general point that arises from such models is that a structural reorganization of the urban space leads to a corresponding reorganization of the mental maps and values of the various actors. Essentially the symmetry breaking properties of self-organization lead to a corresponding expansion of the dimensions of the actors' value space. For example, in the example given above, while initially under circular symmetry, the variables and para-

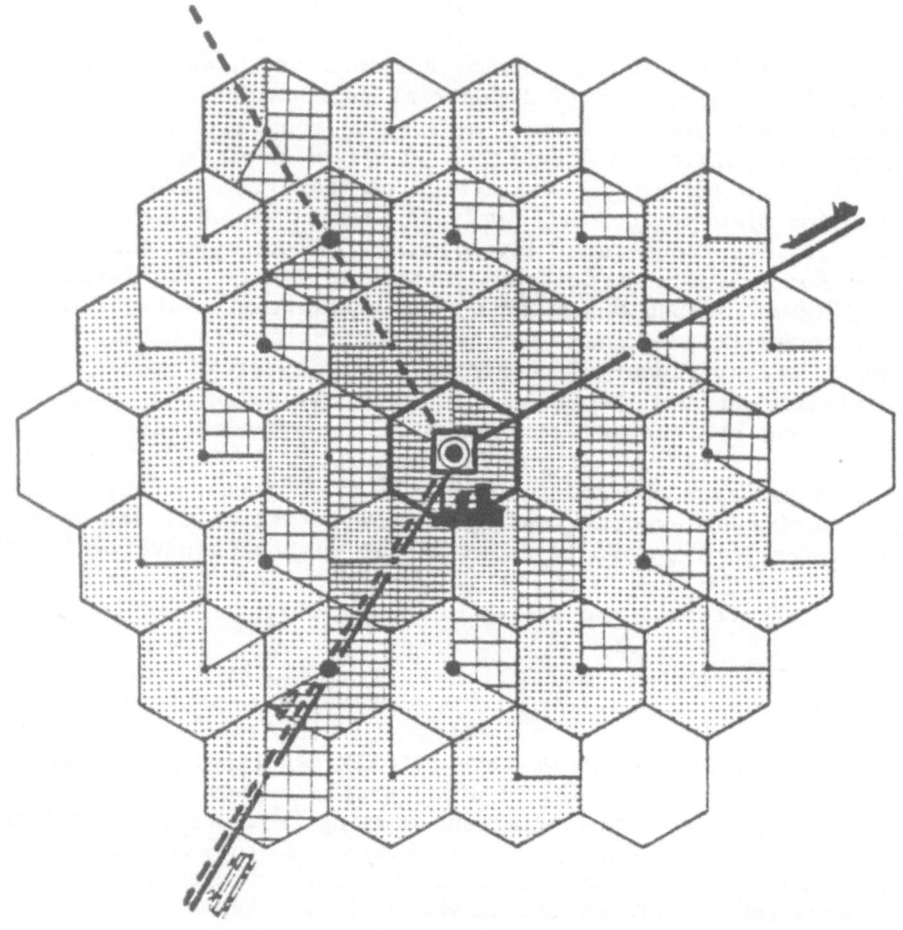

Figure 16 – The Evolution of the Urban System as Modified by the
Construction of a Metro Line. Apart from concentrating
the population densities along its length, causing
changes in land prices etc. as it does, and changing the
residential mix in consequence, it also provokes the
growth of two sub-centres of retail employment at its
outer extremities. This could be very important in
later evolution.

meters of decisional criteria can all be expressed in terms of the scalar distance from the centre, once the symmetry is broken, the value space expands to include all the angle-dependent possibilities. Similarly, for example, when all cars were black, the question or value attached to colour was of no importance. Once the symmetry had been broken however, and cars of other colours appeared, then the new dimension is created in the value space of buyers, and finally it can become an important factor in sales.

Complexity feeds on itself because it creates new situations and dimensions, which widen the experience of people and create new tastes and qualities, leading to new behaviours and to further complexity.

The important point is that fluctuations around "normal" behaviour in the real world, and fluctuations in the mental models of actors, both explore situations which are "richer" than the reduced description of the world which is given by our model. These explorations can be amplified by the non-linearities in the system and lead to a structural evolution of the system. But is all this really true, or is it wild speculation? Where precisely are all these non-linearities which I am suggesting characterize human systems so strongly?

Firstly, there exist purely physical non-linearities in the workings of objects, related ultimately to energy and matter flows, but also to such effects as, for example, surface area to volume ratio. These lead to "optimal sizes" for elements, and give rise to economies and dis-economies of scale, to division of labour, to aggregation and cooperation, pooling of resources, etc. Let us try to imagine a "city" of say a million inhabitants, which has no internal structure. Each small locality contains small units of each type of industry, of all types of shops and services. In such a city there are no "head offices" or central depots because such things already arise because of the spatial self-organization processes of the type I am attempting to imagine absent. Amid all this, we find at each point the same mix of blue collar and white collar residences. Clearly, such a vision is impossible in reality, because there exist real advantages in the functioning of certain size units of each type (due to analogous quesions of internal organization), which means that they obey non-linear laws. Thus, any small fluctuations away from total uniformity will be amplified by the advantages perceived by at least some of the actors. Herein lies the other very strong source of non-linearity in human systems. It is the term

$$\frac{(1/d_i)^I}{\sum_{i'}(1/d_{i'})^I}$$

which we derived from our model of preferences which is inherently and dramatically non-linear. If information is readily available, I is large and even a small change in the advantage offered by a particular choice i can result perhaps in a large change in the population's behaviour. Thus, it is the human capacity to "treat" information and to choose his behaviour which is at the root of much strong non-linearity. However, even in a system with rather poor information concerning the "true" advantages and disadvantages of different behaviour, which I believe is our situation, people's behaviour is largely determined by repeating previous actions which were not calamitous, and when a change is thought desirable, by imitating others. This imitation introduces a strong element of non-linearity into the problem.

Equally true, of course, is the fact that by manipulating information one can change the evolution of the system. Values are, it seems, not the simple rational, even biological certainties which we may have believed. Even such "sure-fire" values as maternal love have recently been revealed to be time-dependent. What we must face is that almost all of our everyday actions are not the expression of an absolute rationality, but result from the dynamic dialogue between "system" and "values", between "supply" and "demand", during which particular bifurcations have occurred. Their "rationality" is simply conferred on them by the society in which they are thought normal, where they evolved, and they can, and will, change. The problem of policy-making in a world with changing values is indeed a fundamental one.

Summarizing the main points made above, then, I believe that the true nature of living systems is only beginning to be understood. The new concepts arising from "dissipative structures" offer us a new basis for understanding, and a much more profound view of human behaviour and rationality. Only models which can take into account this "active" nature of living systems can be really useful in our understanding of the possible evolutions that may occur. They are the necessary prerequisite of any discussion of planning, policy and problem-solving. Indeed, what constitutes a "problem", and what may be a "solution" depend on the particular views of the moment. Our model begins to show us the real difficulties of living in an interdependent society. The evolution of any neighbourhood town or city can never be disassociated from that of the surrounding regions or the included elements, and the decisions policies and plans executed in any sub-unit of the whole will influence the evolution of all the other parts of the system. What liberty should therefore be accorded to the individual, the local community, the region or the nation? At what level should policy decisions be evaluated, and whose money should be used to implement those decisions? Such questions lie at the root of all political debate and, of course, there is no clearly discernable answer to them. But, why should there be? The notion that human systems must be characterized by well-defined "problems" to which there must exist well-defined "answers" is not a result of this new point of view.

What is revealed, however, is a basic paradigm for the creative, self-surpassing evolution that characterizes human systems.

This paradigm is the "dynamic dialogue" that goes on between three levels of the system: the properties of matter and the physical resources of the system, the entities of organization and their roles, which constitutes the type of actors making up the system, and finally, the perceptions, beliefs and values of these actors as they reflect and act on the changing world around them. The presence of irreversible processes leads to the possibility of non-linear interaction which is then responsible for the self-organization and complexity of the system and the possibility of the emergence of new hierarchical levels. The existence of choice and the subtle importance of "timing" and "dynamics" in evolution is a message of real importance for human systems.

Because of the fallacious analogy with the approach to equilibrium of an isolated physical system, it was thought that, in some way, there was an "invisible hand" guiding the evolution towards the equilibrium state, and hence, that the role of "planners" and "governments" was simply to alleviate marginal areas of socially unacceptable hardship, while events moved inevitably towards the "best" (in the circumstances) solution. Even if many of us disavow such a view consciously, nevertheless vestiges of this idea often colour our reactions. The idea that "progress" is necessarily something good is a tenacious one.

Our new perspective tells us that this is not the case, that a choice of futures may indeed exist, and that these possible futures are of different "optimalities", some being more efficient from certain points of view than others. What we must accept is that, knowing this, it becomes our responsibility to abandon ideology and dogma, and to attempt to understand the real choices open to us, and the means to guide the system in that direction, while admitting that this will almost certainly be a direction of discussion and compromise.

APPENDIX I: EQUATIONS DESCRIBING THE EVOLUTION OF THE SYSTEM

The different types of equations expressing the evolution of these variables in each point of the system are

a)
$$\frac{dS_J^L}{dt} = \varepsilon^L S_J^L \left(D^L \frac{A_J^L}{\sum_{J'} A_{J'}^L} - S_J^L \right)$$

with

$$A_J^L = \left[\frac{(1 + \rho^L S_J^L)}{(\mu^L + \alpha_J \phi^L)} \quad x \quad \frac{\tau^L}{(\tau^L + \sum_{k'} \gamma^{k'} x^{k'} + \sum_{L'} \gamma^{L'} s^{L'})} \right]^{co^L}$$

this equation describes the evolution of the employment linked to an external demand which we call the "industry" (S^1) and the "finance" sectors (S^2), $L = 1,2$ respectively.

b)
$$\frac{dS_J^L}{dt} = \varepsilon^L S_J^L \left(\sum_{J'} \frac{\sum_{k'} \beta^{k'} L_x k'}{(\mu^L + \phi^L \delta_{JJ'})} \times \frac{A_{JJ'}}{\sum_{J*} A_{J*J'}} - S_J^L \right)$$

with

$$A_{JJ'}^L = \left[\frac{(1 + \sum_{L'} \rho^{L'} S_J^{L'})}{(\mu^L + \phi^L \delta_{JJ'})} \times \frac{\tau^L}{(\tau^L + \sum_{k'} \gamma^{k'} x^{k'} + \sum_{L'} \gamma^{L'} S_J^{L'})} \right]^{co^L}$$

this equation describes the evolution of the employment related to a local demand which we call the "ubiquitous" (S^3) and "specialized" services (S^4), $L = 3,4$ respectively.

c)
$$\frac{dx_J^k}{dt} = \eta^k x_J^k \left(\sum_{J'L'} \xi^{kL'} S_J^{L'} \frac{R_{JJ'}^k}{\sum_{J*} R_{J*J'}^k} - x_J^k \right)$$

with

$$R_{JJ'}^k = \frac{\nu^k (1 + \sigma^k x_j^k) e^{-b^k \delta_{JJ'}}}{(\nu^k + \sum_{k'} \gamma^{k'} x^{k'} + \sum_{L'} \gamma^{L'} S^{L'})}$$

this equation describes the evolution of the two types of population considered, which we have called "blue collar" residents (x^1) and "white collar" residents (x^2).

APPENDIX II: DEFINITION OF PARAMETERS

The meaning of the parameters in these equations is given as follows:

The $\varepsilon^L, \varepsilon^u$ and η^k characterize the dynamic response of employment L and the population k to external environment.

A_J^L and $A_{JJ'}^L$ represent the attractivity of the economic function at the point J.

D^L is the external demand for the function L that in the following simulations we have kept constant.

ρ^L measures the cooperativity between the different functions.

μ^L is the production cost which contains the input cost for the industrial sector.

$\emptyset^L$ is the transportation cost which can be a combination of time and money.

The α_J parameter represents the access of the point J to the communication axes and in the following simulation it is more especially the access to the canal/railway which is very important for heavy industry.

The τ^L parameter measures the intensity of the crowding supported by the function L, the crowding of a point J takes into account the population $\sum\limits_{k'} \gamma^{k'} x^{k'}_J$ and all the employments $\sum\limits_{L'} \gamma' S^{L'}_J$ at the point J.

$\beta^{k'L}$ is the quantity of function L demanded per individual at unit price.

The parameter CO^L measures the unanimity of the response of the consumers.

e^L is the elasticity of the service L.

The parameter $\xi^{kL'}$ is related to the percentage of people of the type k working in the sector L'.

$R^{K}_{JJ'}$ is the residential attractivity of the point J viewed by someone who is employed at the point J'. It contains the parameters σ^k which expresses the affinity between members of a population of the same type.

ν^k represent the sensitivity to the crowding perceived by the population k. The parameter b^k is related to the ease with which an individual of the type k may commute daily the distance $S_{JJ'}$, which is the distance between his residence located at J' and his work at the point J.

REFERENCES

Allen, P.M., Deneubourg, J.L., Sanglier, M., Boon, F. and De Palma, A., 1979, Report to the Department of Transportation, US, under Contracts TSC-1185, TSC-1460 and TSC-1640, Washington, DC.

Allen, P.M. and Sanglier, M., 1979a, A Dynamic Model of a Central Place System, Geographical Analysis, 11, 256-272.

Allen, P.M. and Sanglier, M., 1979b, A Dynamic Model of Urban Growth-II, Journal of Social and Biological Structures, 2, 269-278.

Allen, P.M., Sanglier, M. and Boon, F., 1980, A Dynamic Model of Intra-Urban Evolution, Second Interim Report to the Department of Transport, US, under Contract No. TSC-1640, Washington, DC.

Badinter, E., 1980, L'Amour en Plus, Flammarion, Paris.

Glansdorff, P. and Prigogine, I., 1971, Structure, Stability and Fluctuations, Wiley Interscience, London.

Nicolis, G. and Prigogine, I., 1977, Self-Organization in Non-equilibrium Systems, Wiley, New York.

Prigogine, I., Allen, P.M. and Herman, R., 1977, The Evolution of Complexity and the Laws of Nature, in E. Laszlo and S. Bierman, (Eds.), Goals for a Global Community, Pergamon Press, New York.

Roy, B., 1968, "Classement et Choix en presence de points de vue multiples (La Methode Electre), Revue Francaise d'Informatique et de R.O., 8, 57-75.

Roy, B., Vincke, P. and Brans, J.P., 1975, Aide a la Decision Multi-critere, Revue Belge de Statistique, d'Informatique et de R.O., 15, 23-53.

ACKNOWLEDGEMENTS

The author wishes to thank Professor I. Prigogine, whose ideas have inspired this approach, for his constant interest and invaluable comments. The urban models were developed in collaboration with Mms. F. Boon and M. Fischer-Sanglier, and owes much to the discussions with and support from D. Kahn, R. Crosby and F. Hassler at the Department of Transportation, US. This work was also supported by the Actions de Recherches Concertees of the Belgian Government, under convention No. 76/81 phase II.3. The author is also most grateful to P. Kinet for drawing the urban simulation figures.

TECHNOLOGY CONSIDERATIONS IN URBAN SYSTEMS ANALYSIS

William L. Garrison

Institute of Transportation Studies
University of California at Berkeley
Berkeley, California, 94720
United States of American

System analyses of urban areas usually include certain features of technologies: the inputs required, steps to transform inputs to outputs and the stochastic properties of those transformations, costs, risks and so forth. But the processes of technological change are not ordinarily incorporated in analysis, and their omission can lead to forecasting errors. More serious matters are the lack of attention to policies influencing technical change and system development and the overlooking of sources of critical problems and inefficiencies.

Having said that, the tasks here must be to identify the technologies and problems of concern, to establish how technology occasions change, and to suggest how the analyst may incorporate technological change in systems models.

ORIENTATION

Our topic is unconventional, orientation to the topic and our approach will be useful.

First, our approach will centre on change in public facility technologies and the organizations that deliver them. It will not deal with the set of issues that arise in the operation of public enterprise organizations and from concerns about public goods. As Turvey (1971), Phillips (1975), and others have shown, there is much wisdom about the operation of public enterprises, given those enterprises. We do not take them as given.

As noted, technology considerations already enter analyses. Incorporation of engineering or economic production functions is quite common. Production processes are ordinarily taken to be fixed. Fixed coefficients are used to translate relations between inputs and outputs, quantities and costs. There can be no quarrel with this procedure for the analysis of stationary systems. But if technological change is or could be an important element in systems evolution, and technology is a target for policy, then it is completely inadequate.

A more dynamic and policy-oriented analysis can be constructed by including time-dependent and policy-dependent changes in the coefficients of production functions. More radical technological change shifts the kinds of inputs and outputs, and the nature of production and consumption.

Over the last several decades, the attention of analysts has drifted from preoccupation with the deployment of the technologies of physical systems -- especially highways, sewage systems and housing -- to the problems of social service organizations. The factors possibly behind this shift of attention include the notion that physical systems are relatively mature, many of them are already extensively deployed, and where planning continues, such as planning in support of investments being made toward the goal of fishable and swimmable streams, planning is relatively routine.

We are dissatisfied. First, analysis protocols for these physical systems give little or no attention to changes in existing technologies, or to new technologies suited to particular environments. Thus, we do not regard the problem of planning the deployment of physical systems as more or less tamed and reduced to the application of techniques.

Our second objection is that while planning has shifted from physical systems to social service systems, they, too, are technologies, and the management of the evolution of those technologies is disregarded.

There is a large literature on technological change. It reveals that the creation of technologies varies by circumstance: it depends on matters such as the complexity of the processes involved, the size of firm and industry concentration, whether the market is highly concentrated or dispersed, regulation, and the interrelations of technologies with each other. The literature is oriented to the private sector. Policy variables are analyzed at the national level: patent policy, taxation policy, educational policies, and policies with respect to monopolistic practices. As a consequence, our knowledge of innovation and technological change is ill-configured to the scales and contents of urban systems analyses.

There is also a close coupling between technologies and the

institutions that control and manage them. Society is a set of organized structures, structures that perform functions. At the heart of these structures or institutions are technologies, hard or soft.

The study of institutions has a rich history, as does the study of technologies; paradigms have evolved for each. To an extent, each has an independent logic responding to the forces working on it and protocols adopted. Yet the analyst's homily that technological and institutional problems are rather different seems not very satisfactory, the connections between technologies and the institutions that operate them are much too close.

Even with this orientation, we still are not at the core of the problem of presenting our topic: it is that we are surrounded by unimagined opportunities. Technological change takes place over decades, and analysts and their clients and sponsors are concerned with today's problems. Also, technological imperatives are so often promoted in the absence of knowledge of their markets and how they might be implemented that analysts shy from them. We, nonetheless, persist in our emphasis on technological change because of its known effects on systems and the leverage for change which it may provide.

WHICH TECHNOLOGIES, WHAT PROBLEMS

It would be desirable to develop a method general to the incorporation of technological considerations in systems analysis. But some technologies are old, others are relatively new; some incorporate complex physical and management rules; others are simpler. Some are at a scale where they form large systems; others are within systems. Diversity is so great that a general method will come hard.

However, the technologies treated in urban systems analysis share some common features. They are spatially interrelated, subjected to close public sector control, and provide goods or services regarded as rights by publics. They form public facilities systems about which generalizations may be made.

Technologies ordinarily provided or tightly controlled by the public sector include water and sewage systems, electrical and other utilities, transportation, and education, police, health and other service systems. No one of these is completely provided by the public sector because private sector equipment suppliers, contractors, etc., are involved. Some, such as electric utilities, have a facade of private enterprise, although they are so completely circumscribed by public sector considerations that they behave as public enterprises.

Maturity

It is fair to characterize these systems as mature and increasingly ubiquitous.

One aspect of maturity is market saturation: all are served who
are likely to be served at current price levels. When there are
large groups that are not served, then public actions may be seen as
necessary to continue to expand service: transit and sewage service
expansions are spatial market examples and lifeline rates for tele-
phones are an income segment market example.

Maturity also has a technological basis: readily obtainable
product and process (of production and distribution) improvements
and economies of scale in the deployment of facilities have been
achieved. The status of electric utilities provides an example:
economies of scale through higher voltage transmission seem largely
to have been achieved, and no technologies seem to be available
offering marked generation economies. Water supply has the same
character. The more desirable storage sites have been developed,
economies of scale in transmission of water and its distribution
have been largely realized.

Growth Dynamic

Maturity is achieved as technology evolves in a particular
form. Early, there is a period in which several forms of an innovation are explored. Second, a predominant technological form emerges
providing a product successful in the market place. Subsequently,
the provision of the product to the market drives rapid growth which
is reinforced by process and product improvements. Finally, the
product reaches market equilibrium. The market is served geographically, technology improvements are no longer so dramatic, and economies of scale are achieved. The system is mature.

In a paper published in 1978, we applied this paradigm to
several public facilities (Garrison, 1978). Later (Garrison,
1980a), we developed it in some detail for transportation systems.
In the automobile-highway system case, the predominant technology
was embodied in the Ford Model T with technological protocols for
highway facilities and traffic control soon following. Continuing
improvements in the vehicle and its production, improved highway
facilities, and improved traffic control devices decreased their
real costs. The product, the automobile-highway system achieved
success in the market place. It was implementable everywhere
because of the availability of the horse, buggy and wagon highway
system. It did take decades, of course, to improve that system so
that it was suitable for automobiles, but now some 3 million of the
3.75 million miles of roads in the United States have been paved,
and they must be judged suitable enough by the populace, for improving the road system has lost its political salience.

Figure 1 compares highway improvements with vehicle sales. A May 26, 1980 article by Edward Lampman in Automotive News projects slowing growth in automobile sales, and ties future sales to a market growing with GNP and shifting due to demographic and location factors. The system has achieved maturity.

The examination of the history of electric utilities, eg., Conniff and Conniff (1978), reveals that their growth dynamic is quite similar in pattern. Markets for production and distribution technologies gradually emerged, first street lighting, then street cars, residential lighting, and the use of electric motors in manu-facturing processes. The predominant technological form coalesced around alternating current and higher and higher voltage transmis-sion lines. In the central power stations, steam turbines emerged around the first of this century as the predominant technology. Urban markets were served first. Later, government programs assisted deployment of service to sparse, rural markets (Brown, 1980). For this technology and its cost and service structure, the system has achieved maturity.

Changing Public Attitudes

James Q. Wilson has recently (1980) edited a volume in which authors discuss the state regulation of electric utilities, the US Environmental Protection Agency, the Food and Drug Administration and other regulatory agencies. Wilson develops a thesis about the origin and behaviour of regulatory agencies: regulatory agencies serve several masters. Regulatory behaviour is the product of events leading to regulation, the continuing source of support for regulation, and the conflicts between interested parties to regula-tion. Our growth dynamic paradigm may add something to this.

During early stages of growth, and while growth is rapid, chief public interest is buying into a system and obtaining its advan-tages. For example, buying into the automobile-highway system required having the funds to purchase an automobile, but it also required that a suitable road system be available and that one could safely operate a car on the system. After one bought in, public support for system development continued, for the system was improv-ing year-by-year as the road system was further expanded, and pat-terns of production and consumption shifted in response.

Buying into the telephone system had similar rewards and sus-tained a growth dynamic (Artle and Averous, 1973), for, once in, service improved year-by-year as others bought it. The same thing can be said of the electric utility systems and other public facili-ties.

In the case of the automobile and telephone, accessibility improved. In the case of electric utilities, the reliability of the system improved, as did its stability. In all facilities, the real

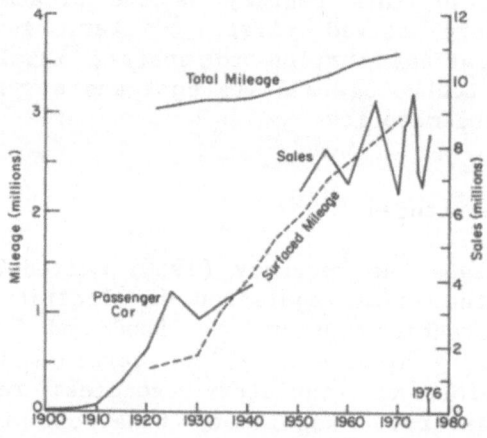

Figure 1-- Comparison of Passenger Car Sales With Total
and Surfaced Mileage of Roads and Streets

Sales shown every fifth year to 1970, annual thereafter.
Sources: Motor Vehicle Manufacturer's Association, 1977,
and the U.S. Bureau of the Census, 1975.

prices paid by consumers decreased as new technologies were introduced or existing technologies improved, and as economies of scale
were achieved.

Access to systems was the matter around which consumers coalesced politically. So it is not surprising that consumers allowed
government-producer alignments, as long as those alignments reinforced services to consumers. Wilson's findings about state utility
commissions and their service orientation are one outcome. Another
is government provision of all or critical parts of supply infrastructures such as highways and water systems. The extent to which
regulation supported the interests of producers versus consumers is
most difficult to analyze. Regulation did provide for monopolies,
providing stability for access for capital. Its emphasis on access
for consumers resulted in cross subsidies divorcing prices from
costs.

As mentioned, expansion of systems to support economic development and/or to provide access continues as a public policy goal.
Today, however, consumers seem no longer content to allow regulatory
agencies and delivery systems to bargain alone, subject to the
constraint that the regulatory agencies assure access. Externalities are less readily tolerated; agency styles, costs and prices are
debated. Growth is questioned.

Our explanation is that mature systems no longer provide
unquestioned, ever-increasing values for consumers. Alternative
explanations, which are not contradictory, include the argument that
the basic needs provided by these facilities are being met, public
attention is elsewhere. This is borne out by the decreasing proportion by the GNP devoted to public works, a matter that has been
extensively studied in the United States (CONSAD Research Corporation, 1980). There also is the argument that the costs of regulation are now better and more widely understood, providing a political base for its demise. Likely each of these factors is playing
a role, along with others.

Economies of Scale

The matter of scale needs to be discussed more sharply to tie
it to technological opportunities, the issues are those of who is
achieving economies of scale, and where.

Four slices are needed. The notion of economies of scale is
usually associated with production organizations, in particular, the
economy of scale of the firm. But when considering networked facilities, it is important to recognize the scale economies that can be
achieved by densifying movements on routes. Following nomenclature
developed by Robert Harris (1977), these may be referred to economies of route density.

There are scale economies achieved by increasing the geographical span of activities, and it is descriptive to term these economies of articulation. In most cases, these economies of articulation are closely related to economies of density. For example,
inter-ties among electrical networks support the robustness or
dependability of systems and transferring power in order to accommodate peaking loads and minimize capacity at generating stations.
Wheeling of power does require transmission lines, and economies of
density in those lines enable the economies of articulation.

In other systems, the economies of articulation also enable
economies of density. In highway networks and rail networks, for
example, the economies of density are achievable because the system
is sufficiently articulated that movements can be grouped on
routes. The economies of articulation and density are supported by
a high level of system standardization and the use of routes for
general purposes (all-comers), rather than specialized purposes
(truck only, eg.).

There are also economies of production associated to what is
put into the system (eg. electricity or the products of farms) and
economies of consumption. The economies of consumption are frequently quality derived. An example is the variety of foods available to the consumer at any time or any place.

Costs

Just as we may measure scale in several frames, costs may be
measured in several frames. It is useful to recognize user costs,
supplier costs, and the costs of access to a system. Excepting
subsidies and externality costs, all costs are user costs, of
course. Those costs are sub-divided here to recognize the first
impact of a technology improvement.

In general, economies of route density are captured by suppliers and passed on to users in some averaged fashion. An urban
freeway is for a highway department a very low-cost method of
providing for a unit of travel. Users have lower costs in situations where freeways are not congested and higher speeds are possible, but direct user costs are not so markedly reduced as are the
suppliers.

The user strives to keep access and user costs down. The
facility supplier is motivated to reduce his cost. Minimum system
or net social cost motivates no actor.

The Question

To draw themes together, our focus is on public facilities systems, such as water and electricity supply and transportation. Most

of these technologies are mature, in the sense that they have saturated markets and, also, in the sense that their technology components are well-established. Component technological improvements have been vigorously sought and achieved; economies of density, articulation, and production and consumption have been sought, now real cost is not decreasing. Now that the technologies are no longer improving, public attitudes toward them seem to have changed from one of supporting development to questioning prices, styles of behaviour and externalities.

For most analysts, this sketch begs the problems to be treated. How to raise the capital to preserve the physical infrastructure is a question of priority for almost every system. Political constituencies no longer form to raise public capital. Lacking productivity improvements, private capital does not find attractive investments.

With maturity, public attitudes have shifted, and traditional regulatory-delivery system alignments are under pressure. These pressures sometimes take the form of more public involvement in decision-making. There also is the rejection of old interest group-government alignments as suggested by deregulation movements.

The notion that the systems are in place may partly undergird interest in conservation, making do with what we have, although conservation priorities also reflect shifting attitudes about resources and their distribution.

Some of the institutions are viewed as mismatching their present tasks, and that sometimes is believed to be the problem. Institutions geared to construction and expanding to serve markets, such as state highway departments and electric utilities, are said not to match the needs that today's situation requires.

These questions all assume that major system improvements have ceased and major changes are not possible. The problem is to keep old systems working.

True, there is a high level of stasis for the systems and their institutions and technologies. That is exactly the problem. The problem is not to keep old systems working. It is to make major changes in them, or to create new systems so that they can again provide goods or services with lower and lower costs.

The question is that of breaking the maturity mold and creating development paths for the systems that maximize ever increasing net social benefits.

TECHNOLOGY AS A CHANGE AGENT

The pattern in the past, as discussed, has been for new technologies and delivery institutions to coalesce following innovations, for economies to be achieved, and for markets to be served in improved ways or for new markets to appear. The new technologies may be system-enhancing, system-changing, or system-creating in scope, although boundaries between these categories are not sharp.

System-Enhancing Technologies

System-enhancing technologies provide marginal improvements in existing systems. They may be forced on a system to control undesirable externalities -- such as catalytic exhaust converters to reduce automobile air pollution, and the replacement of PCB capacitors with mineral oil capacitors in electrical systems. They may be transferred into a system from technologies developed elsewhere, such as the adoption of modern control theory and electronics in numerous systems, or they may be a special-purpose technology created for some particular system function. Rail steel especially developed for use on high density, heavy car rail freight routes provides an example.

As discussed before when the concept of maturity was examined, one feature of mature systems is that gains in productivity attributable to technology are limited. This argument is developed by Abernathy (1978) for the automobile industry and it is one that we have applied to transportation systems generally (Garrison, 1980a), and the highway system in particular (Garrison, 1980b), as mentioned. The extent to which the argument holds needs clarification, for often the generation of system-enhancing technologies is vigorous. For examples, there are lively research and technology development activities within the highway system, in sanitary engineering, and in electric utilities, especially in the US Electric Power Research Institute.

But research and technology managers focus on within systems productivity improvements: longer-lasting asphalt-concrete pavements, more energy efficient pumps for water and sewage systems, and improved stability in electrical networks. Yet system productivity is properly measured on the users gains: more consumption choices from greater accessibility, for example.

Indeed, the pattern is for managers to strive for internalized systems goals, and to have little sense of the larger social and economic purposes of systems.

System-Changing Technologies

The Boeing Commercial Airplane Company's 707 and the Douglas Corporation's Type 8 Aircraft, provide an instructive example of system-changing technologies. The previous air transportation system can be appropriately described as a DC-3 system. The DC-3 aircraft of the early 1930's played a role in air transportation similar to the role of the Model T in automobile transportation, or the turbine generator in electric power stations. The DC-3 had a strong influence on all other parts of the system, airports and terminals, airways, air traffic control, and the dynamic of market acceptance. While larger and longer range aircraft were deployed subsequent to the DC-3, it was not until jet aircraft were deployed that there were major increases in productivity. Other components of the system began to adapt, including patterns of usage. The system changed. Bright (1976) provides a lucid description of this period.

Systems analysis played little or no role in the change in the air transportation system; it may have retarded that change. The systems analyst could count passengers among city pairs, only to find that the daily passengers (between, say, New York and Rome) would hardly warrant high-capacity aircraft and frequent flights. There were uncertainties about seat mile cost and aircraft reliability, lack of suitable terminals and runway strengths and lengths, investment requirements in fuel distribution systems, and air traffic control restraints on climbing and descending rates. The matter was further compounded by the need for a sizable production run of aircraft in order to keep their unit cost reasonable. Analysts were not optimistic. (Lest this seems picayunishly harsh on other analysts, this author had some involvement in these studies.)

In response, the Boeing Company delayed its production decision until competitive and prestige forces forced its hand. Juan Trippe, then Chairman of Pan American Airways, could imagine jet aircraft operating in PAA markets, and he bargained with the Douglas and Boeing Companies to get larger than prototype versions of aircraft in production (Daley, 1980).

Examples similar to this may be drawn from other systems. The break-bulk ship gave way to the containership, and long-distance communication technologies for data transmission and other purposes have pushed aside traditional arrangements. Some years ago, use of new filter concepts (Hazen, 1980), plus strong demands for system improvements greatly changed water supply and sewage systems (Tarr et al, 1980).

Besides the point already made that such changes are difficult for analysis, several other generalizations or patterns stem from examples:

1. There is increased specialization. The longer-range jet aircraft served all-comers, but only between selected city pairs. Corporate jet aircraft specialize to particular purposes, as do many data transmission links.

2. A plethora of industry, trade organization, and/or government regulations and standards, together with existing infrastructure investment and patterns of institutional thinking and

behaviour, all put in place for good reasons, thwart change. The new technology has to be highly productive in order to push those barriers aside.

3.   Productivity gains may appear as sharp cost reductions, however, quality changes and market responses are key.

4.   The user's system access cost may be increased. This may be because the new technological form is restrained to certain geographic markets, and/or because the user has to purchase equipment or undertake activities which would not otherwise be necessary.

5.   If specialization provides quality differences in products or services and the market interrelates dynamically in the growth-sense discussed previously, the ambience of system maturity is broken. There is new or revitalized dynamic growth.

Here are some further examples: The Springfield, Illinois traveller has to access Chicago in order to take advantage of jet flights to London (increased access cost). Data transmission and other communications capabilities enable firms to organize and operate new ways (economies of route density captured by users). Taking advantage of United Parcel's specialized service requires that the shipping dock deal with a specialized trucking firm (higher access cost, change in service quality). A variety of previously not-transported products move in refrigerated containers (quality change in the service). Interchange standards for cars operated on more than one railroad constrain technological change (change thwarted by regulations).

New Systems

New systems represent instances where system change has been thorough. Modern water supply and sewage systems, for example, differ from early methods of procuring water and disposing of sewage. They incorporate and utilize construction, water and sewage treatment, pumping, measurement and monitoring, and other technologies, and they are matched with institutional and fiscal arrangements suited to the technology. The automobile-highway system had its antecedents in the horse-buggy and wagon system.

Telegraph, mail and messenger systems were available prior to the telephone. Steam, water power, animal power and coal gas were doing the jobs that the electrical systems now fill.

In each of these examples, the component parts of the technologies have been thoroughly transformed. Too, the social purposes served by the systems have been transformed -- not by the system change alone, but by combinations of changes that sum to social or cultural change.

Consider the streetcar. Its precursor was the horse-drawn omnibus, and although it resembled the omnibus, it evolved a larger, heavier, more reliable and faster vehicle; propulsion and control protocols were thoroughly changed. Its spatial deployment differed.

Also, its social or market interrelations differed from those of the omnibus, as the social organizations within the city used the streetcar for their purposes. At first, the streetcar was substituted where existing markets enabled the capital investment requirements for electrification. Later, market interactions and land-development decisions made the streetcar much more than a substitute technology for horse-drawn vehicles. Systems and individual streetcar lines interacted with urban development decisions in ways that were not foreseen by early promoters of streetcars.

Generalizing from this example, a change is introduced in an old system. At first, the change may represent only a marginal improvement. Subsequent change in the ways in which the technology is used, coupled with the capability of the technology to evolve, find economies of scale, and serve users, then spawns a new system.

To Summarize

Some systems are changing as actors enhance existing technology. Other systems are changing because some technology and associated institutional development has changed some of their parameters. New systems evolve from old systems through sweeping changes in technology and market relationships.

Institutional and market relationships affect the potential for change, and the nature of change. System-enhancing technologies are within institutional ones. Systems are modified by system-changing technologies -- a part of the system is changed by specialization to markets. In other instances, there are thorough changes in systems and their social relationships.

Change is a matter of degree, of course. Although there are sharp differences between the within-system and system-changing classes, our classification serves no purpose other than to illustrate degrees of change.

IMPLICATIONS FOR ANALYSIS

The question now is how technological-institutional change can be specified and introduced into systems analyses.

Mature Systems

In considering mature systems, such as the highway system, water and sanitary systems, and electrical systems, the analyst has at least these options:

1. Assume no change in the technology; the problem is then incorporating deployment of existing technology in the models.

2. Incorporation of system-enhancing technologies in models.

3. Consider policies that might change the supply of system-
   enhancing technologies.

4. Consider system-changing or new systems technologies thus
   breaking the maturity mold.

The first of these options is conventional and does not require
discussion.

The incorporation of system-enhancing technologies borders on
the conventional, if the analyst simply examines technologies development and incorporates expected changes in parameters in the analysis. An example is the analysis accompanying air pollution control
planning for air sheds. Considerations of the types and quantities
of emissions are typically tempered by projected technology development and expected technology deployment. Regional and urban energy
analyses also typically consider technological changes and incorporate these in projections. The analyst's concerns here are chiefly
uncertainties about technological development, and the rate of the
deployment of technologies.

The analyst may wish to incorporate policy variables bearing on
the development and deployment of technology. Conventional variables include funding, contents of enabling ordinances and standards,
and other conditions for technology deployment. The technology may
require companion infrastructure changes which require planning and
investment. The electric car is an example. Its deployment raises
a series of infrastructure questions, ranging from the availability
of charging facilities to the fuels and facilities used by electric
utilities. These questions, in turn, have extensive policy and
planning implications.

The analyst may expect that feedback from his work will affect
the ways actors within systems are generating new technology. As
emphasized before, however, the work of within-system actors
reflects the existing structures of the systems, the division of
tasks within the systems and among institutions, and peer group
traditions and power positions. So the analyst is not likely to
have much impact on the day-to-day workings of mature institutions.

Turning now to the ways in which a mature system may be transformed into a revitalized growing one serving new or modified purposes, as mentioned, the pattern has been for one of a system's
technological components to be improved with subsequent changes in
other components of the system.

A large improvement in a technology component will not, by itself, assure system changes. There are two necessary conditions. The more important is that improvement be measured on the market; that it occasion major market reactions, for it is the market that is the force for system change. The second condition is that the other components of the system be changeable.

An example will illustrate both points. The diesel locomotive was much more productive than the steam engine, and it replaced the steam engine in a relatively short period of time. Its costs were approximately four times less than that of the steam engine. Other differences included increased reliability, capability to run through longer distances, capability to link locomotives effectively and increase the size of the trains, and improvements in matching the power assigned to trains to specific routes. Consequently, the diesel locomotive had impacts on the quantities of labour required by railroads, the locations of repair and service facilities, and methods of making up trains and managing cars.

But from the point of view of the users of the system, changes were not so marked. Undoubtedly, the use of diesel locomotives reduced the average cost of service, although the adoption of proto-cols of longer trains and increased switching and other work in marshalling yards may have reduced service qualities, including increased delays. Also, the other system components, such as the location and character of guideways, were unchanged.

So, an increase of the productivity of a component of a system may be oriented to supplier costs savings without much impact on markets, other than that of keeping the overall cost umbrella down.

Earlier, distinctions were made among user costs, supplier costs, and access costs, and also among economies of route density, articulation, and production and consumption. The latter four represent situations where technology may offer major productivity (decreased cost per unit of production or change of quality) improvements. Once an improvement is found, then the question is, who gains?

We judge that technologies that decrease supplier costs usually have limited impact. As illustrated by the diesel locomotive, supplier costs savings are averaged over all users, and thus, do not create notable reduction for any consumer and strong feedbacks for continued change. What seems to be required is targeting costs savings to market segments.

Access costs have been identified as a special class of costs. As mentioned, consumer desires for system access have been bargained through regulatory processes, with the result being cross-subsidies to keep access costs low, and the subsidies required for access reductions are one reason why supplier costs savings are averaged across users.

A good example of costs averaging and possible consequences is associated with technological improvements in long-line communications. Costs reductions have been averaged within the general cost structure of the telephone system, so although user's charges for long distance calls were reduced, many of the cost savings went to subsidize other activities. Currently, in the United States, competitive suppliers of long-distance communications links are offering marked price reductions to consumers. There is a very rapid growth of that specialized service with impacts on the telephone companies that traditionally supplied that service, but that have been unable to pass along major savings.

Returning to the issue of technological change, three questions need to be answered in the affirmative:

1.   Is an order of magnitude improvement in a component technology achievable?

2.   Is there some specialized market to whom cost savings will be passed (ie. savings will not be averaged in the general cost structure)?

3.   Are the elasticities in that market such that demands will increase and drive change?

The analyst will find instances where these questions may be answered in the affirmative. One reason is because mature systems with internalized goals have not aggressively explored options supportive of user's goals. The goal of a sewage disposal district, for example, results in large collector systems treating all sewage in centralized plants, whereas specialized small systems might better meet the needs of users. Users in an industrial district, for example, may pay surcharges for the treatment problems that result when their sewage appears at large central plants. The alternative of specialized plants in the vicinity of the origin of their sewage is not considered.

The tests provided by the three questions may dampen technological imperatives or steer those imperatives in a constructive fashion. Technology offers supersonic aircraft; very high-speed trains; economies of scale in the construction of long tunnels; the automation of vehicle control on highways; computer-communications devices at home; point-of-use energy conversion with congeneration of heat, steam, hot water, and electricity; high throughput freight hauling capsules in tubes; and numerous other options. Do these technologies satisfy the questions; might they be configured to do so?

While most public facility systems are mature, several of them such as air transportation, computer and communications services, and perhaps truck freight transportation are in a growth mode. That is, the technologies are continually improving; there is dynamic market response; the market is not yet saturated.

A critical matter is the extent to which systems have achieved the institutional features of maturity, even though they continue their dynamic growth.

This is critical because the typical pattern is for institutional structures to coalesce to deliver successful technologies. Creating the institution stabilizes the situation, the set of options that are available, and the development pathway that the analyst may wish to project. Indeed, if institutions have been coalesced to deliver technology, and if the analyst's objective is to incorporate continuing development, then analysis is straightforward.

It also can be very valuable. It will enable predicting the time when the system will become mature, and the problems that arise because institutions developed to deploy technology must change to manage a mature system. Also, it will enable predicting system problems that result from size impacts. The continuing growth of air transportation, for example, driven mainly by improvements in aircraft and market response, is creating pressures on other components of the system: problems of airport and airways capacity.

A less frequent case is where the institutions have not coalesced, and the analyst may wish to consider options with respect to institutional forms. There is a trade-off. If some desirable system form has emerged, then there is merit to creating institutions, standards, funding and marketing arrangements to stabilize the situation and provide for rapid deployment of the technology. On the other hand, it is highly desirable to maintain flexibility so that entrepreneurs may continue to explore options and markets.

Examples suggest that market and supplier interests very quickly stabilize technology and the delivery system, as was the case with, say, the delivery of television services, and, more recently, stabilization of container sizes and handling protocols.

While the analyst may have some success in alerting policymakers to trade-offs, this is inherently difficult because the futures of technologies and market responses are speculative, and desirable services and products intensely interest today's markets and suppliers.

Perhaps the analyst's most likely contribution can be made by accepting the mandate for early delivery of desirable goods and services, and anticipating options for continuing technology change, adjusting the delivery institutions so that they may later accommodate those options. This is an easy remark to make in the abstract, but it is not one that can be supported by examples and precedents. Rather, leaving options open is not in the interest of suppliers that are creating organizations and committing capital. The task of leaving options open confronts desires for stability.

SUMMARY AND SYNTHESIS

To summarize and provide coherence for the topics touched on in this paper, a short-paragraph summary will now be provided.

The priority of today's problems, and the interests of clients and sponsors of studies have focused the attention of analysts on deploying existing technologies. However, technological change may provide important options for systems development.

Public facilities systems, the concern of much urban systems analysis, have certain features in common. Most public facilities systems are mature: their markets are saturated, or nearly so, technological change is constrained, and public attitudes about them have shifted.

In considering public facilities systems, it is useful to recognize economies of density, articulation, and production and consumption and the association of costs with suppliers, users and system access.

System managers give high priority to technological changes that will lower their costs in pursuit of their goals. Economies obtained from their new technology tend to be averaged and not specialized to particular users.

Technological changes in system components may drive changes in the systems as a whole. If the gains from those technological changes are focused to specialized users, then market response and rejuvenated growth are possible.

Some systems are not mature, their technology is still evolving and their markets are growing and adapting. In some instances, institutions have been created to deliver the technology, and because of the traditionalist characteristics of institutions, the analyst's options may be limited to protections of growth and the anticipation of problems. In other instances, the analyst may consider technological-institutional options that would allow for continuing improvements in technology.

Options for technological change may become evident on close examination of systems. On occasion, these options for change may be in place, although they are not occasioning change because of institutional barriers or because the impacts of change have not been passed to specialized user groups.

REFERENCES

Abernathy, W.J., 1978, The Productivity Dilemma, Johns Hopkins University Press, Baltimore, Maryland.

Artle, R. and Averous, C., 1973, The Telephone System as a Public Good: Static and Dynamic Aspects, Bell Journal of Economics and Management Science, 4, 89-100.

Bright, C.D., 1978, The Jet Makers, The Regents Press, Lawrence, Kansas.

Brown, D.C., 1980, Electricity for Rural America: The Fight for the REA, Greenwood Press, Westport, Conn.

Conniff, J.C.G. and Conniff, R., 1978, The Energy People: A History of PSE&G, Public Service Electric and Gas Company, Newark, New Jersey.

CONSAD Research Corporation, 1980, A Study of Public Works Investment in the United States, Prepared for the US Department of Commerce, Washington, DC.

Daley, R., 1980, An American Saga: Juan Trippe and His Pan American Empire, Random House, New York.

Garrison, W.L., 1978, Thinking About Public Facility Systems, The National Research Council in 1978, National Research Council, Washington, DC.

Garrison, W.L., 1980a, Innovation and the Structure of Transportation Activities, Innovation in Transportation, National Academy of Sciences, Washington, DC.

Garrison, W.L., 1980b, Renewing the Automobile-Highway System, Working Paper, University of California, Institute of Transportation Studies, Berkeley, California.

Harris, R.G., 1977, Economics of Traffic Density in the Rail Freight Industry, Bell Journal of Economics, 8, 556-64.

Hazen, R., 1980, People in Public Works, APWA Reporter, August, 1980, 4-6.

Lampman, E., 1980, Long Term Trend: Slowing Growth in Autos, Automotive News, May 26, 1980, 1.

Motor Vehicle Manufacturers Association, 1977, Motor Vehicle Facts and Figures '77, Detroit, Michigan.

Phillips, A. (Ed.), 1975, Promoting Competition in Regulated Markets, Brookings Institution, Washington, DC.

Tarr, J.A., McCurley, J. and Yosie, T.F., 1980, The Development and Impact of Urban Wastewater Technology: Changing Concepts of Water Quality Control, 1850-1930, in M.V. Melosi, (Ed.), Pollution and Reform in American Cities 1870-1930, University of Texas Press, Austin, Texas.

Turvey, R., 1971, Economic Analysis and Public Enterprises. London, George Allen and Unwin, London.

US Bureau of the Census, 1975, Historical Statistics of the United States, Bureau of the Census, Washington, DC.

Wilson, J.Q., (Ed.), 1980, The Politics of Regulation, Basic Books, New York.

A PLEA FOR PLANNING-ORIENTED RESEARCH

Henk Voogd

Department of Urban and Regional Planning
Delft University of Technology
Stevinweg 1-552, 2600 GA Delft
The Netherlands

INTRODUCTION

This paper will differ from many other papers presented at this conference in the sense that it focuses mainly on the research activity in urban and regional planning, and not on research for urban and regional planning in general. This distinction is, in my opinion, very important, because the investigations done by academics, consultants, etc., will bear on other characteristics and meet other constraints than the investigations performed by civil servants. For instance, the first category often allows a higher level of sophistication than the second category. For people who believe that systematic policy-making and planning can be done adequately, if not perfectly, scientific research will be seen as an important component of a planning process. For others, who see more merit in the traditional bargaining or "muddling through" approach, the role of research will be far more limited. Although the first viewpoint may not be empirically correct, it holds good promise for an improved functioning of urban and regional systems. However, it is also true that a bargaining "commitment generating" approach without research support will almost always result in a satisfactory plan or program for all parties concerned. Nevertheless, there will then be a big chance that the plan or program itself is intrinsically inconsistent. This will undoubtedly be revealed in the implementation phase.

For planners who adopt the pragmatic "bargaining" view on planning -- and I think many practioners do -- research is often seen as just an activity with almost no practical value, but which in a planning process, has to be tolerated. This viewpoint might be true, in so far as it concerns conventional research approaches which are mainly concerned with the investigation of interesting, but rather "academic" and mostly time-consuming WHY questions. In my opinion, it must be possible to come to a more planning-oriented approach to research in which the conventional WHY question is replaced by the more practical HOW question. This type of research will be referred to as "<u>planning-oriented research</u>" (POR). By POR I mean a professional activity in a planning process, based on a scientific methodology and insights, which have the aim in assisting in the preparation and execution of policy proposals. Consequently, POR is only one of many activities in an urban and regional planning process, next to -- for instance -- administering, communicating, bargaining, and so forth.

It is the purpose of this paper to explore the conditions under which POR might take place in practice. Some ideas will be developed on the role of planners and POR in a planning process. It will be argued that if urban and regional policy-making has to be improved, policy-makers require all the assistance possible from advanced planning and problem-analyzing procedures now available. It is asserted that the normative character of POR has to be stressed more in practice, ie. the research-oriented planners should be more aware of the sensitive and highly subjective claims made by or on behalf of various interest groups (including the political executives). Consequently, I emphasize strongly that planning in a pluralistic society means that the planner no longer can conceal his expert advice. He is more and more obliged to relate his work to the various sets of values and norms. It is necessary, therefore, to suggest conditions under which this information of the planner is actually conveyed to those who need it for practical decisions, while being able to check it by both the criticism of other experts and the political responsibility.

In order to do this, it is firstly necessary to explore the role of a planner in a planning process. In the next section it is pointed out that there are a variety of different viewpoints to trace. Therefore, it is impossible to determine "the" role of a planner without considering his planning environment. It is shown that this will have consequences for the interpretation and utility of POR.

In addition, an elaboration is given of the main characteristics of POR, after which the major problems are explored with respect to the integration of research and policy-making. It is argued that POR actually is confronted with two fundamental issues: on the one hand the dilemma which results from the contradiction between the wish to have insight into a "complex reality" and the general desire to gain this knowledge by means of a "simple approach". To overcome these problems to some extent, an organizational framework for planning-oriented research is suggested. Other changes are also needed, not only in methodology, but also in attitude, and this is briefly discussed. Finally, the last section contains some concluding remarks.

THE ROLE OF THE PLANNER

A literature survey will reveal that in publications on planning theory, most attention is focused on "general" planning concepts. The role of the planner himself -- as being an "expert" in the field of urban and regional planning -- has not received proportionately very much explicit attention. However, there are a lot of implicit viewpoints to trace. Some of these have been developed and tested in practice, whereas others more properly can be characterized as theoretical possibilities. Often, these concepts are not related to the environment in which a planner functions.

The role of the planner depends largely on the position he takes in an organization, ie. his function determines the kind of skills he has to be equipped with. A discussion about the instruments of a planner can never be started before a consideration of the position of the planner in the planning environment. The concept of a "planning environment" is used in the sense described by Faludi (1973a).

The function of a planner can be described in at least four dimensions which can be characterized by the words "analyzing", "synthesizing", "visioning" and "administering". Differentiations with regard to the various disciplinary (ie. sociology, economics, etc.) and functional (ie. shopping, recreation, etc.) specializations are neglected for the moment because it is felt that the addition of these "dimensions" will not increase the relevance of the arguments. Each of these dimensions will, to a certain degree, contribute to the definition of the planner's tasks. In fact, this means that, at least in theory, numerous different functions are possible, as are numerous different roles of a planner. Therefore, attention will be limited in this section to the global capacities related to each dimension. By combining these capacities, a systematic overview of the various functions can be given. Figure 1 shows that in total fifteen combinations (ie. viewpoints) are possible. In viewpoint (a), the planner is considered as a technical expert -- an analyst. He is seen as a professional who possesses the capacity, gained through his experience and training, to provide a systematic clarification of the (spatial) problems of the community. Through his analysis he gains knowledge of what the problems are, how they are interrelated, and how the spatial systems function.

An opposite viewpoint is that the function of the planner should be an integrating one. In viewpoint (b), his main task, as being a synthesizer, is not to go in depth, but to assemble parts in new ways so as to create wholes or systems. In other words, under this view the planner is the one who puts the pieces together.

In viewpoint (c), the creative nature is emphasized. The planner is mainly considered as some kind of a visionist or an idea man. His major task is reawakening imagination and encouraging

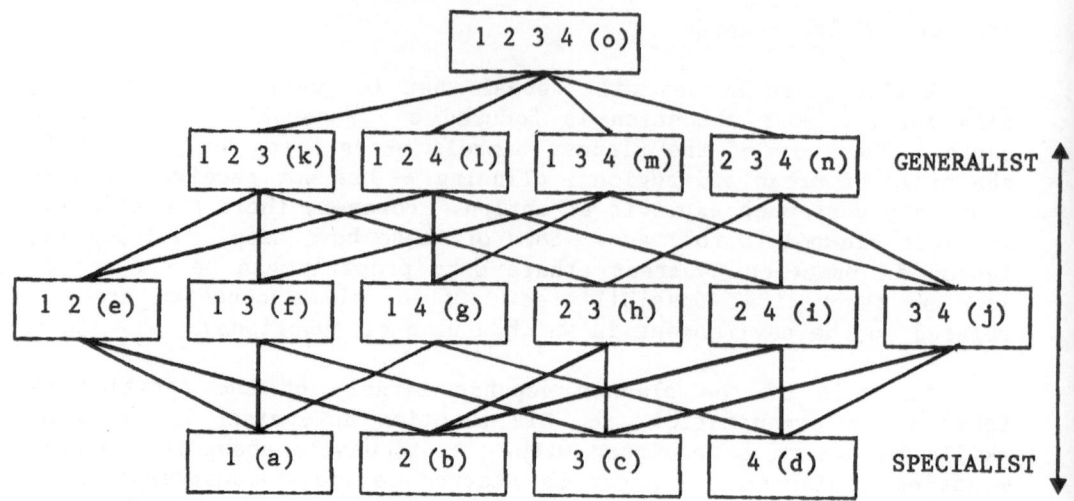

CAPACITIES

1 = analyst
2 = synthesizer
3 = visionist
4 = administrator

Figure 1 – Global Overview of the Various Viewpoints with Respect to
          the Role of a Planner.

open-mindedness to new ideas. This means that he is the person in
government who invokes new arrangements of existing institutions and
creations of new organizational arrangements to achieve plan or
program objectives.

If the work of the planner places great demands on the preparation
and administration of ordinance and regulations, his functioning
corresponds to viewpoint (d). According to this view, the planner
is a civil servant whose major efforts lie in negotiation,
communication and general bureaucratic enterprise.

With a town or regional planner, the job only differs from
other civil servants by having urban and regional problems as the
prime subject of consideration.

The four viewpoints mentioned above all over-emphasize one
particular dimension. This may result in certain shortcomings of
the planner's functioning in a planning process. For example,
sometimes the notion of the planner-analyst has been equated with
the dogma of the "objective professional". This means that, in this
case, not only a level of sophistication of POR is presupposed that
does not exist, but it also fully neglects the normative character

of urban and regional planning. The limitations of an over-emphasizing of viewpoint (b), (ie. planner-synthesizer) are also evident where the quality of the whole depends on the quality of the pieces. Without sufficient insight into the characteristics of parts of the spatial system, the planner works in a vacuum. It is, therefore, necessary that the planner, who stresses the synthesizing function, does not neglect the value of analyzing. If a planner wants to act according to viewpoint (c), he has to become a staff member with access to a strong executive or other official who has the political power to bring his ideas to realization. On the other hand such a planner might become a political activist, using the conflict over interests or issues within a community as a means of welding favourable interest groups into an organization for political action. Viewpoint (d) is perhaps the most realistic one. However, the planner who adopts this role will soon find out that this job seldom lives up to the glowing descriptions he will find in many books on planning theory. Mostly, his work becomes rather routine, carried-on frustration rather than challenge.

Other viewpoints on the function of a planner can all, in a way, be considered as amalgamations of two or more of the viewpoints mentioned before. In viewpoint (e), the planner adopts the role of an analyst who is able to integrate his particular research into a broader framework. This type of planner, who might have different disciplinary backgrounds (eg. sociology, economics, geography), can especially be found in the research units of the larger governmental organizations.

Experience shows that viewpoints (f) and (g) hardly occur in practice. These notions can more properly be considered as theoretical possibilities, which are difficult to realize. On the other hand, planners acting according to viewpoints (h), (i) and (j) can be very well recognized in the daily practice of urban and regional planning. To a certain degree, the same holds for viewpoint (k), whereas (l), (m) and (o) again can be categorized as theoretical possibilities because analyzing capacity and administrative capacity are often very difficult to combine. Finally, viewpoint (n) can be considered as an enriched representation of, especially, the viewpoints labelled under (d), (i) and (j).

In general, it can be said that a planner, who fulfils a more complicated role (ie. the generalist), will be more appreciated in the planning practice than a planner who is specialized in one particular capacity. An analyst, who has absolutely no feeling for the other three "dimensions" mentioned, will encounter many (implicit or explicit) resistances in the discharge of his job. Strangely (or unfortunately?), this mostly is not the case with the other "specialists", who need not be bothered with knowledge of the other "dimensions"!

In theory, all fifteen viewpoints, and even more, concerning the role of the planner are possible. It is evident that this will have consequences for the appraisal of POR. A planner who works according to, for instance, viewpoint (j) will undoubtedly evaluate its usefulness much lower than a planner who meets viewpoint (e). Therefore, it is essential that these differences in tasks and attitudes are taken into account when thoughts are developed concerning the role and function of POR.

ON THE RELATION BETWEEN PLANNING AND POR

The various publications on planning theory show that "planning", as a concept, has no balanced, theoretically consistent, identity (see interalia, Faludi, 1973b). Some authors, for example, consider planning as an activity with some specific characteristics, whereas others see planning more as a framework in which decision-making processes are taking place. It seems to me that the second view holds more promise than the more conventional, first-mentioned point of view. Many actors are involved in planning, and each of these actors are performing certain activities. All these activities have to serve specific purposes, which sometimes might be in contradiction with each other. Because of the independent position of many actors, it is, in my opinion, not very realistic to consider planning as one coordinated activity. It seems more appropriate to favour a notion where planning is seen as a framework, in which the various activities take place with their own specific reference-points with respect to coordination. One of the main purposes of this framework is to guide possible future actions. This is only possible if effective and efficient procedures are available to rule the various processes. Thus, planning can be defined as a set of procedures to guide future actions. A cursory literature review reveals strong evidence to suggest that these procedures at least must have three important characteristics, and these are (1) they should be flexible, (2) they should match the characteristics of the "political ring", and (3) they should be able to integrate research contributions.

To outline in detail the various consequences of this planning notion is a task difficult to fulfil in the context of this paper. Nevertheless, let us explore the characteristics mentioned above more closely.

In my point of view, planning is related to decision-making (cf. Faludi, 1978; Friend and Jessop, 1969; Friend et al, 1974). Planning can facilitate decisions, for instance, by making choices more clear and concrete. It is necessary, however, to recognize that not all decisions are a part of a planning procedure, nor can this be inserted into such a procedure.

There is much to say for the arguments of Lindblom (1968) and others, who asserted that public decision-making processes are more directed to attack present, social imperfections than to the promotion of future social goals. Sometimes the (political) urgency of a decision may be so great that it allows no time for a profound confrontation with planning. This means that planning itself has to be flexible enough to allow these unanticipated decisions.

The uncertainty aspect is also very essential. Planning proposals specified today that are to be taken at some time in the future may no longer appear appropriate when that time arrives (see, for an elaboration of the uncertainty aspect, Friend and Jessop, 1969). Our knowledge about possible actions will change over time. It is, therefore, necessary that the regulating functions of planning are not too heavily stressed, but that instead, possibilities are created to deal with those uncertainties.

All this implies, in fact, a continuous adaptation of planning. I refer to this as a planning process, which can be described as a succession and adaptation of procedures in time. Consequently, I do not associate planning with some kind of control mechanism. Control presupposes strong regulations which might be in contradiction with the "flexibility" intention.

Planning, and thus POR, occurs in a political environment. This environment affects to some degree the types of problems analyzed, how it is done, which groups of persons and organizations are involved and how decisions are implemented. Therefore, both planning and POR must to some extent cope with politics. There has been a growing awareness in recent years that the formulation of any sort of public policy is certain to be accompanied by conflicts among individual and collective values. This has led to the recognition that the political process of dealing with these conflicts must be explicitly incorporated in the urban and regional planning process.

It is important not only to question whether the planning outcomes will reflect politics, but also, whose politics will it reflect? What values, and whose values have been taken into account, and in what way? Clearly, the "best" plan or the "best" action can have no meaning without a normative aim. This implies that planning is of major concern to each actor in a planning process if he is to have any hope of having some of his values and norms fulfilled. On the other hand, this evokes the condition that planning should meet the characteristics of the political ring, which will have consequences for the way planning is organized. I will come back to this point later.

It would be disastrous to conclude that the political nature of planning excludes the research activity. The social system, in which a large number of different kinds of elements interact in a great many ways, is extremely complicated. Systematic analysis is, therefore, for a profound consideration and understanding of this system, inevitable.

Because of the complexity of the spatial-economic system, the conventional paradigm of rational-comprehensiveness seems to be a very valuable one. Critics of this paradigm often argue that in case one wants to pursue this concept, all conceivable alternatives must be identified and evaluated against all relevant objectives and means; otherwise, one could never be certain that the chosen alternative strategy might be "optimal" (see, for instance, Lindblom, 1968). After advancing this statement, they continue by calling this paradigm unrealistic for in almost every situation the establishment of a complete set of alternatives and objectives will be impossible. I do not consider this criticism to be quite fair. Most practitioners of the rational-comprehensive notion undoubtedly will immediately confirm the impossibility to consider all alternatives and all objectives. But this may not be a reason not to try to have an overview of the possibilities.

By only paying partial attention to planning problems, certain areas will be left unconsidered, thus having a fair chance that this evokes unforeseen, and perhaps costly, consequences in the near future. Besides, a comprehensive paradigm offers better possibilities to justify decisions than a "partial" paradigm, because the normative aims behind the reduction of comprehensive set of choice-possibilities can be better outlined, than the, also normative, decision to focus the attention only on a few choice-possibilities.

The most important criticism of the rational-comprehensive planning paradigm is, from my point of view, the lack of sufficient data. This is indeed a very essential issue and, unfortunately, often overlooked in the preparation of research. Perhaps the only way to solve this problem is to use a mixed planning paradigm (Etzioni, 1968), ie. a comprehensive examination of the parts of a system, and an incomplete review of the rest. However, a crucial point in this concept is the decision about which parts should be scanned in detail. According to Etzioni, the level of detail of an examination should depend on the importance of the particular sector. However, how do we measure this importance? Should we take into account the political relevance only or, additionally, the degree of integration (viz. the interactions) with other sectors?

If one should decide to consider only the political relevance of a sector, then this will undoubtedly increase the uncertainty of the examinations, since there is a fair chance that important consequences will be overlooked. Besides, the political relevance of a sector does not automatically coincide with the availability of adequate data. The lack of data will therefore be generally a serious problem in this paradigm too. On the other hand, if one should decide to use the degree of integration as a criterion for the detailedness of scanning, then this presupposes a comprehensive knowledge of the whole system, which in the mixed-scanning paradigm does not exist. In a practical application of this paradigm, however, the theoretical objections mentioned above undoubtedly will seem less serious than is suggested here. Nevertheless, the arguments against it remain valid, which reveals some evidence to suggest a new line of thinking, in which especially the existence of data problems is taken into account. I will come back to this point in the discussion on methodological issues.

SOME FUNDAMENTAL PROBLEMS IN INTEGRATING PLANNING WITH POR

Planners in general, are involved in almost every activity of a planning process. The extent of their involvement varies, however, depending on such factors as the problem itself, the context, the time available for analysis, the state of information, and, last, but not least, the decision-taker. The same factors also determine the use that is made of POR. Planning practice learns that so far only a limited use has been made of the many possibilities of POR. Until now, systematic analysis is mainly seen as a means to advocate a particular position in policy debate, or to help political executives in formulating and implementing their favourite policy. Consequently, planners have to be well aware of the "political costs" of their activities. On the basis of considerations of political acceptability of policy proposals, the planning department and the administrator concerned have to communicate with many social groups and organizations. It is very essential that the political influence of those groups and organizations are taken into account, ie. insight should be created into the probability that their opposition will result in a rejection by the municipal council, parliament or other governmental council, of their proposals. With respect to the implementation of planning proposals, it is also very important that the parties concerned are convinced of the reasonableness of the policy in question.

For these reasons, POR needs, on the one hand, a rather "open" approach, where the cooperation and communication between the planning department and the other groups and organizations mainly will serve the purpose to minimize the political costs. By gauging the social needs and political demands with the help of consultations and public participation, favourable conditions will be created for the acceptance of a certain policy. This is very important, as well as for the planning department and the political executive concerned. It may be a considerable support for a political executive in his negotiations, for instance, with his colleagues.

If a consensus between the department and important social groups and organizations on a certain policy has been achieved, the political executive will hold a strong position in his discussions with fellow-politicians and other administrators.

An "open" approach of POR has, on the other hand, also certain disadvantages. A premature disclosure of research results or policy intentions may, under certain circumstances, be a weapon in the hands of the opposition of the political executives. Certain "weaknesses" might be revealed, which might be detrimental to the negotiations. It is not unusual that planning reports for these reasons often rely rather heavily on vague formulations.

The impression exists that in the practice of urban and regional planning, the arguments against an "open" approach, in general, receive a higher priority than the arguments in favour of openness. This implies that the research activity in planning is often limited in a sense that it must give major allegiance to (the ideas of) the policy-makers, thereby mostly neglecting to assist and inform other social groups and organizations.

For a planner who is involved in POR, this may cause a dilemma. Although he may fully understand the advantages of an analytic approach which takes into consideration various sets of values and norms, at the same time he must face the realities of politics which means advocating the opinions of the political executives. If he should doubt whether the planning proposals actually will attain the goals desired by the administrators, or if he should fear that other harmful consequences might result, his alternatives are limited. He may warn his superiors or resign. In either case, he is expected to refrain from taking his case to the public or to "competing" governmental bodies. It is, therefore, not surprising that the conflict which the planner may feel between professional standards and the demands made on him as a subordinate employee, often will be resolved in favour of the latter.

Another major problem of integrating POR with the planning process lies in the difficulty of combining the planner's advice with democratic institutions. Research-oriented planners have, for instance, often difficulties in communicating properly with policy-makers. As a consequence, politicians and civil servants often make judgements without availing themselves of the knowledge provided by experts in the field, or without making an accurate assessment of it. Apart from possible causes such as lack of time and type of problem, there are two major reasons to be recognized. The first reason can be described as an "unwilling attitude" of both planners and policy-makers towards each other's way of thinking. There are still many researchers left who fulfil their jobs according to a notion of planning and POR in which the normative character is fully neglected. For example, Ter Heide (1977), an acknowledged head of the research department of the Dutch National Physical Planning Agency even stresses the fact that research is not normative! By not taking into account the fact that their "professional assessments" are also based on one particular set of values and norms, for which other sets can be substituted, they isolate themselves from the political platform where planning actually is formed. It is, therefore, not incomprehensible that many policy-makers sometimes reveal a rather sceptical attitude towards urban and regional research. I will come back to this in the next section.

As a second reason for this communication gap, I want to mention the presentation of research results to the policy-makers. It is not unusual that relevant policy conclusions and recommendations are hidden in a huge amount of information. In reading a planner's advice, policy-makers sometimes are confronted with difficult mathematical formulae or equally deterrent set of graphs and figures. It is easy to see that this does not improve the popularity of POR. The only way to overcome this problem is to separate the description of the research efforts from the policy recommendations and conclusions. This conclusion may seem obvious, but many planning reports in the Netherlands (eg. national structure schemes) often exceed the three hundred pages. In this case, a planner's advice will be directed to all those who are involved in a planning process while being able to check it by both the criticism of other experts, and the controls of political responsibility. Professional language, mathematical formulae or complicated graphs are within this context only allowed as a means of mutual communication between planners. This means that I do not agree with those views, in which also the simplicity of the planners' proceedings is stressed. Here, too, a distinction must be made, just as with cooking and baking, between the ingredients and the recipe. One cannot make a cake by putting all the required ingredients in a bowl and just stirring them. The same applies to urban and regional planning. It is meaningless to aim at relevant policy recommendations without a profound and systematic consideration of the planning problem at hand. A systematic consideration in some cases calls for a technical approach. If we neglect this fact, we are bound to get instead of the "expected cake" and indefinable "broth", for which none of the parties concerned ultimately will have any appetite.

Taken together, the previous remarks suggest to us that POR actually is confronted with two fundamental problems: on the one hand, the choice between an "open" or a "concealed" approach, and on the other hand, the dilemma which results from the contradiction between the wish to analyse a "complex reality" and the general desire to do this by means of a "simple approach". Every competent planner will consider these problems as knotty issues, which hardly inspires hope for a compromise. Probably the only solution might be an adaptation of the organizational framework in which POR might operate. This will be discussed in the next section.

TOWARDS A NEW ORGANIZATIONAL FRAMEWORK

Governmental planning departments, at least in the Netherlands, often have to fulfil two different roles. On the one hand, they have to function as "research centres" with the purpose of doing POR on the urban and regional systems with which they are concerned. On the other hand, they are functioning as a "policy service", whose most principal duty is to prepare governmental policy in the field of urban and regional planning. Another way of looking at these two roles is by means of distinction into two specific dimensions of POR, namely the policy exploration and the policy advocacy. By policy exploration I mean the investigation of possible policy-lines and its consequences. Policy advocacy concerns the substructuring, elaboration and advocacy of a proposed line of action of the political executives.

Policy advocacy is often seen as the most important task of a planning department. Consequently, policy exploration is subordinate to policy advocacy, which is mostly justified by a pronouncement like "What is the sense of further research in this direction, if we already know what we want"? In certain circumstances (eg. time pressure), there is much to say for this attitude. However, the disadvantages of this, in my opinion, often too strongly emphasized, pragmatism seem to be ignored most of the time. For instance, some policies might become very dogmatic. There is a fair chance that this attitude results in an insufficient exploration of the possibilities and an insufficient consideration of the various viewpoints (ie. values and norms) and, as a possible consequence, a bad decision.

To show the policy exploration to full advantage, a sharper distinction between policy exploration and policy advocacy is desirable in practice than at the moment is normally usual. However, this statement leads to a crucial question, viz. should this distinction mean a stronger organizational separation, or, on the contrary, a better organizational integration? Both viewpoints have their advantages and disadvantages. A separation might, for instance, result in a mutual distinction. There is, in this case, a fair chance that policy explorative research will match insufficiently the questions which are posed with regard to the policy advocacy.

Champions of the integration of both types of research find, for this reason, a lot of response. However, I want to argue that communication problems, which are often mentioned in this respect, need not easily be solved if it is aimed at a stronger integration of policy exploration and policy advocacy. In general, communications problems will be caused to an even greater extent by the different directions of the various actors (eg. with respect to interest, education, qualities). Besides, I have already mentioned that a (too strong) integration undoubtedly will result in a neglect of the policy exploration, and thus, will lead to a superficially founded urban and regional policy.

For these reasons, it seems reasonable to assume that an organizational separation of the policy exploration and the policy advocacy will lead to an improvement of the POR activities in an urban and regional planning process. Let us consider this hypothesis in more detail.

Not so long ago, urban and regional planning was mainly considered as a "technical" exercise, in which the planner's profession in general was occupied by architects and civil engineers. Their main task was to design a plan, which was then given to the policy-makers concerned who had to take care of the implementation. However, the dissatisfaction with this kind of approach increased gradually because there turned out to be only a minority of plans that had the full agreement of policy-makers, ie. political executives.

Consequently, the call for a better integration of the preparation of these plans and actual policy became increasingly louder, resulting in stronger cooperation between planners and administrators.

Although this development is understandable, I want to argue that it can also be considered as fundamentally wrong. Indeed, the political character of urban and regional planning had been recognized, but at the same time a very one-sided interpretation was given of this political character, namely, only an agreement with the wishes of the political executives was pursued! It was only partially recognized that the pluralistic complexity of spatial planning also encouraged a much stronger integration of the ideas and wishes of other, not directly involved, departments and private groups or organizations. This resulted in a strong commitment of research which satisfied the wishes of the political executives. Unfortunately, it is not unusual for research outcomes to be suppressed from other concerned parties or published incorrectly. POR will, therefore, be very restricted, also because an extensive analysis of alternative options is mostly not allowed or, at least, not encouraged.

An increased emphasis on the policy exploration as a separate activity might offer a fruitful contribution to the solution of these problems. A planning department can, for instance, be split up into one or more exploratory groups, and one or more advocacy groups. The exploratory groups might be involved in substantial analysis of the spatial system and various value systems as well as with the preparation of possible solutions for policy problems at hand. These solutions should reflect the various value-viewpoints as well as possible, and not be curtailed a priori to the opinions of the political executive(s). A publication of these results will enable the political executives to watch the reactions of the various parties concerned, without the need to commit themselves to a premature disclosure of their position. They will only present their proposal if their position is firm; ie. after the advocacy group(s) have studied the publications of the explorative groups and the reactions of the parties concerned. In other words, the advocacy groups are involved in the interpretation of the available information and the rearrangement of this information in such a way that it optimally supports the wishes of the administrators.

As a consequence of this organizational construction, the role of the planner will also be different in both groups. When the planner works in an explorative group, he may consider himself as an agent of a wider society trying to influence the administrators to consider a problem from a different perspective. On the other hand is a planner who works in an advocacy group, more or less morally obliged to help his political executive in its "battles" with the rest of the governmental agencies and other parties concerned.

The latter must thus accept considerable restriction on publication, for in the conflict environment of public policy, any criticism, even the most constructive one, can be used as an argument by the opposition. It is easy to see that the organizational structure in terms of its broad structure, holds for the plan-making phases in a planning process, as for the monitoring phase. Monitoring can be defined as a continuous activity of recording and analyzing changes with regard to their influence in affecting a plan or policy (see for different treatments of the monitoring concept: Voogd, 1980a). Evidently, monitoring requires a systematic selection and scanning of a wide variety of information. This information has to be analyzed and assessed for its significance, eg. in terms of the indicated performance of the policies and programmes. To fulfil these tasks, separate organizational units can be created, which in a number of cases already is done in Great Britain, (ie. the so-called monitoring units). As may be noticed, the aims of these monitoring units correspond strongly to the tasks of the exploratory groups.

Taken together, the advantages of the organizational separation between policy exploration and policy advocacy might be summarized as three major points, and these are:

1. Because of the relatively independent position of the explora-
    tory groups with respect to the political executives, the
    results of POR can be published more openly than at this moment
    is usual. It will be much easier for an administrator to allow
    publication because these publications cannot directly be
    associated with the final policy-choice.

2. The effectiveness of the research activity in planning will be
    increased. Because of the greater public disclosure of POR
    results, the various participants may be better informed about
    the possibilities and/or consequences of policy proposals.
    Consequently, their contributions may be substantially
    improved, which again will enlarge the insight of the political
    executives into the possible outcomes of their proposals.

3. There will be better allocation of tasks in a planning process
    (viz. planner-explorer versus planner-advocate), which enlarges
    the possibilities of arriving at more efficient planning proce-
    dures. Planner-explorers will be better able to exploit their
    full skills, whereas their results can be concealed no longer
    by the administrators as "redundant information" because they
    will also serve other participants in the planning process.

Clearly, this organizational structure might also encounter disadvantages, especially if the separation on the policy explora-tion will lead to an overemphasis of those research activities which can be better performed at academic or specialized research insti-tutes. However, I feel that these difficulties can be prevented by taking care that the exploratory groups are directed by people, who feel themselves mentally responsible for good connections to the advocacy group(s).

TOWARDS AN IMPROVED POR-METHODOLOGY

Urban and regional planning embodies a variety of activities undertaken by various planners (and other participants) under different circumstances. This means that a consideration of the POR-methodology only makes sense if it is related to the particular functions and possibilities of the planners involved. For example, a planner-explorer will act differently in a planning process than a planner-advocate does. Each planner will be guided by "rules" and will use "tools" that he and his fellow-planners within his group are familiar with and trust. In other words: each group of planners will behave according to their own interpretation of rationality. It is evident that this will determine the methodology a group will use.

In the third section of this paper I suggested that planning procedures should be flexible, and be able to match the nature of the "political ring". In addition, they should be able to integrate contributions of POR. These characteristics and the remarks made in the preceding sections reveal some evidence to suggest a number of conditions, which methods and techniques of POR have to meet. These conditions, which may look in part rather trivial, can be summarized as follows: POR methods and techniques:

Must Enlarge Insights into the Solution of the Planning Problem

Methods and techniques must possess the capacity to extend insights into the question "how to cope with the problem at hand". An approach which only serves to "gain insight" into a particular problem without leading to appropriate solutions is often too time-consuming to be relevant in a planning process. Too often, the decisions are already made before such studies have been finished.

Must Be Able to Handle Changes of Circumstances in a Quick and Adequate Way

Urban and regional planning can be associated with continuous deliberations and negotiations. This means that the ideas and insights in the object of planning are constantly changing. Relatively time-consuming methods and techniques are for these reasons less preferable, although they can not be excluded a priori.

Must Give Surveyable and Interpretable Outcomes

Although this might seem a rather trivial condition, it appears to be sometimes of such importance that it cannot be neglected. This is especially true in cases where the results of a particular method or technique play a dominant role in the planning recommendations.

Must Be Based on Easy-to-Illustrate Principles

Although not all the detailed structure of some techniques need
to be (nor can be) explained to non-experts, it is very important
that the basic principles can be explained. If this is impossible,
then the introduction of this technique in a planning process
undoubtedly will encounter considerable resistance.

Have To Be Cheap (Qua Manpower and Finances) in their Application

The more expensive an application of a method or technique is,
the more resistance will exist in applying it. This condition can
be partially relaxed if the method or technique is able to answer
relatively many questions or a very important question.

Must Be Able to Make the Values and Norms Explicit upon which the
Results Will Be Based

Forecasting can, for example, never be based solely on scientific deduction process, but is also the result of value judgements
(viz. assumptions about future characteristics). To make the outcomes debatable, it is therefore necessary that these value judgements be made explicit.

Looking at current planning practices, it can be concluded that
these conditions are not always met (at least not very closely),
particularly conditions (b) and (f) which are often violated. There
are still many methods and techniques which place excessive demands
upon the quality of imputed data. For instance, too many so-called
disaggregated allocation models have been developed without questioning the availability or quality of the necessary data. In the
third section, I have already mentioned the data-situation as a serious problem in POR. Recently some more attention has been paid to
methods and techniques which are able to deal with qualitative or
ordinal information in a theoretically consistent way by using the
principles of ordinal geometric scaling (see, inter alia, Nijkamp,
1979, Nijkamp & Voogd, 1980; Voogd, 1979, 1980b). It appears that
these principles can be extended to various fields of application;
for instance, spatial interaction modelling (see Van Setten & Voogd,
1979).

Therefore, it seems reasonable to conclude that it is meaningful to pay more attention in the near future to the development of
methods and techniques which can treat qualitative information.

TOWARDS A CHANGE OF ATTITUDE

POR has been described in the introduction as a professional
activity based on a scientific methodology, which has an aim in
assisting in the preparation and execution of planning procedures.

Consequently, as has been seen, POR can have a rather powerful (ie. influential) position in a policy-making process. It is able to emphasize certain value preferences, and to neglect others. For this reason, I have argued that the normative character of the research contributions should be more explicitly taken into account than is often done today. This induces both changes in the conditions under which POR is performed, and in research methodology itself.

In the previous sections, I did make some suggestions for improving the research contribution in planning. However, one important aspect has been neglected so far, namely, that a change in the organizational structures and the methodology also necessitates a change in attitudes of the various persons and agencies. This means that the planner who is involved in practising POR has to be more aware of the debatable character of many of his research results. Each individual professional, whether he is a planner or a political executive, has to be open-minded with respect to innovations in their specialties, and be willing to adapt his own way of working (see Hall, 1980).

The daily practice of urban and regional planning learns, however, that this change of attitudes is realized only in a very modest way. One of the most important "obstacles" in the introduction of innovations is the inertia of the human mind. Organizational experts mention, in this case, the inability of many persons to absorb effectively in a short period of time, large amounts of information. Most of them take each day a large number of routine or programmed decisions, and only a very limited number of non-routine or new decisions. Professional activities can mostly be considered as daily customs, and are characterized by the small amount of information that is necessary to act. For instance, one is familiar with the problems one encounters; one knows the persons and the departments with whom one is related; one is familiar with the backgrounds of the persons one is working with, etc.

Non-routine decisions that cause new behavioural patterns, demand more information and psychological energy. Innovations in a profession involve many of such non-routine decisions, which might raise some kind of a psychological stress among the persons concerned.

The danger of mental "overburdening" is, under such circumstances, not purely hypothetical. Well-known symptoms of such an "overburdening" are, for instance, a denial of the reality (details are blown up as if they were very important matters) and an escape into specialization (persons retract in their own specialities and tolerate no interferences of "outsiders").

The reader will undoubtedly recognize one or more of these symptoms, which means that a change of attitudes is perhaps one of the most difficult tasks to fulfil. Nevertheless, if we want to

improve planning in general, and POR in particular, we have to deal
in a proper way with the corresponding need to adjust attitudes.
The "type" and the "strength" of this adjustment depends on the
innovation at hand. It seems to be very essential, however, to
allow the persons concerned to determine their own pace in the initial phase of any introduction of an innovation. Suspicion might be
overcome if persons are able to become acquainted with a new situation and/or "tool" in a very relaxed way. A proper escort of any
innovation by experts in this field is for these reasons inevitable.

In my opinion, this implies for POR that the introduction of
any new elements into this research framework will have the best
chance of success if the persons, who are directly concerned with
these elements, are already a little experienced in it in an earlier
stage. Methods and techniques correspond in a way to automobiles.
If you have not learned to drive them, it might be very dangerous if
you, in spite of this tremendous handicap, want to drive a car your-
self. You (and others!) will be better off if you give the wheel to
a more experienced driver, or if you firstly try to get a driver's
licence. The same situation holds for methods and techniques. If
you are not familiar with them, it will be very difficult, at least
in case of sophisticated methods, to apply them in practical circumstances with tight time constraints. Either you do it incorrectly,
or it takes too much time. In both cases, it will be considered by
the "outside" world as a failure, with all the negative consequences
for the further introduction of this particular method or technique.

FINAL REMARKS

It is obvious that POR alone can never be sufficient to develop
urban and regional policy. There are always considerations that
cannot be handled quantitatively, or even analytically, or systematically, and there may be problems with no solution. Thus, communication and negotiation have been, and will also be in the future,
major pillars on which urban and regional planning rests (see also
Kreukels, 1980). The claims for what one can expect from POR should
therefore be rather modest. It is no more that an adjunct, although
a valuable one, to the judgement, experience and intuition of
decision-makers (see, for instance, Masser, 1979).

Nevertheless, POR can be a useful aid to help in the preparation of planning proposals and decisions, if political executives
are able to appreciate its limitations and know what to expect from
it. However, many administrators do not seem to be very aware of
how POR can really help them, besides as a means to justify a decision already made, or to advocate a particular position in policy
debate. This means that mostly only a limited use has been made of
research. One of the main reasons for this attitude comes from the
political environment of urban and regional planning. Planning
corresponds to politics in a sense that politics also deals with the
judgement of observations. Both notions, however, differ from each

other because politics operates from one particular set of values and norms, whereas in planning one has to deal with various sets of values and norms. It has been argued in this paper that the choice of these sets entirely depends on the position of the POR-activity in the planning process, ie. has it an explorative function or an advocacy function? Since POR is able to emphasize certain value preferences and to neglect others, it is necessary to shape conditions which makes it possible to overcome this intrinsic "unfairness". It has been emphasized in this paper that changes are needed, not only in the organization and methodology, but also in attitude. It is to be hoped that the exploration of new avenues indeed will result in a rapid improvement of urban and regional planning. The present problems of many regions and cities are too important to rely fully for their treatment on conventional insights and negotiations.

REFERENCES

Etzioni, A., 1968, The Active Society: a Theory of Societal and Political Processes, Free Press, New York.

Faludi, A., 1973a, (Ed.), A Reader in Planning Theory, Pergamon Press, Oxford, UK.

Faludi, A., 1973b, Planning Theory, Pergamon Press, Oxford, UK.

Faludi, A., 1978, The Decision-Centred View as a Research Paradigm for Planning, Werkstukken PDI, 18, University of Amsterdam, Amsterdam.

Friend, J.K. and Jessop, W.N., 1969, Local Government and Strategic Choice, Tavistock, London.

Friend, J.K., Power, J.M. and Yewlett, C.J., 1974, Public Planning: The Intercorporate Dimension. Tavistock, London.

Hall, P., 1980, Great Planning Disasters, Weidenfeld and Nicolson, London.

Heide, H. Ter, 1979, Relations between Research, Planning and Policy, in D. Hazelhof, (Ed.), Recent Developments in Planning Methodology, Ministry of Housing and Physical Planning, The Hague, The Netherlands.

Kreukels, T., 1980, Developments in Strategic Planning: an Overview, in H. Voogd, (Ed.), Strategic Planning in a Dynamic Society. Delft University Press, Delft, The Netherlands.

Lindblom, C.E., 1968, The Policy-making Process, Prentice Hall, Englewood Cliffs, New Jersey.

Masser, I., 1979, The Limits to Planning, Working Paper TRP 16, Department of Town and Regional Planning, University of Sheffield, Sheffield, UK.

Nijkamp, P., 1979, Multidimensional Spatial Data and Decision Analysis, John Wiley and Sons, Chichester, UK.

Nijkamp, P. and Voogd, H., 1981, New Multicriteria Methods for Physical Planning by Means of Multidimensional Scaling Techniques, in V. Haimes and A. Kindler, (Eds.), Water and Related Land Resource Systems, Pergamon Press, New York, forthcoming.

Setten, A., van and Voogd, H., 1979, Interaction Modelling under Fuzzy Circumstances, Research Paper, 17, PSC-TNO, Delft, The Netherlands.

Voogd, H., 1979, Ordinal Data Analysis in Urban and Regional Planning. Proceedings Seventh European Symposium on Urban Data Management, Volume 2, Paris, France.

Voogd, H., (Ed.), 1980a, Strategic Planning in a Dynamic Society, Delft University Press, Delft, The Netherlands.

Voogd, H., 1980b, Stochastic Multicriteria Evaluation by Means of Geometric Scaling, Delft Progress Report, 5, 112-125.

CONCLUSIONS

ATTITUDES TOWARDS URBAN SYSTEMS ANALYSIS:

A SURVEY OF THE PARTICIPANTS

Michael Batty and Richard Spooner

Department of Town Planning
University of Wales Institute of Science and Technology
Cardiff, CF1 3NU
United Kingdom

INTRODUCTION

A real sense of community now exists within the fields which comprise urban systems analysis. It has taken twenty years for this community to establish itself, but readers of this book will by now be aware that many of the authors writing here have identified with this area for much of its history. Such a development is perhaps not particularly surprising, although the critiques and prognoses for the field sketched here, emphasize the multitude of difficulties urban systems analysts have faced and continue to face. Nevertheless, a small community of like-minded people now exists and this, it is often argued, represents a fundamental prerequisite for a sustained attack on the major research problems in any field. Commonality of interest and agreement about the key problems facing the field are all important. For example, Fleck (1979) argues that only when such consensus emerges does a scientific community exist, and as the community begins to focus on more detailed problems the differences of opinion become smaller.

It is quite clear that these characteristics apply to the group of participants who attended the NATO Advanced Research Institute on Systems Analysis in Urban Policy-Making and Planning. Indeed, the fact that such a meeting could be arranged at all attests to the fact that a rudimentary organization of interests in this area could be quite easily detected. The evaluation of discussions and recommendations confirms the feeling that the field has acquired coherence and its advocates, a consensus about where it should next be directed. There are other indicators too: the "Urban Modelling List" organized by Michael Wegener from the University of Dortmund was set up some two years or so ago (1979) from the time of writing,

567

with the intention of developing a network of contacts concerning those interested in land use modelling. The Regional Science Association in its British Section has set up a group to look at the history of planning method and analysis. Annual Conferences and meetings on transport modelling have been regular events for a long time, and there is now evidence that this is occurring within less established modelling areas such as land use and public systems modelling.

Just as a good deal of the evidence concerning this emerging community of urban systems analysts has been ad hoc, a good deal of the critique of the field has been somewhat polemical, even rhetorical. Thus, given that there appeared to be consensus in the field over mistakes made in the past, the NATO ARI at Oxford seemed to be an ideal opportunity to attempt some more systematic analysis of the factors affecting the development of the general field. Accordingly, the authors set out to systematically monitor the opinions — attitudes of the participants at the conference, in terms of a limited, reasonably narrow set of issues. With an attitudinal model in mind based on Saaty's (1981) "prioritization" technique, questionnaires were distributed to all the participants early on during the week-long meeting, and this paper presents the results drawn from the completed questionnaires. The results are hardly startling, indeed some might argue that they are so predictable to be trivial and certainly, there are limits to this type of analysis and its applicability. Yet, they are worth reporting if only to show what is possible when a group of highly motivated and committed experts assemble at such a meeting.

There is another, possibly more obscure reason for presenting this work. The major problem pervading the social, in contrast to the physical sciences, is the tautology that "man himself is part of his own field of observation". In other words, the behaviour of say, urban systems analysts, is one small part of the subject matter of urban systems analysis. In short, models of the modelling process — of urban systems analysis — are related to models of the urban system. This inevitable interpenetration of ideas gives the social system its rich and ambiguous flavour, but it also suggests that any model which an analyst uses to model others (including other analysts) must to a degree be applicable to the analyst in question. The ideas which underpin the analysis presented here form a very different corpus of techniques to those in common use in urban systems analysis, and it is this difference which reflects one of the key recommendations from the NATO meeting: that there be a shift in analytic style from quantitative to qualitative. The models proposed here are very definitely qualitative and thus, very much in the spirit of the conclusions reached in the meeting itself.

In this paper, we will content ourselves with providing an overview of the questionnaire we distributed, the model we used for analysis and the essential results. We will not dwell on technical matters, for these are dealt with more thoroughly elsewhere (Batty and Spooner, 1982) and we will not elaborate the analysis into sociological models of the community of urban systems analysts: this we also do elsewhere (Spooner and Batty, 1981).   However, we will attempt to demonstrate how the technique might have been improved, and we hope this will be useful for others thinking about using such attitudinal models.   Accordingly, we will first show how we structured the questionnaire from our knowledge of the field, and then we will recount how we distributed the questionnaire to participants. Then we will launch into an analysis of results — first by outlining the technique itself, and then by concentrating on the results per se.   Finally, we will conclude with some recommendations relating to the use of such techniques.

THE DEVELOPMENT OF URBAN SYSTEMS ANALYSIS AND ITS CRITIQUE

When the idea of this NATO ARI was considered first, it was originally suggested that the meeting be one on technical concerns, continuing the tradition of four-yearly international conferences in the land use modelling field.   However, the NATO Systems Science Panel were more concerned to begin a survey of the applications of systems analysis and thus, it was suggested that the meeting be wider than just land use oriented — be "urban" in focus, that is — and be concerned with applications and their evaluation rather than the development of technique.   After a good deal of useful discussion, it was suggested that the general area could be partitioned into four fields, namely public services planning, physical networks planning, strategic land use planning and corporate planning.   These fields spanned the range of developments during the last twenty years, and thus, it was proposed to draw participants from these four areas, as well as from academia, practice and the political arena.

Early on in the planning of the meeting and in our survey of these fields, it became clear that these four areas had rather different communities of interest.   For example, physical network and land use planning had fairly well-defined communities of analysts, called respectively, transport and land use modellers.   However, in public services planning where the problems were rather tightly defined, the community of interest was a branch of operations research; that is, operations researchers saw the field as an extension of their primary skills.   In corporate planning, the community was even more diffuse with a strong concentration on practice, and with a much wider range of disciplinary skills ranging from public economics to accounting.   These difficulties in identifying relevant communities certainly led to an imbalance in participants, against the corporate and public services planning areas, and for the land use and transport planning areas.

The second influence which was not anticipated in the conference pre-planning was the emphasis on modelling.   Clearly, systems analysis has come to be regarded as modelling by many, but the organizers had no intention of selecting a list of participants who all had technical experience in modelling. The academics and practitioners clearly would have, but in the event the politicians and bureaucrats who attended, also had backgrounds in modelling, through economics and engineering. As pointed out in the previous chapter, this meant that the predominant concern of the meeting was with application of modelling technique as Vickers (this volume) also notes. Before the conference began we had detected this bias, and in preparing our questionnaire we decided to inquire into the attitudes of the participants with respect to the four stated fields, but with the qualification that the focus be on "modelling" rather than "planning". The four fields were thus stated and defined as:

1. Land Use Modelling: involving analyses using operational models of urban spatial structure, location and distribution and relating to the strategic level of planning;

2. Transport Modelling: involving analyses using models of traffic flow, distribution and assignment, embedded within the conventional transport planning process relating to questions of economic and engineering efficiency;

3. Corporate Planning-Modelling: involving analyses of the budgetary process and models of resource allocation, bureaucratic and organizational efficiency, and

4. Public Systems Modelling: involving analyses of localized delivery systems based on models of scheduling, locational optima and public services.

These four fields were relatively well-defined by the time the meeting began. The organizers had continually emphasized the nature and scope of urban systems analysis in their frequent bulletins to the participants. However, the factors which had affected the development of each of these fields was by no means so finely articulated. Indeed, it was the intention of the meeting to inquire into the major factors which had influenced the field in general, although in seeking to inquire as to the attitudes of participants with regard to key issues, a list of such issues had to be prespecified. Of course, as most of the participants were invited for their visible commitment to enhancing the application of urban systems analysis, most were familiar with the major critiques of the area. In fact, it is these critiques which throw up the major issues, and thus, we felt reasonably satisfied that the list of issues we would present to participants would not be in any sense unfamiliar.

We eventually decided that six issues were important. Problems over the size of data bases, the availability of data and computer access, etc., were raised early on in critiques over the development of the area and these seemed worthy of inclusion. Technical skills, their quality and availabiity also figure. There were three other issues of a broader, more general kind: the question of the receptivity of the organizational environment to techniques always seemed to loom large in discussion of where the successes and failures of urban systems analysis had occurred. The relationship of analysis to policy -- the degree to which any model addressed the issues confronted by policy-makers -- was judged important. But the more general issue of the theoretical quality of the model under scrutiny was frequently raised and thus, it was felt that issues concerning the lack or otherwise of social theory in the model should be considered. The list of six issues appeared as:

1.   <u>Data</u>: problems of availability, level of detail and such like;

2.   <u>Computation</u>: problems of the scale and accessibility of the available hardware and software;

3.   <u>Social Theory</u>: problems posed by the absence of relevant social (and economic) theory in models;

4.   <u>Policy Issues</u>: problems involved in designing models to mirror relevant policy issues;

5.   <u>Technical Skills</u>: problems of quality and availability, and

6.   <u>Organizational Environments</u>: problems posed by the lack of receptivity of the organization to the development of models and techniques.

We excluded general issues of cost because we felt these were too pervasive an influence on the field to be usefully distinguished from the six chosen factors. Also, we excluded issues concerning the abuse of modelling, sharp practice and oversell, for these did not appear to be directly comparable with those in the list of six.

From the four fields and the six issues or factors, as we will call them henceforth, we can ask twenty-four questions of the general form:

> "How important has factor x been in affecting the difficulties experienced by field y?"

We can also ask four questions of the four fields:

> "How successful has been the development of field y?"

and six questions of the general form:

> "How problematical has been factor x in the development of urban systems analysis in general?"

Thus, these fields and factors define a rather specific set of issues which we wished the participants to respond to, and as with any well-defined structure such as this one, the particular form of the inquiry depended upon the logic and technique of the method used. To this we now turn.

THE QUESTIONNAIRE AND THE PARTICIPANTS

In eliciting responses to the thirty-four questions posed above, a method for forcing consistency on the judgements sought was adopted. It was always intended to ask participants to judge the relative importance of any pair of questions, and from such estimates (paired comparisons) we would extract a set of weights which approximated the responses in some best possible way. Thus, participants would be asked to examine pairs and to order the comparison of one question with another, according to a given nine-point scale. The technique then used for extracting the weights implied by these comparisons, would be based on Saaty's (1981) prioritization method, which in essence is a type of scaling technique. Inevitably the set of paired comparisons would not be totally consistent, but this would avoid the participant having to arrange and compare a large number of issues simultaneously, and it would avoid ranking. We will discuss the technique in the next section.

However, in asking each participant to make paired comparisons in terms of the given set of questions, this would involve a large number of comparisons; in fact, too many. For example, assuming that the comparison of factor x with factor y is the reciprocal of the comparison of y with x -- a very reasonable assumption -- and that the comparison of x with itself implies a ratio of 1, then for twenty-four questions, two hundred and seventy-six comparisons need to be made. Clearly, as n, the number of questions increases, the number of comparisons increases as $(n)(n-1)/2$. Thus, it is necessary to minimize the number of questions asked. This is where the second aspect of the use of Saaty's method comes in. For example, a list of twenty-four questions is too large, and therefore, it is suggested that the time-honoured principle of breaking down the questions into manageable subgroups be invoked. In Saaty's treatment, he imposes an hierarchy on such a list and employs his paired comparison technique at each level. This will become clear in the sequel when the technical nature of this process is elaborated. In fact, in this context the twenty-four questions referred to above are broken into their four groups of six, each six pertaining to one of the four fields. Thus, the question becomes:

"Given field y, how important has factor x been
in contributing to its difficulties?"

The way in which the overall importance of x in y is calculated can then be approximated by relating these weights to those relating to the importance of each field y, but we are getting ahead of ourselves.

During the NATO ARI, thirty-five people participated, including the secretary, the organizers and the visiting speakers.   On the third day of the meeting, the questionnaires were handed out to thirty-two participants who were then available, and they were asked to return the questionnaire by the last day.   The questionnaire itself was arranged as follows. First, the participant was asked to make six paired comparisons between each of the four fields to eval- uate their relative success.   Then, for each of the four fields, the participant was required to respond to fifteen paired comparisons relating to comparisons between each of the six factors: this sec- tion required a total of sixty responses.   Finally, each participant was asked to judge the overall difficulties posed by the six factors in general, and this also required fifteen comparisons, making a total of eighty-three in all. For each comparison, the participant was asked to identify the one item from the pair which was more important and then estimate the ratio of its importance to the other using a nine-point scale.   The scale begins at one, in which the items compared would be equally important and ends at nine, where one item totally dominates the other in its importance.   Further detail on the appropriateness of the scale can be found in Saaty (1981): suffice it to say that we, like Saaty, have found the scale robust under sensitivity testing.

The responses were judged reasonable in the circumstances. From   thirty-two   questionnaires   circulated,   twenty-four   were returned -- a seventy-five percent response rate.   The non- respondents fell into two groups: those who genuinely found it dif- ficult to answer due to their experience being rather specialist in terms of the responses required, and those who quite definitely were too busy during the week to find the time to complete it. As in all NATO ARI's, the week involved a programme of intensive seminars involving discussion of many of the issues identified in the ques- tionnaire.   In fact, there was very little free time to fill in the questionnaire as participants were required to produce position papers, make short presentations as issues emerged through discus- sion and attend more formal evening sessions/lectures, as well as read and re-read the large set of background papers produced by the participants for the meeting. Nevertheless, the participants were highly motivated and it appeared the questionnaire was quite well- received.

Of the twenty-four who responded, seven completed the question- naire only partially, in that the questions to which they did not respond largely related to their lack of knowledge of certain areas.   In the areas of corporate and public systems modelling, some participants could not respond to the importance of factors in each field, but all twenty-four responded to the importance of each field, and each factor overall.   This bias against corporate and public systems modelling is hardly surprising, given the backgrounds of participants: of the thirty-two who were issued the questionnaire, seventeen were associated with land use modelling, ten with transport, four with corporate and only one with public systems analysis per se.   This imbalance might seem surprising, but it reflects what we were referring to earlier when we noted the difficulties in identifying participants from the more diffuse field of corporate modelling on the one hand, and on the other, the rather narrow field of public systems modelling, a branch of operations research.   Moreover, it may seem curious that some participants were ignorant of some fields within urban systems analysis, but as the previous discussion attempted to imply, the questions asked were extremely specific.

Respondents made a number of comments about the questionnaire. It was very clear that the questionnaire was laborious and difficult to answer, but few hints were available upon how to elicit such information in a less routine fashion.   Furthermore, several participants clearly found it hard to generate the requisite level of discrimination concerning factors in various fields, but this is possibly due to the fact that the questions asked are more appropriate to some fields than others, although ignorance may enter the picture here.   The. scale used was also a little problematic. Respondents tended to use the odd numbers which were specified as defining the range of the scale, rather than the intermediate values, and a few only used the lower range of the scale (1-5, for example).   Nevertheless, the questionnaire was very well-received despite its somewhat experimental nature, and there was considerably greater tolerance for it than expected.   Control over response could have been improved by setting aside time for its completion, and in hindsight, some group response, rather than individual, might have been a fruitful way of entering into discussion about these issues. However, these are all matters for any future development of this technique.

TECHNIQUES OF ANALYSIS

We begin by examining any set of n questions having elicited the nxn pairwise comparison matrix $\underline{A} = \{a_{ij}\}$. We will develop the general technique first before extending this to an hierarchical structure of questions and to an average of weights across a series of hierarchies. If the set of comparisons is consistent, then each comparison a_{ij} is the ratio of the weight of i, w_i to the weight of j, w_j. If we assume that these weights sum to 1, then it is clear that

$$\sum_j (w_i/w_j)w_j = \sum_j a_{ij}w_j = nw_i$$

A similar condition can be placed on the summation over the rows of the matrix, but as we do not use this in the sequel, we can ignore it.

In general, the matrix of paired comparisons will not be consistent, that is, $a_{ij} \neq w_i/w_j$, and thus, equation (1) above will not hold. In fact, what we seek is an equation from which we can recover the set of weights $\{w_i\}$, which are unknown, which is "as near as possible" in some sense to equation (1). Saaty (1981) shows that the appropriate equation to solve is the principal eigenvalue equation associated with a condition on the rows of the matrix $\underline{A}$. Then the equation to solve is

$$\sum_j a_{ij}w_j = \lambda w_i \qquad (2)$$

where λ is the principal eigenvalue of the matrix $\underline{A}$ and thus, $\{w_i\}$ is the right-hand eigenvector associated with $\underline{A}$ and $\overline{\lambda}$. Saaty (1981) also shows that if the matrix $\underline{A}$ is a reciprocal matrix, that is $a_{ij} = a_{ji}^{-1}$, thus $a_{ii} = 1$, $\forall_{ij}$, then the maximum eigenvalue λ is always greater than or equal to n and the difference $(\lambda-n)$ is a measure of the consistency of the paired comparison matrix. Thus, for any matrix $\underline{A}$ satisfying these requirements, the vector $\underline{w}$ can be computed by solving $(\underline{A} - \lambda\underline{I})\underline{w} = \underline{0}$ where $\underline{I}$ is an nxn identify matrix, $\underline{w}$ is an nx1 vector of weights and $\underline{0}$ is an nx1 null vector.

For the problem in hand, it is necessary to find six sets of weights from six pairwise comparison matrices using equation (2); that is, a set of weights referred to in the sequel as $\{w_i\}$, indicating the importance of each of the n=4 fields; four sets of weights, each one dealing with the importance of factor j, contributing to the difficulties facing field i and referred to as $\{w_{ij}\}$; and one set of weights $\{w_j\}$ dealing with the importance of each factor to the overall area of urban systems analysis. Note that it is assumed that

$$\sum_i w_i = 1 , \quad \sum_j w_{ij} = 1 , \quad \forall_i , \quad \sum_j w_j = 1$$

These weights are related as follows. Once the success of each field $\{w_i\}$ has been found, it is possible to compute the difficulties caused by a factor j in a particular field i in an absolute sense, by multiplying w_i by w_{ij}; this gives the overall difficulty caused by factor j in field i in terms of all other factors and all other fields. Now, to predict the difficulty of any factor j in terms of all fields, we must sum $w_i w_{ij}$ across all fields i. This gives

$$\overset{*}{w}_j = \sum_i w_i w_{ij} \qquad (3)$$

where $\overset{*}{w}_j$ is the "predicted" importance of any factor j. As we already have estimated a set of weights $\{w_j\}$ indicating the importance of each factor, we can compare $\{\overset{*}{w}_j\}$ with $\{w_j\}$ as a check on the method.

There are a number of points relating to this application. Equation (3) provides a two-level hierarchy: fields are assessed, then factors in each field, from which the overall effect of any factor can be predicted. We could extend equation (3) to many levels, thus meeting the requirement for data/questionnaire manageability through decomposition. Secondly, in our particular application where responses are missing (in our case where $w_{ij} = 0$), a suitable re-scaling of $\{w_i\}$ enables equation (3) to be applied. Finally, it is a simple matter to extend the computation of consistency indices to equation (3) using a concatenation of the indices for $\{w_i\}$ and $\{w_{ij}\}$ with the same structure as equation (3). All these details are provided in the full paper by Batty and Spooner (1982).

Finally, in the analysis which follows, it will be necessary to predict "average weights" by averaging out the weights over the set of K respondents. We assume that the weight of each respondent is the same due to the fact that all respondents have special (unique) contributions to the field, and it would not make sense to attempt any differential weighting. To develop our notion of "average weights" we must now index the weights by the particular respondent k, thus forming $\{w_{1k}\}$, $\{\overset{*}{w}_{jk}\}$, $\{w_{jk}\}$, $\{w_{ijk}\}$ from the weights defined above. It is a straightforward matter to compute the average weights for fields $\{\overline{w}_i\}$, for factors "observed" $\{\overline{w}_j\}$ and for factors "predicted" $\{\widetilde{w}_j\}$. These are given as

$$\overline{w}_i = \sum_k w_{ik}/K \ ,$$

$$\overline{w}_j = \sum_k w_{jk}/K \ , \text{ and}$$

$$\widetilde{w}_j = \sum_k \overset{*}{w}_{jk}/K \ .$$

The average conditional weight $\overline{w}_{ij}$ is a little more tricky. If we assume that the predicted weight $\{\widetilde{w}_j\}$ can be formed from

$$\widetilde{w}_j = \sum_i \overline{w}_i \overline{w}_{ij} = \sum_{i \ k}(\sum w_{ik}w_{ijk}/K) \ , \tag{4}$$

then a little manipulation of equation (4) gives

$$\overline{w}_{ij} = \sum_k (w_{ik}w_{ijk})/(\overline{w}_i K) \ ,$$

and it can now easily be checked that all these simple averages are consistent (see Batty and Spooner, 1982). Of course, consistency indices can be extended to these average weight sets.

Let us briefly summarize what this technique yields. First, for each individual respondent, we can examine the importance he gives to the fields, then to the factors in each field, and we can predict the importance of each factor overall. However, we have estimated the importance of each factor to urban systems analysis in general and therefore, we can use this as a check on our prediction of this importance. We can also examine these estimates and the prediction for consistency of response. In essence, we can do exactly the same for the average weights. However, there is a complication due to the partial nature of some of the responses. In the case of seven respondents, it is not possible to compute average conditional weights $\{\bar{w}_{ij}\}$, due to the fact that there are some non-responses in this regard. However, it is possible to compute the other averages and therefore, in the sequel, most of the analysis will pertain to the twenty-four respondents, although the seven respondents together with an eighth who attempted to anticipate consistency, but failed, will be excluded from the analysis of average conditional weights.

RESULTS OF ANALYSIS: THE IMPORTANCE OF FACTORS AND FIELDS

We will first examine the individual results, and then present the aggregate analysis, finally summarizing the aggregate response in numerical terms. In fact, a considerable amount of analysis has been undertaken with this data, but here we are content to simply describe the main conclusions without swamping the reader with graphs and tables. This we do elsewhere (Batty and Spooner, 1982; Spooner and Batty, 1981) so our emphasis here is on more substantive conclusions. Before we begin our analysis, two points are worthy of note. First, through the analysis, the inconsistency due to averaging and prediction was well within the bounds set by Saaty (1981). That is, none of the indices exceeded the permissible bounds defined by Saaty, and only in one or two cases is there any indication that the responses should have been re-elicited. Of course, this would have been impossible anyway, but it is encouraging that the results are so consistent. In fact, in the light of using this technique in other contexts, we feel that the results reported here are remarkable in their consistency, and perhaps this implies that our participants "really know their field". Finally on this point, the "predicted" and "observed" weights of factors are close in every case and this further reinforces the high level of consistency.

Second, we have attempted quite elaborate analysis of the variation in this data set; that is, variation between individual respondents, variation of individuals from the average, as well as variation across fields and factors. In general, we find considerable variation with respect to the importance of factors, and of fields, but very little variation of factors between fields, and not much variation between respondents. Thus, our results imply strong consensus among the respondents with agreement that certain factors and fields are predominant. The same type of conclusion emerges from our attempts at extracting the "networks" from such data (see Spooner and Batty, 1981).

First, an examination of the weights of fields, factors in fields, and "predicted" and "observed" factors shows quite sharp differences. For fields, it is clear that most respondents consider that transport and/or public systems models have made the greatest impact on the success of urban systems analysis. In terms of factors in fields, the key factors which have provided most difficulty for each field are lack of social theory on the one hand, and policy relevance on the other. Data problems also appear, but these are of lesser significance. From these results a point of substantial interest emerges, and that is that technically oriented factors -- skills, computation, etc. -- are of lesser importance to the problems facing each field than theory or practice as considered through policy. In fact, this is one of the major themes which emerges from the meeting and which was elaborated in the paper on the discussion, presented as the previous chapter. It is also widely confirmed in more recent critiques of this area (see Breheny, this volume).

Between respondents, this initial picture of substantial consensus over the importance of factors and fields is reinforced. For fields, about half the respondents identify one field as of major importance; the other half identify two fields, and these are both invariably transport and public systems modelling. There are a couple of respondents who identify corporate planning as being the success story of urban systems analysis, but all are agreed that land use modelling has had only limited success. Across factors in fields, the same points made above are emphasized: agreement that social theory and policy relevance pose the main problems for the field with data being problematic for some, and the receptivity of the organizational environment being of occasional significance.

When the average weights are computed, it is necessary to consider the restricted set of sixteen respondents in making a general analysis. In reducing the set of individuals to sixteen, if anything the variation between respondents is reduced. For example, for fourteen of the sixteen respondents, the most important factors in fields measured by $\{w_{ik}w_{ijk}\}$ are restricted to four: social theory and policy issues in transport and public systems modelling. When the average weights are computed, the consistency is found to be still quite acceptable and the conclusions of the individual analysis are reinforced for every set of weights. For example, for the twenty-four factors in fields, the same set of six, identified by scanning the individual responses are significant; these in rank order are -- policy issues in transport modelling, policy in public systems, social theory in transport, social theory in public systems, data in transport and data in public systems modelling.

This paper has not gone very far into the results which have been produced, but in a sense, this reflects the great consensus found among the responses. Nevertheless, it is worth drawing together these ideas and presenting some numerical results for the average weights. First, the average weights for the importance of each field -- the success of each field in contributing to urban systems analysis -- are recorded in Table 1 for the twenty-four, then sixteen respondents. As expected, the rank orders are the same for both sets, and the actual weights are close. As implied previously, the set of sixteen produces a slightly sharper response than the set of twenty-four. With respect to the average weights for factors overall -- predicted and observed -- these weights are very close for both sets of respondents. The same is true for the average conditional weights -- factors in fields. These are all shown in Table 2, and in only one case --that of $\{\bar{w}_{2j}\}$ -- does the rank order differ from: policy issues, social theory, data, organizational environments, computation and technical skills; and this is only a reversal of the rank in computation and technical skills. In fact, the ratios of these weights are all close, whichever way they are taken across this Table -- indeed considerably closer than expected, indicating high consensus in the set of responses once again.

CONCLUSIONS

What then is the significance of these results? There is little doubt that there is massive consensus among the respondents; this also reinforces the agreement which was apparent in discussions throughout the meeting, and reported in the previous chapter. Yet, there is also considerable bias within the data set -- bias which relates to the choice of respondents -- participants in the first place. The classic reaction by non-systems analysts to results such as these would be that the prior selection of participants entirely conditions the types of response, and that the particular responses made are totally predictable, hence trivial. There must be some force to this argument, but there are some unpredictable issues which emerge from these results, indeed from the Institute itself. For example, it was surprising how strongly the participants felt theoretical and policy issues to be in this field. Ten years ago, technical issues would have been to the fore, and we were surprised how little significance was attached to these now. This, we would argue, was not totally predictable before the meeting.

The selection of questions too was rather restrictive, and never entirely clear in meaning. In particular, social theory and policy issues are at a different level from the other factors. Moreover, as noted earlier, the fields are by no means similar in their evolution and present organization. It appears that different levels of response might have been made to different questions, thus destroying comparability. Indeed, it was difficult to control responses, but we feel that any further attempt at such an exercise, should be more tightly controlled. A group response based on group discussion may have been helpful, and it may have been useful to have attempted to rephrase questions to inject more conflict into the responses. Nevertheless, the conclusions do support the rest of those in this book. Readers who wish to examine the way this metho-

TABLE 1

RELATIVE SUCCESS OF EACH FIELD BASED ON AVERAGE WEIGHTS $\{\bar{w}_i\}$

Fields	Weights Based on 24 Respondents	Weights Based on 16 Respondents
Land Use Modelling	0.178	0.155
Transport Modelling	0.394	0.372
Corporate Planning	0.156	0.154
Public Systems Modelling	0.272	0.320

TABLE 2

RELATIVE IMPORTANCE OF EACH FACTOR BASED ON AVERAGE WEIGHTS

Factors	24 Respondents $\{\tilde{w}_j\}$	$\{\bar{w}_j\}$	16 Respondents $\{\tilde{w}_j\}$	$\{\bar{w}_j\}$	Conditional Weights Based on 16 Respondents $\{\bar{w}_{1j}\}$	$\{\bar{w}_{2j}\}$	$\{\bar{w}_{3j}\}$	$\{\bar{w}_{4j}\}$
Data	0.179	0.156	0.176	0.152	0.168	0.198	0.118	0.182
Computation	0.065	0.049	0.068	0.046	0.062	0.087	0.059	0.055
Social Theory	0.251	0.282	0.254	0.286	0.298	0.224	0.280	0.253
Policy Issues	0.331	0.338	0.322	0.344	0.324	0.328	0.350	0.301
Technical Skills	0.081	0.085	0.078	0.076	0.067	0.064	0.078	0.100
Organizational Environments	0.093	0.091	0.102	0.096	0.081	0.098	0.115	0.110

dology can be expanded further should consult the two papers referred to earlier (Batty and Spooner, 1982; Spooner and Batty, 1981).

From this survey, let us re-emphasize future directions for this area. It is never very clear what future directions and priorities should be, for these not only depend on what the present field wants, but on its context, on who joins and inspires it, and on who supports it. But it appears that a greater concentration on theory in terms of social issues and on policy in terms of the relevance of models will be important during the next decade. It might be argued that a change in style is required to capture the essence of these issues. For example, perhaps social psychology should replace social physics as the dominant informing theme in this area and, in a sense, this is already happening. Any subject area which begins to explore its own history and context, to reflect on itself, to turn inwards, contains a sign that the predominant theoretical approach to its own subject matter is changing.

REFERENCES

Batty, M. and Spooner, R.S., 1982, Models of Attitudes towards Urban Modelling, Environment and Planning B, 9, forthcoming.

Fleck, L., 1979, Genesis and Development of a Scientific Fact, University of Chicago Press, Chicago, Illinois.

Saaty, T.L., 1981, The Analytic Hierarchy Process: Planning, Priority Setting, Resource Allocation, McGraw-Hill Book Company, New York.

Spooner, R.S. and Batty, M., 1981, Networks of Urban Systems Analysts, Environment and Planning B, 8, in press.

THE FUTURE OF URBAN SYSTEMS ANALYSIS:

ISSUES, THEMES AND RECOMMENDATIONS OF THE INSTITUTE

INTRODUCTION

Any history of public policy in the Western world during the last 30 years inevitably identifies the conflict between technics and politics, between rationality and ideology, as being central to the way problems have been defined, policies formulated. At the present time, the paradoxes and dilemmas involved in the development of technical approaches to political problems are uppermost in the minds of those involved. Thus it is not particularly surprising that the predominant concern of this NATO Advanced Research Institute has been with non-technical issues, rather than with those technical concerns from which most of the participants originate and with which they continue to identify. The papers in this volume mirror this wider concern and although the Institute was planned and the participants invited to reflect the context of technical analysis rather than analysis per se, it is still somewhat surprising that the flavour of the meeting was and the papers reproduced here are so explicitly non-technical.

Readers unfamiliar with technical issues of systems analysis will already have found that these papers deal with the cultural, organizational, perceptual, ethical, philosophical and theoretical, but definitely not technical, context of urban systems analysis. This was reflected also in the meeting itself. In summing up the work of the Institute, Hoos said:

> "In essence, the questions raised had to do with
> social responsibility and morality in an era
> when public decision-making depends largely on
> "technical" experts using techniques which can

583

be used to "rationalize" and which effectively
divorce the process from social consequence.
Discussion of such matters was an interesting
contrast to the more conventional preoccupation,
at such meetings, with the how, or mechanical
aspects, of model construction.   Indeed in
pondering the whether, why and by whom, this
Institute   scrutinized   the   implications   of
relying on quantitative methods of management
science to deal with such socially-imbedded
matters as transportation and urban planning."

The purpose of this paper is to elaborate and then draw
together the themes and issues which emerged from this meeting.
Many of the themes are woven into the background papers in diverse
ways and thus participants had access to these ideas prior to the
meeting itself.   The ideas to be elaborated here flow from the
meeting as informed by the background papers. As the meeting progressed position papers largely directed to the issues discussed
were produced, and at the end of the meeting each participant produced a paper emphasizing prospects for the field.

This paper attempts to synthesize all this material. As there
are many themes and issues emerging from this material, it is worth
spending a little time discussing how this synthesis should best be
organized. In the first paper in this book which recounts the background to the setting up of the Institute, five subject areas —
public services planning, physical network planning, strategic land
use planning, built-form planning, and corporate planning were contrasted against five methodological themes dealing with aspects of
analysis and planning. This concatenation of the subject and methodology of systems analysis guided the original organization of the
Institute, and the range of questions posed in the first paper
reflects this focus.   In fact, the actual Institute was organized
much more along the lines of diagnosing the nature of the field and
its problems before moving to prognosis and speculations on the
future; this was in the spirit of reviewing the past and the present
contribution of urban systems analysis before moving to future prospects.

It might be argued that the best way to evaluate the work of
the Institute would be to attempt to answer the questions strictly
posed in the first paper, and specifically addressed during the
meeting.   Indeed, this is the way we will begin, but it is not
enough: cutting across these questions are several interacting
themes relating to the application and context of urban systems
analysis — to philosophical issues, to communication between actors
and to societal problems. Furthermore, the meeting was not entirely
devoid of a concern for technics, and various suggestions were made
in this regard which will also be summarized here.

There are other ways of attempting this synthesis.   In the final sessions of the meeting, an attempt was made to impose a rather different structure on the discussions and conclusions. An attempt was made to embed the proceedings into the conventional model of rational decision -- the problem-solving process, but it proved easier to relax such an order, for the discussion during the meeting ranged much further and wider than implied by the conventional image of systems analysis. The fact that the proceedings could not be fitted into any conventional model of problem-solving should not imply any widespread disagreement among participants. Far from it -- quite the reverse: considerable agreement was reached on many issues. But the issues raised were so wide that no preconceived model could encompass this diversity. Thus, we will assume here a rather simple structure. First, the issues addressed will be recounted briefly, and then key themes running throughout the papers and the discussions will be highlighted. These themes which involve both diagnosis and prognosis for the field will then merge into explicit recommendations and conclusions.

THE ISSUES ADDRESSED

The discussion during the meeting was organized according to the six key questions posed in the first paper in this volume, and reflected in different groups dealing with the various subject areas identified. In fact, only three of the five subject areas were emphasized -- public systems or services, transportation and urban systems. There is little point in referring to these subject areas further, for participants organized their discussions in terms of the broader questions addressed rather than specific subject areas and consequently, there was little to distinguish the groups dealing with different subject areas. This may have been due to the fact that most participants came from the transportation-urban systems areas, although we feel that this concern for wider issues reflects more the mood of participants than their particular expertise.

The six questions themselves begin with diagnosis of the field and extend to prognosis, and the meeting was organized to reflect this. The earlier discussions dealing with the past and the present were separated from those dealing with the future by discussion of broader issues -- of the role of science in public affairs -- but these general issues permeated all sessions of the meeting and will be addressed later in this paper. In discussing responses to the six questions posed, it is proposed to broaden the scope of each question in elaborating the discussion that took place, due to the wide ranging nature of the discussion, but also to give the reader a greater feel for the issues dealt with. The first three questions:

What has been the experience in applying rational analysis techniques to urban problems?

Why are the results frequently ignored from the perception of the analyst?

and

> What have we learned from trying to implement
> urban systems analyses?

all deal with the past experience of this field, its specific failures and the learning which has obviously occurred among both
analysts and clients. All these issues are widely accepted as
readers of the papers presented here will be aware, but they will be
addressed rather generally in the sequel. The question:

> Are the difficulties in the urban field simply a
> manifestation of the broader issue about the
> relevance of science to human affairs?

is of more general import and will be picked up as a later theme.
The last two questions:

> What are the current and expected urban issues
> and what processes should be followed to respond
> to these issues?

and

> What should be done and what should the various
> actors do?

pertain to the future. These two sets of questions and discussion
during the meeting addressing them directly, will now be presented.

Past Experiences and Failures

The background implied by these questions and sketched in the
papers presented here in diverse ways was generally accepted by the
participants.

The experience had been characterized by a failure on the part
of analysts to read the context; there had been little sensitivity
to what was possible and impossible with such techniques and the
experience was dominated by over-ambition and naive application.
Opinions in the literature of critique range from deep and substantial criticism of the lack of knowledge of social phenomena by
analysts, and of the blunt ways in which analysts and decision-
makers assume interference in social systems, to evaluations and
critiques acclaiming the experience as salutary, implying modest
achievements. Notwithstanding these latter sentiments, there are
none who see the experience as a totally worthy one.

This was the starting point. In the early discussions, general
contextual issues were immediately raised. The nature of the planning system, the nature of methods for informing such systems, the

ethical dilemmas and responsibilities carried by systems analysts operating in the public arena, the changing nature of urban problems in a turbulent world, and the whole range of communication problems between analysts, professionals, politicians and the public -- these were the types of issues raised. What the discussion was not characterized by was discussion of technique despite the use of technical examples to illustrate various points, and an assumed familiarity with systems analysis on the part of all participants.

It is worth emphasizing the particular flavour of these discussions a little more. There was some discussion of the nature of public and social planning. In particular, the sentiment that public policy does not lend itself to clear decision-making and problem-solving was loudly expressed and indicative of the dilemma faced by using analysis techniques developed for problems with a well-defined structure. The problems of quantifying for quantifying's sake or because no other way is known, were raised and the instability of analysis due to volatile changes in public and professional perception of key problems was identified. Throughout this discussion the perennial problem of a lack of communication between analysts and clients, between scientists and politicians was raised. In one sense, this appeared to be the "clash of cultures" referred to by C.P. Snow, but it was more than this for discussion soon clarified that the difficulties of communicating systems ideology were reflected in a clash of many cultures, rather than just two.

There was also some talk about more modest low level scientific analysis -- about the use of models to impose discipline upon data, and about organizational/institutional difficulties in responding to analytic techniques. There was discussion of standard issues such as the abuse of models and the poor quality of various studies, but by far the most important problem was the perceived lack of understanding which all participants felt with respect to the subtle nature of the planning and policy-making system. These were all pervasive in the discussions and deeply felt by professionals which led to considerable agonizing of a personal nature. Other issues broached were definitional -- what is systems analysis, systems theory and systemic thinking? -- but these will be presented later as part of a more general theme.

Future Images

As discussion progressed, it was increasingly guided by the key issues raised above, and in one sense, this began to determine views of "what should be done?". It was in the latter part of the meeting that the focus changed towards the future. In fact, the spirit of the discussion was optimistic rather than pessimistic. There was a general feeling that much had been learned and that events such as this meeting itself showed how important it was to communicate this learning. This volume can be seen even as a part of the recommendations emanating from the meeting in terms of the need for greater communication of the diverse context of urban systems analysis.

Two main issues were addressed: first, there was an explicit examination of the types of problem which the urban planning and policy-making process would likely embrace during the next two decades. In particular, there was recognition of the increasingly volatile nature of urban society and social change. And second, there was a specific focus on systems analysis techniques, appropriate organizational forms, and the impact of technology on an appropriate organization for the field. The most important conclusion relating to future problems was the need to ensure that systems analysis be "issue" or "reality-driven" in the future. Too often, urban systems analysis was seen as technique in search of a problem; and as the problem changed, technique became increasingly irrelevant and inappropriate.

Another key issue relates to uncertainty and the need to develop more sympathetic and sensitive approaches to social change. This had implications for model structures in which evolution and adaptation played a central role. Moreover, this implied techniques more able to handle micro-issues -- questions of equity, of distribution, rather than of efficiency. A whole range of suggestions with respect to researching more appropriate techniques were made, focusing on the need to develop more supply-oriented, more partial and more economic market models.

Furthermore, the role of aggregate modelling was broached. The most exciting prospect for such aggregate work which, in some respects, represented the heritage of urban systems analysis, was in developments in self-organization theory and catastrophe theory. Such models represented the long-awaited shift from quantitative to qualitative in terms of structural change, as well as a new vision of urban dynamics. Yet, despite the evident excitement for such work, the possibilities here were largely conceptual rather than operational in image and thus, represented somewhat bold, if useful, speculation.

Finally, as in the earlier discussions of experiences, organizational issues were raised, particularly those dealing with communication -- with the need for a more fruitful dialogue between analysts and clients, between scientists and politicians. Many other more detailed issues were broached and several suggestions at a technical level made. Some of these will be picked up later, but before we do so it is worth examining some of the surprises of the discussion where potential issues turned out not to be issues at all.

Non-Issues

We have already emphasized that the whole question of analytical and modelling technique was not regarded as a central preoccupation of the conference. Moreover, this was in stark contrast to earlier critiques of the field of urban systems analysis where technical issues were regarded as rather central to the difficulties facing the field. However, what is perhaps a little more surprising is that technical issues are no longer seen as important in the future of the field. For example, computation, so important in the early days of systems analysis is no longer regarded as an issue. This is all the more curious, given the massive changes taking place in this area. The impact of the microprocessor is more and more visible in society-at-large, and a session of the meeting was devoted to such developments. All the obvious issues concerning micro-processors were listed — its widespread availability aiding learning and familiarity of technique, coupled with its possible widespread abuse in fostering unthinking application. New developments in graphics aiding communication and the help such developments would give in disciplining data collection and management, the shift from quantitative to qualitative information, were all recognized, but these issues were generally held to be separable from the wider non-technical issues affecting systems analysis. Indeed, it is interesting to speculate that this separability of social from the physical domain represents a much wider influence in modern society than has been hitherto recognized or accepted.

The other issue which did not receive much attention was the notion that optimization might contain the way forward in thinking about more realistic applications of systems analysis. It was felt that formal optimization techniques were far too narrow to be of other than conceptual use in planning, but Harris did argue in his usual persuasive fashion for the use of these ideas to illuminate the problems of applying systems analysis in general. However, there remained an implicit view that social systems were so infinitely complex that optimization could not take place, indeed that optimization was not a particularly useful concept — in short, that optimization was a dysfunctional concept in systems analysis. However, this argument was not really pursued to any depth during the meeting, although it remained significant.

THE EMERGING THEMES

Already several themes have been identified from the discussions summarized above, and these can be grouped into three main areas. First, and by far the most dominant during the meeting, were those themes which pertained to the nature and scope of systems analysis, and the consequent suggestions which flowed from an appreciation of the wide role of systems analysis in social action. These we will develop as the three related themes concerning the nature of systems analysis, professional and ethical responsibilities, and problems of communication and participation.

The second main area related to a concern for the system itself
and its analysis and planning. Two clear themes emerged here: the
whole question of substantive versus procedural approaches, and the
questions surrounding the emergence of new problems in an unstable
future. Finally, and by far the most contained, were suggestions
for future technical work with models and techniques, and although
largely separable from the previous two areas there is some prospect
that in the future, a deeper concern for context and emergent problems will guide technical work. We will elaborate these below.

The Nature of Systems Analysis

Although conscious of the wide and diverse characteristics of
systems analysis, we (the Editors and Organizers) kept its definition fairly narrow, yet implicit, in our preparation for the
meeting; only in our selection of participants did we condition the
type of systems analysis we were appealing to. Thus, at the outset
it was inevitable that there would be considerable discussion of the
scope of the meeting in terms of a definition of urban systems analysis. Our implicit definition from the background papers viewed
urban systems analysis as the application of analytical ideas, typically involving mathematics and computers, to problems involving
understanding, design and action, typically problems of policy-
making and planning in the urban domain. Such a definition is
inevitably narrow for it begs the question as to how it relates to
systems theory, and more generally to logical, rational, systemic
thinking.

A good deal of time was spent by more reflective participants
musing upon the legitimate concerns of the meeting in these terms.
For example, Savas in an early discussion defined the type of analysis in question as carrying "...the connotation of careful thinking,
as opposed to relatively unthinking reliance upon technique and a
premature emphasis on quantification". This seemed to represent the
dominant view although the experience of participants was rooted in
technique despite the widespread expression of a broader interest.
In fact, the participants were characterized in these terms more by
their intentions than by their actual experiences, skills or roles.

This point was detected quite early in the meeting by Vickers
who argued that the question of defining systems analysis was more
than just a matter of semantics for the way words like system,
model, rationality and ethics were used by many participants portrayed the narrower view of technique rather than the broader view
of systemic thinking. In a sense, this was represented by the
considerable confusion in discussion over a suitable and agreed use
of the term "urban systems analysis", and over the implications
which flowed from this. Indeed, it is fair to say that the absence
of any agreed and explicit theory of social action embracing such
analysis, was not felt to be a problem, but the fact that these con-

cerns were not really articulated demonstrates more the origins and concerns of the participants rather than any comprehensive assessment of the field. In fact, this illustrates the partial nature of the discussion of systems analysis, although, nevertheless, one of the first such explorations to be conducted by technical experts of their wider role. And for this, the approach taken is significant.

Professional and Ethical Responsibilities

The concern over the nature of systems analysis was also paralleled by a concern for the legitimacy of the systems analyst. However, there was little discussion of any grandiose theory linking positive to normative knowledge, or positivism to realism and idealism, or of theory to practice, possibly due to the professional-practical orientation of the participants despite their academic origins. There was much more discussion concerning the practical difficulties encountered in the role of systems analyst, in terms of the actual execution of the task. These difficulties were raised from actual experience, not in most cases from considered reflection on the theoretical origins of such difficulties.

The obvious tendency in the field to formulate problems in terms of known, "off-the-shelf" techniques rather than any problem-oriented approach, was a constant source of worry. From this arise the problems of quantifying the unquantifiable and the whole range of abuses which spin off from this — "modelitis" as Ida Hoos so graphically portrayed it, usually requiring "radical modelectomy". However, there was a large measure of agreement about the fact that systems analysts were often the victims of circumstance — "hired guns" rationalizing the covert actions of public or other agencies. There was also a good deal of tacit recognition that systems analysis may well be part of the problem addressed, but few took the argument this far. Occasionally, there were forays into more considered philosophy through such notions as: "...analysts should be seen as a part and apart from the problem..." and such like.

In one sense, these issues raise the age-old clash between ideology and rationality which although rarely articulated here in such terms, seemed to be at the root of much of the confusion and agonizing. Yet, the strategy for exploring these issues further was well-contained; few if any attempted to see the resolution of these issues in terms of any wider social philosophy. In fact, the predominant response involved suggestions for resolving the conflicts over professional responsibility through advocacy on the one hand where values could be laid bare, and the laying bare of assumptions on the other; the former implying a more radical tendency than the latter. However, both these responses were elaborated in the wider theme of enhanced communication and participation in systems analysis to which we now turn.

Communication and Participation

Throughout the meeting, one of the continued pleas was for more openness and transparency in the planning and policy-making process, consistent with the "laying bare of assumptions" and the quest for legitimacy from a public wider than the client per se. There were suggestions for increased participation at various stages of the model-building and analysis process, but these can be viewed as much as the desire to reinforce and legitimize the conventional planning process as to gain any deeper understanding of the process through radical alternatives.

One key problem which was clearly identified as leading to the need for greater communication was the confusion over the nature of the public policy-making and planning process. It was widely recognized that the process was not one of clear decision-making or -taking, that it was much plagued by non decision-making and high uncertainty and thus, it was not amenable to classical methods of problem-solving. The various portrayals of social problems as wicked, squishy, ill-defined, ill-conditioned and so on, is consistent with such confusion. Moreover, the analyst is never in command of the process — his skill always comes too late after the problem is formulated, and this skill is never used to assess the implementation of the proposed solution.

An obvious response to such issues is increased communication in an effort to get increased understanding between relevant actors, analysts, bureaucrats and politicans, and also to gain more information about the problem. Such calls for communication also betray a recognition that the nature of the problem-solving process is intrinsically political. Communication also helps in times of increasing instability when responses are required more quickly. The need for dialogue was voiced many times, but perhaps such pleas were a little unselective. These are obvious points at which increased communication may help resolve conflict and gain new information, but they are all rooted in the conventional model of systems analysis, and essentially extend the role of analyst to that of catalyst. Nevertheless, much of the discussion was consistent with the general need to broaden the outlook of both analysts and clients, to embrace a wider system of interest, to shift from quantitative to qualitative, to engage in systemic thinking and to get to grips with the inherent contradictions of public policy-making.

There is one final point concerning communication. The call for simple, robust techniques, for small, intelligible models perhaps conflicting with one another, for a plurality of technique is quite consistent with the need for enhanced communication of ideas. Moreover, this discussion very much implies a more humble, more modest role for the systems analyst. This is laudable, but perhaps to be expected, in an era when government, professionals, science and technology are all held in some disregard in Western society.

Substantive and Procedural Approaches

One of the dominant concerns in planning theory at the present time is the distinction between approaches which address substantive issues concerning the type of urban future to be desired, or ways in which we might proceed towards it. The last 30 years has increasingly seen emphasis on procedures, on ways to make decisions, on means rather than ends, on the assumption that there is no optimal plan or if there is, it cannot be found. Hence, the only feasible approach has been to optimize the process. This emphasis is now changing.

Dramatic changes in public mood and in the perceived importance of urban problems, reflected in changing approaches to urban planning dominated by material rather than rational concerns, have fostered these changes, and there was some considerable evidence at this meeting that these concerns had influenced systems analysts. For example, the generally felt requirement that approaches should be "issue or reality-driven" rather than technique-dominated was an ever present concern. Some, such as Harris, sketched the way in which the present procedurally-dominated techniques could be matched with such approaches. He argued that the realism of any problem should be construed in terms of three interacting worlds — functional realism concerning models of the appropriate behaviour of actors in the system, value realism concerning the motivation and processes of goal formulation, plan-design and evaluation, and political realism concerning the constraints and influences on feasible social action. This type of sketch is in some ways a turning point, for few if any discussions of systems analysis hitherto have been pitched in these terms.

A related point concerned the extension of systems analysis to embrace the design of systems traditionally assumed to act as a constraint on analysis. Typically, the extension of analyses to deal with the design of institutions and the assessment of new technologies was discussed, particularly by Garrison who argued that: "...the analysis should be configured in order to evolve new institutional-technological forms, and provide for their evaluation and implementation". This sentiment echoed much of what had been discussed earlier in analyzing the failure of urban systems analysis in terms of "organizational disharmony" and lack of communication, although there were few proposals for suggesting how to implement such ideas in terms of operational-usable models and techniques.

Such concerns did not explicitly address these issues in terms of substance or procedure. In fact, there were few who broached the issue head-on, but where it was discussed there seemed little agreement on future approaches. Indeed, for those who addressed the issue the predominant concern was that procedural issues be handled in a new light, and some implied that there had been too much emphasis on substantive questions. For example, Couclelis argued that there had been less emphasis on processes of planning than on system models with a substantive content and that "...the balance should be redressed". However, she also argued that the present procedural emphasis on the rational decision model of planning and decision-making had shown that such a model was a most unsatisfactory representation of the behaviour in a planning system. She then suggested:

> "There should be more emphasis on the development of procedural methods that are compatible with incremental, non-synoptic, non-comprehensive, non-centralized modes of planning behaviour, as most real forms of planning seem to be these days. This would imply more research into institutional and decision-making structures and the information flows (and blocks) inside these. Properties of these structures such as resilience, inertia, response to uncertainty, adaptive capacity etc., should be better explored. Models of decision-making should expand to include representations of collective or decentralized decision."

In fact, these suggestions relate very strongly to some of the substantive concerns identified above, and although the meeting was not primarily addressed to the search for new organizational forms, there was considerable agreement on these points, implying a convergence of substantive and procedural issues. Contained herein was one of the most exciting issues addressed by the meeting relating technical, behavioural and political issues, and this related to "getting-to-grips" with the urban crises involving the seeming instability of present social life.

The Unstable Future

Since systems analysis began to be applied in any significant way to urban planning and policy-making -- since the mid- to late-1950's -- Galbraith's image of an "Age of Uncertainty" has become sharp reality. The decade of the 1960's now looks like a golden age compared with the beginning of the 1980's. Opinion is divided upon the generic reasons for the present crises -- some argue the political and economic order contains the seeds of its own collapse, others argue that it is more local issues of management, conservation and control which are the cause, but whatever the blend, the most significant issue for urban systems analysts is the increasing volatility and turbulence of urban issues due to the inordinate complexity of modern life. Moreover, the increasing pressure on resources, the collapse of old and the opening up of new economic markets -- perhaps a sign of the population explosion -- has made a mockery of existing approaches. In short, current images and models of urban systems and problems are grossly inadequate, the product of an earlier age dominated by the assumption that the world was a stable, simply ordered, certain place. Nothing could now be further from the facts.

Throughout the meeting, there were continual suggestions for approaches to be rooted in real problems, rather than techniques, despite the obvious and perhaps inevitable conclusion that problem-oriented approaches might become totally inadequate and inappropriate as problems change. New syntheses of these difficulties were urgently required and the one which found the greatest favour was the emergent theory of social and physical dynamics based on self-organization theory — in one version pioneered by Prigogine, called "order through fluctuations" but linked to other related views of change from Boulding's "Eco-Dynamics" to Rene Thom's "Catastrophe Theory". What made such approaches attractive to systems analysts was the fact that they are strongly rooted in systems thinking, and systems theory which stress notions about adaptation, evolution, disequilibrium, discontinuity and structural change.

Couclelis's suggestions for a change in emphasis on the procedural basis of systems analysis, given above, reflects these concerns. Moreover, these approaches, although largely of conceptual interest, have been developed in highly operationalized forms for physical systems. Indeed, there are examples such as those due to Allen where these ideas have been applied to urban systems. There was considerable excitement during the meeting that such approaches might contain hints on how to address a range of problems — how to understand adaptive and evolutionary behaviour giving a richer insight into system behaviour, how to understand discontinuity, indivisibility and crisis, how to understand qualitative change, technological change. Yet it would be wrong to give the impression that there was other than tacit agreement that such approaches presented new and fresh ways of conceptualizing urban systems, problems and processes. It is perhaps significant that other new approaches which view cities as material, structural entities were not engaged here, thus indicating once again the origins, concerns and vantage point of those making up the Institute. This is not a criticism, but an acceptance that bias in outlook is the prerogative of any group reaching conclusions about a set of issues.

Technical Work

Out of the Institute came many detailed suggestions for refreshing and enriching the technical effort in urban systems analysis. There were several general proposals to concentrate more technical work on new approaches such as those developed by Allen, or on extending models to deal with institutional-organizational concerns. But few of these were followed up in detail. However, a number of significant substantive and methodological suggestions were made and these are worth presenting.

There was a general quest to deepen understanding of the system of interest in an effort to gain more knowledge of the system being interfered with.   In fact, this was rather surprising in that the majority of participants felt quite strongly that there should be much more work into the understanding, rather than the design of systems. This was not rooted in any strong theoretical concern, but in the inadequacies of available knowledge felt in a more practical context. This was the traditional quest in this field, and participants elaborated their concern for a deeper understanding in several ways. There was a professed need to examine micro-behaviour, especially the economic behaviour of urban markets and institutions, the concept of market failure and the whole economic dimension to urban systems in both their positive and welfare aspects. There was the need to disaggregate models to deal with individuals and groups, while there was a general feeling that aggregate approaches were of limited use. Indeed, the paradox between easy-to-operationalize, but often irrelevant aggregate models and hard-to-operationalize, but often relevant disaggregate models was often broached, but never resolved.

A key focus, as well as that on markets, was on questions of mobility. It was felt that existing models did not address such issues in any depth and there was an increasing need for model structures able to handle redistributional issues.   Indeed, the whole question of the growth-oriented bias of conventional techniques was raised several times, and there was tacit agreement that models should be developed to mirror the mechanisms of decline and stagnation. There was also general concern over the need to shift from quantitative to qualitative. Systems analysts, accused of quantifying the unquantifiable, desperately need tools to handle qualitative issues. Two sorts of extension were envisaged: first, there were the qualitative issues relating to new approaches to systems analysis in which qualitative shifts are modelled as structural changes -- this is the "order through fluctuations" approach involving shifts in modes of system structure rather than system values. The second approach is more modest and involves shifts from continuous quantitative to discrete quantitative -- from continuous to ordinal -- from quantitative to semi- or "poor" quantitative. Such tools are useful in articulating certain discrete situations. Many of these issues were permeated by more general concerns to build models dealing with distributional issues -- of equity rather than efficiency, and of institutional rather than activity/land use change.   Such concerns do have technical ramifications but merge quickly into the broader themes already discussed.

Methodological issues were much more contained.   There was a general view that simpler, partial models, more transparent models were required.   These were best summed up by Goldner who said: "Somewhere between these data hungry modelling monsters and penscratching on the back of a used envelope are simpler policy-oriented models". There was a general quest to find these, but with the condition that such partial models be tailored in modular fashion. To build from the ground up, sub-system by sub-system, is a time-honoured principle of design, and as Harris stressed, this should become the widely adopted principle of model design by urban systems analysts. What were not addressed in this Institute were particular technical strategies of model design such as calibration or optimization or model specification. Such issues as Hoos remarked in the quote at the beginning of this paper were never broached here. The attitude of participants in this regard was summed up by Schussmann:

> "Techniques, however important they are in this context, do not bother me very much. I take it for granted that they have already reached a very high level and they will improve very likely in the future and that they will never be perfect. That means that the questions we are confronted with in this context are the limits to model use."

RECOMMENDATIONS AND CONCLUSIONS

The broad conclusion of the Institute was that systems analysis has had a fairly limited impact on urban affairs in that the problems addressed were little influenced in their definition by systems analysts, and certainly, their solutions proposed were never implemented or adapted using such expertise. Yet, despite this inferred political ineffectiveness of systems analysts, the participants were agreed that the conventional mode of systems analysis and planning ranging from problem definition through to alternatives generation and plan-evaluation, was also a highly limited, somewhat artificial and sometimes counter-productive vehicle for analysis.

There was also substantial concern for the mismatch between systems analysis techniques and currently perceived urban problems. A major reason for the mismatch seems to be due to a combination of difficulties in defining the problem to be resolved and the availability of techniques to enable systems analysis to proceed. Falling back on technique to the detriment of good problem formulation had characterized systems analysis and this had not been aided by adequate communication between analysts and clients. As a result, most studies had developed a very limited range of options with most alternatives representing quantitative variations on one theme, rather than qualitatively different alternatives. Some participants felt that organizational structures and constraints were primary contributors to these narrowly defined problems, although there was a general feeling that it was the techniques-based bias of systems analysis which was the greatest bind on realistic planning and policy-making.

Some suggested that the limited ranges of policy responses considered in many studies could be traced to the incompatibilities that exist between analytic and creative capabilities. Most systems analyses have been dominated by staffs with strong analytic capabilities, but rather weak creative abilities. The mental maps of the systems analysts tended to be dominated by the analytical skills and current "models of the world" of those disciplines dominating the studies. Most felt that the deficiencies of current analytical techniques were not major contributions to the ineffectiveness of systems analyses. While a number of participants expressed the view that modelling techniques could be improved, there was little overall feeling that better models would lead to significant improvements in the effectiveness of systems analysis. However, some expressed the view that models that improved our understanding of urban phenomena and the response of these phenomena to changes in public policy would yield better problem definitions and through this, better policy responses.

The evaluation phase of the systems analysis process was also identified as a major source of inadequacy. Most participants argued that urban affairs tended to be dominated by equity issues rather than by issues of economic efficiency, and urban systems analyses have contributed little to the evaluation of alternative courses of action based on distributional concerns. There were also suggestions that many systems analysts paid little attention to the problem of implementing solutions. The comprehensive nature of many systems analyses produced comprehensive long-term solutions which did not recognize the rapidity of change in social, economic and political environments.

Reference was made to the lack of ability of many systems analysts to respond to changing environments. Many studies required periods longer than the response time of the key systems issues addressed, before recommendations were produced and during this period substantial changes had occurred in the system and its environment. Substantial investments were made in computer models which in many cases had been tailored to a specific issue. Issues changed and it was difficult to adapt the available analytical infrastructure to these new issues. Most participants felt that the rigidity of the systems analysis process was one of its principal difficulties. In conclusion, it was felt that successful applications of systems analysis occurred in situations where a strategic direction had first been identified by a number of actors and that systems analysts had only been used to detail the tactics for achieving the overall strategy.

Let us attempt to sum up these diverse issues and themes and present them in the form of a small number of positive recommendations for the field. Although all these points were discussed and agreed during the meeting, it is important to note that these points have been synthesized by ourselves (the Editors) from the variety of material presented and as such there is bound to be argument over our interpretations. Nevertheless, let us reiterate the fact that in terms of future work, the participants agreed that further technical work on models and analysis techniques which followed the existing conventional wisdom, would be of much lesser importance than work on how to relate systems analysis to its context; that is, the interface between urban systems analysis and the wider environment in which it is used — between the problem-solving process and its variants and how these relate to the institutions in which they exist and the problems to which they are addressed. Urban systems analysis as currently practised is postulated according to a relatively stable and unchanging context, but it is clear that as the world is perceived to have become more unpredictable, the match between urban systems analysis and its problems has become poor. It may even be that this match was never any good in the first place, but as problems have changed, systems analysts have failed to adapt.

Participants were agreed that it was the area of the use of systems analysis and its relevance to issues and problems addressed and their relations to organizational environments, that required research. Technical issues were rarely raised and there seemed to be a tacit assumption that further elaboration of existing techniques would lead to nothing. In Cole's phrase, there appeared to be a consensus, that "...putting models into society rather than society into models" should be the dominant thrust for the future. The concern over substantive or procedural approaches alluded to earlier suggested a blurring of these issues, in that analysts were now more familiar with the notion of procedures representing substance, and substance-containing procedures. In a sense, this parallels the quest to "explore the interface between urban systems analysis and its context, and with the oft-quoted shifts from hard to soft, quantitative to qualitative and such-like, which were expressed.

All these notions can be encompassed by the general plea that there should be a shift from the modelling dimension of systems analysis to a theoretical dimension — from analysis to theory, from systems modelling to systemic thinking.

As a conclusion, let us now list ten key recommendations which emerge from this discussion in the hope that these might act as a guide or perhaps a starting point for future work: it is recommended that efforts be made

> to broaden the current conventional wisdom of systems analysis: to embrace systems theory and systemic thinking.

> to explore the institutional context of urban systems analysis in terms of its operation and use in appropriate organizational environments.

> to make urban systems analysis much more receptive to currently perceived problems and policies: to anticipate future problems.

to shift from quantitative to qualitative: to
quantify qualities and to model structural
change.

to recognize the limits of formal systems analysis and the context in which it should or should
not be applied: to extend systems analysis to
help in exploring solutions rather than in
choosing them.

to explore radically different and conflicting
methodologies of systems analysis.

to explore the central questions of philosophy
and ideology pervading systems analysis: to
broach ethical, moral, logical and epistemological issues.

to explore methods for capturing distributional
issues — of equity rather than efficiency.

to engage in more intelligible communication
between systems analysts and other communities
with whom they interact.

to examine the professional role of systems analysts, to improve their credibility and to
enhance their educational opportunities in the
light of rapidly changing circumstances.

THE AUTHORS

PETER M. ALLEN is a Research Fellow, Service de Chimie-Physique, Universite Libre de Bruxelles, Belgium. He is currently working on the application of the theory of dissipative structures to urban phenomena and human systems.

PETER W.J. BATEY is a Lecturer in the Department of Civic Design, University of Liverpool, United Kingdom. Prior to this he worked in local government in Northwest England where he was involved in developing and implementing land use and transport models. His present research is concerned with the integration of analysis in strategic land use planning and demographic-economic forecasting frameworks. Between 1975 and 1980, he served as Secretary of the British Section of the Regional Science Association.

MICHAEL BATTY is a Professor of Town Planning at the University of Wales Institute of Science and Technology, Cardiff, United Kingdom. He has worked on the development of operational land use models and their application in the development of strategic land use-transport plans in the Nottinghamshire-Derbyshire and Reading areas of the United Kingdom. He is the author of a book on urban modelling and of many papers in the fields of urban modelling and design theory.

JOHN A. BONSALL is Assistant General Manager of the Ottawa-Carleton Regional Transit Commission, Ottawa, Canada. Prior to this he was responsible for all transportation planning activities in the Regional Municipality of Ottawa-Carleton including long range planning for the official plan, project planning, neighbourhood traffic plans as well as the planning and scheduling of the public transit system.

MICHAEL BREHENY is a Lecturer in Planning in the Department of Geography, University of Reading, United Kingdom. Prior to joining the University of Reading, he worked for eight years in strategic planning in British local government and has written and edited a number of publications on the application of models in British planning. He is currently Secretary of the British Section of the Regional Science Association.

SAM COLE is a Senior Research Fellow in the Science Policy Research Unit at the University of Sussex. He was a member of the Sussex Policy Research Unit forecasting team and co-editor and author of several major publications resulting from this study. He is a member of the Technical Working Group of the ACC Task Force on long-term Development Objectives. He is a consultant to the UNESCO Medium Term Plan, a Fellow of the United Nations Institute for Training and Research, co-director of the UNITAR Project on Technology, Domestic Distribution and North-South Relations, and is an associate member of the Centre for Environmental Studies Ltd. He is currently preparing a Long and Medium Term Development Plan for The Netherlands Antilles.

HELEN COUCLELIS is a Scientific Advisor in the National Council for Physical Planning and Environment, Ministry of Co-ordination, Greece. She has worked on a variety of national planning and development problems and has written several papers which deal with the philosophy of policy analysis and modelling.

JOHN W. DICKEY is a Professor in the Center for Public Administration and Policy, Virginia Polytechnic and State University, Blacksburg, United States of America. He is the author of three books and a number of papers in the areas of land use, transport and other urban sectors. He has worked for the Division of Building Research, CSIRO in Australia, and the US Agency for International Development and serves as a consultant to the World Bank.

MARCIAL ECHENIQUE is a Reader in the Department of Architecture, University of Cambridge, United Kingdom. He is the author of many papers in the area of urban modelling and has served as a consultant on transport and land use models to a number of agencies in South and North America, Europe and Asia.

WILLIAM L. GARRISON is Director, Institute of Transportation Studies, University of California at Berkeley, United States of America. He has served on the faculties of several American universities and as a consultant to a number of agencies and organizations. He has worked on transport problems, urbanization, and regional and economic development issues.

PEDRO GERALDES is an Associate Professor, Department of Highways and Transport, Technical University of Lisbon and a researcher at LNEC, Portugal. He has worked for some years on the development of land use and transport models for the Lisbon and Bilbao regions, and on computer packages for transport engineering design.

MICHAEL A. GOLDBERG is a Professor of Urban Land Economics and Associate Dean, Faculty of Business Administration, University of British Columbia, Canada. His present interests are in urban housing, land market analysis and policy studies. He is currently involved in comparative studies of Canadian and US metropolitan areas and the policy implications of the observed differences and similarities, as well as modelling neighbourhood change in Canadian and US cities.

WILLIAM GOLDNER is a Visiting Research Economist, Institute of Governmental Studies at the University of California at Berkeley, United States of America. He has worked for some fifteen years guiding the design and implementation of the Projective Land Use Model (PLUM) in the San Francisco Bay area. He has also contributed significantly to the study of regional economic structure in the Bay area and has authored a number of publications dealing with the office industry, tourism, the service sector and trade.

BRITTON HARRIS is UPS Professor of Transportation and Public Policy and Dean of the School of Public and Urban Policy at the University of Pennsylvania, Philadelphia, United States of America. His major interests are in transport and land use modelling; planning, design and optimization theory; and in computer applications. He has participated in numerous studies in the US and abroad and has published widely in a variety of fields.

IDA R. HOOS is a Research Sociologist and heads the Social Sciences Group at the Space Sciences Laboratory, University of California at Berkeley, United States of America. She is the author of a book on systems analysis and public policy and a number of papers in the areas of futures forecasting, technology assessment and the methodology of decision-making. She has served on a variety of US Government tasks forces and advisory groups.

BRUCE HUTCHINSON is a Professor in the Department of Civil Engineering, University of Waterloo, Canada. He has worked in the areas of transport engineering, land use-transport planning and cost-benefit analysis. He has published a book on urban transport planning and a number of papers. He has served as a consultant to a variety of Canadian and Australian transport and regional planning agencies.

ROBERT LAURINI is a Senior Lecturer at the Institut National des Sciences Appliquees in Lyon, France. His current interests are a study of local planning practice in France, the development of computer tools for practising planners, and in decision-making, public participation and the use of computer graphics as a communications aid.

IAN MASSER is Professor of Town and Regional Planning at the University of Sheffield, United Kingdom. He is the author of a number of books and papers which deal with the development and application of planning methodology. Prior to his current position, he taught at the Universities of Liverpool and Utrecht in The Netherlands. He has been involved in a variety of planning projects in the United Kingdom, The Netherlands, Germany, Malta, Brazil, Korea and Uganda.

JANET ROTHENBERG PACK is an Associate Professor in the School of Public and Urban Policy at the University of Pennsylvania, Philadelphia, United States of America. She has studied the institutional settings of urban modelling studies in the US and has published a book and a number of reports and articles on the findings of these studies.

JURI PILL is Executive Director of Planning for the Toronto Transit Commission, Toronto, Canada. Prior to his current position he worked with the Metropolitan Transportation Plan Review developing the policy evaluation methodology and financial analysis. He has worked as a consultant in urban planning and authored a book on the Plan Review experience. He is currently responsible for all operational and strategic planning with the Commission.

TONY M. RIDLEY is Managing Director (Railways), London Transport Executive, London, United Kingdom. He has served in executive positions with the Greater London Council, the Tyne and Wear Passenger Transport Executive in Northeast England and the Hong Kong Mass Transit Railway Corporation.

STEVE SAVAS is a Professor of Public Systems Management and Director, Center for Government Studies, Graduate School of Business, Columbia University, New York City, United States of America. Prior to this he served as First Deputy City Administrator of New York City and Manager of Urban Systems for the IBM Corporation. His current interests are in the areas of urban services delivery, productivity, performance, privatization, institutional arrangements and solid waste management. In February 1981, President Reagan nominated him as Assistant Secretary for Policy Development and Research, US Department of Housing and Urban Development.

KLAUS SCHUSSMANN is Director of Population and Employment Studies for the City of Munich, West Germany. He has interests in the philosophy of urban systems analysis, and the development of urban economic analysis where he has published a number of papers.

RICHARD SPOONER is a Research Assistant in the Department of Town Planning, University of Wales Institute of Science and Technology, Cardiff, United Kingdom. His research interests are in housing policy, and in the development of computer graphics for planning.

GUNDUZ ULUSOY is an Associate Professor in the Department of Industrial Engineering, Bogazici University, Istanbul, Turkey. His current interests are in urban transport, network optimization and mathematical programming.

GEOFFREY VICKERS was educated at Merton College, Oxford University and was admitted as a solicitor in 1923. He served in World War I and was awarded the Victoria Cross in 1915. In World War II he served as Deputy Director General, Ministry of Economic Warfare in charge of economic intelligence and was knighted in 1946. He served as legal adviser to the National Coal Board and as Board Member in charge of manpower, training, education, health and welfare until 1955. He has served as a member of many public and professional bodies and is the author of some fifty papers and six books on the application of systems theory to management, government, medicine and human ecology.

HENK VOOGD is a Lecturer in the Department of Urban and Regional Planning, Faculty of Civil Engineering, Delft University of Technology, The Netherlands. Prior to this he was associated with the Research Centre for Physical Planning TNO where he was involved in several research projects for various governmental agencies. His current interests are in the areas of multicriteria planning evaluation and process monitoring and review.

MICHAEL WEGENER is a Research Fellow, Institute of Urban and Regional Planning, University of Dortmund, West Germany. He currently heads a simulation project at this university and prior to this worked for the Battelle Institute in Frankfurt and was involved in the development of models for urban and regional planning.

FRANS WILLEKENS is Deputy Director, Netherlands Inter-University Demographic Institute, The Netherlands. He worked for some years as an agricultural economist at the National University of Zaire, at the International Institute for Applied Systems Analysis in Laxenburg, Austria, and as Director of Research at the European Research Institute for Regional and Urban Planning. His current interests are migration studies and multi-regional demography.

DOUGLAS T. WRIGHT is currently President of the University of Waterloo, Canada. He trained originally as a structural engineer and was Dean of Engineering at the University of Waterloo. From 1970, he has worked in various senior executive positions with the Government of Ontario in a variety of sectors including education, social and medical services, and latterly, culture and recreation where he held the post of Deputy Minister.

PARTICIPANTS IN THE INSTITUTE

The authors listed above participated in the Institute, together with Bulent Bayraktar, Secretary to the NATO Special Programme Panel on Systems Sciences; Donald Clough, Professor of Management Sciences, University of Waterloo, Canada, and Member and former Chairman of the Special Programme Panel on Systems Sciences; Jose J. Dias Coelho, Associate Professor in Operations Research at the University of Lisbon, Portugal; Ismet Kilincaslan, Associate Professor in the Faculty of Architecture, Istanbul Technical University, Turkey; and Georgio Leonardi, Research Fellow in Public Facility Location Models at the International Institute for Applied Systems Analysis in Laxenburg, Austria.

Standing (l-r)

Steve Savas (US), Peter Allen (Belgium), Mike Goldberg (Canada),
Richard Spooner (UK), Michael Wegener (West Germany), Bill Garrison
(US), Mike Breheny (UK), Giorgio Leonardi (Italy), Peter Batey (UK),
John Bonsall (Canada), Ian Masser (UK), Henk Voogd (The Netherlands),
Marcial Echenique (UK), Helen Couclelis (Greece), Ismet Kilincaslan
(Turkey), Britton Harris (US), Don Clough (Canada), Robert Laurini
(France), Geoffery Vickers (UK), Juri Pill (Canada), Ida Hoos (US),
Gunduz Ulusoy (Turkey), Bruce Hutchinson (Canada).

Kneeling (l-r)

John Dickey (US), Klaus Schussmann (West Germany), Frans Willekens
(The Netherlands), Jose Coehlo (Portugal), Pedro Geraldes (Portugal),
Michael Batty (UK), Doug Wright (Canada), Bill Goldner (US).

Abercrombie,P.,90,91,104
Abernathy,W.J.,534,542
Adorno,T.W.,141
Agarwal,S.,156
Albers,G.,127,137,145
Albert,H.,141,145
Alexander,C.,430,445
Alonso,W.,129,145
Allen,P.M.,470,514-517,524
Allison,G.T.,372,375,443-444
Ancot,J.P.,158,166
Archibald,A.,19
Arch+,139,145
Aron,R.,468,471
Arrow,K.J.,442,445,463,471
Artle,R.,529,543
Ash,D.A.,51
Ashby,W.R.,428,461,471
Atherton,T.,59,66
Atkin,R.H.,469,471
Averous,C.,529,543
Ayeni,M.,59,66

Badniter,E.,517,524
Ballard,K. & Assoc.,230,239
Bartholmai,B.,180,190
Barras,R.,79,84,112
Batey,P.W.J.,69,85
Bather,N.,85,119
Battelle,Institute,128
Baumeister,R.,135,145
Bayley,D.H.,385,388
Batty,M.,7,50,59,66,78,107,109
    110,114,190,272,361,434,440
    441,449,568,576,577,581

Beer,S.,428-430,433-435,442
Bell,D.,364,375
Bellush,J.,48
Beltrami,E.,7
Ben-Akiva,M.,59,66
Berlinski,D.,426,445
Berry,B.J.L.,49,466,471
Bishop,Y.,238
Blakemore,C.,436
Blum,H.,146
Boguslaw,R.,403,426,466
Bohr,N.,470-471
Bolan,R.S.,103,104
Boltan,R.,229,239
Bonsall,J.A.,328,340
Boulding,K.E.,437
Box,G.E.P.,436,438,445
Boyce,D.E.,8,71,92,104,112
Brady,R.H.,368,375
Brass,W.,235,236,240
Breheny,M.J.,69,75,79,85
Bretschneider,M.,180,190
Brewer,G.D.,8,110,120,426,446
Bright,C.D.,543
Bright,J.R.,535
Broadbent,T.A.,79,112
Brooking,C.G.,237,240
Bryson,J.M.,104
Buchanan,C.,25
Buckley,W.,451,458,471
Buhr,W.,49
BURISA,99

Camhis,M.,108,109,112,117,120,
    450,453

Castells,M.,101,105
Castro,L.,236
Cater,E.,98,105
C.E.T.,263,269
Chadwick,G.,8,92,93,105,453
Chamberlain,K.J.,86
Charles River Assoc.,64
Churchman,C.W.,129,146,439,444
Clarke,M.I.,60
Coale,A.,236
Coalho,J.D.,161,166
Coniff,J.C.G.,529,543
CONSAD,531,543
Couclelis,H.M.,462,469,471
Courbis,R.,230
Crenson,8
Crissy,8

Dahan,Y.M.,203
Daly,A.,64,66,535
Daland,R.,320,326
Dantzig,G.B.,434,445
DATUM,129,147
Davidoff,P.,101,105
Davis,R.C.,217,226
Day,N.,8,71,85,92,104,112
Dekker,F. & Mastop,H.,96,105
de Leon,P.,48
De Leuw Cather & Co.,331,340
Delbecq,A.L.,104
Demeny,P.,236
Diamond,367,375
Dix,M.C.,66
Doat,P.,193,203
Domencich,T.,59
Donnea,F.X.,59
Dorsch Consultants,159
Downs,A.,441
Drake,M.,8,113,120
Dror,Y.,370
Duc,T.,85
Dziewonski,K.,235,240

Eichperger,Ch.L.,229,240
Ellul,J.,372
Emplasa,262
Enns,J.,48
Etzioni,A.,144,147,552,563

Faludi,A.,97,103,105,450,467,
    547,550
Fehl,G.,129,138,139,144
Fester,M.,144
Flaming,K.H.,48
Fleck,J.D.,369,375
Fleck,L.,567,581
Fletcher,S.M.,366,375
Flore,G.,187,190
Flowerdew,A.D.J.,250,270
Floyd,M.,98,100,105
Foley,D.L.,467,471
Forget,J.P.,203
Forrester,J.W.,209
Francis,K.,85
Franke,D.,131,180
Friedlander,D.,238
Friedmann,J.R.P.,112,120,144,
    443
Friedrich,P.,49
Frienburg,S.,238
Friend,J.K.,73,74,94,101,119,
    550,551

Galin,D.,26
Gall,J.,434,446
Garrison,W.L.,528,534,543
Geraldes,P.,161,165,166
Gibberd,R.,234,240
Gillard,A.A.,80
Glansdorf,P.,495,524
Glickmann,N.,232,240
Goldberg,M.A.,49,50
Goldman,R.E.,275,289
Goldner,W.,50,276,289
Goldstone,S.,190
Goodman,L.,238
Gordijn,H.,231
Gould,P.,470
Graybeal,R.S.,275,289
Greber,J.,331
Gutch,R.,112,120

Habermas,J.,141,147
Hake,D.A.,237
Hall,P.,49,561
Hansen,N.M.,49
Harris,A.J.,59,66,139,443

Harris,B.,120,272,289,446
Harris,R.G.,105,531,543
Hausknecht,M.,48
Hazen,R.,535
Heclo,H.,367,375
Hegemann,W.,135,147
Heggie,I.G.,66
Heida,H.,231,240
Hickling,A.,74,87,95
Hightower,H.C.,450,471
Hill,M.,72
Hirsch,F.,374
Hoberg,R.,131
Hoffman,S.R.,274,289
Holland,P.,238
Holling,C.S.,452,466,472
Hoos,I.R.,8,129,148,435,426
Hudson,B.,101,102,112
Hueftlein,L.D.,51
Huff,D.L.,51
Hutchinson,B.G.,7,59,61,66
Hujer,R.,139,148

Jaffe,M.,112,120
Jefferson,R.,77,86
Jessop,W.N.,119,550,551,563
Jones,P.M.,60

Kade,G.,139,148
Kane,J.,437
Kanter,J.,118,120
Kash,D.E.,380,388
Kavalsky,B.,156,166
Keyfritz,N.,234-235,240
Klitzing,130
Koestler,A.,119,121
Kop,A.G.,136,145
Krieger,E.,130
Kruekels,T.,562
Kubatzky,U.,184,190
Kuhn,T.S.,402,426,436,453,472
Kunzmann,K.R.,131

Lampman,E.,543
Langley,P.,72,85
Laurini,R.,193,203
Lee,D.B.,Jr.,94,109,190,129
   453,471

Leicester & Leicestershire,71
Leigh,C.M.,51
Lembke,R.G.,274,289
Lendaris,G.G.,426,446
Lerman,S.R.,59,67
Levin,P.H.,105
Lilienfield,R.,403,426,435,438
Lindblom,C.E.,95,364,467,550,
   552
Linstone,H.A.,426,446
Lithwick,N.H.,49
Lowry,T.S.,275
Luckman,J.,73
Luhmann,N.,138,144
Lyden,F.J.,365

Mackett,R.,59
Manski,C.F.,59,66
March,L.,7
Martin,Voorhees & Assoc.,160
Martin,L.,7
Masser,I.,103,106,562
Masters,R.,86
Mattenberger,P.,159
McCreery,P.,86
McCurly,J.,543
McDonald,C.,8,71,85,92,104,112
McDougall,G.,73,86,112,121
McFadden,D.,59,66
McLoughlin,J.B.,71,85,92
Meadows,D.,388
Meintz,Th.,180
Meise,J.,128
Mengden,J.,131,149
Meyer,K.,136,149
Milgram,S.,429,446
Miller,E.G.,365,427
Miller,R.W.,375
Morel,J.P.,203
Morrison,H.,375
Morrison,W.I.,118,120 369
MTRPA,358
Mumford,L.,90

Narr,W.D.,138,149
Naschold,F.,138,139,144
Neutze,M.,49
Nicolis,G.,495,524

Nieuwenhuis,W.P.,229,240
Nijkamp,P.,96,106,238,560
Nonnenmacher,W.,131
Nowak,J.,190
Nowlan,D.,365,375
Nutall,R.L.,103,104

Offe,C.,139,144
Openshaw,S.,80,85
Osborn,F.J.,49
Ostermann,J.,131

Pack,J.R.,8,112,113
Paelinck,J.H.P.,157,158,166
Palen,J.J.,48
Parker,J.,320,326
Parry-Lewis,J.,112,121
Pearce-Williams,L.,113,120,121
Perry,H.A.,85
Phillips,A.,525,543
Pill,J.,321-323,326
Pittenger,D.,229,241
Polenske,K.,232,241
Popp,W.,130,190
Popper,K.R.,443,470
Por,A.,233,239,241
Power,J.M.,94,101,105,136
Prest,A.R.,365
Prigogine,I.,495,524
Prognos,A.G.,186,191
Putnam,S.H.,50,67,285
Pyhrr,P.A.,368,375

Rapp,M.H.,159,167
Raquillet,R.,239,241
Rautenstrauch,L.,137
Rees,P.,232,241
Reginal Munic.,331
Richardson,L.F.,444
Ritchie,R.S.,370,376
Rittel,H.W.J.,96,138
Roberts,A.J.,71,75,79,116
Roberts,N.A.,86
Robertson,A.,86
Ronge,V.,139
Rogers,A.,232,234,236,426
Rose,R.,368,376
Rosenhead,J.,466,472

Roy,B.,502,524
Ruiter,E.,59,67
Ruppert,W.R.,130

Saaty,T.L.,568,572,575,581
Said,G.,59
Sanglier,M.,466,470,517
Sayer,R.,108,112,121
Savas,E.S.,384,388
Schaffer,F.,49
Scheurwater,J. & Masser,I.,99
Schroeder,D.,186,191
Schuon,K.T.,139,150
Schussman,K.,131,186
Schwoerer,I.,131
Scott,D.,98,105
Sener & Preyser,158
Senghaas,D.,138,144,150
Senior,M.L.,58,67
Sessions,V.S. & Sloan,L.W.,226
Shelsky,H.,140
Sherman,L.,59,66
Shubik,M.,435
Siebel,W.,137,150
Sikdar,P.K.,61,67
Simmie,J.M.,468,472
Simon,H.A.,112,423,428,465
SISTRAN,260,270
Sitte,C.,135
Smith,D.L.,72,86
Smith,D.M.,86
Smith,D.P.,61,62,66
Smith,W. & Assoc.,278,290
Snow,C.P.,313,326
Solesbury,W.,83,87
Sorgo,K.,128,150
Spooner,R.S.,568,576,577,581
Stahl,K.,131,180
Steiss,A.W.,209,226
Stopher,P.R.,59
Stradal,O.,128,150
Stuebben,J.,135,151
Sutton,A.,74,85,119

Tanner,J.C.,59
Tarr,J.A.,535,543
Taylor,B.,326
Tenbruck,F.H.,138,151

Ter Heide,H.,130,229,235,254
Thorburn,A.,73,87
Trussel,T.J.,236,240
Turvey,R.,365,375,525

Van Delft,A.,96,106
Van Hamel,B.A.,229,240
Van Setten,A.,560,564
Verlinsky,I.B.,437,445
Vickers,G.,27,436
Vidal,J.,203
Volwahsen,A.,130,151
Vooga,H.,558,560,563

Wade,B.F.,77
Wakeland,W.,426,446
Warren,R.M.,323
Webber,M.M.,112,118,121,442
Webber,R.,121
Webster,N.,369,375
Wedgewood & Openheim,F.,99,
    100,106
Wegener,M.,128,144,151
Wendt,P.F.,275
Whittick,A.,49
Wildavasky,A.,7,366,367,374
    381,388,467
Williams,C.,85
Williams,H.W.C.L.,58,66
Willikens,F.,233,235,238,239,
    241
Willoughby,C.R.,156
Wilson,A.G.,50,51,87,167,241,
    472
Wilson,J.Q.,69,158,232,466
Winch,P.,451
Wiseman,C.,95,106
Whitehead,A.N.,326
Wuerdemann,G.,130

Yewlett,C.J.,94,101,563
Yosie,T.F.,543

SUBJECT INDEX

accessibility model,185
airport development impact,298
airport system study,275
analysis styles,434

basic employment model,282
Bayesian analysis,399
Belgium,491
benefit-cost analysis,191,131
     364,394,396,450,489
Bilbao,160
Brusselator,494
budget review techniques,367
built form planning,5,7

Canada,13,14,31,33,53,56,62,
     317-325,327-332,370
Club of Rome,409
cohort-survival method,230,301
Cologne,129
communication
  between planners and
    decision makers,554
  of planning results,77,144,589
  of results,592
  with decision makers,207
comprehensive auditing,368
computer-aided planning,128
  technology,42,286
control-engineering,322
  theory,27
corporate planning,6,7,14,309,
    316,320,324,365,370,570
cost-effectiveness analysis,394

criteria for good models,42,
  60,438,593

decision maker requirements,
  466
decision making context of
  models,6,45,56,71,101,114,
  550
decision making environment,
  of systems analysis,6,17
  in planning,73,138,195,312
  550
  in public sector,379
  using model information,220
definition of problems,17,20
  76,319,438,476
demographic accounting models
  232
demographic models,14,131,229
  criteria for models,233
  data requirements,238
  disaggregation,234
  discrete multivariate
    analysis,238
  implementation,237
demographic patterns,36
demographic policy analysis,
  236
demographic scenarios,236
demographic theory,235
demographic trends,35,36,220,
  413
dissipative structures,493
domain of urban planning,5

economic base-analysis, 278
  studies, 229
economic models, 229
economic evaluation of plans,
  161
economies of scale in urban
  systems, 531
education sector, 209
employment allocation model,
  186, 282, 301
  shift analysis, 187
energy planning, 286
energy policy, 400
environmental indices, 252
environmental issues, 34
equity in public policy, 596
ethics of system analysts, 587,
  591
evaluation of development
  policies, 272
evaluation procedures
  benefit-cost, 91, 250, 254, 365,
    394
  goal achievement, 91
  multi-criteria decision
    analysis, 96, 449

France, 14, 193-203, 230
French planning, 414
French planning practice, 193

goal achievement analysis, 91
government organization, 370
Greece, 449

Hong Kong, 348
housing market model, 130, 180
  policy variables, 184
housing policy, 34, 35
housing market simulation,
  Munich, 183
housing supply, 133

information systems, 170
  for urban areas, 216
  for urban planning, 177, 178
  for in-planning, 75, 130

infrastructure planning system
  186
institutionalized policy
  review, 369

land market simulation, 262
land use model, 108, 272
land use modelling, 7, 570
land use planning
  data bases, 280
  environment of, 133
  historical development in
    West Germany, 135
  information systems, 99
  institutional context, 77
  institutional framework, 459
  legislative context, 273
  legislative context in
    France, 193
  monitoring, 98
  organizational inertia, 78
  planning environment, 101
  polictical context, 102, 125,
    305
  structure plans, 70
land use planning process, 475
land use-transportation models,
    14, 58, 128, 165, 243, 296
  Atlanta, 299
  Miami Valley, 300
  Portugal, 153
  Puget Sound, 296
  San Diego, 276, 300
  San Francisco Bay Area, 275,
    278
  Sao Paulo, 246
  Spain, 153, 160
  Twin Cities, 301
  Washington, D.C., 302
  West Germany, 128-130
  transport planning, 71, 160, 161
  transport planning
    institutional context, 162
  transport planning process,
    246
  transport studies forecasting
    frameworks, 91

land use-transportation models,
  (cont.)
  transport study,93,99
large scale models,407,480
legislative context of systems
  analysis,193,273,324,344
limitations of theory,415
limits of growth,411
Lisbon,159

management by objectives,368
management science,423
management theory,363
market failure,596
Marxist theory in planning,
  139
Marxist tradition in planning,
  460
mental maps,26,459
Metropolitan Toronto,317
Miami,209
Miami Valley,300
military systems analysis,435
model buiding myth,382
modelling styles,479
models
  future development of,417
  logical structure of,436
  theoretical basis,479
monitoring of planning,98,99
  208,215
monitoring of urban systems,215
multi-criteria decision
  analysis,96
multiregion economic studies,229
multiregion input-output
  model,232
multiregion population studies
  231
Munich,129,177-190

New York City,307,384
normative issues in planning
  478
normative planning models,453
nuclear accidents,397
Ontario Government Organization
  369

Oporto,159
organizational complexity,426
  context,593
  response,430
Ottawa-Carleton,327-332

Pentagon,393
physical network planning,5
plan making process,74,112,197
planning data bases,479
planning issues,101,455
planning organization,555
planning process,15,43,71,89
        91,97,112,138,198,246,
        272,318,330,456,462,
        466,478,545,553,593
planning process improvements,
  559
planning process
  characteristics,312
planning theory,134,137,383,
        454,460,475,545
policy analysis,19,207,218,236,
        254,295,319,363,475,526
policy problems in planning,133
policy research institutes,370
policy variables,480
political context of planning,
  551
political theory,140
pollution regulation,395
pollution allocation model,184,
  301
population forecasting 129,231
population forecasting model,
  180
population models,227
Portugal,14,154-165
positivist approach in
  planning,460
PPBS,365
problem solving approaches,425
  myth,380
  styles,454
process planning,97,103
production functions,526
public participation,592
  in planning,126

public participation (cont.)
  policy analysis,336
public policy,363
  analysis issues,588
  future development of,372
  interactions between
    decisions,487
  issues,175,207,218,385
  role of systems analysis,363
  variables,273,304
public policy problems
  definition of 583-586
  nature of,4
public services
  planning,5,185
  modelling,570
public transport
  finance,351
  legislation,358
  planning,332,341
  Bay Area Rapid Transit,278
  institutional context,278
  policies,322,324,347
  subsidies,321
public utilities,532
Puget Sound,296

qualitative information inputs
  to models,215
qualitative inputs to planning
  analysis,560
qualitative inputs to models,
    596

rational analysis,19,23,364
  in planning,333,552
  techniques in planning,424,450
rational behaviour,457,463
rationality,26,591
  of planning,468
  in public policy,126,136,395
  in urban planning,451-2
regional demographic models,228
regional land use planning
  San Diego,276
residential location choice,506
role of planner,312,546-8

San Francisco Bay area,273
Santo Andre,161
Sao Paulo,243-264
scientific method,310,453,485,
    492
shopping centre impact models,
    416
shopping centre impacts,298
simplified analysis techniques,
    592
simplified models,596
Sines,161
social impact assessment,365
social planning,209,393,476
social policy,366,385
  analysis,425
social responsibility of
    systems analysts,583
social systems,425
solid waste disposals,386
Spain,14,154-165
spatial interaction models,238
strategic-choice,94,97,119
  land use planning,5,8,69,74
    89,93,153,300
strategic land use planning
  decision making,97
  decision making context,117
  institutional context,95,114
  organizational inertia,116
  plan evaluation,93
  political context,109
  process planning,97
  role of systems analysis,107
  strategic choice,94
strategic monitoring of
  planning,100
strategic urban and regional
  planning,134,158
structure plans,112,193
systematic methods in planning,
    13,69,70,72,76,107,424
systematic planning
  shortcomings,75
systematic thinking,17
systems analysis
  assessment of,424

systems analysis (cont.)
  availability of information,
      170
  conceptual framework,24
  critique,392,437
  culture of,13,313,443,452
  definition of,3,18,108,309,
      590
  education for,170,386,393
  effectiveness of,8,109,
      371,399
  environment of in Turkey,169
  future developments,173,469
  in planning: assessment of,
      110
  institutional inertia,171
  limitations,14,17,25,334,
      359,371,380,391,399,401
      407,586
  myths,387
  organizational context,586,
      588
  philosophy of,6,8
  political context,385
  procedure aspects,109,112
  social consequences,401
  social context,391
systems concepts
    bifurcation,493
    definition of,19,309,364,
        428,486,537
  hierarchical structure,428
systems definitions, open
  systems,19
systems dynamics models,14,
      177,209,213,409
systems engineering,127
  instability,23
  regulation,20,21,24
  theory,426,428,433,440
  thinking,13,462
  understanding,25

technological assessments,365
  change,413,526
  impact,525
technological impacts on
  systems analysis,527

technology as agent of change
      533
The Netherlands,99,229,370,545,
      554
Toronto,317-325
Toronto Transit Commission,320
transport impact study,284
  model,158
  modelling,570
  multi stage aggregate model,
      159
transport planning,7
  decision making process,342
  definition of problems,335
  energy planning,332
  institutional context,337
  institutional constraints,338
  legislative context,331,334
  objectives,335
  process,330
  studies,54,90,246,318,343,348
transport plans
  Metropolitan Toronto,317
  Sao Paulo,256
transport policy,25,34,243,327
      341,528
  current issues,65
  evolution of,53,243,247
  equity in,258
  historical development,52
  Hong Kong,341,348
  impact of,258
  Metropolitan Toronto,317
  options,54,254
  Ottawa-Carleton,327
  policy analysis,56
  policy responses,54
  Sao Paulo,246
  Tyne and Wear,341,343
transport studies,90
transport systems analysis,
      13,14,53
  criteria for good models,60
  disaggregate models,57,59,63
  economic evaluation,250
  effectiveness,61,334,359
  evaluation procedures,254
  future development,65

transport systems analysis
    (cont.)
  household based activity
    models,59
  models,159
  simplified methods,318
transport system impacts,299,
    303
transport system models
  aggregate,57,58,158,159,247
  disaggregate,59
  institutional context,62
  network equilibrium,285
  multi stage aggregate models,
    57,247,279
transport systems planning,319
    327
  institutional context,329
Turkey,14,169-176
Tyne & Wear,343

uncertainty in decision making,
    397,452
  in land use planning,95
  in planning,73,162,476,551
unintended consequences of
    planning,477
United Kingdom,13,35,69-78,89,
    93,107-119,343-348,410
  in land use planning,95
urban and regional planning,13,
    545
  role of systems analysis,72,
    109,125
urban and regional policy
    making,546
urban development
  growth control,35
  patterns,32,132,253,327
urban evolution,507
urban land use planning,193
  planning issues,33,38
urban planning
  historical development,72,
    125,127,132,134
  role of systems analysis,126
  planning process,43,193,440
    449

urban planning (cont.)
  strategic,69,89,153,158,331
  structure plans,70,112,160
urban policy analysis,271,293
    303
  requirements of,311
  making,8,32
  problems of,441
U.S.A.,13,14,31-34,53-56,71,91
    112,117,129,207-225,
    273-286,296-307,383
    393-401,409,480,525,528
urban systems,31
  complexity of,488,492
  equilibrium in,492
  growth control,132
urban system analysis,142,178,
    440,449
  assessments,117,131
  attitudes to,567,577-579
  context,40,44,56,154,584
  critique,569
  development patterns,33,253,
    327,507,528
  future development,44,143,127
  growth control,35,507
  impact of infrastructure
    technology,525
  legislative context,157
  uncertainty,45
urban system models,13,31,36,
    79,130,153,185,265,271,
    293,491
  achievements,41
  behavioural components,484,
    502
  communication with decision
    makers, 165
  complexity,461
  consistency between sub-
    models,79
  criteria for good models,42,
    208,224,490
  decision making context,224
  different modelling styles,
    274
  disaggregation of,481

urban system models (cont.)
  effectiveness of,131,188,293
       296,304
  future developments of,163
  future directions,40,163,587
  historical development,31,37-
       39,275
  impact of increased travel
    costs,514
  institutional context,305
  interactions between sub-
    models,187
  interactions with decision
    makers,45
  limitations,39,110,224,293,
       307
  logical consistency of,484
  model of San Francisco Bay
    Area,275
  necessary structural character-
    istics,164
  parameter stability over time,
       287
  policy inputs,218
  policy relevance,39,204
  policy variables,295,510
  preparation of data input,283
  requirements for good model,39
  Sao Paulo area,243
  self-organizing,499
  shortcomings,80
  simplified analysis techniques,
       165
  simplified approaches,482
  spatial choice mechanism,504
  structural re-organization,513
  theoretical basis,81,493,521
  use by planning agencies,294

Valencia,160
very large scale models,483

Washington,D.C.,302
water resources planning,277
West Germany,14,125-143,177-190
world population model,412
World War I,21

zero based budgeting,368,450